Functional morphology of the evolving hand and foot

Functional Morphology

of the

Evolving Hand and Foot

by

O. J. LEWIS

Professor of Anatomy,
St Bartholomew's Hospital Medical College,
University of London

CLARENDON PRESS · OXFORD

1989

Oxford University Press, Walton Street, Oxford OX2 6DP

Oxford New York Toronto
Delhi Bombay Calcutta Madras Karachi
Petaling Jaya Singapore Hong Kong Tokyo
Nairobi Dar es Salaam Cape Town
Melbourne Auckland

and associated companies in
Berlin Ibadan

Oxford is a trade mark of Oxford University Press

Published in the United States
by Oxford University Press, New York

British Library Cataloguing in Publication Data
Lewis, O. J.
Functional morphology of the evolving
hand and foot.
1. Man. Evolution
I. Title
573.2
ISBN 0–19–261684–6

Library of Congress Cataloging in Publication Data
Lewis, O. J. (Owen John)
Functional morphology of the evolving hand and foot.
Bibliography Includes index.
1. Hand—Morphology 2. Hand—Evolution. 3. Foot
—Morphology. 4. Foot—Evolution. 5. Mammals—
Morphology. 6. Mammals—Evolution. 7. Anatomy,
Human. 8. Human evolution. I. Title. [DNLM:
1. Anatomy, Comparative 2. Evolution. 3. Foot
—anatomy & histology. 4. Foot—physiology. 5. Hand
—anatomy & histology. 6. Hand—physiology.
WE 830 l675f]
QL950.7.L48   1989   611'.98   88–25495
ISBN 0–19–261684–6

Typeset by Cotswold Typesetting Ltd, Gloucester
Printed in Great Britain by Butler & Tanner Ltd, Frome

Preface

In less than one hundred million years evolutionary diversification has produced a remarkable array of mammalian species, including some very bizarre forms, but on close examination it is apparent that these transformations have almost entirely affected the head region and the distal parts of the limbs. The trunk, and its contained organs, have remained really quite conservative.

Excellent insights into the evolution of the head region are readily available from published material. For the limbs, however, despite a vast monographic literature on muscles, bones and so on, there has been virtually no serious attempt at an evolutionary synthesis. This book attempts to rectify that. It represents an entirely personal viewpoint, and results from firsthand familiarity with the material; it is not an abstract compilation from the literature. Some of the chapters are based upon papers published by me over the last twenty years, but much is entirely new.

It is my deeply held conviction that without the evolutionary perspective the study of human anatomy, even in its most utilitarian form, lacks an important dimension. The comparative viewpoint provides a new plane of understanding and heightens observation—variations and anomalies take their place in a logical framework, and subtle morphological features, that have often been disregarded, spring into focus.

It is hoped therefore that this book could be read with some profit by anyone wishing to achieve a deeper understanding of the complex anatomy of the human limbs. It is not a text for beginners, however, and readers will need to be reasonably familiar with traditional descriptions of human anatomy.

This is not primarily a book of fossils. This is partly dictated by space, but also by the restricted availability of much of the relevant material. However, it is hoped that it does recount remarkable natural experiments in evolution of the most complex kind. It is humbling to realize how rudimentary are our present ideas about the mechanisms of evolution at this detailed morphological level. There is a great challenge here for the future.

I would like to thank Mrs Jean Politi for the highly efficient way in which she has produced the typescript for this book. Mr Roger Ivings and Mr Philip Hazell provided skilled technical and photographic assistance. I should also like to thank the staff of the British Museum (Natural History) for courteously allowing me to examine material in their care.

London OJL
April 1988

Contents

1

Evolutionary theories and comparative anatomy

In the two thousand years before Darwin popularized the idea of evolution as the source of biological diversity, a considerable body of morphological knowledge, some monographic but much of it truly comparative, had already accumulated. The concept of comparability between the structures and organs of different creatures, or homology, was already firmly established, but belief in the biblical account of special creation hampered philosophical interpretations of this, and some sort of explanatory framework was generally sought in notions of basic archetypal designs manipulated by a divine architect. Dimly perceived suggestions of historical relationships and of a progressive elaboration of form and structure, however, repeatedly appeared.

By the first half of the nineteenth century the biblical story of creation was being openly challenged. Lamarck's publications, although often rather unfairly misrepresented (Cannon 1958), provoked a series of important British contributions (Darlington 1974), and ideas of evolutionary change were being quite freely bandied about even with some insight into mechanisms of inheritance, and vaguely suggesting a possible role for natural selection. Thus the stage was set for intervention by Darwin. With the publication of *The Origin of Species* in 1859 belief in evolution became firmly established and catalysed a great flurry of morphological investigation; Darwin himself, however, had relied on detailed comparative anatomy to only a very limited extent. Whilst belief in the reality of evolution thus firmly took hold, strident controversy developed about the possible underlying causative mechanisms. Darwin's main theme was that the prime directional force in evolution was selection acting on small random variations, but in later editions of his book he increasingly had

recourse to Lamarckian ideas which invoked direct effect of the environment, or of use and disuse—what has been called 'soft inheritance'.

The discovery of Mendelism at the beginning of the century ushered in a turbulent period of controversy which lasted until the early 1940s when a broad consensus was reached. This rather shameful period of academic antagonism and woeful lack of communication between different disciplines was followed by a reconciliation which has recently been chronicled, often by the principal protagonists (Mayr and Provine 1980). The early Mendelians challenged the whole fabric of Darwinism, allowing selection only a minor role in establishing variation within species (microevolution); all advance above the species level (macroevolution) was credited to mutation pressure, produced by the sudden emergence of major genetic changes. Such attacks on Darwinism were mounted in various countries. Particularly in Britain, Germany, and Russia the Darwinians fought back and neo-Darwinism, purged of its defensive appeal to 'soft inheritance', took firm root and produced a great stimulus to morphological enquiry. In France, however, Darwinism never did become established, and neo-Lamarckian schemes and other viewpoints, such as, so-called 'creative evolution', flourished. The latter, based on a belief in a predestined order of progress, unrelated to selection, and incurred by the appearance of new genes, has its adherents to the present day (Grassé 1977). Neo-Larmackian theories, stressing the response of organs to use and disuse, and to new needs triggered by environmental changes, coupled with the inheritance of these acquired characters, proved seductive to a powerful school of American palaeontologists and to a number of morphologists.

Meanwhile the increasing sophistication of the population geneticists (neo-Mendelians) narrowed the gap between their views and those of the neo-Darwinians, leading to the reconciliation known as the 'Modern Synthesis'. Perhaps the most influential and earliest statement of this viewpoint, in the West at least, was by Huxley in 1942, followed rapidly by important works by Mayr, Simpson, and Dobzhansky. Recombination, both chromosomal and genetic, was emphasized as the source for much of the variation, which is the raw material acted upon by selection. The common occurrence of the pleiotropic effects of genes was recognized, as was the significant role of 'rate genes' which controlled a number of different developmental processes. These two effects supplied the solution to Darwin's recognition of the common occurrence of correlated characters, where a structure with self-evident adaptive significance is often linked with another lacking any such clear-cut role. The synthesis presented a scenario of a steady, majestic progression of evolution in entire large populations by the accumulation and integration of small genetic changes: phyletic gradualism! Species formation by branching was conceived as a mere fringe effect, a biological luxury, as was the possible role of genetic drift in small populations. When Mayr (1942) highlighted the importance of geographical isolation for speciation, this idea was merely subsumed into the synthesis.

Morphology figured in only a minor way in the construction of the fabric of the modern synthesis in the West. Morphology, as the cornerstone to evolutionary theory was, however, being pursued to considerable effect in Russia and an earlier, and possibly more refined, synthesis with the maturing science of genetics was achieved than in the West. This was soon snuffed out by the machinations of the charlatan, Lysenko.

Embryology, like morphology, played no great part in the early stages of formulation of the modern synthesis, apart from the recognition of the significant role which genes have in controlling the rate of developmental processes. De Beer (1951) refined and elaborated these ideas and devised a comprehensive theory to embrace the varying relationships between ontogeny and phylogeny. Central to his scheme was the concept of heterochrony—the realization that certain genes alter the rate at which structures appear, although the effects may be either retardation or acceleration. Thus changes, or indeed reversals of the order of development, can take place. The retraction of adult characters into younger stages of development with addition of new stages at the end of the life history, giving a pattern of recapitulation, is by no means a universal occurrence. These ideas dealt the final death blow to Haeckel's (1910) Biogenetic Law, 'ontogeny is a recapitulation of phylogeny', and interest waned in embryological studies which had a specifically phylogenetic bias. In a scholarly account, Gould (1977) has revived interest in the relevance of acceleration and retardation (particularly of retardation of somatic development, or neoteny) making the telling point that a small genetic change in such regulatory systems may produce profound phylogenetic changes even where the difference in structural genes is minimal, as it is between humans and chimpanzees. He predicted that renewed interest in these mechanisms would become a major issue of evolutionary biology and would perhaps provide insights into the mechanisms of macroevolutionary change and of the emergence of evolutionary novelties.

Some of the embryological deficiencies in the synthesis were remedied by Waddington (1957), who demonstrated the mechanism of 'genetic assimilation'. Where environmental change acts to induce non-heritable modifications in the phenotype, selection will favour those variants showing optimal response to the new conditions; development is thus canalized to the point where the well-adapted phenotype is 'assimilated' into the genome, giving the impression of inheritance of acquired characters, for example, in the case of inherited skin callosities.

The modern synthesis, devised in the 1940s and subsequently further refined, has had a relatively short tranquil period of widespread acceptance. Now, however, the theoretical world of evolutionary studies is in tumult again over the promotion of the paired concepts of a punctuational model of macroevolution and the analytical tool of cladistics. To read the popular press one would conclude that the whole neo-Darwinian fabric of conventional evolutionary belief is again under attack. A more balanced view would be that we are entering a further phase in refining and clarifying evolutionary mechanisms—a process started by Darwin and continued by the architects of the modern synthesis.

Cladistic analysis

Following the English translations of the works of Hennig (1965, 1966) 'cladism' has emerged in the last decade as a dominant theme in much of the writing on phylogeny and taxonomy. The essence of cladistics lies in the recognition of specialized, derived, or apomorphic characters and their distinction from primitive or plesiomorphic ones. The emergence of new apomorphic characters, coupled with the branching of lineages or clades, (which as a methodological device is always considered as dichotomous although in nature it may not necessarily always be so) gives rise to sister groups sharing these derived characters or synapomorphies. In this way a branching hierarchy is derived, consisting of a nested set of synapomorphies uniting sister groups: a cladogram (Fig. 1.1A). A consequence of this methodology is that all known species, including fossils, are located on terminal branches, never at the nodal points which would rather represent hypothetical ancestral morphotypes. The rationale behind this is that it is supposed that it can never be fully possible to establish that fossils which are usually fragmentary, might not possess some unrecognized apomorphic characters which would thereby exclude them from ancestral status. Theoretically then, cladograms are rigorously objective and are moreover falsifiable in the best traditions of the scientific method.

The theories of cladistics and punctuated equilibria share the notion that evolution progresses by the branching of lineages, or speciation, rather than by the transformation of existing species, which is known as 'phyletic evolution'. However, the idea that ancestral forms cannot be identified, is at odds with the very basis of the theoretical model of punctuated equilibria which holds that new species arise from peripheral populations, isolated from the main body of the parental species which may itself persist unchanged. The reluctance to identify ancestral species has sparked off the most vehement opposition to cladistics and its use in, for instance, recent exhibitions at the British Museum (Natural History); the charge has even been made that there is an underlying political (Marxist) bias in this (Halstead 1980). Yet when the palaeontologists in the past have proposed ancestral status for a particular taxon, they have often done this in the full realization that it should only be viewed as a model, ancestral only in those features of mosaic evolution which are preserved (Delson 1978).

Moreover, when a cladogram depicts a split into two sister groups, it is by no means improbable that one or the other may remain unchanged (Delson 1977; Eldredge and Tattersall 1975), converting the relationship in reality into a parent–daughter one (Eldredge and Tattersall 1975). The cladogram merely avoids commitment on which of the possible ancestor–descendant relationships exists.

Thus cladograms, true as far as they go, and also testable, are but one stage in the reasoning leading towards the more speculative procedure of construction of phylogenetic trees which then include a reasoned assessment of ancestor–descendant relationships and also of stratigraphic order (Tattersall and Eldredge 1977).

New taxonomies may also be derived from cladograms. These often differ radically from the traditional ones which are based not only upon phylogenetic branching but also upon overall phenetic resemblance, and which have done good service as an acceptable and stable information-retrieval system, although not necessarily completely mirroring phylogeny (which may be quite impracticable) but not seriously conflicting with it. Doctrinaire application of cladistics to taxonomy has generated much of the acrimonious debate surrounding the technique. Branching points are taken as irrevocably determining the categorical rank of subsequently evolving taxa: a single branch or a whole nested set of branches must be given similar taxonomic rank. As Mayr (1974) has pointed out the whole enormously diverse bird radiation would then have to be accorded the same categorical rank as a restricted sister group such as the crocodilians.

There is little doubt that such revolutionary classifications tend to create great confusion and indeed often seem to offend simple common sense. If this controversial aspect of cladistics is set aside however, there is widespread agreement that the

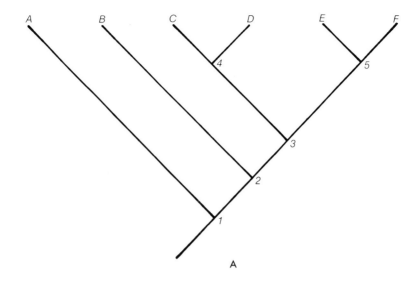

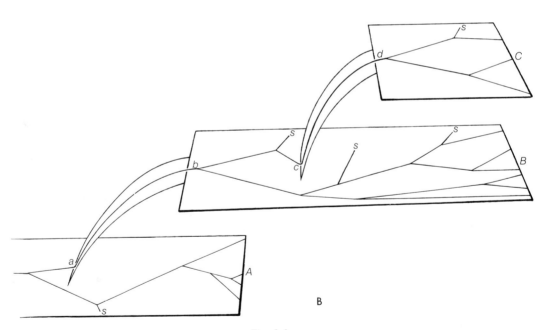

Fig. 1.1
A, a typical cladogram showing the relationships of six known phyletic groups, *A–F*. The nodes *1–5* identify successive morphotypes defined upon the basis of the establishment of new derived character states which are shared as synapomorphies by all the groups beyond that node. *C* then is the sister of *D* since both share the apomorphic characters defined by node *4*; in the same way *A* is the sister group of the whole clade *B, C, D, E. F*. In a sense the nodes correspond to hypothetical ancestors. B, an interpretation of the evolutionary view of Sewertzoff (1929). The dichotomous branching on the surfaces *A, B, C* represents idioadaptation; great declinations (*S*) mark specializations. Aromorphosis is indicated by *ab* and *cd*.

approach has had a beneficial catalytic affect on phylogenetic studies by the way in which it has concentrated attention in a disciplined way upon the recognition of apomorphic characters (Mayr 1974). Yet this really only represents the culmination of a changing emphasis in evolutionary studies. At one time it was fashionable to stress the value of non-adaptive, 'sheltered', or 'palaeotelic' characteris in determining phylogenies (Gregory 1910). Gradually, however, there has been a trend towards the recognition of the overriding importance of adaptive characters (Clark 1959) also known as evolutionary novelties (Mayr 1960).

Cladistic methods, however, are not a universal panacea for solving the problems of evolutionary relationships, and often they merely provide a spurious veneer of scientific impartiality. Thus, Gardiner (1982) applied this method to tetrapod relationships and came up with the startling conclusion that birds are the sister group of mammals and cynodonts are only distantly related to mammals, even more distantly than crocodiles and dinosaurs. As Cox (1982) has pointed out, the flaw in this analysis is the inadequacy of the morphological data used. A similar notorious example linked monotremes with marsupials on the basis of a single synapomorphy (Kuhne 1973) and for the author this represented 'the end of the argument'.

Relying on the vagaries introduced by the use of too few characters is only one of the pitfalls awaiting the unwary in cladistic analysis. In a transitional series of states of one character complex it is of fundamental importance to establish which of the states is primitive, and which derived: the morphocline polarity. Moreover failure to recognize parallelism or convergence may give rise to false indications of affinity. Szalay (1981) has rightly stressed that the most potentially useful characters are those complex ones which can be related to some major function such as feeding or locomotion, and which have been subjected to detailed functional analysis.

Attempts to devise methods for weighting such characters and establishing correlations help the credibility of the analysis. This approach contrasts with the view of post-Hennigean purists who scorn any attempt to incorporate functional analysis or weighting. At the end of the day, cladistic analysis is only as good as the quality and sophistication of the functional morphological markers utilized.

Punctuated equilibria and macroevolution

The central dogma of the synthetic theory is that phyletic evolution (or phyletic gradualism) is the prevailing mode of evolutionary change. Of course, Darwin and his later disciples recognized that splitting, or speciation, must occur or there would be no increase of taxa; but this splitting was regarded as essentially incidental and proceeding by the same mechanisms of selection of small variations. Indeed, Darwin believed that species were progressively modified across their population range, but he did recognize that the rate of change might fluctuate widely. Moreover he was well aware of the significance of geographical isolation as a factor in evolutionary change (Rhodes 1983). When, rather belatedly, the importance of geographical barriers was appreciated more precisely this produced no particular conflicts within the synthetic theory and the notion became recognized as the prime mode of speciation by most biologists; geographic speciation was seen as a slow, steady transformation of two separate lineages, essentially two cases of phyletic gradualism progressing in tandem. Nor did any special explanations seem to be required to explain evolution above the species level (macroevolution); all that was needed was extrapolation of the same process to a higher level. Clearly the fossil record should reveal this majestically gradual transformation but it has become abundantly clear that it does not. Refuge has usually been sought in the facile appeal to imperfections in the palaeontological record, yet this is far from convincing. Species of one geological bed often differ widely from the preceding ones, even where there is no stratigraphic gap. This, together with the widespread acceptance that allopatric speciation is of fundamental importance, has seriously undermined the whole concept of phyletic gradualism.

In response to this the alternative idea of punctuated equilibria was formulated (Eldredge and Gould 1972) and has been progressively refined (Gould and Eldredge 1977; Stanley 1979). In reality the germ of the notion was nascent in the synthetic theory: the concept is less revolutionary than is sometimes portrayed. The norm for a well established species with a large population is stasis with only minor progressive changes, producing a superficial veneer of modification of the basic morphology in factors such as body size. In essence this is just phyletic gradualism, and change is buffered by stabilizing selection which eliminates genes which might disrupt canalized developmental pathways and by the homeostatic mechanism produced by gene flow (Charlesworth and Laude 1982). It is suggested in the punctuated equilibria theory, however, that small peripheral isolates of the main body of the parent species commonly occur and that these rapidly experience developmental instability leading to the eventual acquisition of morphological distinctions and genetic isolating mechanisms. No revolutionary new mechanisms are proposed for the genesis of this divergence, but the emphasis on the various mechanisms recognized in the synthetic theory is rather different. It is known that phyletic evolution in domesticated animals, and in the 'Drosophila cage', can occur at rates corresponding to the geological 'instants' which are envisaged in the punctuational theory for divergence of these geographical isolates. Rather more significance, however, can probably be attributed to the founder effect, evolutionary drift, and the enhanced potentiality for fixing chromosomal changes in these small populations; what Stanley (1979) has termed 'high-amplitude' sources of variability. The role of regulatory or rate genes, which, with a miniscule change in the genome can initiate the rapid appearance of evolutionary novelties, is perhaps of especial significance. The real point is that within this great proliferation of diverging species, the changes are essentially random with regard to the mainstream of evolutionary trends in the parent population. Fortuitously, one of these peripherally isolated species may be especially adapted to colonize preferentially the ancestral range and in so doing to cause the extinction, or the emigration, of the parent species. Thus, the stratigraphic record of the ancestral range will show a break, but it is a real discontinuity, not an artefact of the record. Clearly the search in the fossil record should be for geographic rather than stratigraphic variation, although the actual chance of finding the location of the emergent new species is obviously quite improbable. The geological record then should show long periods of stasis interspersed with quite rare punctuational events of virtually 'instantaneous' change.

Speciation in the random isolates is envisaged as a 'boom or bust' phenomenon with small populations technically attaining the status of species with considerable frequency in these 'experimental' forcing grounds. Their subsequent survival or not, with extinction being the most likely outcome, is subject to drastic pruning. Thus selection is envisaged as occurring at a hierarchical level above that of the individuals of a local population. This is species selection: whole species take the place of individuals, extinction substitutes for birth and death, and speciation provides the variability, just as mutation and recombination do in local populations. The speciation itself has a strong random element, even if natural selection is the main factor in moulding new species in their isolated environments. Species selection provides the directional element. In this way it is said that macroevolution is decoupled from microevolution: a new direction is introduced into the ponderous and restricted progress of phyletic evolution. Great proliferations in numbers of species, with subsequent extinction of many, is a well recognized feature of the fossil record, and the idea of species selection is really as Darwinian in essence as is the usual concept of selective establishment of genes in local populations.

The suggestion that macroevolution has been decoupled from the steady stream of phyletic change has provoked some of the most adverse reactions to these theories of punctuational evolution. They are often depicted as representing something like a resurgence of the ideas of saltative evolution and 'hopeful monsters' promoted by the mutationists. However, in essence the punctuational model only invokes mechanisms recognized in the modern synthesis. In the lengthy perspective of geological time the punctuational mechanism presents a disjointed picture but if the time scale is changed to a finer one there is a continuity of descent differing from the synthetic view largely in emphasis. Thus a comprehensive theory seems to be emerging which accommodates punctuational ideas, but it is a far less radical departure from convention, than is sometimes supposed.

What appears to be a convincing palaeontological documentation of this process of punctuated equilibrium in action has been given by Williamson (1981*a*) for molluscs in the Turkana basin. It has been pointed out (Jones 1981) that the rates of change in these peripheral isolates (5000–50 000 years) are quite compatible with the neo-Darwinian mechanisms operating in domesticated breeds and in the *Drosophila* population cage but this, of course, is an integral part of the view of the punctuationists, as pointed out by Williamson (1981*b*). Yet very real questions remain to be answered about the validity of this apparent demonstration of punctuated equilibrium (Mayr 1982).

Cronin *et al.* (1981), in an important survey of hominid evolution, have failed to find anything in it compatible with the model of punctuated equilibria. Indeed, it seems probable that the evolution of the genus *Homo*, characterized as it is, in essence, by increased brain size, is truly an example of phyletic gradualism, and, of course, it has never been denied that this mechanism does operate on occasions. Even these authors concede the possibility, even likelihood, that speciation occurred earlier in small populations at the pongid–hominid divergence.

Comparative anatomy and evolutionary theories

It has been mentioned above that detailed comparative anatomy played only a minor part in forging evolutionary theory in the West. It was not so in Russia, where a surprisingly sophisticated synthesis was achieved before that in the West (Adams 1980). The central figure was Sewertzoff (also spelt Severtsov) whose writings were almost entirely in Russian although there is one major survey in English (Sewertzoff 1929). It is startling to see how he pre-empted, and even improved upon, a number of the ideas which are currently being debated (Fig. 1.1B). He was well aware that evolution proceeded by speciation, depicting this as a dichotomous branching, and he realized that this must proceed by the natural selection of those derived characters which are adaptive to changes in the environment. Moreover, he distinguished changes which merely adapted organisms to their own environment (idioadaptations) from those which produced a quantum leap to another level of organization (aromorphosis). His diagram neatly

portrays the idea that macroevolution is decoupled from the underlying phyletic trend. Thus he was familiar with the essence of cladistics—emergence of apomorphic characters—but saw the importance of distinguishing the most important of these (aromorphosis); the idea of weighting characters is only now receiving due attention in cladistics. Moreover, Sewertzoff realized that periods of aromorphosis were probably of short duration, leaving few transitional fossil forms, whilst idioadaptive periods might be very long; he had thus perceived the basic features of the evolutionary model of punctuated equilibria. This brief period of enlightened morphological enquiry waned and comparative anatomy languished in Russia, as in the West.

There has been something of a revival in comparative anatomy with the trend away from sterile purely descriptive studies, to those of functional anatomy (Bock and Wahlert 1965). The emphasis has moved to a search for what have been called 'habitus' features, that is those genetically controlled characters correlated with a particular way of life, which have a trigger or threshold effect such that under favourable circumstances they may result in a period of explosive evolution (Davis 1954). These characters are often very few in number (Schaeffer 1948). Indeed, much of my own work, which will be recounted in later chapters, has been concerned with identifying just such derived character states in limb morphology, where they have a clear-cut functional and adaptive role.

So it was that even before the term 'apomorphic' came into fashionable usage, morphologists were aware, at least intuitively, of the evolutionary importance of such derived states. Current debates about cladistics and macroevolution have brought such studies into the limelight. The Primates, and this book is largely about Primates, might seem unfavourable material for such studies. Their apparent essential primitiveness, and that includes man, is a byword among morphologists. Thus Davis (1954) has bemoaned the fact that Primates, 'almost alone among the large and successful orders of mammals, are not characterized by an obvious and clear-cut major adaptation.' As will be seen this is an illusion: at an appropriate level of resolution of the anatomical detail, helped by the enormous data base provided by human anatomy, they will be seen to provide a most instructive field for such enquiries.

Evolution in the mammals has primarily involved modifications in the head and in the limbs, in particular in the distal segments of these. No satisfactory attempt at evolutionary synthesis of the morphological data for the latter has been produced, and the material to be presented here, restricted to the emergence and subsequent evolution of the primate limbs should provide a fertile field for the analysis of evolutionary principles as they are currently being debated. Particular attention will be paid to the recognition of apomorphic characters, what their functional meaning might be, how they might therefore be weighted, and what evolutionary mechanisms might have brought them about.

In limb evolution morphological change can be convincingly correlated with function. Subtle changes in the form of joint surfaces can produce disproportionate modifications in the whole pattern of movements and these are matched by appropriate ligamentous restraints and suitably changed muscle attachments. Could such correlated complexes of adaptive structure really be moulded by a neo-Darwinian process involving the gradual incorporation of minor variants each with its own genetic determinant? Certainly, a significant number of practical morphologists have felt that this strains credibility, and have been seduced by Lamarckian ideas invoking the inheritance of acquired characters. The solution to this dilemma seems to be gradually emerging, as experimental embryologists continue to explore the problem (Bonner 1982). It seems that genes specifying particular protein molecules must be at a number of removes from the actual developmental processes unfolding the overall integrated morphology. The middle ground between genome and environment seems to involve supragenomic, or epigenetic mechanisms where inbuilt morphogenetic rules hold sway. These are apparently often controlled by regulatory genes whose roles may be quite non-specific, yet they set off accumulating and cascading effects during development. Some of these processes will be explored in Chapter 3.

2

The origin, evolution, and geographical deployment of the Primates

It is well known that the Primates, and even man, have retained a remarkably primitive mammalian structure in many aspects of their anatomy. For this very reason perceptive comparative anatomists have long realized, at least intuitively, that marsupial anatomy can yield useful insights into primate morphology. The morphological strategy used in this book will repeatedly have recourse to consideration of marsupial and even monotreme anatomy in order to elucidate emergent primate functional complexes. To vindicate this approach it is necessary to survey briefly the 'dark ages' of the evolution of the mammals—this comprises two thirds (140 million years) of their history, from their origin in the latter part of the Triassic, to their extensive radiation at the beginning of the Paleocene.

The mammals in the mesozoic

The fossil record covering this phase of mammalian history has been greatly expanded in recent years. Coupled with the revolution in the geographical sciences, providing information on the changing position of the continents and oceanic basins, a clearer realization is beginning to emerge about the main trends characterizing this early phase of mammalian evolution. The published data about these fossils is almost entirely restricted to observations on dental and cranial morphology and inevitably this greatly limits speculation about the probable ecology and likely behavioural repertoires of the fossils. Yet quite a large number of postcranial specimens (even apparently whole skeletons) have been recovered, but largely remain undescribed; the expertise and the background data of functional anatomical

knowledge necessary for meaningful assessments have simply not been available.

However, the emerging, although blurred, picture derived from the fossil record, provides a framework for hypotheses developed in this book; perhaps then the conclusions arrived at can be fruitfully applied to the postcranial fossil record in order to validate or falsify these ideas.

Throughout the Triassic, and extending into the early Jurassic, the continents as depicted on modern maps were confluent forming one supercontinent—Pangaea. During the early part of this period the predominant land vertebrates were mammal-like reptiles, in particular therapsids, including both herbivorous and carnivorous varieties. In the latter part of the Triassic the therapsids were largely eclipsed by the rise of the Archosaurs or ruling reptiles. The dinosaurs, both saurischian and ornithischian forms, were most prominent amongst these and were to hold sway until the latter part of the Cretaceous when they too became extinct. However, two very mammal-like groups of therapsids, the trithelodontids ('ictidosaurs') and tritylodontids (both now regarded as members of the Cynodontia) did survive the extinction of the remainder of the therapsids and lingered on into the early Jurassic. More significantly at some time during the Triassic the first mammals were derived from one of the groups of carnivorous cynodonts, perhaps from early representatives of the trithelodontids (Kemp 1982).

These first mammals were tiny, varying from shrew to rat size. They were insectivorous and probably nocturnal, and were apparently highly active probably at the interface between the arboreal and terrestrial habitats (Crompton and Jenkins 1979). It is generally assumed that they were egg-laying as were their reptilian precursors;

they were the first 'Protorherians'. The situation in which they found themselves was ripe for their diversification: small-bodied species, morphologically unspecialized and with limited mobility and a coarse-grained perception of the environment, would readily segregate into isolated subgroups and would thus tend to speciate according to the theory of punctuated equilibria (Gould and Eldredge 1977).

These 'Protorherians' soon diversified into three orders, the Triconodonta, the Docodonta, and the Multituberculata which were widely dispersed throughout the confluent continental mass of Pangaea. The parent or stem order from which the others were derived was apparently the Triconodonta (and in particular the family Morganucodontidae). These 'Protorherians' formed the major part of the mammalian fauna (although they themselves were completely overshadowed by the dinosaurs) until the late Cretaceous, when 'Therians' emerged to become the dominant mammals. The docodonts died out at the end of the Jurassic (Kron 1979). The multituberculates, however, flourished (Clemens and Kielan-Jaworowska 1979) in their primary role as herbivores (the 'rodents of the Mesozoic') and in fact showed a considerable expansion in the middle Cretaceous, with the origin and spread of the flowering plants or angiosperms. They reached their greatest diversity in the Paleocene, but finally succumbed in the Eocene in the face of the overwhelming competition posed by emergent placental herbivores.

The Triconodonta (Jenkins and Crompton 1979) however, diversified into three families, and survived until the end of the Cretaceous. The stem family of this minor radiation, and the earliest to appear was the Morganucodontidae; almost at the same time as morganucodontids appear in the fossil record so too do the first 'Therians' (*Kuehneotherium*). These belonged to the earliest representatives of the pantotheres, the Symmetrodonta, and it is probable that they were derived from early morganucodontids (Cassiliano and Clemens 1979). It is generally considered that these first 'Therians', unlike the 'Protorherians', were viviparous and produced altricial young. Thus it is supposed that there was a very early dichotomy in mammalian evolution into the basic two lineages, 'Protorherians' and 'Therians' but it is now widely held (Crompton and Jenkins 1979) that mam-

malian origin was monophyletic as stated here: in the past a polyphyletic origin for the different mammalian groups, supposedly derived independently from the cynodonts, was commonly canvassed. Yet the early stages of this dichotomy are blurred and the terms 'Protorherian' and 'Therian' are used in only an informal taxonomic sense (Lillegraven 1979*a,b*). This blurring is emphasized by the way in which certain 'Protorherian' groups, although without surviving descendants, demonstrated a latent potentiality to evolve what are usually thought of as therian postcranial features (Jenkins and Crompton 1979). The multituberculates in particular evolved hindlimb characters, particularly in the tarsus, with quite advanced mammalian characteristics. It has even been suggested that some of the later representatives had independently evolved viviparity, producing altricial young (Kielan-Jaworowska 1979). From this palaeontological perspective the two genera of highly specialized living monotremes (platypus and echidna) are usually considered as surviving 'Protorherians' although no clues about the likely origin among fossil 'Protorherians' have been forthcoming (Clemens 1979*a*).

By the mid-Jurassic the Symmetrodonta, destined to die out themselves at the end of the Cretaceous, had given rise to the more advanced Eupantotheria; these two groups are considered to be orders belonging to the infraclass Pantotheria. There is apparently an undescribed eupantothere skeleton and the suggestion has been made that they were possibly arboreal (Kraus 1979).

By the late Jurassic the northern group of continents had been separated from the southern plate (Gondwanaland) by the Tethys sea, although there were probably intermittent connections between Europe and Africa (Lillegraven *et al.* 1979). Pantotheres, especially Eupantotheres, certainly occurred at this period, on both sides of this divide, in what was to become North America, Europe, Africa, and South America. The Eupantotherians were destined to be the source of the more advanced Therians; the symmetrodonts, the other subdivision of the Pantotheria, were doomed to extinction by the end of the Cretaceous.

The more advanced Therians made their appearance in the so-called 'middle' Cretaceous—an informal term for the end of the Early and the beginning of the Late Cretaceous. This was a time of commencing continental break-up both north and south of the Tethys sea; Eurasia was separating from North America and Africa from South America. It was also the time of a beginning incursion of epicontinental seas: the Mowry sea, to be succeeded by the Western Interior seaway, between Eastern and Western North America, and the Turgai strait or Obik sea between Europe and Asia. There was, however, at least partial or intermittent connection across these waterways for a time (Lillegraven *et al.* 1979). It is likely that these factors favoured geographical isolation and speciation. This was also the period of spread of the flowering angiosperm flora with its attendant pollinating insects. In these circumstances the more advanced therians, the so-called Theria of Metatherian–Eutherian grade, characterized by a tribosphenic dentition (Bown and Kraus 1979), suited to an insectivorous diet were derived from the Eupantotheria. They were the precursors of the marsupial and placental lineages. This important phase in Therian evolution is unfortunately only documented by scraps of jaws or teeth from Asia and North America, and in England from rather earlier in the Cretaceous (Kielan-Jaworowska *et al.* 1979*a*).

The late Cretaceous was characterized by further barriers to dispersal and at this time Metatherians (Marsupialia) and Eutheria were derived from the Theria of Metatherian–Eutherian grade. There is little doubt that these latter already possessed many of the characters usually considered as diagnostic of the marsupials, and in these the marsupials are therefore merely primitive (Clemens 1979*b*); it has been suggested also that they had reproductive systems of essentially marsupial type and were viviparous in that characteristic fashion (Lillegraven 1975, 1979*b*). Similarly the early derived Eutheria inherited many 'metatherian' characters, destined to be lost in later Caenozoic placentals.

It seems likely that the Metatheria evolved initially in Western North America. These primitive didelphid-like marsupials in the late Cretaceous negotiated what was probably a 'sweepstake' route of island chains to reach the separate South American continent. Some of these primitive didelphids ranged further reaching Australia, via what is now Antarctica, apparently in the early Tertiary. Here they give rise to the extensive marsupial radiation on that continent. There is

fairly general agreement on the broad outlines of this scenario, although various points are still disputed. Authoritative reviews have recently been given by Clemens (1971, 1977), Lillegraven (1974), and Keast (1977). Marsupials lingered on in North America (and in Europe which they reached in the Eocene) as a very minor element in the mammalian population until the Miocene. Late in the Pleistocene the opossum *Didelphis* invaded North America from South America across the Panamanian isthmus formed at the end of the Pliocene, thus reestablishing a marsupial presence in North America (Keast 1977; Clemens 1977). Recently fragmentary putative marsupial fossils have been described from Africa and Asia but this does nothing to invalidate this general story since it is likely that they were far reaching wayward members of the Eocene incursion into Europe (Benton 1985).

In contrast to the Metatheria, it seems likely that the Eutheria emerged from the Metatherian–Eutherian stock in Asia. These earliest Asian placentals retained many features usually considered as 'metatherian', and there are indications in the pelvis that the gestation period was very short and of marsupial type (Kielan-Jaworowska 1975). These emergent Eutheria belonged in the families Leptictidae, Paleoryctidae and Zalambdalestidae, and might be considered as belonging in the 'wastebasket' order Insectivora (Kielan-Jaworowska *et al.* 1979*b*).

Eutherians of this type presumably reached North America by a 'sweepstake' route involving intermittent connections or island chains across what is now the Bering Strait. It is noteworthy that no marsupials apparently made the reverse passage, perhaps because of the marine circulation limiting the direction of rafting (Clemens and Kielan-Jaworowska 1979). At first placentals must have been very rare mammals in North America, but by the end of the Cretaceous they were becoming more numerous and diverse: members of the order Condylarthra had appeared (*Protungulatum*) and what is supposedly the earliest Primate (*Purgatorius*). It has been suggested that the whole of the succeeding Paleocene eutherian fauna of North America and Europe (collectively known as Laurasia) could have been derived from *Procerberus* (a leptictid), *Protungulatum*, and *Purgatorius* (Clemens *et al.* 1979).

Relationships of living monotremes, Metatheria and Eutheria

Huxley (1880) who coined the terms Prototheria, Metatheria and Eutheria, used them to represent stages of evolution, and only with reservations did he include the modern monotremes in the first and the living marsupials in the second. This was because the living monotremes are clearly highly aberrant and specialized, and even the most generalized modern marsupials present quite extreme specializations in, for example, dentition and reproductive tract. Above all, in the minds of most comparative anatomists these three mammalian groups are separately categorized by their widely different reproductive strategies, yet the gulf is much more apparent than real.

Reproductive strategies in the three mammalian grades

Oviparity in monotremes is more advanced than that in, say, chelonians or in birds, for the intra-uterine period of development is prolonged and both embryonic nutrition and respiration are carried out by the shelled egg in this environment, with considerable growth of the embryo up to a size of 18 or 19 somites. In a morphological sense marsupials show only a slight advance on this and marsupials are strictly ovoviviparous: the embryo completes two thirds of its intra-uterine development surrounded by the shell membrane and floating freely in maternal fluid, and there is only a brief period of choriovitelline placentation. There is also a striking similarity in developmental state between newborn marsupials and newly hatched monotremes, and their homologous pattern of differentiation of fetal membranes supports the notion that the ovoviviparity of marsupials was derived from a system of oviparity comparable to that of monotremes (Luckett 1977).

Certain authors who are convinced of the unique distinctiveness of marsupials, and who regard Huxley's idea that the Metatheria were a grade on the way to the Eutheria as a baneful influence on understanding mammalian evolution, believe that viviparity must have evolved independently in these two groups. Yet there is ample evidence that this is not so (Lillegraven 1975) and that the marsupial system is phylogenetically ancestral to the eutherian reproductive strategy.

Marsupials are born at a very immature stage, roughly equivalent to that of a ten-day mouse or rat embryo, and much of the process of organogenesis occurs after birth. Continued development is ensured by the embryo transferring to the nipple and effecting a seal by closing its lips around it; a pouch is not invariably present to protect it here— it is lacking, for instance, in the American genus *Marmosa*. Similar epidermal seals close and protect the eyes and ears in this exposed extra-uterine situation. Yet similar transitory closures occur intra-uterinely even in nidifugous Eutherians. Moreover in some nidicolous insectivores and rodents lip seals are not merely a fleeting developmental phenomenon, for the young, little more advanced in development than marsupials, may show extensive attachments to the nipple and be dragged about by the mother, in marsupial fashion. There is thus persuasive evidence indicating that the immediate ancestor to marsupials and placentals was viviparous in the manner in which marsupials are today. With the 'invention' of trophoblast by the eutherians, which conferred immunological protection and the capacity for a lengthened gestation period, an entirely new evolutionary vista of heterochrony and anatomical 'experimentation' seems to have been opened up.

Origin of the monotremes

It has long been recognized that monotremes share a wide variety of character complexes with the supposedly much more advanced Therians, and this even prompted Gregory (1947, 1951) to propose his radical palimpsest theory which proposed that the monotremes were, in effect, backsliding derivatives of the Australian marsupial phalangeroid stem. This idea has been resurrected by Kuhne (1973) who on the strength of a cladistic analysis using only a single apomorphic character, maintained that monotremes have their nearest relationship with marsupials and that the dichotomy of these two lines must have occurred after the cladistic separation of placentals and marsupials, therefore not earlier than 'middle' Cretaceous. Few would now subscribe to such a radical view which has the inherent difficulty of explaining why the monotremes should have abandoned a marsupial reproductive strategy for a return to oviparity. Yet the theory did focus attention on the anatomical similarities between the two groups. Jenkins (1970*b*) in a wide-ranging review of the anatomical and physiological characteristics of monotremes convincingly argued that they possess a wide array of characteristics which are shared with the Theria and which in all probability must have been inherited from a common ancestor—his 'prototherian level of organisation'! A major stumbling block to the idea of any relatively late derivation of the monotremes was the belief that the cavum epiptericum of the skull had been enclosed quite differently in monotremes and in Therians, leading to apparently irreconcilable differences in the morphology of the side wall of the brain case. The essence of this conflict has now been resolved (Presley 1981; Kemp 1982). Kemp (1982) has suggested that the monotremes diverged from an early (eupantothere) part of the therian stock. This suggestion circumvents a major difficulty with the notion of an earlier prototherian derivation: monotremes possess ear ossicles closely resembling those of Therians, yet in the Mesozoic Prototherians and the earliest Therians (for example, the pantothere *Kuehneotherium*) both reptilian and mammalian jaw joints were present and post-dentary bones had not realized their new sound-conducting role in the middle ear (Crompton and Jenkins 1979). Those who argue for a prototherian origin of monotremes necessarily invoke parallel evolution of a quite improbable scale, to explain this. The idea that monotremes had a therian origin has received a considerable boost from the discovery of an early Cretaceous monotreme fossil in Australia. Archer *et al.* (1985) in describing this specimen even suggested that it had tribosphenic teeth indicating a very late therian derivation. This has been refuted (Kielan-Jawarowski *et al.* (1987), yet it is agreed that the fossil was derived from well along the therian line, probably in the late Jurassic. The only real conflict is with current taxonomy which is, in any case, in a state of flux. It is clear that the earliest members of the 'Theria', before the branching off of the monotremes, must then have had an oviparous mode of reproduction making inappropriate their present therian designation (which is

based entirely on cranial and dental criteria) with its implications of viviparity.

Eutherian–Metatherian relationships and the arboreal theory of the origin of mammals

The detailed studies of Bensley (1903) (based on earlier work by Huxley and Dollo) have been widely accepted as virtually proving that marsupials were descended from arboreal ancestors. Bensley showed that the primitive marsupial foot was of the prehensile type, similar to that now possessed by certain of the Didelphidae, and having a widely divergent opposable hallux, adapting it to a clinging, perching grasp in the trees. A more cursorial locomotor pattern in derivative forms induced changes in this structure involving loss of prehensilism, a reduced hallux, and elongation of the foot, producing a pedal mechanism adapted to quick balanced running in the trees or on the ground. Even in the most primitive and arboreal of the Dasyuridae, the phascogale, these trends are firmly established. In the more terrestrial members of the family they are more advanced (Figs 2.1A, B, C, D). The Phalangeridae retain a foot of essential arboreal type, except for the specialization of syndactylism of the second and third digits; again, their terrestrial descendants, the Macropodidae show specialization of the pes involving elongation.

Matthew (1904) developed this idea and sparked off what has become a longstanding controversy by suggesting that not only marsupials, but also placentals, had arboreal ancestors. It is often suggested that this idea was initiated by Matthew, only to be later refuted by Gidley (1919) but this is a simplistic view of the real progress of the debate. In fact, neither of these authors provided firm evidence either for or against the proposal. Matthew almost entirely confined his locomotor speculations to the forelimb suggesting that the pollex was primitively opposable and he merely mentioned opposability of the hallux as a late arboreal adaptation. This was unfortunate, since it is the foot which is primarily concerned with grasping in arboreal locomotion; the hands are more involved in feeding and grooming. Gidley, however, mounted his attack on Matthew by correctly pointing out the lack of opposability of the hallux in tree-living rodents and

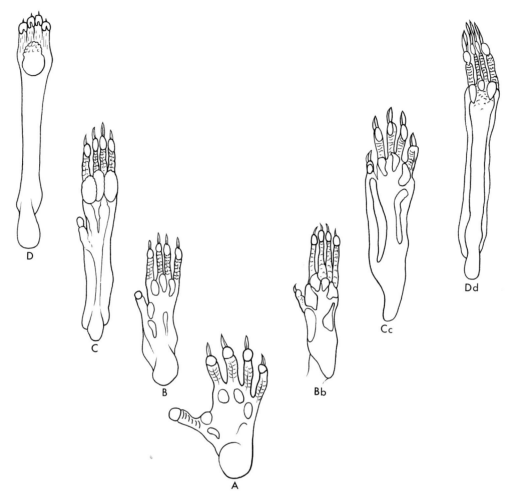

Fig. 2.1

A–D, modifications in right feet showing increasing terrestrial commitment in a series of marsupials, in part after Bensley (1903). A, a primitive didelphid; B, *Phascogale*; C, *Sminthopsis*; D, *Antechinomys*. Bb, Cc, Dd, comparable modifications in selected eutherian mammals. Bb, *Ptilocercus*; Cc, *Tupaia*; Dd, *Macroscelides*.

insectivores and even in the supposedly primitive marmosets among the Primates. Yet Gidley did concede that the primitive Eutherians must have possessed a divergent hallux, one which was readily convertible to an opposable one—his criterion for arboreality. Haines (1958) also rejected Matthew's theory finding no evidence of ancestral arboreality in the hands of placental mammals. It is noteworthy that these rejections of the arboreal theory took little note of foot architecture.

The discussion on reproductive strategies above makes it quite feasible, even likely, that the antecedents of the Eutheria had an essentially 'metatherian' reproductive pattern and presumably other aspects of their anatomical organization were similarly 'metatherian'; it would, however, be inappropriate to name these common ancestors Metatheria. Moreover, the feet of extant Eutheria having a varying cursorial and terrestrial commitment parallel in form those of comparable marsupials in a striking way (Figs 2.1Bb, Cc, Dd); it would be not unreasonable to suppose that they too were derived from an ancestral prehensile pattern like that of the stem marsupial (Fig. 2.1A). This does not presuppose, however, that the most committed arboreal Eutherians, the Primates, had merely preserved the foot structure of that stem form. In fact, evidence from the anatomy of the joints of the ankle and foot, which will be considered in Chapters 12, 13, 14, dictates otherwise. The following hypothesis, based on these findings, was originally proposed by myself some years ago (Lewis 1980a).

An hypothesis of the role of arboreality in the evolution of Metatheria and Eutheria

It is probable that the last common ancestors of the marsupials and placentals, the advanced Therians of the 'middle' Cretaceous, were in many features of their locomotor anatomy (particularly in the hindlimb) similar to generalized living marsupials and were well adapted to a scansorial or arboreal life. This does not necessarily mean exclusive canopy-dwelling; it could equally well have included the spatially complex interface between arboreal and terrestrial habitats represented by the forest floor and margins, with its tangle of roots, vines and bush-like growth. It is likely that the emergent marsupials became particularly committed to arboreal life, whilst the placentals pioneered the more terrestrial niches. This marsupial–placental dichotomy could have been related to the spread of the flowering plants occurring at that time. As Cartmill (1974) has pointed out the accompanying insect fauna are concentrated in two main strata: the canopy on the one hand, and the forest floor and bushy forest margins on the other. The marsupials may have occupied the first zone and the placentals the latter. The Primates were derived, in a manner as yet unclear, from these early placentals. A good case has been made (Jenkins 1974) for the fact that these stem placentals occupied the three dimensionally complex forest floor niche, including both 'terrestrial' and 'arboreal' features, the habitat occupied by some living treeshrews. As Jenkins (1974) then says: 'the adaptive innovation of ancestral Primates was, therefore, not the invasion of the arboreal habitat, but their successful restriction to it.' On the basis of this idea that 'terrestrialism' and 'arborealism' are not discrete phenomena the old standing controversy about an arboreal or terrestrial origin of placental mammals becomes to some extent irrelevant.

It should be emphasized that treeshrews are not really adequate locomotor models for the stem form from which the primate lineage might have been derived, although they are only too often supposed to be. The treeshrew pes is lengthened and the hallux reduced, convergently resembling the more terrestrial dasyurids (Fig. 2.1B). Jenkins (1974), however, has stressed that even this reduced hallux shows strikingly effective independent action on uneven or arboreal substrates. It seems much more likely that the placental ancestor of the Primates inherited a relatively unreduced opposable hallux, more comparable to that in living didelphids or phalangers, even if other internal structures such as the ankle joint were remodelled away from the marsupial pattern.

Moreover this secondarily arboreal primate lineage include members where the pes has undergone further lengthening and reduction of the hallux in response to a return again to a terrestrial niche. As described by Morton (1924) this has occurred in the marmosets among the New World monkeys and among the Old World monkeys the Patas monkey shows a marked degree of terrestrial modification. Thus, it is not a question of the Eutheria having either an arboreal or a terrestrial ancestry; in a sense they have had both.

Szalay (1984) has vigorously attacked this hypothesis but his case contains many distortions, misunderstanding of published anatomical data, and quotations taken out of context. His partisan viewpoint is exemplified in the title of the article: 'Arboreality: is it homologous in metatherian and eutherian mammals?' It should be clear from the above that there was never any suggestion by me that it was homologous. Szalay's article is based almost entirely upon foot anatomy and some of his points will be considered further in Chapter 17.

The first primates—Cretaceous and Palaeocene

The first Eutherian mammals of the late Cretaceous are usually referred to the order Insectivora. This is a paraphyletic or 'wastebasket' order consisting of a diverse assemblage of primitive placental mammals with uncertain affinities, although the Leptictidae are usually considered to be central to this evolutionary radiation (Van Valen 1967). The condylarths (*Protungulatum*), the ancestors of the ungulates had separated from this group by the end of the Cretaceous, as had the creodonts, the ancestors of the carnivores. Many believe that the tupaiids (treeshrews) and Primates were also independently derived from some leptictid insectivore such as *Procerberus* (McKenna 1966; Van Valen 1965).

Until quite recently it has been common practice to include the treeshrews in the order Primates but they are now almost universally excluded. The status of the treeshrews has recently been reassessed in cladistic terms in a multiauthor volume; the general conclusion (Luckett 1980) was that the tupaiids do not share any uniquely derived features with living or fossil Primates which would warrant their inclusion together in a monophyletic single order, and indeed the available evidence rather suggests that the treeshrews have evolved independently since the early Tertiary. Taxonomically they might be included in the paraphyletic order Insectivora or perhaps better in a new order, the Scandentia. Their superordinal affinities remain unclear but some of the immunological data in this review volume and a contribution by Szalay and Drawhorn on tarsal anatomy provided a minority viewpoint that the Scandentia, the Primates and the Dermoptera should be lumped together in a superorder Archonta (this resurrected term had been used originally for these groups together with the Macroscelidea, the elephant shrews). Despite their separate taxonomic status, however, the treeshrews still can provide some insight into the heritage characters of the emerging Primates.

The first supposed Primate, *Purgatorius ceratops* appears in the late Cretaceous of Montana and is said to be a paromomyid (a member of the Paromomyiformes) with resemblances to condylarths and leptictid and erinaceoid insectivores (Van Valen and Sloan 1965). These primitive presumptive Primates (the Paromomyiformes) formed a radiation of four families (Fig. 2.2) in the Palaeocene: Paromomyidae, Plesiadapidae, Carpolestidae, and Picrodontidae (Savage 1975). It has been generally assumed that the suborder was arboreal and that the central family was the Paromomyidae. The better known Plesiadapidae were quite aberrant cranially. These paromyiforms

dispersed even to Europe bridging across the North Atlantic via Greenland. It is, however, arguable whether the Paromomyiformes were really arboreal and even whether they should be considered as Primates (Kay and Cartmill 1977).

The radiation of the prosimians

Undoubted prosimian Primates were widespread in Laurasia, the continental mass north of the Tethys sea, in the early Eocene. It is likely that North America was the centre from which they dispersed, and that they reached Europe, with the great influx of advanced placentals in the early Eocene, across a North Atlantic land bridge including Greenland (Russell 1975; Savage 1975). Access to Asia was also possible perhaps from North America across a circum-Arctic route where now lies the Bering Strait. The initial immigrants to Europe included both lemuriform adapid Primates (Northarctines) and primitive tarsiiforms (Omomyidae); most believe that the latter were derived from early members of the former (Szalay and Delson 1979; Szalay 1975). It is probable that the adapids and omomyids reached Africa from Europe via a land bridge at Gibraltar (Campbell and Bernor 1976) in the early Eocene but the poor Paleocene and Eocene fossil record makes this speculative (Hofstetter 1974) although there is circumstantial evidence that this was feasible for rodents similarly colonized Africa from Europe at the same place and time (Lavocat 1980). Probably about this time adapids entered Madagascar (Walker 1972) from Africa probably by a filter bridge route. As Madagascar subsequently drifted away into increasing isolation, the colonizing lemuriforms underwent a remarkable radiation giving rise to the extant Lemuridae and Indriidae. The most popular view concerning the origin of the Lorisidae (Szalay and Katz 1973; Hofstetter 1974) is that they were derived secondarily from the Cheiragaleinae, the little mouse lemurs. The lorises, derived in (or entering) Africa gave rise to the extant forms, and fossils are known from the Miocene. Apparently the lorises entered and populated Asia, just as hominoids did, via a land bridge at the beginning of the middle Miocene. An alternative, but less popular view, (Charles-Dominique and Martin 1970) derived the lorises

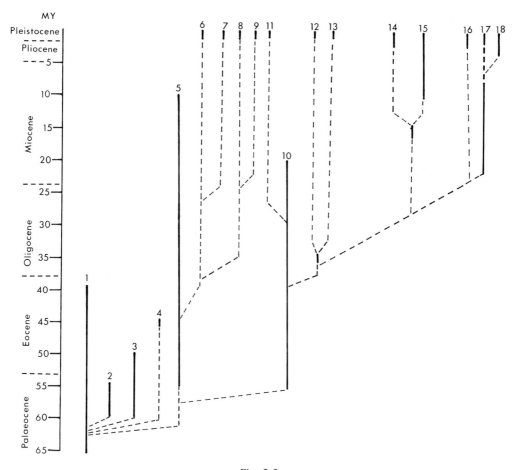

Fig. 2.2
Hypothetical relationships of primates at the family level, (and in a few cases subfamily level) redrawn, simplified and in some cases modified from figures by Szalay (1975) and Szalay and Delson (1979). The scale on the left is in millions of years before present. Heavy solid lines indicate known ranges and broken lines define relationships. The resolution, being only at family level, is insufficient to indicate the special relationship between *Pongo* and the other great apes, and of the Hominidae to the African great apes. Numbers indicate taxa given in the synoptic classification at the end of this chapter.

and lemurs independently from a primitive primate stock, and sees the galagines and cheirogaleines respectively, as the most primitive, and so most similar, members of those diverging lines.

Europe received a second invasion of more advanced prosimians from Asia, presumably arriving there from America via the Bering strait, at the end of the early Eocene, (Russell 1975) although this migration subsequently was effectively halted by the transgression of the epicontinental Obik sea (Turgai Straits) which isolated Europe from Asia in the middle Eocene. These late immigrants included

advanced, and so more specialized, members of both the lemuriforms (adapines) and tarsiiforms (microchoerines).

This scenario for the origin and dispersal of the undoubted Primates of modern aspect—the Euprimates, which excluded the paromyiforms—is very plausible, but, in fact, is based upon quite flimsy evidence, and has not gone uncontested. The discovery of a primitive Euprimate in the middle Palaeocene of Asia (Szalay and Li 1986) raises the alternative possibility that this continent may have been the original source for a widespread northern dispersal event. Africa has even been proposed as the source (Schwartz and Tattersal 1979), but with the lack of any substantial Palaeocene fossil record from that continent, this remains an unsubstantiated hypothesis.

Origin of the higher Primates

There can be little doubt that the higher Primates (monkeys and apes) evolved from some prosimian stock. A plausible scenario for this dichotomy has been suggested by Cartmill (1980) and runs as follows. Prosimians are fundamentally nocturnal insect eaters. A shift to diurnal fruit-eating, with opportunistic predation on insects, perhaps characterized the emergence of a Tarsius–Anthropoid clade, which included synapomorphies such as development of a retinal fovea, postorbital septum etc. Cartmill suggested that the living *Tarsius* is a secondarily nocturnal survivor of the early pioneers of this line. Fundamental to this idea is the separation of the Primates into two natural groups: Haplorhini (higher Primates plus the tarsiiforms) and Strepsirhini. This presupposes the origin of the higher Primates (Anthropoidea) from early tarsiiforms, and most likely from the Omomyidae. This idea has been widely espoused in recent years (Szalay 1975; Szalay 1977; Szalay and Delson 1979). A different but less popular view holds that the Anthropoidea, together with the lemurs and the lorises, were derived from Adapidae, but the tarsioids were independently linked to earlier plesiadapiform Primates; the basic dichotomy then according to this view is between the Lemuriformes, Lorisiformes, and the Anthropoidea on the one hand, and the Plesiadapidae and Tarsiiformes on the other (Gingerich 1975; Gingerich and Schoeninger 1977). In an update of this theory Gingerich (1981) inclined to the view that the tarsiiforms (omomyids and *Tarsius*) were derived from the Adapidae, eliminating plesiadapid Primates from this lineage, yet continuing to deny any place for the tarsiiforms in the ancestry of the higher Primates.

The radiation of the higher Primates

The earliest catarrhines and platyrrhines

Undoubted higher Primates are found in the Oligocene of the Fayum region of North Africa and already they were quite diversified. The earliest platyrrhine in South America (*Branisella*) is also from the Oligocene. The origin of the Anthropoidea, whether it be adapid or omomyid, although the latter is probably more likely, was thus presumably in the later Eocene, but the locale of the transition remains debatable. It might of course have been within Africa; both omomyid and adapid stocks had already entered there, just as rodents did from Europe. Wood (1980) however has suggested that the rodents entered Africa from Asia Minor, and the implication is that the first higher Primates could have been immigrants using the same route. Indeed Gingerich (1980) has suggested that *Amphipithecus* and *Pondaungia* of Burma were transitional 'protosimians' of ambiguous adapid/simiiform status which entered Africa by this route across a narrow Western part of the Tethys Sea; *Oligopithecus* of the Fayum has been said to be of similar grade. An Asian origin for the Anthropoidea (but from an omomyid stock) is also suggested by Delson and Rosenberger (1980) and Ciochon and Chiarelli (1980). This notion hinges on the real status of the Burmese fossils, but this is controversial, since they are known only from a few jaw fragments. Whether as immigrants, or evolved *in situ*, it is clear that protoanthropoids were present in Africa in the latter part of the Eocene. For long it had been accepted that the Platyrrhini and Catarrhini had evolved in parallel from separate prosimian stocks. Tantalizing resemblances of the Fayum Primates (undoubted catarrhine precursors) to the platyrrhines (Conroy 1976), however, has prompted speculation on an African origin for the New World monkeys (Hoffstetter 1974; Szalay 1975, 1977).

The origin and radiation of the platyrrhines

The idea that the New World monkeys were initially derived from Africa is gaining respectability and support. A recent multidisciplinary review (Ciochon and Chiarelli 1980) came out strongly in favour of the view that the Anthropoidea represent a monophyletic group, whose ancestry extends back into the Eocene, and that the Platyrrhini and Catarrhini are two separate and equal divisions of this group, each being strictly monophyletic in its own right. This same review concluded that palaeogeographic evidence favoured an African origin for the platyrrhines; comparable arguments favour the origin of the caviomorph rodents of South America from an African stock. The apparently insuperable barrier posed by the South Atlantic has been seen to be less daunting, as understanding has increased about the processes of continental drift. In the Early Tertiary, up to and including the Eocene–Oligocene boundary, it seems probable that oceanic islands, some hundreds of square kilometers in size, existed offshore from both Brazil and West Africa and that the mid-oceanic rise was also exposed for much of this period (Tarling 1980). Deep oceanic waters probably rarely exceeded 200 km and the stepping stone route which was thus uncovered seems to have been favoured by east to west currents.

In contrast the Caribbean separation between South America and Central America (which was part of the North American plate) in late Eocene–Early Oligocene times seems to have exceeded 1200 km. Although chains of volcanic islands may have existed here, the oceanic currents would seem to have favoured drifting from South to North America and so inhibit southerly migration (Tarling 1980). Nevertheless, there are still proponents backing the origin of both South American Primates and rodents from the North by this route (McKenna 1980; Wood 1980; Orlosky 1980; Gingerich 1980; Delson and Rosenberger 1980). Even this view of the direction of influx of the South American Primates is not, however, incompatible with the concept of the Anthropoidea as a monophyletic group (Rosenberger 1986) since at least some of these workers suggest that the North American protoplatyrrhines were derived from an Asian stock which had entered North America via a Bering bridge connection, and which may also have entered Africa across the Western Tethys Sea, there to evolve into catarrhines.

Cartmill *et al.* (1981), basing their views on a study of the temporal bone have also rejected the idea of an African derivation of the platyrrhines. They did, however, support the concept of monophyly for the Anthropoidea and suggested that the ancestral anthropoid was a mid-Eocene inhabitant of either Eastern Asia or Western North America and that dispersal from one hemisphere to the other was probably via Beringia. They differed

from other authors, however, in stating that the ancestral anthropoid was presumably derived, neither from known adapids nor from omomyids, but from an undiscovered group of Eocene prosimians. Despite rejecting the omomyid view of ancestry for the Anthropoidea, they did propose that *Tarsius* is the sister group of the anthropoids— this is the so-called tarsiid–anthropoid hypothesis.

All in all, however, the balance of evidence seems to favour an African origin for the platyrrhines, even though the known Fayum fossils are probably too advanced and too committed to the catarrhine path of specializations, to have themselves provided that stock. The earliest platyrrhine fossil is *Branisella boliviana* from the early Oligocene which dentally is intermediate between omomyids and cebids. Other fossils of Oligocene and Miocene age from South America are known and all are cebid-like. The extant genera *Cebus* and *Saimiri* appear to be closest to the basic platyrrhine type, and the Callitrichidae, far from being primitive as is often assumed, seem to be derived forms (Rosenberger 1977).

The origin and radiation of the catarrhines

The mid-Oligocene Primates of the Fayum are the earliest known undoubted Anthropoidea (higher Primates). They include two relatively large species, *Propliopithecus* and *Aegyptopithecus* which have been described as ape-like; indeed *Aegyptopithecus* and *Propliopithecus* have been described as the ancestral apes (Simons 1972, 1978). This view is increasingly discounted in favour of the idea that the suite of supposedly 'hominoid' features which they have been said to possess, are in reality just primitive catarrhine characters which were, however, retained in the true apes which emerged later (Delson and Andrews 1975; Delson 1975a, 1977; Andrews 1985).

A number of authors, however, persist in including these fossils in the Hominoidea (Fleagle and Simons 1982; Fleagle and Kay 1983, 1985; Simons 1985) despite their acceptance of the primitive character of the Fayum Primates, and the likelihood that they are ancestral to both cercopithecoid monkeys and the apes. Two smaller marmoset-sized fossil Primates, *Parapithecus* and *Apidium*, are also found in the Fayum and have been considered for their part as ancestral cercopithecoids. More likely they were merely the ecological vicars of monkeys among the early catarrhines of the Fayum (Delson 1975*a*). The conservative nature of these two genera is emphasized by their possession of three premolars and this suggests their possible sister group relationship to the platyrrhines. It seems unlikely that they are related phyletically to modern catarrhine monkey species (Delson 1975*a*).

From a pool of undifferentiated catarrhines such as those represented in the Fayum must have been derived both the apes and cercopithecoid monkeys.

The emergence of the apes   Fossils of undoubted apes appear first in early Miocene interrift localities of East Africa such as Songhor and Rusinga Island, and also other sites such as Koru, which are currently being opened up (Andrews 1981*a*; Harrison 1981). All these sites were in forested areas (Andrews and Van Couvering 1975; Evans *et al.* 1981) and there are indications that these emergent apes were frugivorous and arboreal (Kay 1977*a*). Six species of these apes are members of the Dryopithecinae (Andrews 1974; Delson and Andrews 1975), and it is now probably preferable to allocate them to three genera (Andrews 1977)—*Proconsul major*, *Proconsul nyanzae*, *Proconsul africanus*, *Rangwapithecus vancouveringi*, *Rangwapithecus gordoni* and *Limnopithecus legetet*. Also found here is *Limnopithecus macinnesi*, now assigned to the genus *Dendropithecus* (Andrews and Simons 1977); this last genus was originally allocated to the Hylobatidae, but it is now more usual to consider it a conservative leftover from the Oligocene catarrhines (Andrews 1981*b*, 1985) which in a few ways has converged on the gibbons, but whose gibbon-like dentition is merely primitive; and it lacks both clear-cut hominoid and cercopithecoid synapomorphies. *Micropithecus clarki* from Uganda has been said to be a related genus (Fleagle and Simons 1978). It has been suggested that *Proconsul major* was directly ancestral to *Gorilla*, and *Proconsul africanus* to *Pan*, but most now consider this unlikely, and that resemblances are more probably the results of ecological and size related convergence (Delson and Andrews 1975).

About 16MY (million years) ago a land bridge was completed between the Afro-Arabian plate and Eurasia, an event which is taken to herald the beginning of the middle Miocene (Andrews and Van Couvering 1975; Thomas 1985). The collision of Africa and Eurasia greatly reduced the Tethys sea, resulting in a cooler and drier climate, with the conversion of closed forest into open forest-woodland (Laporte and Zihlman 1983). Apes, perhaps related to *Proconsul* (Andrews and Tobien 1977) crossed and dispersed into Eurasia along this wooded corridor and there evolved into at least two species of *Dryopithecus* and the two genera *Ramapithecus* and *Sivapithecus* which are closely related to one another; there is a tendency recently to subsume *Ramapithecus* under *Sivapithecus* (Andrews 1982). It is possible that the true apes were accompanied by, or preceded by, survivors of the early conservative catarrhine radiation which then gave rise to the European genus *Pliopithecus*, often, like *Dendropithecus*, mistakenly said to be a gibbon relative (Delson and Andrews 1975). The fossil history of the gibbons is in fact shrouded in mystery and neither *Pliopithecus* nor *Dendropithecus* are likely candidates for ancestry. A slightly better case may be made for the African *Micropithecus* and similar fossils to it have been found in China; at least these would be in the right place at the right time (Fleagle 1984).

Ramapithecus is known from teeth and jaws of middle to late Miocene (15–7MY) age in Pakistan, China and West Eurasia (Delson and Andrews 1975; Simons 1981). The earliest specimens appear to be from Pasalar in Turkey (Andrews and Tobien 1977), and related fossils, inducing the usual proliferation of dubious taxonomic names, have been found in Greece and in Hungary. At the *Ramapithecus–Sivapithecus* sites there appears to have then been an environmental switch from forest to open woodland interspersed with grassland although not actually savannah. *Ramapithecus–Sivapithecus* were apparently not restricted to Eurasia but also populated sites in Africa at middle Miocene age (about 15MY) which were also open woodland (Evans *et al.* 1981; Shipman *et al.* 1981). If it is true that the initial ape diversification took place outside Africa then the specimens at these sites must represent repatriated migrants from Asia.

Accompanying *Ramapithecus–Sivapithecus* in Asia, but of later Pliocene and Pleistocene age, is the larger *Gigantopithecus*. All these fossils show dental resemblances to hominids, and this dental morphology has been interpreted as a specialization correlated with ground foraging for seeds (Jolly 1970) the implication following from this being that these apes were becoming terrestrially adapted.

The stage was thus set for the claim that *Ramapithecus* was the ancestral hominid and this view was promoted with considerable conviction (Simons 1972, 1977, 1978). Simons (1978) proposed a seemingly convincing scenario: that the middle Miocene apes were confronted by regions with climatic variation so that the successive ripening crops of fruits enjoyed by their early Miocene predecessors in the tropical forests might not always have been available enforcing increased forays to the ground for omnivorous feeding. Kay (1981) queried the accepted view on the nature of the dental specializations of *Ramapithecus* suggesting instead that they are adapted to a diet of hard fruit and nuts, thus throwing doubt on the evidence for terrestriality.

Support for the hominid affinities of *Ramapithecus* has now waned and the idea is being increasingly canvassed that the *Ramapithecus–Sivapithecus* group is ancestral to *Pongo* (Pilbeam 1982; Lipson and Pilbeam, 1982; Andrews 1982; Pilbeam 1985). This view seems to have much to recommend it. A minority viewpoint, which has not received wide acceptance is that the 'ramapiths' had their origin in Africa, spread thence to Eurasia, and were the stock from which all great apes and hominids were derived (Laporte and Zihlman 1983).

In the late Miocene, continued cooling and desiccation of the Mediterranean region brought about the extinction of the hominoids in southern Eurasia (Laporte and Zihlman 1983).

The emergence and diversification of the catarrhine monkeys   It is a seeming paradox that the forests of the early Miocene were dominated by apes and not monkeys (Andrews 1981b). How then, and when, did the Old World monkeys arise? The view that the cercopithecids were derived from parapithecids of the Oligocene Fayum (Simons and Delson 1978) now receives little support and the parapithecids, *Apidium* and *Parapithecus*, are rather being viewed as merely primitive catarrhines, showing some convergent resemblances to monkeys. The idea that the so-called apes from the

Fayum, *Propliopithecus* and *Aegyptopithecus*, are really only primitive catarrhines and may form the source of both true apes and cercopithecids is also gaining ground (Delson 1975a; Kay et al. 1981; Fleagle 1983). The earliest indisputable fossil cercopithecids are known from the Early Miocene of Egypt (*Prohylobates*) and Uganda (Delson 1975a, 1979; Delson and Andrews 1975). Their emergence seems to have been correlated with an ecological experiment in supplementing the primitive frugivorous diet with leaves in times of shortage. With increasing dependence on a folivorous diet it seems that a bilophodont dentition was gradually established. These emergent cercopithecids seem to have been essentially arboreal or perhaps semi-arboreal quadrupeds inhabiting deciduous seasonal forests. Kay (1977b) has dissented from this view, and considers that the ancestral cercopithecids were semi-terrestrial frugivores, and Delson (1979) also admits to such a possibility.

However this may be the earliest catarrhine monkeys soon split into two stocks, the Colobinae and Cercopithecinae. The colobines were aboreal, possibly secondarily so, and folivorous, retaining an essentially primitive skull form; the cercopithecines were semi-terrestrial with an eclectic diet and face lengthening (Delson 1975a). The first evidence of these emerging specializations is shown in the fossil record by the two species of *Victoriapithecus* (Koenigswald 1969) from the middle Miocene of Maboko island in Kenya. It seems that certain of the colobines became more terrestrial by the late Miocene (11MY ago), left Africa via a wooded corridor, and colonized Europe where the fossil genera *Mesopithecus* and *Dolichopithecus* are found; apparently they reached Asia at the end of the Miocene where the extant genera are *Presbytis*, *Nasalis* and *Pygathrix* (Delson and Andrews 1975; Delson 1975a,b). The more arboreal colobines remained confined to Africa where the fossil *Libypithecus* is found in the latest Miocene of Egypt and *Paracolobus* and *Cercopithecoides* in the African Plio-Pleistocene (Leakey 1969); the living representatives belong to the genus *Colobus*.

The Cercopithecinae, in contrast to the colobines seem to have been basically semi-terrestrial and by 10–12MY ago had formed two divergent stocks. One, the tribe Papionini was basically terrestrial and includes the baboons, geladas, macaques and mangabeys, the last two, however, including some quite arboreal species. They appear to have been confined to Africa and perhaps did not leave with the colobines because they were too terrestrially committed to navigate the partly forested corridors to the north, but when the Mediterranean became dessicated at the end of the Miocene (6MY ago) macaque-like representatives apparently entered Europe and are known in Southern France from the earliest Pliocene (Delson 1975*a*). Colobines were by then the dominant European monkeys, and these two monkey lineages filled the place now vacated by the extinct European hominoids. In the later Pliocene and Pleistocene the colobines were completely supplanted in Europe by macaques, which had by then also reached Asia (Delson 1980).

The other divergent tribe of the Cercopithecinae was the Cercopithecini: the guenons. This tribe appears to have been essentially a secondarily arboreal stock, perhaps derived from early mangabeys and *Cercocebus torquatus* and *Cercopithecus nigroviridis* may be living transitional forms (Kingdon 1971). These guenons re-invaded the arboreal milieu where they there came into competition with the folivorous colobines and the various frugivorous apes, largely superseding the latter (Kay 1977*a*, *b*; Andrews 1981*b*).

The evolution of the hominids

The time, place and mode of origin of the Hominidae, the family of man, remains something of an enigma. Even whilst *Ramapithecus* was strongly touted as an early hominid, there remained the so-called 'Pliocene gap', (Simons 1981) in the fossil record from 8MY to almost 4MY ago, at which time the earliest specimens of *Australopithecus*, the first undisputed hominid, appear in the African fossil record. Now that for most people *Ramapithecus* has been effectively eliminated from a role in hominid ancestry, the gap stretches back to the early Miocene. There is, however, little doubt that the apes of the African early Miocene were the ultimate precursors of all the living hominoids including man. Also, abundant anatomical, biochemical and karyological evidence links *Homo* with the African apes, *Gorilla* and *Pan*, and perhaps particularly the latter.

Precisely when the hominid lineage diverged from that leading to the African apes, however, remains undocumented in the fossil record. There is little doubt that unequivocal hominids (leaving out of consideration *Ramapithecus*) emerged and adaptively diversified in the forest margins and the savannah of Africa. The taxonomic status and phyletic relationship of the many fossils now found in the Great Rift valley, from Ethiopia in the north to the Transvaal in the south, however, provides endless and often acrimonious debate.

Until a decade ago a fairly convincing scenario (Tobias 1973) reflecting a widely held but not universal view, ran as follows. The emergent African hominids were believed to have been lightly built small-brained gracile bipeds, first appearing in the fossil record in significant numbers about 3MY ago; they belonged to the species *Australopithecus africanus*. The majority of the fossils were known from two South African sites, Sterkfontein and Makapansgat. This basal stock was believed to have undergone an adaptive radiation, still confined to Africa, just before 2MY ago. A heavily built, muscular robust stock with an impressive armamentarium of jaws and teeth, seems to have then been derived from the gracile precursors and dispersed widely along the rift valley. Separate South African and East African species—*Australopithecus robustus* (earlier called *Paranthropus robustus*) and *Australopithecus boisei*—are commonly recognized. These robust australopithecines were doomed to extinction, but they survived well into the Pleistocene living sympatrically with other more advanced evolving hominids. The juvenile of Taung, the first autralopithecine found, and paradoxically the type specimen of *Australopithecus africanus*, now seems likely to be in fact a juvenile of the robust form and dated at just under 1MY ago (Tobias 1978) although this dating is seriously challenged (Bishop 1978).

Apart from the robust forms a further advanced stock, characterized by brain enlargement, seemed to have been derived cladogenetically from *Australopithecus africanus*, which species perhaps persisted little changed in co-existence with it for a brief but indeterminate time. The first of these relatively 'brainier' fossils, known as *Homo habilis*, appear in the fossil record at about 2MY ago (or a little less), coincident with the first clear evidence for stone tools (Isaac 1981). A specimen of *Homo habilis* from East Rudolf, was initially thought to be about 1MY older (Leakey 1973), owing to an inaccuracy in dating (Hay 1980). It seems that these hominids were but little advanced in structural grade from *Australopithecus africanus*, apart from an increase in brain size, having means between 494 cm^3 and 656 cm^3 (Tobias 1968, 1971), and there have been heated debates as to whether they should really be classified in the genus *Homo* or merely as another species of *Australopithecus* (Reed 1967). However this may be, the scenario leaves little doubt that these ape-men were transitional forms leading to the emergence of *Homo erectus* at about the beginning of the Pleistocene. For a time *Homo erectus* coexisted with the last remnants of the doomed robust australopithecines, and ultimately evolved into *Homo sapiens*.

This fairly clear-cut scenario was thrown into confusion by the discovery of abundant australopithecine fossils from Hadar, in Ethiopia (Johanson and Taieb 1976) of an age older (about 4MY ago) than South African *A. africanus*. Similar fossils, even somewhat older, were recovered from Laetoli in Tanzania (Leakey *et al.* 1976). A new taxon *Australopithecus afarensis* was created for these earliest australopithecines by Johanson *et al.* (1978). Johanson (1978) and Johanson and White (1979) postulated that *Australopithecus afarensis* represented the basal source of two divergent hominid clades; one leading to the robust australopithecines, with *Australopithecus africanus* as a transitional form, and the other leading to *Homo*. Tobias (1981) was strongly critical of this idea on the grounds of doubtful taxonomic validity, and also whether one particular geographical variant of *Australopithecus africanus* (as he interpreted it) should be considered as ancestral to *Homo*. Yet the proposal is gaining widespread acceptance and gains considerable support from the finding by Rak (1983) that there are significant synapomorphies in the facial architecture of *A. africanus* and *A. robustus*.

A sound working hypothesis seems to be that *Homo habilis* was derived from *A. afarensis* and that *H. habilis* in turn gave rise to *Homo erectus*, all of which occurred in Africa. From about 2MY ago the geographical range of *Homo* (certainly *H. erectus*, and perhaps possibly *H. habilis*) was extended beyond Africa to Eurasia reaching from Germany

to Java (Howells 1980) where the most extensive sequence of fossils has been found (Jacob 1973; Santono 1975), extending back to about 1.9MY. Claims that certain of the Javan fossils, known as *Meganthropus*, are australopithecine (Koenigswald 1973, 1975) are not however generally accepted (Isaac 1978).

The subsequent transition between *H. erectus* and archaic and modern *Homo sapiens* on the worldwide stage is still a matter of strenuous debate but has no particular relevance to this book.

This general theme for hominid evolution, although having wide appeal has not gone unchallenged. One alternative scenario has as its essence the notion of an independent *Homo* lineage, paralleling that of the australopithecines, which are thus deprived of any direct link in human ancestry. This view has been championed by Richard Leakey. The story presented is that a basic *Ramapithecus* stock had split by the end of the Miocene (6MY ago) to give rise to three new lineages: *Australopithecus africanus*, *Australopithecus boisei* and *Homo* (*Homo habilis* and later *H. erectus*). As the story goes, up to the Pleistocene these three hominid lines co-existed, at least at East Rudolf (Lake Turkana), although only *Homo erectus* was destined to survive (Leakey 1974; Leakey and Lewin 1977; Walker and Leakey 1978). This tale seemed to draw sustenance from the apparently great age (over 2.6MY) of *Homo habilis* specimens at Lake Turkana (Leakey 1972); later dating has shown that in fact they are about 1MY younger. The idea also appeared to be supported by the initial assignment of the 3.7MY old specimens from Laetoli to the genus *Homo* (Leakey *et al.* 1976); later the true nature of these specimens was realized and they were included in *Australopithecus afarensis* (Johanson *et al.* 1978). The main weak point in the story, however, is the dubious evidence for the survival of *Australopithecus africanus* into the Pleistocene at Lake Turkana. Indeed this rests on the interpretation of one skull (KNM-ER1813) and there is considerable indication that this specimen should really be included with the early *Homo* at this site (Cronin *et al.* 1981; Dean and Wood 1982). This would then leave only two contemporaneous hominid species in the early Pleistocene: early *Homo erectus* and *Australopithecus boisei*, and indeed the scenario has been presented in this light (Leakey and Walker 1976).

If the fossil evidence alone is considered there seems to be no persuasive reason for abandoning the more likely majority viewpoint given earlier.

There are also various other fringe viewpoints. Yet another theory, aptly called 'devolution from giants' by Simons (1978), considers the Asian *Gigantopithecus* to be the ancestral hominid (Robinson 1972; Frayer 1973; Eckhardt 1975). This viewpoint has been demolished by Simons (1978) who has pointed out that a form has to be older than its descendants in order to qualify as an ancestor; *G. bilaspurensis* can scarcely be older than 5MY. The evidence, in fact, is persuasive that *Gigantopithecus* was a giant open-country, terrestrial, gorilla-like member of the *Sivapithecus–Ramapithecus* group which found it possible to flourish in Asia at so late a date because of the absence there of other more advanced hominoids in its particular niche. However, regardless of the unlikely ancestral status of *Gigantopithecus*, the theory has been developed by Robinson (1972) and this has had a wider impact. He held that, after *Gigantopithecus*, the African hominid radiation was derived from an early '*Paranthropus*' which he conceived as a woodland dweller, incompletely adapted to erect posture, still to some extent arboreal, and existing as a specialized herbivore with a dental apparatus adapted to powerful crushing and grinding; the robust forms were envisaged as persistent relicts of this ancestral type. He suggested that a further lineage, all allocated by him to the genus *Homo*, was derived from these and consisted of *Homo africanus* (the gracile australopithecines), *Homo erectus*, and *Homo sapiens*. This later lineage he considered to be fully adapted to the erect posture and to a life of hunting on the open plains with an omnivorous diet, and enjoying an emergent culture. Despite the unlikely features of this theory the idea of dietary differences leading to niche specialization between the sympatric gracile and the robust australopithecines has had considerable impact on ideas about early hominid evolution.

In complete contrast to all these views is the so-called single species hypothesis, which views human evolution as proceeding in four broad stages: australopithecine, pithecanthropine (*Homo erectus*), neanderthal, and modern. At any one time there would be only one species, albeit a highly variable one. This is the view of Brace *et al.* (1971). Wolpoff (1978) also counsels caution on assuming the presence of two species at any one time, and throws doubt on the dietary hypothesis.

The most radical proposal of all (Schwartz 1984) would seek to divorce the African apes from any particular close relationship to the hominid line and to replace them instead with the orang-utan, also incidentally reinstating *Sivapithecus* to hominid ancestry. The study was essentially a cladistic one, relying only on apomorphic characters. Yet, as so often in these studies, it was in the restricted choice of characters where the fallacy probably lay. As this book progresses innumerable morphological features will be seen to link man inextricably with the African apes, and in particular with *Pan*. Moreover, there seems to be abundant molecular and biochemical evidence cementing this relationship.

Molecular evidence for hominoid evolution

When the idea of 'molecular clocks' was first floated, and it was suggested that the hominid lineage diverged from the African apes as recently as 3.5–5MY ago (Sarich 1968; Wilson and Sarich 1969), the whole approach was greeted with scorn. Simons (1981) discussed this so-called 'chimpanzee theory' with a scathing attack on 'simplistic cultural assumptions and simplistic biochemical assumptions'.

Controversy has now simmered down and the variety of techniques based on characters at the protein or molecular level have assumed their rightful place in unravelling primate evolution. Andrews (1986) has recently given a balanced review of the state of the art. In general it may be said that the molecular evidence confirms the evolutionary tale unfolded in this chapter. The timing of evolutionary divergences, in particular those derived from DNA hybridization, in general agree with those shown in Fig. 2.2. The following times of divergence (millions of years ago) are given by this technique: Old World monkeys, 27–33MY; gibbons, 18–22MY; orang-utans, 13–16MY; gorillas, 8–10MY; chimpanzee from man, 6.3–7.7MY.

All this seems very satisfactory but a word of caution is needed. Molecular methods do not form an infallible panacea for resolving evolutionary dilemmas. The techniques have their own shortcomings, which have been highlighted by

Andrews (1986) and also by Schwartz (1984). In particular the timing of the molecular clocks is decidedly suspect, and methods which purport to show a closer relationship of man to *Pan*, rather than just to *Pan–Gorilla*, may lack the resolving power to make the conclusions entirely convincing. In the last analysis it comes back to the functioning anatomy of the animals—morphology, in effect.

Classification of the Primates

As Simpson (1945) says, a classification can never fully repeat phylogeny, especially changing viewpoints about it, but should be a stable means of communication. Bearing this in mind, the following synoptic classification of the order Primates is largely based upon that of Simpson, but has been modified to include various changes which have become common usage. These changes draw on certain widely acceptable features of the rather radical new classification proposed by Szalay and Delson (1979) and various aspects of those by Simons (1972) Delson (1975a) and Hofstetter (1974).

In the conventions of taxonomy the names of higher ranks are derived from names of valid included genera by adding stylized endings to the stem form of a given generic name, as follows: Superfamily -oidea (-oids); Family -idae (-ids); Subfamily -inae (-ines); Tribe -ini (-ins). In brackets are given the vernacular endings for these same higher taxa.

The basic unit for discussion has long been recognized as the genus and under each grouping are shown representative genera. These, in fact, are the genera referred to at later points in the book, so that by referring to this scheme when a particular genus is described, its place in the taxonomic scheme of things may be determined. The numbers after the various groups refer to their place in the suggested evolutionary scheme shown in Fig. 2.2.

Another classification not shown in the scheme, is that the Lemuriformes and Lorisformes are often grouped together as the Strepsirhini, and the Tarsiiformes with the Anthropoidea as the Haplorhini; this reflects the likely evolutionary relationship described above.

Order Primates

Suborder Prosimii

Infraorder Paromomyiformes
    Family Plesiadapidae   (3)
      Plesiadapis
    Family Paromomyidae   (1)
      Purgatorius
    Family Carpolestidae   (4)
    Family Picrodontidae   (2)

Infraorder Lemuriformes
  Superfamily Adapoidea
    Family Adapidae   (5)
      Subfamily Adapinae
        Adapis
      Subfamily Notharctinae
        Notharctus
  Superfamily Lemuroidea
    Family Lemuridae   (6)
      Subfamily Lemurinae
        Lemur
      Subfamily Cheirogaleinae   (8)
        Cheirogaleus
    Family Indriidae   (7)

Infraorder Lorisifomes
    Family Lorisidae   (9)
      Subfamily Lorisinae
        Loris, Nycticebus, Perodicticus
      Subfamily Galaginae
        Galago

Infraorder Tarsiiformes
    Family Omomyidae   (10)
    Family Tarsiidae   (11)
      Tarsius

Suborder Anthropoidea

Infraorder Platyrrhini
    Family Cebidae   (12)
      Subfamily Aotinae
        Aotus
      Subfamily Pithecinae
        Pithecia
      Subfamily Alouattinae
        Alouatta
      Subfamily Cebinae
        Cebus, Saimiri
      Subfamily Atelinae
        Ateles, Brachyteles, Lagothrix
    Family Callithricidae   (13)
      Callithrix, Leontocebus, Saguinus, Cebuella

Order Primates *contd*

Infraorder Catarrhini
  Superfamily Cercopithecoidea
    Family Cercopithecidae
      Subfamily Cercopithecinae   (14)
        Tribe Cerocopithecini
          Cercopithecus, Erythrocebus
        Tribe Papionini
          Papio, Cercocebus, Macaca,
            Theropithecus
      Subfamily Colobinae   (15)
        Colobus, Presbytis, Pygathrix, Nasalis
  Superfamily Hominoidea
    Family Hylobatidae   (16)
      Hylobates
    Family Pongidae   (17)
      Subfamily Dryopithecinae
        Dryopithecus, Proconsul, Sivapithecus
      Subfamily Ponginae
        Pongo, Pan, Gorilla
    Family Hominidae   (18)
      Australopithecus, Homo

3

Evolutionary principles applicable to the limbs

For centuries comparative anatomists have dissected vertebrate animals and recorded how diversity in overall form is produced by variation in corresponding component parts. This was usually seen as evidence of purposeful design in creation but the broad comparability of structure between even the most diverse species was not generally attributed to any sort of historical relationship. Owen (1848) conceived the idea of a basic vertebrate body plan, the archetype, which had been modified or varied by a 'Divine Architect', with elaboration or diminution of various parts, so giving rise to the bewildering array of living creatures. He popularized the use of the term homology to signify the idea of equivalence of component parts, in particular of their similarity in position in the basic plan. But Owen was not simply a believer in special creation; he acknowledged some examples of evolutionary change, for instance in the horse lineage, but again under the guiding hand of some divine power.

The determination of homology, which is at the core of the comparative method, might on the face of it seem to be an uncomplicated procedure but in practice it can be dauntingly difficult, as will soon emerge. Haeckel really introduced the term to evolutionary studies, identifying organs as homologous when their presence is due to common ancestry (Szarski 1949). Gegenbaur (1870) added a further dimension to the term by stressing not only community of phylogenetic descent, but also development out of the same anlage. Thus, with the increasing sophistication of evolutionary theory, homology has come to mean not merely similarity of position in a body, that is of anatomical relationships, but also derivation by a similar pathway of development, controlled by genealogically related genes (Hass and Simpson 1946; Roth 1984).

Apparently by analogy with segmented invertebrates Owen (1848) introduced the idea that the vertebrate body consisted of a repeated series of segmental archetypal body plans. Corresponding parts in this series, for example forelimb and hindlimb, would then be serial homologues or homotypes. With the passage of time there has been increasing awareness that the basic similarity in structural plan and pattern of development shared by both pairs of limbs indicates some underlying common genetic control, even though divergent specializations are superimposed on this.

Serial homology and the ontogeny and phylogeny of mammalian limbs

Attempts to rationalize the complex morphology of the upper and lower limbs of man, be devising elaborate schemes of serial homology, have long provided a speculative diversion for human anatomists. Detailed plans have been proposed, embracing bones, muscles, nerves etc. based almost entirely on the position, relations and attachments in the human body and paying scant regard to the facts of comparative anatomy or of palaeontology.

Historically this approach has its roots in the tradition of depicting the human body in the so-called anatomical position, erect and with the forearm supinated, a stylized viewpoint which was used even by Leonardo da Vinci and Vesalius. Springing from this convention is the simple idea, repeated almost invariably in textbooks, that during development the upper limb has been externally rotated through 90° whilst the lower limb was internally rotated by a comparable amount. By the simple device of reversing these dissimilar rotations—internally rotating the upper limb and externally rotating the lower—the limbs could be placed in corresponding positions and deductions made about the equivalence of parts. This was done with great enthusiasm, a typical version being that appearing in *Quain's Elements of Anatomy* at the end of the last century (Schäfer and Thane 1892, 1899). Deductions about matched counterparts were made not only for the many osteological features but even for all the individual muscles, all this being based almost entirely upon human anatomy. It might be thought that this approach would now be only of curious historical interest; surprisingly it has been revived recently in even more elaborate detail (Brindley 1969*a*, *b*).

This idea of a simple, single rotation affecting each limb has produced considerable confusion, although many have noted that the attitude of the forelimb in the anatomical position is an unnatural one. Indeed Humphry (1858) stated that, 'pronation should be regarded as the ordinary or natural position, because it is the most easy, and the one which we assume when placed upon all fours . . .'.

An alternative scheme of serial homology represents the two appendages as effectively mirror images of one another. Humphry had noted that the limbs were surmounted by morphologically similar girdles, though located at opposite ends of the trunk, and that there was correspondence between the exposed surfaces of the upper segments of the limbs: the anterior surface of the forelimb related to the neck and the posterior surface of the hindlimb adjoining the tail. Indeed, it was usual to consider that the deltoid muscle was the serial homologue of gluteus maximus together with tensor fasciae latae and the intervening gluteal fascia. But the mirror image hypothesis as elaborated by Parsons (1908) and Geddes (1912) went much further and postulated a complete 'looking-glass' symmetry between bones and muscles of the two limbs such that tibia should correspond to ulna, olecranon to patella and thumb to the fifth toe.

Absurd though it may seem, this idea has appeared repeatedly and does seem to get some support from the course of the so-called musculo-spiral nerves (Ruge 1878; Brooks 1889) in the two limbs; in modern terminology these are the radial nerve in the upper limb and common peroneal in the lower (Fig. 3.1). In the hindlimb the musculo-spiral winds around the fibula, whereas in the

forelimb it circumscribes the radius. Moreover, in a primitive reptile such as *Hatteria (Sphenodon)* or in the monotreme mammal *Ornithorhynchus* the radial nerve gives off a branch, the nerve to anconeus, passing deep to that muscle to reinforce the lower part of the musculospiral (posterior interosseous nerve) in the supply of the extensor muscles of the forearm. This nerve of course persists even in man, although here it is totally expended in supplying the anconeus.

The situation in the hindlimb is rather different but there are tantalizing similarities. Whereas in the forelimb the nerves to the extensor muscle mass of the proximal limb segment (triceps) are bound up in the musculospiral and the combined nerve enters the limb from behind the girdle, in the hindlimb the nerve to the comparable muscle mass (quadriceps) enters independelty in front of the limb girdle as the anterior crural (femoral) nerve. Significantly in *Ornithorhynchus* this nerve is continued into the leg, just as the nerve to anconeus is in the forelimb, contributing to the innervation of the extensor musculature of the leg. This continuation is lacking in therian mammals but the lower part of the primitive anastomosis may be represented by the recurrent branch of the anterior tibial nerve.

The patterns of nerves in the two limbs have a striking looking-glass quality (Fig. 3.1) and this factor, together with the muscular arrangements at the roots of the limbs, might appear to give some credence to the idea of mirror-image homology. There is little doubt, however, that these arrangements, far from being indicators of an all-pervading system of mirror-image homologies in the limbs, represent specializations, proper to the divergent functional needs of the proximal parts of the individual limb and are projected onto a common basic plan in which preaxial borders and postaxial borders (terms introduced by Huxley) are comparable: tibia equates with radius and great toe with thumb.

Phylogenetic and ontogenetic rotations in the limbs

It is well established that the primitive attitude of tetrapods was a sprawling one (Fig. 3.1) with the three limb segments, stylopodium, zeugopodium,

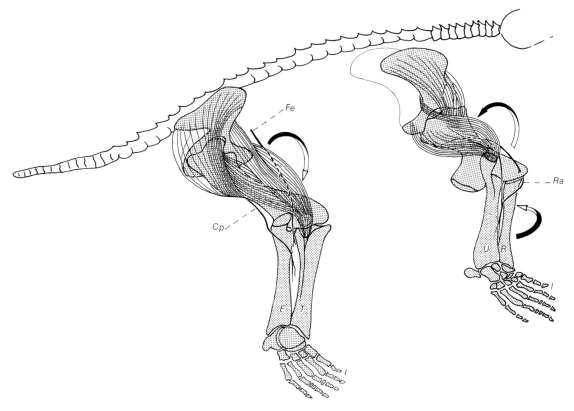

Fig. 3.1

The attitude of the limbs in a primitive tetrapod, with the extensor musculature of hip and thigh lightly indicated (after Romer 1922) and with the inferred primitive course of the nerves supplying the extensor musculature of thigh and leg also shown (after Ruge 1878*a*; Brooks 1889). The rotations effective in converting this limb posture to that typical of mammals are indicated by the arrows. *R*, radius; *U*, ulna; *T*, tibia; *F*, fibula; *I*, first or preaxial digit of each limb; *Ra*, radial nerve; *Fe*, femoral nerve; *Cp*, common peroneal nerve.

and autopodium, acutely bent at their junctions. The stylopodium, incorporating humerus or femur, projected almost horizontally and the autopodium (manus or pes, including the carpus or tarsus) was implanted flat upon the ground. This is the posture of extant urodele amphibians and most reptiles, and the characteristic gait used is a symmetrical one in which one forelimb moves with the contralateral hindlimb, these pairs alternating in body support. Long hindlimbs and shorter forelimbs seem to be a prerequisite for such a gait, by enabling the advancing pes to be placed lateral to the homolateral manus. Such limb disproportions are then also pre-adaptive for a more specialized running bipedal type of gait, which has been achieved in reptilian lines several

times in parallel. The major motive power for sprawling quadrupedal locomotion is achieved by retraction of femur and humerus, accompanied by a rotary motion of these bones about their long axes (Rewcastle 1981). Such serial homology as these limbs possess seems to be clearly one in which the preaxial borders correspond. The musculature clothing the upper aspects of stylopodium and autopodium, and the outer aspect of the zeugopodium is extensor, that on the inner concealed aspects is flexor.

The reptile–mammal transition was accompanied by lifting the trunk from the ground by shifting of the limbs into a more vertical orientation. The classical abstraction, however, of the 'fully mammalian' type, with parasagittal femoral and humeral movements, and feet placed close to the mid-line beneath the trunk, is characteristic only of the specialized cursorial mammals. In more primitive non-cursorial mammals ranging through Echidna, opossum, treeshrew, rat and ferret much of the primitive sprawling attitude is retained and the femur and humerus remain abducted from the parasagittal plane by anything from 10° to 90°. In all these the femur and humerus function in positions nearer to the horizontal than to the vertical (Jenkins 1970, 1971). This change in posture to a mammalian attitude is effected by rotation, or twisting, of the stylopodium—external rotation for the upper limb and internal rotation for the lower—so that the elbow comes to be directed backwards and the knee forwards (Fig. 3.1). In the forelimb, however, compensatory rotation of the zeugopodium in the sense of pronation, which is achieved by a partial crossing of radius over ulna, ensures that the palm remains applied to the substrate in a comparable position to the pes, with the preaxial digit in each case being medial. These rotations are recapitulated during mammalian ontogeny.

The limbs in non-cursorial mammals then provide propulsion in a manner directly derived from that of their sprawling reptilian ancestors. The femur and humerus move backwards in arcs somewhat approaching the horizontal, whilst at the same time rotating about their long axes. Owing to the changed disposition resulting from the limb rotations, these propulsive movements, pushing the body forward, must now be anatomically designated as adduction for the humerus and abduction for the femur.

These same rotational contortions affect the limbs of the developing human. Assumption of the erect posture, however, has superimposed additional extension at hip and shoulder, accompanied by straightening of the elbow and knee. The unnatural 'anatomical position' can then be assumed by unwinding the natural forearm pronation.

Homology and limb osteology

Throughout the phylogenetic history of the tetrapods, the homology of the single bones of each stylopod, or of the paired bones of each zeugopod, is never in any real doubt. This applies even when one of the zeugopodial bones, such as the fibula, is reduced to a mere nubbin, or fused to its partner, the tibia. Even significant morphological features of the bones show a fairly clear-cut genealogical homology, although the pervasive basic form is progressively reworked by evolutionary forces.

The situation is dramatically different in the autopod, with its aggregation of small bones forming the carpus or tarsus, and with radiating rays each composed of successive bony segments. Here reductions and fusions of individual bones, and even the evolutionary emergence of new bony elements, produce a wide spectrum of phylogenetic variation. The true nature of the mechanisms operating may be difficult to unravel and the determination of homologies may be far from easy. The crux of the problem is to try to elucidate how the orchestration of a basic developmental plan has been changed during phylogeny. Haeckel's biogenetic law—ontogeny recapitulates the adult stages of phylogeny—falls far short of providing an adequate interpretive framework. Largely owing to the work of De Beer (1951), it has been realized that the various parts of a master plan of development may be activated at different times and rates: the rate of elaboration of some structural features may be slowed or retarded relative to other parts even leading to paedomorphosis or neotony; alternatively development of other features may be accelerated or condensed (even to the point of elimination) often with the terminal addition of new states, leading to clear-cut recapitulation of at least one feature. Thus different developmental processes may get out of phase by alteration in their rates of expression, blurring any stylized picture of recapitulation. These altered time relationships in development, presumably controlled by 'rate genes', collectively represent the phenomenon of heterochrony.

A confusing variety of complex classifications has been applied to the various aspects of heterochrony expressed in these altered time relationships. Gould (1977) in an excellent review of the history of these ideas, has pointed out that there is an underlying simplicity in all this, which can be explained by the twin mechanisms of developmental acceleration or retardation. These rate processes, apparently controlled by a small package of regulatory genes, explain the paradox of how different genera (such as *Pan* and *Homo*) can be almost identical in structural genes, yet differ markedly in morphology. Gould suggests that these processes of heterochrony are probably the main agent of macroevolutionary change, reshuffling and re-utilizing the structural material of essentially similar genomes, to produce manifestly different final products. The applications of these principles to limb morphology promises to be a fruitful field for the understanding of evolutionary processes.

Heterochrony

Consider the case in Fig. 3.2 where two contiguous bones, *a* and *b*, in a certain species *A*, remain separate throughout life. They appear initially as individual chondrifying elements, each of which resolves itself out of an indeterminate mass of condensed mesoderm in which separate elements are not clearly distinguishable. These two cartila-

ginous precursors ossify and grow in complete independence. In the course of phylogeny, a trend towards coalescence may manifest itself, presumably meeting some new functional requirement. The period of independence of the two elements may progressively be condensed with acceleration of the time of fusion. So that in species *B* fusion may occur in late adult life, in *C* shortly before or after birth, in *D* and *E* at the cartilaginous stage (although two ossific centres develop in *D* but are superseded by a single one in *E*) whilst in *F* only a single chondrified centre appears representing the two previous bones which have now effectively become one. Selection acts on the developing organism just as it does on the adult, and clearly in the case of *F*, no selection would be operating to ensure the persistence of two centres of chondrification. No clue to this phylogenetic history is presented in *F*, even during its ontogeny. Extant species related to the intermediate phylogenetic stages *B–E* may, however, persist and examination

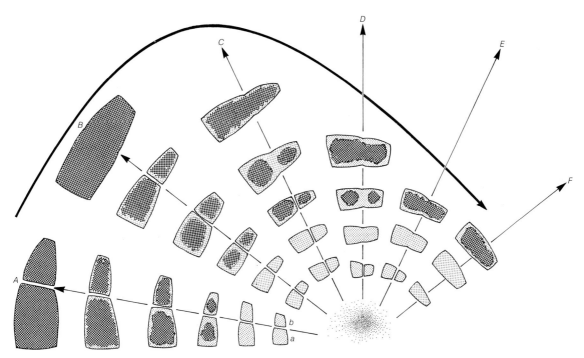

Fig. 3.2

Heterochrony affecting the fusion of two contiguous bones, *a* and *b*. Cartilage is shown in light stipple, and bone is darkly stippled. The ontogenies of a phylogenetic series *A–F* are represented by the radial arrows; only the earlier stages of the last four are figured, since the later stages are identical to those of the precursor species. The course of phylogeny is shown by the circumferential arrow. See text for detailed description.

of young and embryonic stages in the relict forms may clarify the mechanism which converted *A* into *F*.

Or take the case illustrated in Fig. 3.3 where the growth of bony element *b*, one of two contiguous bones *a* and *b*, is retarded progressively within a phylogenetic series *A–D*. Thus in *B*, element *b* is represented only by a much reduced bone. In *C* a chondrifying centre may still appear only to fail to keep pace in growth with its partner, perhaps not ossify, and then be resorbed. Finally in *D* no trace of *b* ever appears: this bone has been erased from the picture. This may be looked upon as retardation, in the sense that in species *B*, element *b* is smaller than in its ancestor *A*, and in fact corresponds in size to an earlier development stage in that precursor. Exactly how this comes about developmentally remains obscure. It could be that in stage *B* a smaller chondrogenic element *b* initially appears in the mesodermal condensation and this reduced structure then grows isometrically with *a*; that is, *b* remains proportional to *a*. Alternatively, and more probably, the rate of growth of *b* is really slowed relative to *a*; in this case we are observing an example of developmental allometry: the expo-

nential growth rates for these two structures have become different. In any case, as has been pointed out (Maderson 1982) heterochrony and allometry appear as two sides of the same coin and may combine to produce changed patterns of structural proportions. Whatever the ultimate truth about the developmental mechanics of the change there can be little doubt that it is under the control of a set of regulatory or 'rate' genes. Fossil specimens recording the end results of this series may be found, and moreover species related to these ancestors may still be extant; in this way the phylogenetic mechanism may be elucidated.

Significantly in the examples shown in both Fig. 3.2 *B–F* and Fig. 3.3 *C, D* the place of two formerly independent phylogenetic elements is now taken by one. In Fig. 3.2 this single element is the conjoined homologue of *a* and *b*; in Fig. 3.3 it is the homologue of only one (*a*). Without knowing the developmental history (not practicable for fossil specimens) it would be impossible to determine which of the two possible homologies is true; only if relict extant species, fairly closely related to the fossil phylogenetic stages persist, whose developmental stages can be studied, could a truly

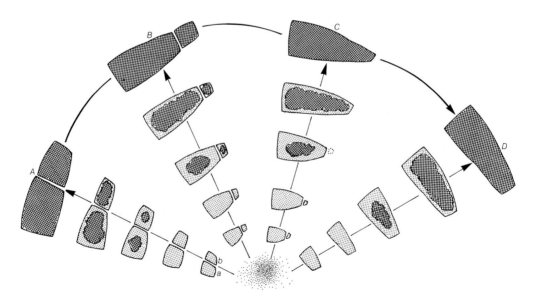

Fig. 3.3

Alteration in the growth rates of two contiguous bones, *a* and *b*, in a phylogenetic series *A–D* (represented by the circumferential arrow). The ontogenies of these successive species are shown by the radial arrows, cartilage being lightly stippled and bone shown in dark stipple. Bone *b* is progressively reduced to the point of extinction, but the increased rate of growth of *a* effectively usurps its position.

rational judgment of homology be made. Interestingly the trend to acceleration or retardation may be reversed at least partly in certain individuals of a given species giving further insights into homology; examples of this will be seen in carpus and tarsus. The way in which atavisms, echoes of distant ancestors which have long lain dormant in the genome, may be re-activated is quite remarkable (Gould 1982).

An example (Fig. 3.4) will show how these uncertainties are responsible for constraints in determining homologies, particularly across wide taxonomic divides. A represents an aggregation of bones in the autopod of an emergent tetrapod which has its counterpart B in the corresponding limb region of a generalized mammal, which is thus far removed from it phylogenetically. C, D, and E are three interpretations of the arrangements in an advanced therapsid mammal-like reptile, at least close to the precursor of the mammal B; these three versions would be indistinguishable as adult fossils but entail theoretically different phylogenetic and ontogentic processes responsible for the transition from A to B. C represents loss of bones *r, i,* and *u* by retardation of the type seen in Fig. 3.3; D represents comparable loss of some members of the *c* group and of 5. E represents one way in which fusions achieved by acceleration, as shown in Fig. 3.2, could convert A into a plausible precursor for B. Of course, other schemes of coalescence and extinction of the primeval morphological elements shown in A could be devised, similarly leading to a suitable mammalian ancestor. All these possibilities, however, must remain hypothetical since no close relatives of the therapsids have survived and hence there is no possibility of using embryological criteria to determine which of the patterns of acceleration or retardation has been operative. In the same way no surviving representatives exist of the pelycosaurs, the precursors of the therapsids, nor are any of the extant amphibia of great use in interpreting the morphology of the labyrinthodont amphibia which were ancestral to the reptiles. Clearly caution is necessary in establishing homologies in such cases. Attempts have been made to identify homologous representatives of all the components of a primitive amphibian such as A, in an autopod such as B, on the basis of indefinite mesodermal condensations described in embryonic haematoxylin and eosin stained wax plate sections. Indeed complex schemes of phylogenetic transformation have been based on this approach, which is deeply influenced by Haeckel's doctrine of universal recapitulation. These claims are now viewed generally with considerable scepticism for the poor definition and doubtful independence of the condensations makes it highly unlikely that they have any real genealogical relationship with the elements in the limbs of their amphibian forebears. More sophisticated methods reveal that, at least for the developing chick limbs, the approach is invalid. Hinchliffe (1977) applied an autoradiographic method to the study of the developing chick carpus and tarsus, labelling the chondroitin sulphate component of cartilage with $^{35}SO_4$. Chondroitin sulphate synthesis is initially widespread in the limb mesenchyme but is progressively localized into a pre-chondrogenic pattern of areas representing the future cartilaginous elements. In both carpus and tarsus the pre-chondrogenic condensations are at first not clearly distinct from one another, progressively resolve themselves as independent structures, and are not more numerous than the definitive number of cartilaginous elements. Thus in the chick limbs, at least, the carpus and tarsus do not pass through a primitive amphibian stage, with all primitive elements present.

Given the inadequacy of the methods available to determine the actual phylogenetic fate of all the elements in the ancestral amphibian carpus, or tarsus, it seems best to establish homologies in a stepwise fashion down the phylogenetic ladder and this can be done with reasonable conviction.

It is clear that elements *R, I, U* in Fig. 3.4 C, D, E are the precursors, and thus homologues, of *S, L, T* in Fig. 3.4 B, whatever their ultimate derivation. Similarly within the mammalian radiation derived from B, clear-cut homologies based on the principles of heterochrony shown in Figs 3.2 and 3.3, can be convincingly established. Filling in gaps in the fossil record can also narrow down the possibilities for far removed homologies. For instance, the presence of a stage between A on the one hand and C, D or E on the other, showing persistence of more than one of the elements *c* would tend to favour option D. In fact, the hypothetical case shown in Fig. 3.4 closely corresponds to the actual phylogeny of the carpus, and

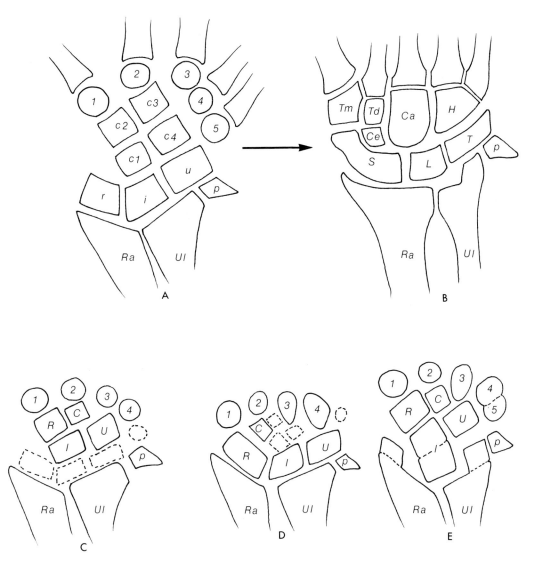

Fig. 3.4

Evolution of the carpus, as described in the text, from a primitive tetrapod A to a generalized mammal B. C, D and E represent possible intermediate stages, at the advanced therapsid level, immediately ancestral to the mammals; all three of these possible intermediates would appear similar at the adult stage (and in the paleontological record) but they differ widely in how they are derived from the primitive tetrapod, either by coalescence or by erasure of individual elements. In all figures: *Ra*, radius; *Ul*, ulna. The carpal elements shown in the primitive tetrapod A, following traditional nomenclature: *r*, radiale; *i*, intermedium; *u*, ulnare; *c1, c2, c3, c4*, the four centralia; *1, 2, 3, 4, 5*, the five distal carpals; *p*, pisiform. The derivative carpal elements shown in the mammal B are: *S*, scaphoid; *L*, lunate; *T*, triquetral; *Ce*, centrale; *Tm*, trapezium; *Td*, trapezoid; *Ca*, capitate; *H*, hamate; *p*, pisiform. The therapsid carpal elements in C, D, and E, following traditional nomenclature are: *R* radiale; *I*, intermedium; *U*, ulnare; *C*, centrale (more than one is present in more primitive mammal-like reptiles); *1–5*, distal carpals; *p*, pisiform. The elements in C, D and E may be similarly named to components in A but are not necessarily homologous with their like-named counterparts.

the nomenclature is that used in palaeontology. More detail will be added in the next chapter.

Supernumerary bones

Considerable arrays of supernumerary ossicles have been recorded in the hand and foot of man and diagrams specifying their location usually pay tribute to the monumental work of Pfitzner in documenting them; a formidable terminology usually complements the description. Pfitzner's work was carried out on a remarkable collection of macerated human specimens and only rarely did he make forays into the comparative field in the search for comparable ossicles. These few cases will be given especial treatment in later chapters. For the great majority of these anomalous human ossicles there is no evidence that they represent the reawakening of any past primordial components by a reversal of the processes of heterochrony shown in Figs 3.2 and 3.3.

Carpal and tarsal bones, of course, ossify in a manner quite comparable to that of mammalian epiphyses. Calcification and degeneration of the core of the cartilaginous model occurs and is replaced by bone receiving its blood supply from adjacent cartilage canals. Growth then occurs by proliferation of the covering investment of cartilage in advance of the radiating expansion of the ossific centre. Supernumerary centres of ossification may occur in processes or apophyses projecting from the main mass of cartilage. These may fail to fuse with the primary centre thereby giving rise to supernumerary ossicles as for example with the os hamuli proprium, a separated hook of the hamate, or the os styloideum, a separated styloid process of the third metacarpal (Pfitzner 1900*a*, *b*). At its more extreme expression this process may lead to dual centres in a carpal or tarsal bone where normally there is one. Thus we may find a bipartite carpal scaphoid or triquetrum, or a bipartite tarsal medial cuneiform. None of these anomalous bones seem to have phylogenetic significance and need to be distinguished from those that do; this point will be made at a number of places in this book.

Epiphyses and sesamoids

Parsons (1904) introduced the commonly used

classification of epiphyses: pressure epiphyses—at the weight-bearing articular ends of bones; traction epiphyses—at the attachment of muscles or tendons; atavistic epiphyses—fused, phylogenetically reduced skeletal elements.

Pressure epiphyses are not ubiquitous at the ends of long bones in tetrapods and so are clearly not essential for bone growth. For instance, amphibia have no pressure epiphyses although they are found in some reptilian bones. Birds have a single one, at the upper end of the tibia (Parsons 1905). They are, of course, almost universally found at the ends of mammalian long bones, but not quite. Where they are absent growth occurs in a manner not basically dissimilar from that of a mammalian long bone equipped with an epiphysis, for the diaphyseal ossification simply invades the growing cartilaginous cap at the bone end whose only modification may be some superficial calcification. Indeed bones which have perhaps historically possessed true epiphyses may revert to this type of growth. This is regularly seen at the head of the human first metacarpal bone, and the bases of certain of the other metacarpal bones where osteogenic tissue from the shaft extends into the cartilaginous extremity and expands there in ballooning fashion as a so-called 'pseudoepiphysis' demarcated by radiographically obvious 'notching' from the shaft (Lee and Garn 1967); at these same sites true independent epiphyses sometimes occur, presumably representing the primitive condition.

Moreover, long bones may grow by a combination of these processes with true epiphyses capping only a portion of the articular end of a growing bone. This is seen especially clearly at the upper extremity of the tibia of birds and at the same site in at least some lizards. Pathologically also, apparently due to abnormal pressure effects, the epiphysis may cap only part of the ends of human long bones as in Madelung's deformity of the radius, tibia vara, and some types of scoliosis (Rang 1969). Significantly in these disorders the metaphysis grows more slowly where it is uncapped by the epiphysis perhaps indicating that the vascular epiphysis typical for mammals promotes growth by increasing the nutritive supply to the underlying proliferative zone of the growth plate. These variations of pressure epiphyses are of significance in interpreting certain following points.

Traction epiphyses have fuelled a considerable controversy. Pearson and Davin (1921*a, b*) sparked this off by the suggestion that sesamoid bones in tendon or muscle arise by the separation of a bony process, in effect a traction epiphysis. Parsons (1904, 1908) turned this argument on its head by contending that at least many traction epiphyses result from the fusion of sesamoids which were present at an earlier phylogenetic stage; Barnett and Lewis (1958) elaborated this theme which seemed to explain in a satisfying way many morphological features. I would not now subscribe to the more extreme view taken then but it still seems virtually certain that in some few cases sesamoids have really fused to form integral parts of related bones, traction epiphyses in effect; awareness of this is central to the understanding of a number of features of limb morphology.

The patella is a sesamoid in the extensor musculature over the knee and has an ancient tetrapod history (Fig. 3.1), sometimes being only cartilaginous but often ossifying in reptiles and invariably so in birds and mammals. There is little doubt that in some birds it can fuse to the tibia to become a large traction epiphysis (Barnett and Lewis 1958); apart from this oddity (which, however, establishes the possibility of sesamoid amalgamation with adjacent bones) the patella, of course, preserves its separate integrity. A comparable sesamoid may be found in the extensor musculature of the forelimb (triceps) where it overrides the elbow, but in the forelimb this muscle group inserts to the postaxial bone, the ulna (Fig. 3.1). This sesamoid, sometimes known as the ulnar patella, is frequently found in reptilia and even in some amphibia (Parsons 1904). These arrangements were a persuasive factor in prompting Parsons (1908) to espouse the idea of looking-glass symmetry in the limbs. Significantly in lizards the pressure epiphysis caps the ulna asymmetrically, rather like the comparable upper tibial epiphysis of lizards or birds. In Primates Cave (1968) has shown that two ulnar epiphyses occur: a pressure one (his 'beak epiphysis') and a traction one (his 'scale epiphysis'), the latter apparently representing the 'ulnar patella'. The fact that the pressure epiphysis only occupies part of the articular surface should occasion no surprise in the light of what has been said above. A dual epiphysis at this site then seeming to be the likely primitive

condition for mammals, but mammals other than Primates have a single one, although often this shows clear indications of a bipartite nature. Cave suggested that in these mammals the pressure epiphysis had been suppressed since the trochlear fossa is purely diaphyseal, but it seems much more likely that it has been incorporated in the single one, lack of contribution to the articular surface being an extreme example of the asymmetrical growth noted above. Other putative cases of sesamoid fusion such as the tubercle of the tibia (Lewis 1958), I now consider to be more likely explained as examples of the intratendinous ossification centres sometimes found at insertions (Haines 1942), rather than as relics of formerly independent sesamoids.

Although the sesamoid ancestry of the traction epiphysis of the ulnar olecranon may be equivocal, this can certainly not be said about the arrangements at the comparable site in the hindlimb. The parafibula is an ossicle, commonly found in marsupials, articulating with the upper end of the fibula, connected ligamentously to the lateral condyle of the femur, and giving origin to the lateral head of gastrocnemius and the plantaris; recognition of its relationships is central to the understanding of the phylogeny of the crural and pedal flexor musculature in the hindlimb. In occasional specimens of the wombat and the Tasmanian devil the parafibula may fuse with the fibula forming a large fibular crest or process, in effect, a traction epiphysis mimicking the olecranon, and giving origin to these same muscles. Such a fibular process, with comparable muscular origins, is a constant feature of both the living monotremes, the echidna and the platypus, and in both the fibular process, as would be expected, ossifies independently. Strangely, Pearson and Davin (1921*b*) reasoned that the monotreme condition must be primitive and that a phylogenetic break-up of this process had given rise to a pair of sesamoids, the lateral fabella in gastrocnemius and the cyamella in the popliteal tendon. This is a classical example of mistaken morphocline polarity and the more rigorous approach dictated by cladistics leaves little doubt that the fibula crest of monotremes is an apomorphic character in these otherwise most primitive of living mammals. Presumably Pearson and Davin were influenced by the generally lowly grade of the monotremes, but this is no guarantee of a pervading primitiveness in all characters. There is little doubt that Pearson and Davin were wrong in considering that the marsupial parafibula was a complex of both cyamella and lateral fabella. Rather, the reduction of this structure in Eutheria gives rise only to the fabella and of course it is often eliminated entirely; as will be seen, the cyamella is a quite different structure arising as a lunula in a knee joint meniscus. The parafibula as a free bone is not restricted to mammals but exists also in some reptiles (Pearson and Davin 1921*b*) but never as a fibular crest. There is, therefore, little doubt that the parafibula is plesiomorphic and the fibular crest apomorphic. For these reasons the parafibula is shown in Fig. 3.1 as a characteristic of the ancestral tetrapod limb, along with the 'ulnar patella'.

Sesamoids and lunulae

It is well recognized that seamoids exist in tendons (and occasionally in ligaments) where they rub around bony protuberances and are subjected to pressure. They may present as hyaline cartilaginous nodules (hemisesamoids) within the tendon, which may subsequently ossify producing true bones (orthosesamoids). Hemisesamoids would then be orthosesamoids in the making and often exist in this state at birth. Pearson and Davin (1921*a*) could not accept this functional explanation for the appearance of sesamoids, for it seemed to them to require a process of inheritance of acquired characters, and this prompted them to propose their alternative theory holding that sesamoids were vestiges resulting from the fragmentation of bony processes. In fact, the appearance of these structures before birth seems to be but one more example of genetic assimilation. Sesamoids are also found in tendons at nodal points where they break up into portions diverging to different attachments. An analogy for their functions here would be the tying of a knot in a rope where it is frayed out into strands passing to different anchorages.

French anatomists (Testut 1904) have long distinguished between 'sésamoides intra-tendineux' and 'sésamoides peri-articulaires'. What has just been described are the former. This distinction has been surprisingly unrecognized in British textbooks with the result that the metacarpophalangeal (and interphalangeal) sesamoids are usually supposed to be of the intratendinous variety, located within the tendons of insertions of intrinsic hand muscles; in fact, they belong to the latter variety. The volar capsules of the metacarpophalangeal joints and interphalangeal joints are formed by massively thick fibrocartilaginous 'glenoid plates' deeply grooved for the long flexor tendons and forming channelled guides controlling the trajectory of the long flexor tendons which are bound into the grooves by the fibrous flexor sheaths. These structures are strikingly developed in marsupials where they represent what is apparently a primitive mammalian feature. The wedge shape of the margins of these guides is sustained by ossific nodules, the peri-articular sesamoids, which are developed by endochondral ossification of cartilaginous precursors. It is true of course that certain of the intrinsic muscles, most notably the thenar muscle tendons, may secondarily attach to the surface of these sesamoids.

These ossific nodules are thus dissimilar in location and function to intra-tendinous sesamoids and are in fact, quite analogous to lunulae which are similar strengthening modifications developed within the thickest portions of wedge-shaped intra-articular fibrocartilaginous menisci (Barnett 1954).

Homology and limb myology

A vast monographic literature dealing with the myology of the limbs was accumulated in the nineteenth century. The authors were seldom concerned with the niceties of the meaning of homology, yet nevertheless they did recognize the 'same' muscles in widely different species. This work was usually deeply rooted in the traditions of human anatomy and indeed the nomenclature used was generally derived from that of the exhaustively studied human cadaver. The names given to muscles commonly suggested their functions, and since action is largely consequent upon the location of tendinous insertions this feature, at least implicitly, became established as having overriding importance in determining the corresponding identity of muscles in different species. It is true, however, that some authors appeared to despair at the apparent lability of muscular attachments, almost as though they were devoid of phylogenetic meaning.

Even more so than with the bones, attempts to determine muscle homologies beyond the confines of single vertebrate classes can be unrealistic and even positively confusing. A consideration of the emergence of the mammalian pattern on the forearm flexors, although somewhat simplified, will demonstrate the problems. The work of McMurrich (1903) is frequently cited in this

regard, and much of the data shown in Figs 3.5 and 3.6 are from that source.

In urodele amphibia (Fig. 3.5A) the forearm flexor musculature is triple layered: a deep layer, pronator quadratus, joins ulna to radius; an intermediate layer, palmaris profundus, descends obliquely from the ulna and is delaminated into several subdivisions; a superficial layer, palmaris superficialis, radiates obliquely into the forearm from the medial humeral condyle. This last

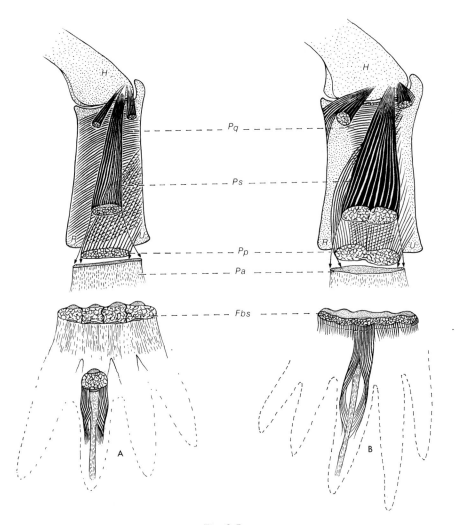

Fig. 3.5

Evolution of the flexor musculature of forearm and hand, as described in the text, in a urodele amphibian (A), and a reptile (B). *H*, humerus; *R*, radius; *U*, ulna; *Pq*, deep pronator layer; *Pp*, palmaris profundus layer; *Ps*, palmaris superficialis, intermediate segment (this is flanked at its origin by the unlabelled radial and ulnar segments, the former achieving a partial radial insertion and the latter a partial ulnar insertion in the reptile); *Pa*, palmar aponeurosis; *Fbs*, flexor digitorum brevis superficialis. (Data from McMurrich 1903.)

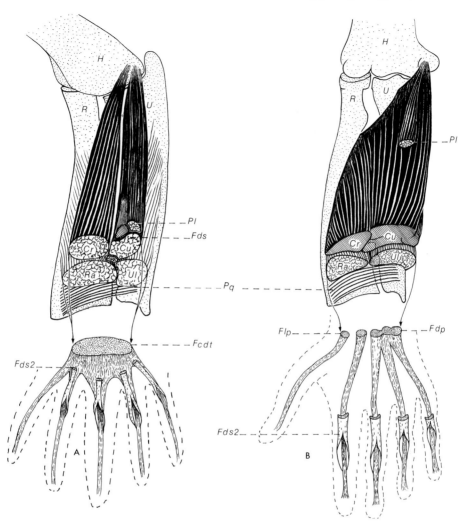

Fig. 3.6

Evolution of the flexor musculature of forearm and hand in a marsupial, the opossum (A), and in a human embryo (B). *H*, humerus; *R*, radius; *U*, ulna; *Pq*, pronator quadratus; *Ra*, radialis; *Ul*, ulnaris; *Cr*, condylo-radialis; *Cu*, condylo-ulnaris; the centralis (unlabelled) is shown between *Cr, Cu, Ra* and *Ul* in A and by the broken line imperfectly separating it from *Cu* in B; *Pl*, palmaris longus (its distal prolongation as the palmar aponeurosis is not shown); *Fds*, flexor digitorum superficialis, variously derived in A and B, is shown cross-hatched and its distal perforatus tendons are indicated (*Fds 2*, that to the index); *Fcdt*, flexor communis digitorum tendon plate in A, giving rise distally to the perforans tendons of the digits; *Flp*, flexor longus pollicis; *Fdp*, flexor digitorum profundus. (Data from McMurrich 1903.)

superficial layer subdivides into three main portions, two marginal parts attaching to the carpus, in part comparable to the mammalian flexor carpi ulnaris and flexor carpi radialis, and a central part contributing to the long flexors; only the last portion concerns us here. The central part of palmaris superficialis when traced distally becomes continuous with a strong palmar aponeurosis whose deep surface receives the attachment of palmaris profundus. Traced into the palm this aponeurosis splits into two layers with another muscle, the flexor digitorum brevis superficialis, sandwiched between them. Further layers of muscle lie deep to this musculo-aponeurotic sandwich but treatment of these will be deferred until the chapter dealing with the intrinsic palmar musculature. As the flexor brevis superficialis traverses the palm it splits into four, each destined for one of the four urodele forearm digits. Carried down distally on both dorsal and ventral surfaces of each flexor brevis superficialis is a thickened prolongation, or slip, of each of the layers of the palmar aponeurosis. At the entry to the digits each subdivision of the flexor brevis superficialis splits to embrace the deep slip, which is then joined by the superficial slip, and the composite flexor tendon thus formed passes on to insert on the terminal phalanx; the flexor brevis superficialis slips insert on the other phalanges. In this way the whole flexor mass activates the digits.

The reptilian condition (Fig. 3.5B) represents a simple elaboration of this scheme. The palmaris profundus has a more extensive origin from the ulna and has become somewhat longitudinally orientated so that its more lateral part overlies the radius; it shows an incipient subdivision into two essentially longitudinally orientated blocks, one related to each of the forearm bones. The central portion of palmaris superficialis is similarly partially subdivided into two portions. All four of these muscular subdivisions insert into a thick 'palmar aponeurosis' just as did their amphibian homologues. The distal attachment and relationships of the palmar aponeurosis are recognizably derived from the amphibian grade. A flexor brevis digitorum superficialis is again present with subdivisions entering each digit where they embrace slips, or tendons, derived from the deep layer of the palmar aponeurosis, which pass to the distal phalanges; the flexor brevis superficialis slips attach to the other phalanges. Note, however, that most extant reptiles have lost the superficial palmar aponeurotic layer, but it seems likely that this is a derived condition which was not present in the reptilian ancestors of the mammals. No muscles comparable to the mammalian palmaris longus or the flexor digitorum superficialis were recognized by McMurrich in amphibia or in reptiles. The radial subdivision of the palmaris superficialis layer, which in amphibia inserts only in the carpus as a flexor carpi radialis has in reptiles attained an additional attachment to the radius,

thus heralding the emergence of a separate pronator teres. The ulnar subdivision inserts to the pisiform but also has a small forearm attachment to the ulna. The subdivision of the flexor mass (leaving out the marginal parts) into four major components, which was incipient in reptiles, has progressed to complete separation in mammals (Fig. 3.6A).

These basic mammalian flexor masses are radialis (*Ra*) and ulnaris (*Ul*) derived from the palmaris profundus, and condylo-radialis (*Cr*) and condylo-ulnaris (*Cu*) derived from the palmaris superficialis, together with a small isolated slip enclosed between them, the centralis. Moreover, a palmaris longus is separated from the palmaris superficialis and attaches distally to a palmar aponeurosis, representing the superficial lamina of the amphibian aponeurosis, which as has been noted is lost in most extant reptiles. Moreover, a flexor digitorum superficialis is present, variably split from the common flexor mass (*Ra, Ul, Cr, Cu* and centralis, collectively called flexor communis digitorum) and this muscle belly has established continuity with the bifurcating slips of the flexor brevis digitorum superficialis, which by conversion to tendon, have become the characteristically split perforatus tendons inserting on the middle phalanges; phylogenetic transformation of muscle belly into tendon, as will be seen, is not unusual. The remainder of flexor communis digitorum muscle inserts into a thick communis tendon, homologue of the deep layer of the amphibian and reptilian palmar aponeurosis, which splits up into the individual perforantes tendons passing to the terminal phalanges of the ulnar four digits, together with a long flexor to the thumb.

There is a clear progressive trend in mammalia for more and more of the flexor communis digitorum mass to be allocated to the flexor digitorum superficialis. At one extreme, as in the opossum *Didelphys virginiana* (Fig. 3.6A), a small sequestered part of condylo-ulnaris furnishes the slender superficialis tendons to digits two, three and four whilst the other tendon to digit five is, at least in some specimens, derived from part of the muscle belly designated as palmaris longus. At the other extreme is man (Fig. 3.6B), where centralis furnishes the perforatus tendon to the index, condylo-radialis (with a newly acquired radial attachment) that for the middle finger, and condylo-ulnaris those for ring and little fingers. Between these extremes, in the mouse, condylo-ulnaris furnishes all four of the perforatus tendons. In terms of their amphibian ancestry, the flexor digitorum superficialis and flexor digitorum profundus of different mammalian species are fashioned out of varying precursor muscular components, and are thus not strictly homologous. Yet to the morphologist restricting his attention to the mammals, the flexor digitorum superficialis of opossum and man should surely be considered as counterparts; they are homologous muscles. The evolutionary trend shown is that the independent superficialis is progressively enlarged by extension of its domain, thus appropriating more and more of the developing common flexor mass, at the expense of that part linking to the profundus tendons.

Apparently such myological evolution proceeds by a resetting of the fundamental pattern of subdivision of a major muscle group as each major taxonomic category is established and this new pattern is then exploited in all its variant forms. The strategy of the morphologist must be to establish the basic ancestral myological stem pattern of the order or class, which can then by known mechanisms of phylogenetic change give rise to the spectrum of modification observed in various species. This stem pattern may be conserved in some extant species or it may only be possible to deduce hypothetically what the ancestral arrangement was likely to have been. In the case of the forearm flexors the opossum matches well the pattern which was probably characteristic of early therian mammals. Such an approach was well established by the classical morphologists (Huntington 1903) and in a modern context this procedure is essentially an inductive cladistic analysis (Gingerich 1984). Understanding of these phylogenetic transformations provides a logical framework for interpreting the many minor variations in muscular attachments, and also the great majority of muscular anomalies; most of these are reversionary, although a very few are what Huntington (1903) called fortuitous or progressive variations. In practice therefore it is necessary to establish muscle homologies for the taxonomic group under consideration: this procedure is followed in a stepwise fashion through successive evolutionary grades, establishing a new set of homologies at each grade of major reorganization. Attempting to homologize by a leap-frogging process between widely separated taxonomic groups may give a misleading notion of the real evolutionary path taken, as for example comparisons of the long flexors of different mammals directly with amphibia. However, in broad terms such wide ranging comparisons do have some value. For instance, the case cited allows one to appreciate how the mammalian perforatus tendons have been derived from the flexor brevis digitorum superficialis or how the common tendon of flexor profundus digitorum owes its existence to derivation from the amphibian palmar aponeurosis.

Muscle nerve specificity

A central dogma dominating the efforts of the classical comparative myologists was the belief that there was an indissoluble link between the source of nerve supply and the identity of a muscle. Indeed, the muscle was originally considered as the end-organ of the nerve (Ruge 1878*b*), and it was believed that there was actual protoplasmic continuity between nerve and muscle, an idea long since disproved. The belief that determining nerve supply provides a universal key for unlocking homologies has been amply discredited (Haines 1935), yet the confusions introduced from this era linger on.

It is now well established that as muscles undergo phylogenetic migrations to new locations in the limbs, so they may appropriate a new and more convenient nerve supply from the nearest available source in a quite opportunistic way. This is analogous to the way in which a denervated muscle may be provided with a new and unusual motor nerve by surgical implantation.

Muscle–tendon relationships

The dense collagenous tissue of tendon, and the fibres of striated skeletal muscle are strikingly dissimilar tissues histologically yet in phylogenetic terms it would appear that comparable mesodermal anlagen possess the dual potentiality to switch their differentiation into either of these develop-

mental paths, according to functional requirements. Indeed what is a muscle belly in one species
may be represented by a tendon in another. Even
whole muscles may become transmuted phylogenetically into ligamentous cords. For instance, in
ungulates the interrosseous muscles are converted
into remarkable ligamentous slings encompassing
the fetlock (metacarpophalangeal) joints, which
by their elastic properties (despite their collagenous structure) store strain energy during gait
(Hildebrand 1982; Alexander and Dimery 1985).

There is a precise adjustment between the
relative sizes of a muscle belly and its tendon, both
during development and during phylogeny. The
extent of contraction of a muscle is related to the
required range of movement and is determined by
the lengths of its fasciculi and the angle they make
to the long axis of the belly. Tendon eliminates the
need for any unnecessary length of muscle
between origin and insertion, and removes the
bulk of a muscle from the vicinity of a joint over
which it acts, whilst concentrating the pull onto a
circumscribed area of bone. Experimental alteration of the extent to which a muscle must
contract in order to be fully effective, by operative
alteration of direction of pull and so on, is followed
by compensatory alteration of the muscle–tendon
ratio, provided that this is done during the growth
period. This correlation is brought about by
changes in the growth rates of muscle and tendon,
the latter growing interstitially particularly near
the muscle–tendon junction (Elliott 1965). This
constant relationship between the length of
muscle fibres and the distance through which
origin and insertion must be approximated, is also
seen in congenital abnormalities such as clubfoot,
where a diminished range of movement is accompanied by a compensatory shortening of the
muscle bellies (Haines 1932). The interdependence of muscle–tendon length and the attuning of
this ratio to functional requirements is apparently
an innate property and is seemingly projected to
the phylogenetic level by some such mechanism as
genetic assimilation.

Migration of muscle origins

It is commonly accepted that certain muscles,
which are usually considered to be homologous,
may vary widely in origin between species. It is this
variability in origin which generates controversy
about muscle homologies. It is manifest that
muscles may expand their area of origin by
appropriating developmental material which in
other species would contribute to a neighbouring
muscle belly, and that by this means the distribution of power to different terminal insertions is
reallocated in response to new functional
demands. A classical example of this has been
described above, demonstrating how the flexor
digitorum superficialis of the forearm, trivial in
importance in some mammals, has been transformed into a powerful human muscle by the
incorporation of muscle material otherwise allocated in other mammals. Moreover, the muscle
has acquired a new radial origin. Thus muscles
may show territorial expansion of their origins by
extending onto adjacent unoccupied areas of bone,
or even ligaments or the fascia covering adjacent
muscles. Such a secondarily colonized origin, if
more distally located may eventually become the
primary one: the muscle has shown descent.
Muscles may also easily exhibit apparent migration down the component segments of a limb by
replacement of tendon by muscle belly, which then
adheres to bone or fascia in the new situation, thus
establishing a more distal origin.

It has been suggested, however, that muscles do
not undertake the converse proximal migration
across joints (McMurrich 1904). This contention
is, however, not entirely true although as a broad
generalization it has some substance. A muscle
most assuredly can sometimes show phylogenetic
encroachment into the territory of a neighbouring
group and thus 'ascend' from, for example autopod
to zeugopod; examples of this will be described,
most notably in consideration of the evolution of
the cruropedal extensors. An alternative, although
unusual mechanism of ascent may involve the
bridging mediation of an intra-articular meniscus.
For instance some intrinsic zeugopodial muscle
substance may adhere to a meniscus located in the
joint between stylopod and zeugopod, which is
tethered to the upper bone by a ligament. Degrada-
tion of the meniscus (perhaps with a contained
lunula) converts it into an intra-articular tendon,
maybe still sesamoid-containing, which has thus
transferred origin to the bone above. This mechanism will be described in connection with the
evolution of the muscle popliteus.

The relative infrequency of muscle migration directly up over joints is possibly a consequence of the fact that an important purpose of tendons is the removal of the bulk of a muscle from the joint over which it acts. There is little scope for phylogenetic conversion of tendon into muscle belly, proximally extending the origin of a more distal muscle.

Migration of muscle insertions

While homologous muscles may show a wide range of shifting origins their insertions tend to remain relatively constant. Yet these terminal tendinous attachments can also change phylogenetically, but only by the mediation of a limited range of morphological devices (Fig. 3.7). There are usually clear anatomical indications of the change that has occurred, and the morphocline polarity is seldom in doubt. Tendons may appropriate some of the connective tissue of adjacent or overlying layers of fascia thus radiating their influence more widely (Fig. 3.7A). This is quite in accord with what is known experimentally: even a strip of loose connective tissue may become transmuted into tendon if subjected to tension (Elliott 1965).

Tendons may similarly become adherent to overlying or adjacent ligaments and so have their insertions widened to include other bones. The differing trajectories of the two component groups

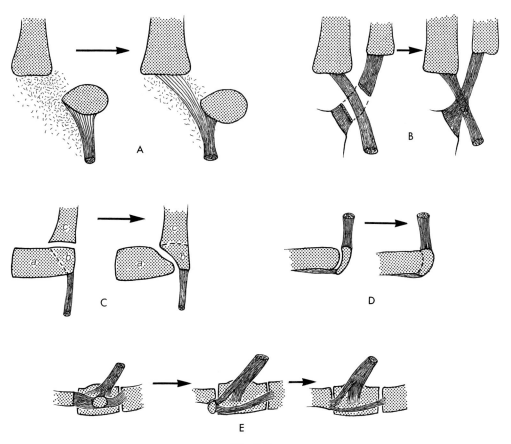

Fig. 3.7

Mechanisms effecting phylogenetic shifts in tendinous insertions (see text): A, by appropriation of adjacent fascial layers; B, by adherence to overlying ligaments; C, by rearrangement of bony fusions where bones have a composite derivation; D, fusion of a sesamoid to form a traction epiphysis; E, partial fusion of a tendon to an overlying sesamoid bone.

of fibres incorporated in the composite tendon termination, usually bear witness to the mechanism involved (Fig. 3.7B).

Where a bone has a composite phylogenetic and developmental origin a tendon may primitively attach to only one constituent part (Fig. 3.7C). This insertion-bearing anlage may in other phylogenetic lines coalesce developmentally with a neighbouring bone to produce a different composite. If the development history of these bones *ab* and *bc* is not recognized it may appear that the tendon has arbitrarily shifted attachment from one to the other; indeed, such apparent shifts may provide the clue to the osteological changes.

Fusion of a tendon sesamoid to form the traction epiphysis of a neighbouring bone can transfer the tendon insertion to that composite bone, leaving the former distal prolongation of the tendon as a ligament or dense fascial sheet (Fig. 3.7D). Tendons may fuse, not only to overlying ligaments but to residual sesamoid bony elements contained within ligaments (Fig. 3.7E). Migration of the bony element may then carry a slip of the tendon to an adjacent bone where it may achieve a definitive attachment either by dissolution of the sesamoid, or incorporation of it into the related bone.

Ontogeny and phylogeny in myological evolution

Unequivocal examples of the phylogenetic trends described above will be dealt with in subsequent chapters. The real crux of the matter, however, is how regulatory genes might produce these changes by appropriately orchestrating the developmental plan. Woefully, information here is very meagre but some insights are emerging (Shellswell and Wolpert 1977) most of it derived from experimental work on developing chick limbs. There is little doubt that at least in birds myoblasts

in the limb are derived from an immigrant cell population from the somites, whilst tendons and the connective tissue framework of the muscle are formed *in situ* from the mesenchyme which is of somatopleuric derivation. Moreover, the development of tendons and muscles are autonomously and independently specified: indeed, experimentally one may develop without the other or the two components may be induced to link up in unnatural ways. If the principle of two independent cell populations with different potentialities applies to mammals how can this be reconciled with the apparent interchangeability of muscle and tendon in phylogeny? There seems to be no particular problem here, as the conversion of muscle to tendon or vice versa could be realized by variation in the rate of proliferation and migration of myoblasts through the mesenchyme responsible for the production of both the tendon and the fascial framework of the muscle; this could be envisaged as an effect produced by rate genes. Little is known about the way in which the dorsal and ventral muscle masses are split up into individual named masses, but the splitting apparently follows a binary pattern and the sculpting appears to be a property of the mesenchymal non-muscle cells. The phylogenetic enlargement of one muscle at the expense of another could again result from differential rates of proliferation of the myoblastic rudiments specified for individual muscles.

Some hint of the underlying genetic mechanisms and of the importance of timing in development stems from the interesting observation that in human trisomies 13, 18, and 21 muscular atavisms are quite common. Moreover, the persistence in these aneuploids of non-human primate characteristics in their musculature appears to be largely generated by a general developmental retardation. (Barash *et al.* 1970; Asiz 1981).

4

The skeleton of the hand

It is widely believed that the skeletons of both the hand and foot in tetrapods share a basically similar primitive pattern: a proximal carpal or tarsal row consisting of an intermedium flanked either side by a preaxial bone (radiale or tibiale) and a postaxial one (ulnare or fibulare); an intermediate centrale (or centralia) and a distal group of five carpalia or tarsalia, bearing the metacarpals or metatarsals. The foot skeleton in mammals is drastically modified from this simple pattern; in contrast the mammalian hand retains the essentials of this basic primeval pattern.

Fin to limb—the embryological approach

A great literature has grown up on the hand skeleton which hints at a far from simple history, much more complex than that outlined above. The root of the problem has been a proliferation of theories aimed at interpreting the tetrapod limb in the light of its supposed fin-like precursors. Hypothetical schemes supposedly demonstrating the skeletal pattern in the early lobe-fin air-breathing fish which emerged on land in the Devonian have been directly utilized in attempting to interpret developmental patterns in the hand of man, and other mammals. Implicit in this approach is a belief in stereotyped recapitulation, which has little regard for the niceties of the relationship between embryology and morphology which was dealt with in the previous chapter.

There is quite general agreement that the tetrapod fore-limb was in fact derived from the fin of a lobefin fish at least similar to the Devonian *Eusthenopteron*. In such a muscular dual purpose appendage the cartilaginous rays, or radials, are packed together and consolidated into plates at the constricted attachment of the fin, especially at the anterior end, producing a longitudinal axis of skeletal elements embedded in the body wall. Freeing of the posterior part of the fin by an indented notch peels away this axis so that the fin projects outwards from a constricted base. At this transitional stage the consolidated cartilaginous elements are then located postaxially, forming the axis of the emergent limb, whilst the residual rays are preaxially arranged. There is little doubt that at this emergent stage there were seven rays (Jarvik 1980). This aggregation of cartilaginous elements formed the building blocks for the stylopod, zeugopod, carpus, and metacarpus of the emergent tetrapod limb. Although opinions differ about the exact branching pattern of the rays from the axis, a commonly held view is that shown in Fig. 4.1A. An alternative view (Gregory 1951) is that the original axis down the postaxial border became shifted radially in the autopod, either by new formation of postaxial rays (digits) at the postaxial border, or by their migration there from the preaxial margin in the earliest tetrapods.

Such a groundplan has been applied directly to the interpretation of development in the human embryo. The most recent of these exercises (Fig. 4.1B, C) which, in fact, synthesizes and reconciles many of the results of earlier studies as well as reviewing them, is by Cihak (1972).

The strategy employed in this essentially embryological approach highlights the dangers mentioned in the previous chapter, of attempting to establish homologies across wide classification gulfs. Such a strategy overlooks the fact that progressive evolution erases, or at least masks, part of the developmental 'blueprint' by the process of heterochrony. However, by establishing homologies in an escalating step-by-step fashion, from

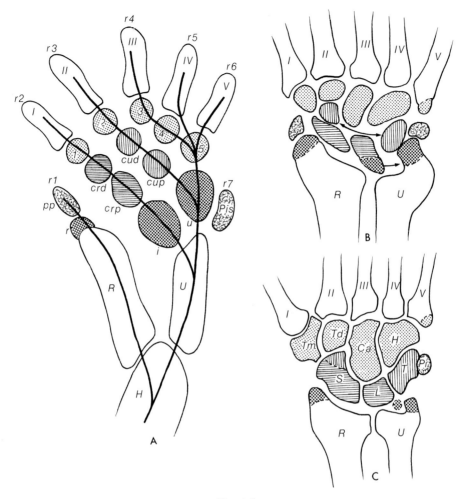

Fig. 4.1
A, scheme of the emergent tetrapod carpus, redrawn from Cihak (1972) and incorporating the views of Steiner. *H*, humerus; *R*, radius; *U*, ulna; *r*, radiale; *i*, intermedium; *u*, ulnare; *pis*, pisiform; *pp*, prepollex; *crp*, centrale radiale proximale; *crd*, centrale radiale distale; *cup*, centrale ulnare proximale; *cud*, centrale ulnare distale; *1–5*, carpalia distalia; *I–V*, the metacarpals; *r1–r7*, the primitive seven rays. B, identification of the same elements, identified by similar shading, in the scheme of the human embryonic hand, according to Cihak. C, identification of the same elements in the human adult hand according to Cihak, shaded as before. *S*, scaphoid; *L*, lunate; *T*, triquetral; *Pi*, pisiform; *H*, hamate; *Ca*, capitate; *Td*, trapezoid; *Tm*, trapezium.

evolutionary grade to grade, this pitfall is escaped. The results achieved by Cihak (1972), which are in at least some features rather in line with the conclusions of Holmgren (1952), are shown in Fig. 4.1B, C. The problem confronting the workers here is that there are more elements in the groundplan (Fig. 4.1A) than there are separate chondrifying carpal elements in the embryonic human hand. In order to reconcile this discre-

pancy mere mesodermal concentrations are portrayed as representing elements from the groundplan; often these condensations are no more than outcrops at the periphery of expanding centres of chondrification. Thus, by this dubious pretext, it is maintained that the radiale becomes the styloid process of the radius, the ulnare becomes the ulnar styloid process, and the intermedium by migration becomes the so-called triangulare; on this interpretation, it would seem that the original proximal carpal row has effectively disappeared, to be replaced by a trio of centralia.

The whole scheme, corresponding broadly to option E in Fig. 3.4, founders when the basic principles governing the embryology–phylogeny relationship are appreciated. Moreover, it will be shown that the so-called 'triangulare' or 'intermedium' of the human embryo is not even a representative of the primeval carpus but is a late acquisition added uniquely in the Hominoidea. These views, although certainly untenable, have been influential is needlessly complicating what is really quite a simple issue.

The evidence from anomalous ossicles

Confusion has also resulted from the elevation of any supernumerary ossicle in the carpus to the status of a primitive carpal element. Pfitzner (1900*a, b*) recorded thirty-three of these anomalous bones in the human hand and accounted all of them as primordial carpal constituents. This was only a vaguely voiced theoretical assumption, lacking any convincing comparative, embryological or palaeontological substance. In fact, of course, the great majority of these ossicles are merely supernumerary ossifications in apophyses located on the true primordial elements and thus have no particular morphological significance at all. A handful, however, do have a credible phylogenetic history and will be dealt with later.

The evidence from the palaeontology of tetrapods

Despite these unwarranted diversions into needless complexity Broom's (1904) statement still seems to hold good: 'While the mammalian carpus has become very slightly specialized and is hence of little service in guiding us to the mammalian ancestor, the specialization of the tarsus is so peculiar . . .'. In fact, the basic components of the tetrapod carpus enumerated in the first paragraph of this chapter have merely acquired new names: the proximal row becomes lunate flanked by scaphoid and triquetral, the centralia are reduced to one, and the five carpalia are represented by

trapezium, trapezoid, capitate, and hamate; the last may represent fused fourth and fifth carpalia, or only the fourth, the fifth having disappeared. At the margins of the ancestral mammalian carpus are the pisiform postaxially and the prepollex preaxially. Whilst the homology of any of the major components with particular cartilaginous masses in any known fossil lobefin appendage may be equivocal, in general principle derivation from a pattern such as that in *Eusthenopteron*, seems valid. There is little doubt that the transitional appendage was seven-rayed (Jarvik 1980) and that the marginal rays have contributed the prepollex and pisiform (Fig. 4.1A), but there is no evidence that these were ever true digits as such. The basic tetrapod hand is therefore pentadactyl. The digits themselves are presumed to have arisen by further segmentation of the cartilaginous radials; the rays of fish are known to have a propensity for breaking up into a number of pieces. Despite its patchy nature the fossil record gives broad confirmation of this view (Fig. 4.2) which in effect corresponds to option D in Fig. 3.4. It must be remembered also that none of the fossils described are likely to be on the direct line of descent to the mammals and moreover they are usually incomplete, with the presence of certain elements shown in reconstructions inferred from their known presence in other more or less closely related specimens of approximately similar evolutionary grade.

Yet certain trends seem quite clear-cut. There is a progressive reduction in the number of centralia from four to one either by actual elimination or by fusion; in the absence of developmental evidence about the chondrifying elements the precise mechanism is likely to remain obscure. The apparent reduction of the intermedium in the cynodonts, and its even more pronounced diminution in other related therapsids, the gorgonopsians, has led to a suggestion that the lunate in mammals was actually derived from a centrale, with the 'intermedium' represented in developing mammals as another small cartilaginous nodule and named the intermedium antebrachii; in fact, as noted above, this latter is a newly acquired structure unique to hominoids. Furthermore as described in Chapter 5 the lunate in certain mammals (the marsupials *Pseudochirus* and *Phascolarctos*) can undergo a reduction comparable to that of the therapsid intermedium (Fig. 5.2B). It is likely then that this

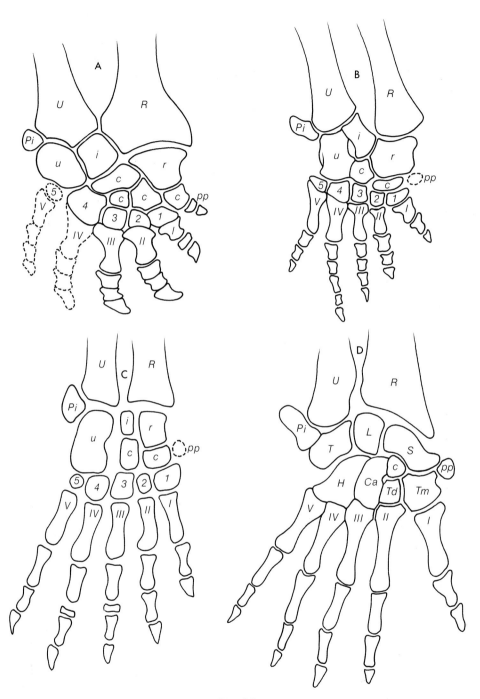

Fig. 4.2
The inferred transformation of the carpal architecture of tetrapods, based upon descriptions and interpretations by various authors, from a primitive labyrinthodont amphibian (A), to an early mammal-like reptile (pelycosaur) (B), then an advanced mammal-like reptile (cynodont) (C), and finally a typical mammal (D). Labelling is as in Fig. 4.1.

condition in some therapsids is a derived one, having no particular relevance to mammalian origins, and that the mammalian lunate is indeed the reptilian intermedium.

There are clear palaeontological indications that the mammalian phalangeal formula of 2-3-3-3-3 was derived from the primitive reptilian one of 2-3-4-5-3, by the reduction of the supernumerary bones to disc-like vestiges which then presumably disappeared (Fig. 4.2).

In cynodonts the fifth carpale is quite small leading to the supposition that the hamate of mammals is representative only of the fourth bone of this carpal series, the last having disappeared. The argument lacks conviction for in other therapsids (gorgonopsians) there clearly appears to have been fusion of the fourth and fifth carpalia to form a compound bone analogous to the hamate.

Embryological evidence confirms this view of the groundplan of the mammalian carpus, since all the individual elements make their separate appearances as chondrifying centres. The subsequent history in different mammalian clades varies owing to certain fusions characteristic of the particular clade; for instance in a number of groups scaphoid and lunate fuse to form a scapholunar bone. Although in a few mammals two centres of chondrification appear for the hamate (Emery 1897; Ratjova 1967) in most, including man, there is only one centre; nevertheless these findings suggest the evolution of the hamate from the fourth and fifth carpalia and are consistent with the doctrine of heterochrony.

The evidence from comparative anatomy

Pfitzner, in his infrequent and limited excursions into comparative anatomy had some inkling about a phylogenetic background for several of his recorded ossicles. These were the so-called os intermedium antebrachii, the prepollex, and the centrale.

The 'os intermedium antebrachii'

Pfitzner (1895, 1900*a*, 1900*b*) noted three differently named ossicles at the ulnar extremity of the proximal row of the carpus. These were triangu-

lare (or intermedium antebrachii), triquetrum secundarium and pisiforme secundarium, most probably varying manifestations (differing slightly in position) of an intrameniscal lunula evolved at a relatively late stage at this site during hominoid evolution. This will be described in some detail in Chapter 5.

The prepollex

A chondrifying element (or elements, for it is sometimes double), which later ossifies, representing the prepollex is the general rule among mammals including the Primates. Its great enlargement in certain mammals with an extremity bearing a pad, nail or even claw (Bardeleben 1894) is doubtless secondary and is no indication that the early tetrapod limb was ever other than pentadactyl. For Bardeleben, however, this was an indication that the primitive tetrapod limb had been heptadactyl with a similar postaxial digit, the postminiminus (pisiform). There are no convincing indications that the primitive tetrapod limb was other than pentadactyl and the most reasonable assumption is that the prepollex is the remains of a rudimentary marginal ray. When greatly reduced the prepollex may have the appearance of a mere sesamoid bone but this is also secondary and reflects the fact that these marginal ray vestiges (including that at the postaxial border) have tendinous insertions and muscular attachments rather like metacarpals in the five complete rays. These same remarks apply to the pisiform which typically has a heel-like character and a terminal epiphysis, and underlies the hypothenar pad. When greatly reduced it assumes a sesamoid-like configuration, as in, for example, man. Its heel-like function, however, should not be taken to imply serial homology with the cancaneal tuberosity, an absurdity apparently introduced by Owen (1866) and repeatedly cited since.

Pfitzner (1900*b*) flirted with the idea that his radiale externum represented the prepollex although he totally rejected the Bardeleben theme that it had ever been a fully equipped and functioning digit. In the very next section Pfitzner (1900*b*) described an ossicle which he named the epitrapezium, and significantly noted that an exactly similar bone was found in the great apes. There is little doubt that it is this ossicle which is the prepollex. Unlike other Primates it is generally believed that no vestige of the prepollex is present at any development stage in man (O'Rahilly *et al.* 1957) but this single case shows that rarely it can be atavistically reinstated in man.

The muscular attachments of the prepollex to be recounted in subsequent chapters allow its homology to be pinpointed precisely and there is no doubt that the ape 'epitrapezium' is, in fact, the prepollex; by implication so is the human ossicle.

It has been noted (Lewis 1977) that in the gorilla this prepollex fuses with the trapezium in older specimens to form a massive bony apophysis, the functional implications of which will be described in Chapter 6. Pfitzner (1900*b*) similarly noted this fusion of the 'epitrapezium' in his gorilla specimens.

The name prepollex, although stripped of any idea that it represents a previously functioning digit will be retained throughout this book—it is descriptively apt and avoids confusion.

The centrale

Particular interest attaches to the fate of the centrale in hominoids. A great number of studies from Rosenberg (1876) to Olivier (1962) have established that a chondrified centrale is found in the human hand during the second and third months, finally disappearing by fusing with the scaphoid. Just where it is represented in the human adult hand is, however, far less clear but the comparative history in other Primates elucidates this and provides a most instructive example of heterochrony in action during primate evolution.

Characteristically, in all those Primates where the centrale is an independent element, it is intimately bound to the scaphoid and carries on its distal aspect the articulation for the trapezoid, although it often also (*Ateles, Lemur, Cebus*) encroaches marginally to articulate somewhat with the trapezium, whose main (or entire— *H. symphalangus*) articulation is with the scaphoid tuberosity (Etter 1974). It has long been known (Rosenberg 1876) that the transient free cartilaginous centrale of man is similarly the structure proximally related to the trapezoid; its morphological homology with the bone in other extant adult Primates is thereby confirmed.

The history of the centrale in hominoids has been well documented by Schultz (1936). In the Asiatic apes (gibbons and orang-utan) the centrale is typically free and carries the articulation of the trapezoid. In aged specimens, however, it may fuse with the scaphoid proper. In chimpanzees and gorillas such fusion is normal but it occurs much earlier, either at the end of fetal life when both elements are cartilaginous or probably more usually in the infantile period when both components have developed their own ossific centres. In man, of course, the fusion is even earlier (third month of prenatal life) and usually only one centre of ossification appears for the composite structure. This is then a classical case of heterochrony with the independence of the centrale being progressively truncated in evolution until it is confined in man to a transient early phase of fetal life. In this way a carpal element which is originally independent is progressively assimilated into the developmental ground-plan; presumably the final phase, not yet achieved in man, would be that even a separate centre of chondrification for the centrale would fail to develop and it would have been effectively erased from the ground-plan (Fig. 3.2).

The question that immediately poses itself is whether it is practicable to identify any homologue of the centrale in the normal adult carpus of *Homo sapiens*. At the outset it should be emphasized that suggestions appearing in textbooks (for example, Jones 1949) maintaining that the whole distal half of the scaphoid including the tuberosity represents the centrale, are mere supposition and totally without foundation. Yet it should be possible to delineate that part of the normal human scaphoid which is of centrale derivation. Moreover, on general principles it would seem not unlikely that the original unabbreviated developmental programme could be replayed on occasion in man, effectively putting heterochrony into reverse, with the reappearance of a free os centrale in the mature carpus. There is, in fact, little doubt that this can occur.

Reports (Pfitzner 1900*a, b*) have suggested that a free os centrale may occur in man, the frequency being given as anything from a fraction of 1 per cent up to 3 per cent. Many of these structures, as described in the literature (Virchow 1929) and often figured in textbooks are quite diminutive ossicles located dorsally where scaphoid, lunate, capitate, and trapezoid are all adjacent. Indeed,

Forster (1933) considered that all such ossicles were in reality not the os centrale but the epilunatum (one of Pfitzner's supposed primordial carpal elements) and, on the strength of reports in the literature of a second chondrifying 'centrale 2' in the human embryonic hand, suggested that this was their source; subsequent detailed study on the stages of chondrogenesis have failed to identify any second chondrifying centrale element, however. There seems little doubt that these ossicles are in reality merely the result of adventitious supernumerary ossific centres in one of the adjoining bones.

Forster (1934) did however, describe a 'true os centrale' in an adult human hand. This was a separated portion of the scaphoid, bearing the whole of the trapezoid facet and, therefore, by all criteria quite homologous with the free carpal element of other Primates. The limits of this free os centrale in man, as described and figures by Forster, are comparable to the area delimited by the broken line in Fig. 4.3A. In fact, on the particular bone illustrated here a shallow groove partly delimited an apparently fused centrale and a radiograph indicated a contrastingly different texture of this part of the bone. Forster noted the occasional presence of such linear demarcations of the original os centrale and even of quite pro-

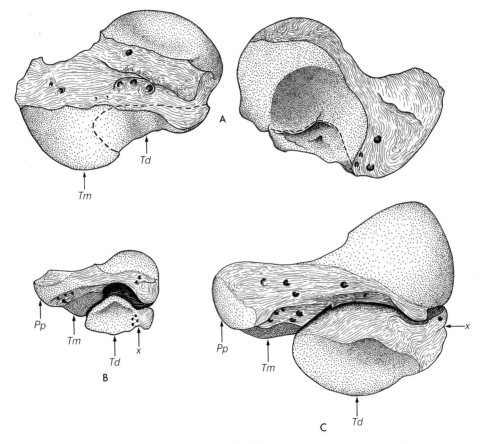

Fig. 4.3

A, a human left scaphoid viewed from the dorsal (left) and palmar aspect (right); in this specimen a groove on the palmar aspect partially demarcated the centrale element whose presumed extent is indicated by the broken line. B, a right scaphoid of *Hylobates lar*, viewed from the dorsal aspect, together with the centrale, which shows a tongue-like ulnar prolongation demarcated by a groove *x*. C, a left scaphoid of *Pongo pygmaeus* viewed from the dorsal aspect, together with the centrale which shows a groove partially demarcating an ulnar portion *x*. The articular surfaces for the prepollex (*Pp*), trapezium (*Tm*), and trapezoid (*Td*) are indicated.

nounced clefts into the scaphoid at the dorsal extremity of the delimiting line. There is, therefore, little doubt that on occasion in man the os centrale may separately ossify, very rarely remain free, but considerably more often fuse with the scaphoid although traces, more or less marked, of the line of fusion may sometimes remain.

This large segment of the scaphoid is quite unlike the tiny dorsal ossicles usually described in the human carpus as representative of the os centrale. It has been noted that Forster instead considered these as representing another supernumerary ossicle, the epilunatum. Above I have suggested that they belong among the category of supernumerary adventitious ossifications, and in fact, they may be such centres in the dorsal tongue of the true centrale. Forster (1933, 1934) indeed noted that this portion of the centrale (x in Fig. 4.3B, C) was demarcated by a groove in *Hylobates*, and he even noted it as a free ossific nodule in a specimen of *Pongo*. This then may be the site of the supernumerary ossific centre producing the 'epilunatum' of man. As noted above, however, recent embryological studies have provided no conclusive grounds for believing that it represents a second centrale, with the implication that the centrale carpal element of mammals in reality resulted from the merger of two reptilian bones of the central series.

The homology of the centrale, even after coalescence with the scaphoid, can be confirmed by fundamental ligamentous connections. The free centrale is dorsally attached to the triquetral by the posterior centrale–triquetral ligament (Figs 5.1A, B) and its free margin is commonly linked to the waist of the capitate by a centrale–capitate ligament (Figs 6.16, 6.18). When amalgamated with the scaphoid these same ligaments pinpoint the centrale component; the centrale–capitate ligament, however, is inconstant.

5

The joints of the wrist

The small bones of the carpus, with their complex pattern of intervening synovial joints, provide a mobile linkage allowing for positional adjustments between the forearm bones and the plane of the hand. These joints are therefore key determinants of locomotor habits and positional behaviour.

In mammals the carpal bones are generally marshalled into two distinct rows. The proximal row consists of the scaphoid (with the centrale firmly articulated with it or fused to it), lunate, triquetral, and pisiform; the distal row consists of hamate, capitate, trapezoid, and trapezium. Each row functions as a unit, the intervening synovial joints between component bones permitting only minor adjustments. The distal row articulates with the metacarpals by a complex system of joints which will be considered in Chapter 6. Between the two rows lies the midcarpal joint, phylogenetically the focus of most movement at the wrist. The proximal row articulates with the forearm by the antebrachiocarpal joint, or wrist joint proper. This is the radiocarpal joint of the human anatomist but in phylogenetic terms this nomenclature is inappropriate for primitively it was an articulation of relatively restricted mobility, with ulna and pisiform participating in the joint. What follows is largely based on a number of papers which I published over a period of years and which are included in the list of references. Here will be found more extensive descriptions and a discussion of the relevant literature.

The prosimian joints

Lemurines

In lemurs (Fig. 5.1A) the elongated pisiform,

supporting the 'heel' of the hand articulates with the palmar surface of the triquetral and the two bones combine to form a concave receptive cup for the lower articular extremity of the ulna. Proximal to this the ulna presents a fusiform swelling, traversed by the lower epiphyseal line in young specimens, and united by an interosseous ligament to the radius. Sometimes a small synovial cavity may be found within the substance of this ligamentous mass and here tiny cartilaginous facets may be found on the exposed epiphyses of the two forearm bones. There is, however, no true synovial inferior radio-ulnar joint nor does the ulna have a 'head', in the sense of human anatomy, and the actual distal extremity is clearly important in weight transmission to the carpus. This ulnocarpal joint is completely separated from the remainder of the antebrachiocarpal joint, the radiocarpal portion, by a thick fibrous septum attaching above to the ligamentous union between the forearm bones and particularly in its more anterior part to the ulna. Below, the septum attaches to the partly exposed ulnar face of the lunate, which is somewhat elevated proximal to the triquetral. In the radiocarpal compartment of the joint the radius articulates with both lunate and scaphoid, and when the joint becomes close-packed in full extension, the convex dorsal articular margin of the radius engages in a hollowed posterior part of the scaphoid. A strong intracapsular ligament descends obliquely from the front of the radial styloid process to the lunate and this palmar radiocarpal ligament has as its counterpart the most anterior part of the longitudinal septum, effectively a palmar ulnocarpal ligament, converging from the ulna. These ligaments clearly are taut and provide support when the joint is extended, which is of course the weight-bearing position in

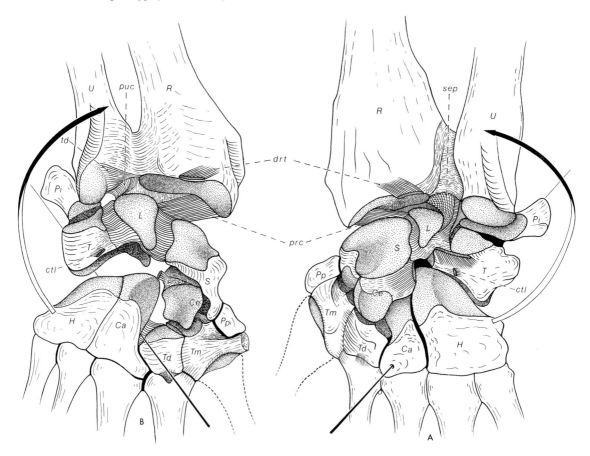

Fig. 5.1 (A, B)
Dorsal views of the left wrist articulations of *Lemur catta* (A) and the right wrist of *Cebus capucinus* (B), with a steel rod transfixing the wrist in each case along the midcarpal axis; the direction of habitual movement about this axis into extension, ulnar deviation and rotation in the sense of pronation is indicated in each case by the arrows. Articular surfaces are shown stippled. *R*, radius; *U*, ulna; *Pi*, pisiform; *T*, triquetral; *L*, lunate; *Ce*, centrale; *S*, scaphoid; *Pp*, prepollex; *Tm*, trapezium; *Td*, trapezoid; *Ca*, capitate; *H*, hamate; *ctl*, centrale–triquetral ligament; *drt*, dorsal radiotriquetral ligament; *prc*, palmar radiocarpal ligament; *puc*, palmar ulnocarpal ligament; *sep*, septum within the antebrachiocarpal joint; *td*, triangular articular disc.

the quadrupedal posture. The posterior aspect of the joint capsule is reinforced by a relatively unobtrusive oblique posterior radiotriquetral ligament.

The proximal carpal row embraces the dome-like conjoined heads of the capitate and hamate to form the functional hub of the midcarpal joint. The centrale and the tubercle of the scaphoid, articulating with trapezoid and trapezium form an auxiliary part of the joint. The centrale is insinuated between scaphoid and capitate in such a way that it forms much of the receptive cup for the head of

the latter bone and it is linked across the back lip of the proximal carpal row by a centrale–triquetral ligament. The prepollex seals off the lateral end of the joint line, articulating with both scaphoid and trapezium, but apparently it has little influence on the mechanics of the joint. Anteriorly the midcarpal joint line is crossed obliquely by an anterior capitate–triquetral ligament.

Movement at this complex midcarpal joint occurs about an axis running from dorsally on the radial side of the carpus with a proximal angulation towards the volar aspect on the ulnar side

(Fig. 5.1A). Determining an axis on such a complex aggregation of small bones is technically difficult and it is not suggested that there is any great precision about the position of the axis depicted, or that the axis does not show some change in orientation during movement. Nevertheless it is a useful device aiding the visualization of the movements.

The habitual movement at the wrist, bringing it into the close-packed maximally congruent position, is centred upon the midcarpal joint and is a combination of extension, ulnar deviation and rotation in the sense of pronation. In this attitude the triquetral has slid down the somewhat spiral facet on the hamate to become lodged in a hollowed cup at its lower extremity. Dorsally the centrale becomes wedged between the two carpal rows which are then firmly compacted together.

This midcarpal movement is complemented by much more restricted action at the antebrachiocarpal joint which occurs about a comparable sort of axis; the movements at the two joints are therefore congruent. The everted dorsal articular lip on the radius impacts on the hollowed posterior concavity on the scaphoid and the extremity of the ulna bears down upon the receptive cup formed by the triquetral and the heel-like pisiform which now especially carries the compressive stresses.

As will be seen, there can be little doubt that in overall structure and function lemurs retain the basic essentials of the morphology of the wrists of the emergent therian mammals. Morever, the lemur provides a good representative model for the ancestral primate pattern.

In contrast, in the other great subdivision of extant prosimians, the lorisines, an apomorphic overlay to the primitive pattern of wrist architecture has been evolved, geared to their specialized locomotor requirements.

Lorisines

Nycticebus coucang (Fig. 5.2A) In this Asian lorisine the lower end of the ulna bears an asymmetrically swollen head proximal to its carpal extremity. This contrasts with the conservative prosimian condition, as seen in, for example, *Lemur*, where the bone adjoining the epiphyseal line merely presents a uniform spindle shaped enlargement. This neomorphic lorisine ulnar head

participates in a synovial inferior radio-ulnar joint which is largely bounded inferiorly by a sloping bony shelf projecting from the radius. The distinction between the antebrachiocarpal joint and the midcarpal joint is blurred in the slow loris and the complex of joints at the junction of forearm and hand can only be viewed as an integrated whole. The antebrachiocarpal joint retains a primitive bipartite character with a thick fibrous longitudinal septum intervening between its radiocarpal and ulnocarpal subdivisions. The bones of the classical proximal carpal row have adopted a stepped configuration with the lunate well elevated proximally into the radiocarpal cavity, taking it out of strict linear alignment with the triquetral, and with its ulnar face instead firmly united to the side of the longitudinal septum. This proximal protrusion of the lunate, which is foreshadowed in *Lemur*, is accompanied by an absence of the usual interosseous ligamentous union to the scaphoid, which is the typical primate condition. The triquetral is thus rather isolated medially, belonging properly neither to the proximal nor to the distal carpal row. It bears a cartilage clothed facet proximally, but in the neutral position (Fig. 5.2A) this does not contact the ulnar styloid process. Instead, the inflected flat cartilage-clad termination of the latter abuts against the side of the longitudinal septum.

The pisiform although somewhat reduced still retains a definite heel-like form but has effectively been displaced distally with the triquetral. The pisotriquetral joint cavity nevertheless retains continuity with the ulnotriquetral joint beneath the crescentic margin of the common articular fibrous capsule. In one specimen, which I have examined, a lobulated fat-filled synovial fold depended from this arcuate fibrous margin; this is not especially noteworthy, for fatty interarticular folds are common findings in diathroses, but in this case such features have been misinterpreted.

The habitual movement at the wrist, bringing it into the close-packed position is centred upon the midcarpal joint, and as in *Lemur* is a combination of extension, ulnar deviation and rotation in the sense of pronation (Fig. 5.2A). In the slow loris, however, this movement is exaggerated and the hamate bears a spiral facet prolonged anteriorly as an excavated cup, within which the triquetral is lodged at the extreme of the movement. Moreover,

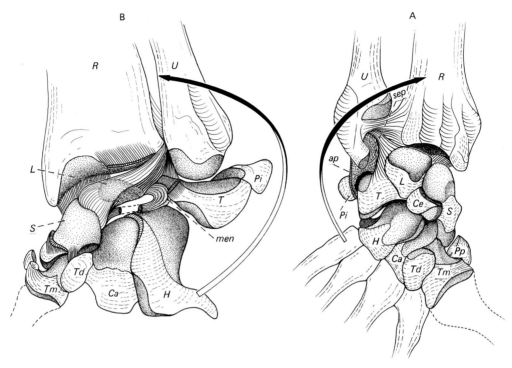

Fig. 5.2 (A, B)
Dorsal views of the right wrist articulations of *Nycticebus coucang* (A) and the left wrist of *Pseudochirus laniginosus* (B) both shown in the neutral position; the direction of the habitual movement into extension, ulnar deviation and rotation in the sense of pronation, is indicated in each case by the arrows. Articular surfaces are shown stippled. *ap*, aperture between ulnotriquetral and pisitriquetral joint cavities; *men*, meniscus-like derivative of the interosseous triquetrolunate ligaments; other labelling as in Fig. 5.1. (From Lewis 1985a.)

when this attitude is assumed the articular end of the ulnar styloid process slides down the longitudinal septum to articulate directly with the triquetral. It is then in very close apposition to the base of the pisiform which is hidden beyond the arched opening of the fibrous capsule, although the latter bone bears no actual articular facet. Although it is difficult to determine axes experimentally in such small specimens it is clear that the pattern of movement is quite comparable to that depicted for *Lemur* in Fig. 5.1A. The arc depicting this movement in Fig. 5.2A would then encompass an axis of similar orientation to that of *Lemur*.

The pronatory component of the movement seems to be slightly transmitted to the radius, and this restricted motion about the ulna is accommodated at the synovial inferior radioulnar joint. More importantly, this joint facilitates distraction of the ulna from the radius, with the head of the former bone sliding down the bony shelf projecting

from the radius and which walls the joint inferiorly. As will be seen the functional needs met by this lorisine inferior radioulnar joint are quite unlike those provided for in hominoids, or even the more restricted movement occurring in some monkeys. Forearm pronation/supination, which involves rotatory movement of the radius carrying with it the whole carpus, is thus quite tiny in amount.

The resultant movement grotesquely splits the hand, with the ulnar three digits, accompanied by the markedly reduced index, becoming offset almost at right angles to the forearm. This powerful lobster-claw like pincer is admirably adapted for securing a vice-like grip on branches as this cautiously climbing creature reaches ahead to bridge gaps in the foliage.

Perodicticus potto In this African lorisine the index finger is reduced to a diminutive nubbin and thus the ulnar limb of the split hand consists effectively of only the ulnar three digits. In form and function the joints between forearm and hand differ from *Nycticebus* in only minor details. For instance it seems that in some specimens at least a minor articular contact between ulnar styloid and pisiform is achieved at the close-packed extreme of the habitual movement. Highly suggestive hints about the true nature of the lorisine adaptations have long been available in the literature. Forster (1933), noting the extreme degree of ulnar deviation possible in the potto, correlated it with 'compression' of the carpus on the ulnar side and modification of the midcarpal joint so as to favour torsional movements, thus compensating for restricted forearm pronation–supination.

The ancestral form and function of the wrist articulations in therian mammals

There is little doubt that extant lemurs conserve a wrist morphology relatively unchanged from that of the ancestral primates. Moreover, it seems clear that in structure and function there has been little modification of the prototypal pattern evolved in the earliest therian mammals.

Marsupials such as the opossums *Didelphys marsupialis* and *Caluromys lanatus*, retaining a generalized clasping type of hand, with all the digits evenly spaced, exhibit the essence of this primitive mammalian pattern in the wrist articulations (Fig. 5.3B, D). The lunate is an integral part of the proximal carpal row and has a tough longitudinal septum attached to its ulnar margin against which the triquetral articulates. The latter bone and the heel-like pisiform form a weight-bearing cup for the articular extremity of the ulna. The Australian brush-tailed possum, *Trichosurus vulpecula*, again with a clasping type of hand, shows comparable structure but the lunate is reduced in size, although retaining its usual location within the proximal carpal row.

The functional significance of the structural features of the wrist articulations becomes apparent when one considers the gait of *Didelphys marsupialis* (Fig. 5.3A, C). This marsupial, an almost unchanged survivor from the Cretaceous, retains much of the primitive sprawling attitude of its reptilian precursors and shares this posture with non-cursorial therian mammals. No longer can it be considered that the characteristic and primitive posture of mammals involves vertically orientated limbs, undergoing parasagittal excursion, but rather this is a specialized derivative attitude and a cursorial modification (Jenkins 1971*b*). Jenkins (1971*b*) cineradiographic study of *Didelphys* shows that at each step the outstretched humerus is retracted (or adducted) and medially rotated pulling the body forward over the hand. As this occurs the changing relationship between the forearm bones and the hand requires that the latter, to maintain its grip on the substrate, must take up a position of ulnar deviation, extension and pronation; this is strikingly suggested by the illustrations in Jenkins' study although not specifically commented upon. Manipulation of the wrist of a wet specimen of *Didelphys marsupialis*, or *Caluromys lanatus*, shows that this is indeed the characteristic habitual movement, and that it occurs largely at the midcarpal joint, about an axis running from dorsally on the radial side of the carpus, with an angulation proximally towards the volar aspect on the ulnar side (Fig. 5.3B). This is similar to the axis of movement in *Lemur catta* (Fig. 5.1A), although the movement in *Lemur* is more restricted in amplitude than in *Didelphys*.

In the quadrupedal position, of course, as shown in Fig. 5.3 the midcarpal axis runs backwards, downwards, and laterally and as the stance phase of gait progresses the forefoot is effectively everted whilst the forearm is medially rotated about this oblique axis. It will be shown later that this midcarpal axis strikingly parallels the subtalar axis in the primitive mammalian hindlimb and that the subtalar axis functions similarly, allowing for angular and rotational adjustment between the implanted and thrusting foot and the crural bones as the femur is retracted (in this case abducted). Here the pes takes up an everted position comparable to the attitude of extension, pronation, and ulnar deviation of the manus. This might appear to be a striking example of serial homology but, in fact, the morphological components pressed into service in the two limbs are in no way comparable. It thus seems reasonably established that the obliquity of the midcarpal axis as described, is a symplesiomorphic feature, originally evolved to satisfy the requirements of the primitive type of mammalian gait. This habitual midcarpal move-

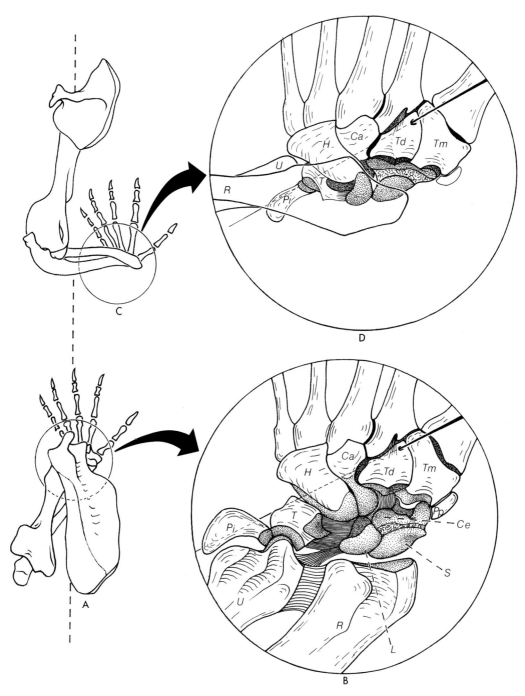

Fig. 5.3 (A–D)
The attitude of the left forelimb bones of *Didelphys marsupialis* at the earlier (A) and later (C) parts of the stance or propulsion phase of locomotion are shown, as demonstrated by cineradiography (redrawn from Jenkins 1971b). The interpretation described in the text of the way in which movement about the oblique midcarpal axis maintains the contact of the manus with the substrate during these movements is indicated in B and D. Labelling is as in Fig. 5.1.

ment at the primitive therian wrist, combining extension, ulnar deviation and pronation, has however been grossly exaggerated by convergent evolution at least three times to meet particular functional requirements. The specialized overlay to the primitive pattern of wrist structure and function which was evolved in lorisines has already been described. Analogous changes have occurred in other mammalian orders.

Certain Australian marsupials have evolved a specialized grasping modification of their hands known as schizodactyly of zygodactyly, in which the second digit together with the thumb forms one arm of a pincer, the other part being formed by digits three–five moving together. This produces a manus somewhat convergent in form to that of lorisines, except for the way in which these primates incorporate the second digit or index in the ulnar arm of the pincer; the resemblance is enhanced, however, by the reduction (*Nycticebus*) or even virtual elimination (*Perodicticus*) of the index. Schizodactylous hands are found in the ring-tailed possums and the koala which are often grouped together in the subfamily Phascolarctinae and also in the cuscuses.

It is reasonable to assume that these marsupials might show derived features in the joints between forearm and hand providing the basis for exaggerated movement of the ulnar limb of the pincer, and analogous at least to the morphology of lorisines. Such is, in fact, the case and the striking nature of the specializations further illuminate the interpretation given above of the significance of the lorisine adaptations. In the ring-tailed possum (*Pseudochirus laniginosus*) the lunate is much reduced in size. It has already been noted that there is a diminished lunate in *Trichosurus vulpecula* and this reduction seems to be a particular feature of the Australian marsupial radiation. In *Pseudochirus*, however, the reduced lunate has been lifted right out of the canonical proximal carpal row and is incorporated in a massive ligamentous band traversing the wrist joint cavity obliquely from scaphoid to attach in the cleft between radius and ulna (Fig. 5.2B). This fibrous tract is clearly a derivative of the primitive longitudinal septum of the antebrachiocarpal joint and incorporates at its lower end the interosseous scapholunate ligament. A thin fibrous sheet, also part of the longitudinal septum attaches to the triquetral, thus sealing off an independent ulno-

carpal joint cavity containing the articulation of the ulna with the triquetral and pisiform.

This disruption of the proximal carpal row has established communication between the radiocarpal and midcarpal joint cavities and the triquetral is left suspended as a relatively independent element, tethered mainly by the dorsal radiotriquetral ligament, and articulating with the side of the oblique fibrous tract containing the lunate. From the radial margin of the triquetral projects a tough, translucent, fibrocartilaginous meniscus whose location proclaims it to be derived from, and thus homologous with, the interosseous triquetrolunate ligament. This meniscus encircles the communication between the radiocarpal and midcarpal joint cavities. The triquetral, with the conjoined meniscus and bearing the articulating pisiform, is thus intercalated as a mobile mass between the ulna and the distal carpal row.

The habitual movement is, in its essence, not unlike that occurring in *Nycticebus* and occurs about a similar axis. Capitate and hamate, bearing with them the three ulnar digits, perform a composite movement, consisting of ulnar deviation, extension, and rotation in the sense of pronation (Fig. 5.2B). The hamate thus comes to lie almost transversely, compressing the triquetral against the ulna, and with its articular head, surrounded by the encircling meniscus, snugly abutting against the lunate. The whole wrist complex is then in close-packed position, and the schizodactylous form of the hand is realized.

The koala (*Phascolarctos cinereus*) shows an even more extreme degree of zygodactyly and it would be expected that the derived features seen in the ring-tailed possum would be even more emphasized here. Examination of ligamentous preparations indicates that this is indeed so. The triquetral is mobilized out of the proximal carpal row and by its tethering to the radius by the dorsal radiotriquetral ligament has effectively become part of the forearm. A tough oblique intracapsular ligamentous band passes from the scaphoid to the radius but no trace of a lunate seems to be discernible within this ligamentous mass. The antebrachiocarpal joint freely communicates with the midcarpal cavity and the aperture is partly encircled by a meniscus-like rim projecting from the radial border of the triquetral. *Phascolarctos* apparently represents the culmination of a progressive trend, the lunate having suffered complete dissolution leaving only a ligamentous connection between scaphoid and radius. The absence of a separate identifiable lunate has led to the erroneous idea that in this marsupial species the carpus includes a scapholunate.

Two-toed sloths, of the order Edentata demonstrate a mobilization of the triquetral out of the proximal carpal row rather similar to that exhibited by *Nycticebus*, and especially *Pseudochirus*. In this case the specialization also provides for a greatly increased range of ulnar deviation.

The joints at the wrist in monkeys

Both the superfamilies of monkeys, the Ceboidea (New World monkeys) and Cercopithecoidea (Old World monkeys) retain the essentials of the basic mammalian quadrupedal type of wrist morphology, modified in only quite small details. The apomorphic overlay, superimposed on a morphology similar to that seen in *Lemur*, or even *Didelphis*, is relatively minor. Despite the widely differing locomotor habits of monkeys ranging from restricted arborealism to committed terrestrialism, their wrist morphology is quite conservative. Even the so-called semibrachiators, both the New World and Old World varieties retain this plesiomorphic structure, and only the terrestrial baboons show any significant departure from the basic pattern and this is really quite trivial in scope.

The inferior radio-ulnar joint in monkeys may retain the primitive mammalian character (as seen, for example, in the marsupials *Trichosurus vulpecula* and *Didelphys virginiana*) and be a syndesmosis—a joint consisting of a firm ligamentous bond without elaboration of a synovial cavity. There is, however, a tendency leading to some modest refashioning of this joint into a diarthrosis, but this is apparently always imperfect. I have observed a simple syndesmosis in *Cebus*, and *Procolobus* but an incipient synovial cavity—a mere bursa deep within the ligamentous union between radius and ulna—may occur in *Ateles*, *Lagothrix* and *Colobus*. A considerable synovial cavity intervening between a cartilage-clothed ulnar head and a less perfectly elaborated articular surface on the radius has been encountered in *Cercopithecus*. In general, the Cercopithecoidea

appear more likely to possess a fairly well formed diarthrosis. It has already been noted that even *Lemur* possesses a bursal cavity in this situation whilst *Nycticebus* has elaborated a substantial synovial joint here. Clearly this derived morphology has been acquired in parallel several times.

The distal part of the ligament uniting radius and ulna forms an intermediate segment of the proximal articular surface of the wrist joint and when a well developed inferior radio-ulnar synovial cavity is present the ligamentous sheet is identifiable as the homologue of the hominoid and human triangular articular disc. This is unlike the arrangement in *Nycticebus* where the inferior radio-ulnar joint is largely walled by a bony shelf from the radius. The triangular articular disc then is a derivative of the original syndesmosis and there is no foundation for the alternative view that it has been derived from two ligaments one palmar and one dorsal (Corner 1898; Mörike 1964).

A derived feature of the monkey antebrachiocarpal joint is that the primitive partitioning into separate ulnar and radial compartments is no longer retained, at least in the adult. Indeed the only monkey in which a longitudinal septum has been observed is the spider monkey *Ateles*. In all other monkeys the septum has been resorbed, at least in its more dorsal portion; its most ventral part apparently persists as the palmar ulnocarpal ligament. The arrangement in *Cebus* (Fig. 5.1B) is particularly suggestive of this derivation of the ligament. Invariably, from monkeys to man, the palmar ulnocarpal ligament is closely associated with the triangular articular disc and protrudes back into the synovial cavity. Perhaps this secondary association led to the belief that the disc was phylogenetically formed by the fusion of two ligaments.

In the ulnar part of the antebrachiocarpal joint the ulna articulates with a concave cup formed by the pisiform and triquetral. The ulna has a constricted neck proximal to this articular extremity, which is the forerunner of the styloid process, and above this presents a variably developed lateral protuberance or emergent head (Figs 5.1B, 5.4A). To avoid confusion, the terms of human anatomy—styloid process and head, respectively—will be used in the following descriptions for the original carpal extremity and for this newly evolved enlargement. It has already been seen that a comparable ulnar head has been evolved in parallel in *Nycticebus*. The facet on the ulnar styloid process is borne upon its convex distal extremity and on the surface facing the interior of the joint (Figs 5.1B, 5.4A). By contrast, the exposed surface, facing laterally in the habitually pronated quadrupedal posture, is non-articular and the disposition of the styloid process is analogous to that of the lateral malleolus of the hindlimb. In the dorsiflexed and pronated palmigrade attitude the pisiform projects back into the heel of the hand, acting as a biomechanical counterpart to the heel in the hindlimb, and forming the bony basis underlying the hypothenar skin pad, which is thus related to the lower extremity of the ulna, extending in effect onto the forearm; this is in contrast to the more distal location of the pad in the Hominoidea. This distinction between articular and non-articular areas on the ulnar styloid process is always quite obvious in the Ceboidea but in a few species of Old World monkeys the dry bones may not truly reflect the arrangements known to hold in the intact state for the exposed surface, known to be non-articular, may appear to be quite smooth. In the radial part of the joint, the convex surfaces of scaphoid and lunate articulate with the radius. The scaphoid articular surface, however, bears a posterolateral concavity (Fig. 5.1B), recalling that seen in *Lemur* (Fig. 5.1A) and even *Didelphis* (Fig. 5.3B), and similarly acting as an articular stop locking onto the convex dorsal border of the radius in full extension. The joint is here strengthened in its volar aspect by a prominent intracapsular palmar radiocarpal ligament joining the front of the radial styloid process to the lunate. Posteriorly the antebrachiocarpal joint is crossed by an oblique dorsal radiotriquetral ligament which is little more than a capsular thickening (Fig. 5.1B). These three ligaments, intracapsular palmar ulnocarpal and radiocarpal ones, and the thinner dorsal radiotriquetral ligament comprise the essential ligamentous apparatus of the mammalian antebrachiocarpal joint and are clearly attuned to its functional requirements of the quadrupedal posture. The extensor carpi ulnaris tendon crosses the dorsum of the wrist between the head of the ulna and its styloid process and then spirals in a volar direction around the exposed non-articular part of the styloid process towards its insertion; the tendon

pursues a similar course in *Lemur* (Figs 1A, 1B).

Variations from this basic joint pattern are only rarely encountered among either New World or Old World monkeys. Reports in the literature that in *Ateles* the pisiform is distally displaced and takes no part in the antebrachiocarpal joint do not accord with my experience. In both a specimen of *Ateles geoffroyi* and one of *Ateles paniscus* the typical monkey ulnopisiform contact was retained, although in the latter it was quite tiny. It does, however, indeed seem true that in a least some specimens the pisiform in *Ateles* is quite distally located and it is likely that then contact with the ulna is lost. Thus, no striking specializations are featured in the antebrachiocarpal joint morphology of even those monkeys exhibiting at least a fairly substantial change in emphasis from arboreal quadrupedal towards suspensory locomotion. Conversely, invasion of a terrestrial habitat has been accompanied by some change in structure. Jones (1967) has described a semilunar meniscus in the dorsolateral aspect of the wrist joint of *Papio ursinus*. I have observed what is doubtless the same morphological entity in *Papio papio*: the lateral part of the dorsal radiotriquetral ligament, which is additionally tethered to the scaphoid in this species, is interposed between the dorsal concavity of the scaphoid and the radius during extension and is here modified to form a somewhat rudimentary meniscus. This is a relatively minor modification of a pre-existing structure, and the remainder of the joint is unremarkable.

The dry bones of monkeys bear the clear imprint of this morphology (Fig. 5.4A). The triquetral is a relatively massive block of bone, rather transversely disposed and bearing a non-articular tuberosity at its medial end. Its proximal surface bears a shallow concave facet which is confluent with a second facet on the anterior surface for the articulation of the pisiform, completing the receptive cup for the ulnar styloid process. The elongated pisiform, with a terminal epiphysis apparent in young individuals, is vertically compressed at its carpal extremity where it presents a terminal facet for the triquetral and a small facet, often restricted to the medial part of the upper surface, for the ulnar styloid process.

The scaphoid is a squat bone, only slightly constricted between tuberosity and body by a

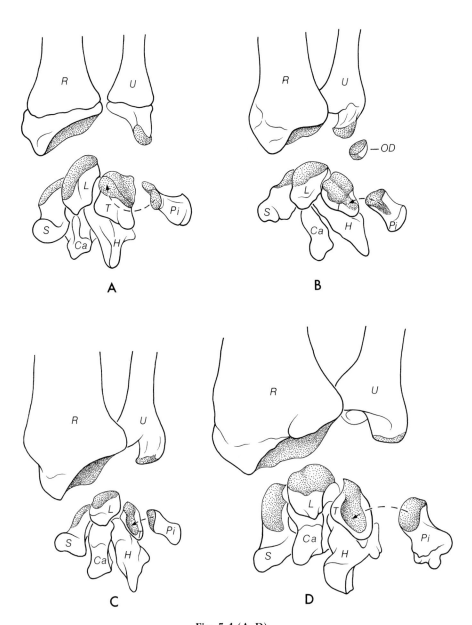

Fig. 5.4 (A–D)
The palmar aspects of the bones entering into the right antebrachiocarpal joints (with the articular surfaces participating in this joint stippled in each case), together with the capitate and hamate in *Cebus nigrivittatus* (A), *Hylobates lar* (B), *Pan troglodytes* [C], and *Gorilla gorilla gorilla* (D). *OD*, os Daubentonii; other labelling as in Fig. 5.1. (From Lewis 1972b.)

groove for the intracapsularly situated palmar radiocarpal ligament, and it is firmly bound to the centrale posteriorly by ligaments; the two bones, however, have a diathrodial articulation between them where the centrale is insinuated between capitale and scaphoid. Posteriorly the centrale is joined to the triquetral by a centrale–triquetral ligament as in *Lemur*. The keystone of the proximal carpal row is the lunate with a form befitting its name, and the whole row embraces capitate and hamate at the midcarpal joint, with the centrale bearing the articulation for the trapezoid and an adjoining part of the trapezium, whose other support is the scaphoid tuberosity (Fig. 1B).

The broadly wedge-shaped hamate is disposed so that its midcarpal articular surface faces rather proximally, forming the gently convex medial part of the midcarpal joint line and providing a stable platform for the triquetral. The convex surface for the triquetral is prolonged as a synovial-lined pocket adjacent to the base of the fifth metacarpal and here the bone may show a concave ventro-medial extension of the triquetral facet. This articular area on the hamate thus may present a gently spiralling appearance. This particular description has led to some confusion and it might be better to say that the surface is twisted concavo-convex, for it is unlike the spiral trough found in hominoids. The blunt apex of the hamate takes up much of the midcarpal surface of the lunate, often virtually excluding the capitate from contact with that bone.

The head of the capitate is small and sharply cut off posterolaterally (Fig. 5.1B) where the centrale contacts it and is wedged into the carpus dorsally between capitate and trapezoid. Here, in a number of monkeys, but not all the centrale is joined to the 'waist' of the capitate by a slender centrale–capitate ligament (Fig. 6.6). This relatively diminutive capitate head sits astride the broadly flared distal portion. In the terrestrial and habitually digiti-grade cercopithecines—*Papio, Mandrillus, Erythro-cebus*—the head of the capitate shows a medial expansion at the expense of the hamate, thus attaining a more extensive contact with the lunate, and imparting an obliquely truncated form to the apex of the hamate; the capitate then has a rather centrally constricted or 'waisted' form but this is quite dissimilar to the waisting which will be described in hominoids. Anteriorly the capitate is joined to the triquetral by an anterior capitate–triquetral ligament obliquely linking the two carpal rows (Figs 6.6, 6.7). This ligament was noted in *Lemur*; it is part of the ancestral primate ligamentous apparatus.

Midcarpal movement in monkeys has the same essential pattern as that found in prosimians, which in itself is merely a primitive therian inheritance, correlated with the quadrupedal method of locomotion. Motion is about a comparable axis (Fig. 5.1B) to that noted in prosimians, or for that matter *Didelphys*, and at the full extent of the movement of extension, pronation, and ulnar deviation, the joint is in its position of maximum congruence. The triquetral has then moved down and forward on the convex head of the hamate and so become lodged in a synovial or cartilage lined pocket adjacent to the base of the fifth metacarpal. Close-packing is achieved at the radial part of the monkey joints, by wedging of the centrale from dorsally into the embrasure between capitate and trapezoid, whilst its concave margin locks against the posterolateral part of the 'neck' of the capitate. The extensor carpi ulnaris must be a prime mover involved in this motion and the spiral course of its groove is clearly correlated with the characteristics of the movements. Movement to the other extreme about the axis involves flexion, radial deviation and supination. When the midcarpal joint is extended and close-packed in the quadrupedal posture, a congruent movement also involving extension, pronation and ulnar deviation close-packs the antebrachiocarpal joint, and the rolled over dorsal border of the radius abuts upon the posterior concavity of the scaphoid. In baboons the close-packed position is here brought about at an earlier point by the presence of the 'meniscus' interposed between the approximating dorsal parts of radius and scaphoid. This appears to limit extension in baboons. Particular interest attaches to midcarpal function in atelines. These New World spider monkeys indulge in a form of suspensory locomotion variously known as bra-chiation, or semibrachiation. This activity however, differs from that of apes in that a third support, the prehensile tail, is utilized.

Commonly the overhead grasp involves taking hold of the support with a contralateral grip (for example, the right hand grasping the left side of the support) and in this case the hand is hyperpronated, as in quadrupedal locomotion, and substantial wrist rotation is not a requirement for progression. Alternatively, an ipsilateral grip is employed and of necessity rotation of the body about the grasping hand occurs. However, this rotation is of only about 90°, substantially less than the 180° of brachiating apes, and the body at both extremes faces half forwards, supported behind by the prehensile tail. I have pointed out that spider monkeys lack the antebrachiocarpal joint modifications characteristic of hominoids, and suggested that this factor limits their bodily rotation.

Jenkins (1981) in an important cineradiographic study of brachiating spider monkeys focused attention on the midcarpal joint and maintained that the limited bodily rotation is largely (70°) accommodated here, the remainder

(20°) being contributed by forearm or radial supination. He went on to suggest that the critical morphological indicator of brachiation is elaboration of a midcarpal joint with substantial rotatory capacity. I have analysed the joint movements and disposition of the forelimb bones depicted by Jenkins during brachiation employing an ipsilateral grip on wet specimens of *Ateles* (Figs 5.5A, B). What was not made clear in Jenkins' study was the quite trivial extent of the movement possible at the midcarpal joint from the neutral position to full supination of the distal carpals. It seems quite clear that the greater part of the movement accommodating the restricted range of bodily rotation, is produced by forearm twisting or supination. In suspensory mode this is, in effect, rotation of the ulna (carrying with it the

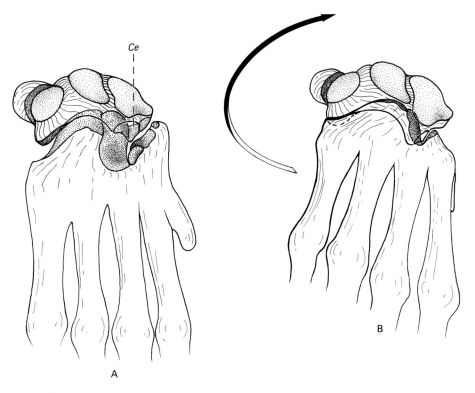

Fig. 5.5 (A, B)

A dorsal view of the left carpus and hand of *Ateles paniscus*. A, the distal carpal row, carrying with it the hand, has been moved into the position of flexion, radial deviation and supination. B, the distal carpal row (and hand) have been moved into extension, combined with ulnar deviation and pronation and the joints are close-packed. The direction of this latter movement is shown by the arrow above, and if a single axis was defined it would be circumscribed by this arrow and would traverse the carpus from the dorsal aspect with an angulation proximally, ulnarly and volarly, as in Fig. 5.1B. *Ce*, centrale. (From Lewis 1985*b*.)

body by virtue of its non-rotatory articulation at the elbow) about the fixed radius and grasping hand. That this is the really significant movement is deducible from Jenkins' radiographs where the rotation of the ulnar head and styloid process relative to the rest of the wrist is clearly shown. This is not especially surprising since monkeys are generally capable of about 90° of pronation–supination. It is likely that diminution (or absence) of ulnopisiform contact in *Ateles* facilitates this but to nothing like the extent attained in hominoids where, as will be seen, the combined gliding and rotational movement of the ulnar styloid process in relation to the carpus is realized by the interposition of a meniscus. Thus *Ateles* species seem to be utilizing a fundamentally conservative monkey morphology for their semibrachiating activities and indeed other monkeys without any claimed uniqueness in wrist structure, including some Old World colobines, are also semibrachiators. Moreover, the ipsilateral grasping attitude is only part of the locomotor repertoire of spider monkeys; as Jenkins' (1981) noted they regularly use a contralateral grip and in this case the hand is initially hyperpronated—in effect it is a quadrupedal attitude of the hand but used to grasp an overhead support. This highlights the other extreme of the movement which is of much greater amplitude. The hand is readily manipulated into a position of marked ulnar deviation, extension and pronation (Fig. 5.5B), a movement similar to that occurring at the midcarpal joint in other monkeys or prosimians, and clearly about a comparable axis running from the dorsal aspect of the carpus proximally and volarly and towards the ulnar side. Movement to the other extreme about such an axis involves flexion, radial deviation, and supination. It is this movement which is somewhat enhanced in *Ateles* by easier access of the centrale into the somewhat widened cleft between capitate and trapezoid.

The ape joints

Analysis of the evolutionary changes in hominoid wrist joints provides a striking example of the way in which comparative studies can sometimes shed new light and understanding on the bald facts of descriptive human anatomy.

For a century it has been recognized that apes have an increased range of hand mobility when compared with other Primates and that this has been effected by two major correlated skeletal modifications. Firstly, the ulna has lost primitive articulation with the pisiform and triquetral, its original carpal extremity thereby regressing until it became the comparatively insignificant styloid process, whilst a neomorphic ulnar head developed and became incorporated into a perfected synovial inferior radio-ulnar joint. Surprisingly, the mechanism effecting the exclusion of the ulna from the antebrachiocarpal joint excited no curiosity, despite the presence of well documented suggestive clues. In the gibbon carpus it was known that a supernumerary carpal element—the bony os Daubentonii (Fig. 5.4B)—occupies the interval created by retreat of the ulna (Daubenton 1766; Leboucq 1884; Kohlbrugge 1890). The gibbon carpal element, and a similarly situated cartilaginous nodule in the human foetus were interpreted as serially homologous with the pedal os trigonum (Leboucq 1886), and just as the os trigonum was said to be derived from the os intermedium tarsi, so this carpal element was interpreted as an os intermedium antebrachii. Since it is obvious that an os Daubentonii is correlated with the retreat of the ulna from its direct articulation with the carpus, it is tempting to suppose that the ossicle is in some fashion directly associated with this retreat.

I was already aware that the free so-called 'os intermedium tarsi' of the marsupial foot (held to correspond to the human os trigonum) is nothing more than a lunula, an ossification within the thickest part of an intra-articular meniscus, (see Chapter 11); the very foundation of the postulated homology of the os trigonum with the os intermedium was thereby refuted. This finding raised the possibility that the supposed comparable carpal homology might similarly be faulty and that the so-called os intermedium antibrachii, and therefore the os Daubentonii, might also be a lunula. This hypothesis presupposed that separation of the ulna from the carpus had been effected by the appearance of a meniscus in the interval. Indeed, it did turn out that in the hominoids the ulna has been excluded from the antebrachiocarpal joint by a substantial fibrocartilaginous meniscus containing a lunula in gibbons, but with this ossicle

generally regressing in other hominoids. Furthermore the great apes show a progressive trend
towards amalgamation of this meniscus with the
triangular articular disc, thus becoming an integral part of the proximal articular surface of the
wrist joint, and leaving the ulnar styloid process
isolated within a proximal diverticulum of the
joint.

The gibbon joints

Gibbons possess fully elaborated synovial inferior
radioulnar joints incorporating a well developed
ulnar head and with a large triangular disc
completely separating its cavity from the antebrachiocarpal joint. Associated with the anterior
margin of the disc is a palmar ulnocarpal ligament
passing from the base of the ulnar styloid process to
the flexor aspect of the lunate bone. This ligament
is so intimately merged with the triangular disc
that it is effectively one with it, giving the
impression that the disc is partly attached to the
anterior aspect of the lunate bone. Rather surprisingly in some gibbons a vertical partition subdivides the antebrachiocarpal joint, passing from the
inferior surface of the triangular disc, adjacent to
its radial attachment, down to the interosseous
ligament between triquetral and lunate; I have
observed such a subdivision in a specimen of
Hylobates lar and in one siamang (*H. syndactylus*)
but it is not a constant gibbon feature. As has been
described above, such septa are primitive mammalian features, persisting in prosimians but
apparently only present among monkeys in *Ateles*.
Their occasional persistence in gibbons therefore is
unexpected. A large fibrocartilaginous intra-
articular meniscus intrudes into the joint, intervening between the cartilage-clothed ulnar styloid
process and the triquetral, adjacent to its articulation with the pisiform. At the anterior margin of
the meniscus the pisotriquetral articular cavity is
in free communication with the radiocarpal joint
cavity proper and also extends proximally as a
short blind diverticulum related to the periphery of
the meniscus; in effect, therefore, the pisiform
articulates largely with the meniscus (Fig. 5.6).
The meniscus effectively excludes the ulnar styloid
process from contact with the pisiform but not
entirely from the triquetral. Thus, part of the ulnar
styloid process is exposed through the wide

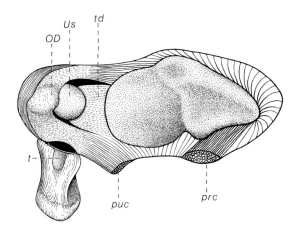

Fig. 5.6
The proximal articular surface of right antebrachiocarpal joint in *Hylobates lar*. The rounded articular
extremity of the ulnar styloid process (*Us*) is encircled by
a meniscus containing a lunula, the os Daubentonii
(*OD*). The pisiform bears a facet (*t*) for articulation with
the triquetral and the blind diverticulum of the pisotriquetral joint cavity between pisiform base and meniscus
is shown. Labelling is as in Fig. 5.1. (From Lewis 1969.)

opening encircled by the meniscus and participates
in the antebrachiocarpal joint. The posterior horn
of the meniscus inserts on the radius behind the
inferior radio-ulnar joint and the anterior horn
attaches to the lunate, in association with the
palmar ulnocarpal ligament. The thick medial
portion of the meniscus contains an ossified
lunula, the os Daubentonii. This supernumerary
carpal bone may be so large that it comprises much
of the mass of the meniscus, and the base of the
pisiform then effectively articulates with it. In
juvenile specimens it is represented only by a
nodule of hyaline cartilage. The os Daubentonii
appears to be a constant feature of gibbons with
one possible exception, the rather specialized
siamang, which indeed has often been placed in a
separate genus, *Symphalangus*. In a juvenile siamang which I have examined there was no trace of
a lunula, not even a chondrified nodule. Similarly,
Schreiber (1934) failed to find an os Daubentonii
in *H. syndactylus*, yet Kohlbrugge (1890) records
its presence in this species; it seems likely, therefore, that the ossicle is inconstant in siamangs, in
contradistinction to other gibbons. The palmar
radiocarpal ligament is thick, essentially intracap-

sular, where it grooves the scaphoid, and attaches proximally to a large triangular depression on the anterior aspect of the radial styloid process; distally it shows a partial subdivision into two bands passing to the lunate and capitate respectively. This extension on to the capitate across the midcarpal joint line is an advance on the monkey morphology; from the same situation on the capitate an anterior capitate–triquetral ligament passes back across the joint line. Dorsally the capsule is reinforced by a rather flimsy dorsal radiotriquetral ligament, and a centrale–triquetral ligament links those two bones.

The dry bones of the gibbon wrist (Fig. 5.4B) present several apomorphic features clearly associated with the elaboration of the meniscus, together with other plesiomorphic aspects clearly recalling the more primitive pattern found in monkeys. In this amalgam of primitive and progressive carpal characteristics the gibbons occupy an intermediate position between monkeys and the other hominoids.

The form of the ulnar styloid process reflects its new role, that of providing an articular surface for the meniscus whilst retaining a reduced contact with the triquetral through the aperture encircled by it. The meniscal articular facet is borne upon the distal and exposed (ulnar) aspect of the process; that for the triquetral, more or less delimited from the former, occupies the tip. Thus, the major articular area on the process is located on its exterior surface, whilst in monkeys it is on the interior one. Again, unlike monkeys, the groove for extensor carpi ulnaris courses straight down the back of the process and does not spiral around onto the exposed surface.

The triquetral, inappropriately named just as it is in monkeys, retains the essentially cuboidal form found in monkeys but is compressed into a rather flatter plaque. The pisiform facet is displaced distally and medially almost onto the tuberosity and here the pisiform articulates by a small concave groove on its dorsal surface. The changed disposition of this articulation, when compared with that of monkeys, results in a distal angulation and displacement of the gibbon pisiform into the palm, and the broad proximal surface of the bone articulates with the periphery of the ossicle-containing meniscus. This reorientation of the pisiform, is the causative factor underlying the

apparent distal migration of the hominoid hypothenar pad.

The scaphoid is more deeply waisted than that of monkeys, consequent upon widening and deepening of the groove for the intracapsular palmar radiocarpal ligament, which in gibbons is bifascicular and attaches not only to the lunate but also to the capitate. The proximal articular surface retains a suggestion of the dorsal concavity typically found in monkeys which locks in extension upon a complementary convexity on the radius. The centrale, usually independent, may fuse to the scaphoid in old individuals, but even when independent it is firmly bound to the scaphoid and forms an integral part of the proximal carpal row, providing most of the articular cup for the head of the capitate.

The lunate is unremarkable in form and its midcarpal articular surface is largely occupied by the hamate, almost to the exclusion of the capitate, just as it is in monkeys.

The hamate is more narrowly wedge-shaped than in monkeys and together with the adjoining head of the capitate may give the impression when seen in profile from behind, of an almost hemispherical articular ball entering into the midcarpal joint. But this is a spurious impression for clearly it cannot function to provide simple rotatory midcarpal movement. Jenkins (1981) indeed suggested this again concluding, as for spider monkeys, that here was located the pivotal point for bodily rotation during brachiation. Significantly, however, he noted that gibbon hands could not be manipulated into a hyperpronated posture as in monkeys. In fact the triquetral articular surface of the gibbon hamate clearly precludes simple rotation and shows incipient signs of a rather subtle transition from monkey to ape morphology, the most significant point being that its orientation is changed from the shallow incline typical of monkeys to the more vertical disposition, characteristic of hominoids. In the extended close-packed position the triquetral rides onto the upper and dorsal part of the hamate where there are indications of a spiral furrow, and is not impacted close to the base of the fifth metacarpal as in monkeys. In fact, manipulation of gibbon and siamang wrists makes it clear that in the habitual movement of extension the midcarpal joint is brought into a close-packed position by a composite movement involving extension, radial deviation, and supination, which is, in fact, the mirror image of that in monkeys and prosimians. The centrale then is locked about the slightly bulbous head of the capitate; here, however, there is a lack of deep undercutting of the capitate head, and so none of the markedly 'waisted' effect characteristic of chimpanzees. The centrale articulates with the 'waist' of the capitate posteriorly, where there is a continuation down of the articular surface of the head to become confluent with the facet for the trapezoid, just as in monkeys (Fig. 6.16). But because of the changed rotatory character of movement in gibbons, the centrale also contacts the capitate anteriorly; here the articular surface of the head is confluent below with the anterior facet for the second metacarpal. Perhaps for this reason there is no centrale–capitate ligament. Gibbons have already initiated the dramatic change in midcarpal morphology and function which characterizes the hominoids and the changed disposition of the groove for extensor carpi ulnaris is an indicator of this.

The chimpanzee joints

The synovial inferior radio-ulnar joint is fully elaborated and is walled inferiorly by a triangular articular disc which is broader transversely in *Pan* than in *Hylobates*, and the palmar ulnocarpal ligament associated with it is a more obvious entity, bulging as a synovial-covered protrusion into the joint cavity. A meniscus (Fig. 5.7), in this case lacking a lunula, but having similar attachments to those found in *Hylobates*, is a characteristic and striking feature of the ulnar part of the chimpanzee joint. Again the pisotriquetral cavity communicates with the main joint cavity around the arcuate anterior border of the meniscus and adjacent to this the pisiform base retains a restricted contact with the periphery of the meniscus, where there is a small blind recess of the pisotriquetral cavity. The cartilage-clothed ulnar styloid process retains some contact with the triquetral through the aperture bounded by the meniscus. The proximal compartment of the joint approached through this aperture extends some way up the anterior aspect of the styloid process and usually contains a prominent aggregation of synovial fringes. The extent of this proximal

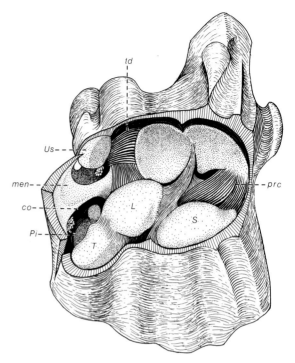

Fig. 5.7

The right wrist joint of *Pan troglodytes* opened dorsally and flexed. The semilunar meniscus (*men*) has been detached posteriorly from the radius and pulled forward from its contact with the cartilage-clothed ulnar styloid process (*Us*), thus revealing the proximal compartment of the joint cavity; this diverticulum extends mainly anterior to the ulnar styloid process (where it contains an aggregation of synovial villi). The anterior margin of the meniscus exhibits a free, crescentic indentation; here the wrist joint cavity communicates (*co*) distally with the pisotriquetral cavity which extends as a short, blind proximal recess anterior to the meniscus. Also illustrated are the synovial-covered palmar ulnocarpal ligament anterior to the triangular articular disc (*td*), the thick intracapsular palmar radiocarpal ligament (*prc*), and a synovial fold attaching to the radius between its facets for the scaphoid and lunate. Labelling is as in Fig. 5.1. (Redrawn after Lewis 1969.)

diverticulum is strikingly demonstrated in arthrograms (Fig. 5.8).

The ulnar part of the antebrachiocarpal joint is confluent with the radiocarpal part and no septum intervenes. A thick palmar radiocarpal ligament which is essentially intracapsular, grooves the scaphoid during its intracapsular course, and is partially subdivided into two bands attaching to

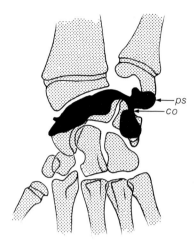

Fig. 5.8
A tracing of an arthrogram of the left wrist joint of *Pan troglodytes* the same specimen as is illustrated in Fig. 5.7. The radio-opaque medium (black) in the main articular cavity is continuous with that in the proximal compartment (*ps*) where it is seen as a prominent ballooning mass investing the ulnar styloid process. The pisotriquetral cavity has also been filled via a communication (*co*) with the main wrist joint cavity. (Redrawn after Lewis 1969.)

lunate and capitate just as it is in gibbons; it arises from a large impression on the anterior aspect of the radial styloid process. The distal band is reflected back across the midcarpal joint as an anterior capitate–triquetral ligament.

In some specimens the morphology is modified and the meniscus may be well incorporated into the proximal articular surface of the joint, merging with the triangular articular disc (Fig. 5.9). This composite fibrocartilaginous structure then forms the deeply concave ulnar part of the proximal articular surface, and is perforated by a small aperture, bounded by the thin sharp margin of the meniscus homologue, through which the ulnar styloid process is exposed. In such specimens the ulnar styloid process is therefore more effectively segregated within its own quite capacious proximal synovial compartment.

The chimpanzee joint presents a suite of more advanced apomorphic features than gibbons, where some resemblance to monkey morphology is retained despite the elaboration of a meniscus. In *Pan* the trend towards opening up of the ulnar side of the joint cavity, incipient in gibbons, is accentuated and major remodelling of the proximal carpal row has led to its quite wide separation from the ulna.

The ulnar styloid process presents an articular surface for the interposed meniscus on its external aspect and retains a more or less distinguishable facet at the tip for the triquetral (Figs 5.7, 5.4C). The extensor carpi ulnaris tendon courses straight down its posterior aspect, grooving it (Fig. 5.7).

The triquetral differs markedly from the rather cuboidal bone found in monkeys and gibbons and justifies its name. It is remodelled into the shape of a triangular pyramid with the edge between palmar and dorsal surfaces (Fig. 5.4C) rounded off proximally to form a smooth facet playing upon the inferior meniscal surface and variably articulating with that exposed part of the ulnar styloid process encircled by it. This facet is confluent below and in front with a sizeable shallowly concave articular excavation receiving the pisiform, and occupying most of the palmar surface of the bone. It is the elaboration of this excavation which has been the major factor in remodelling the bone, and the siamang even gives a suggestion of the emergence of this morphology. On the posterior surface the bone presents a semilunar crevice, shown in Figs 5.10 and 10.1D, where the fibrous capsule, and with it the periphery of the meniscus, are attached. The extensor carpi ulnaris tendon here is in intimate contact with these structures, grooving them, as it proceeds from its straight passage down the back of the ulnar styloid process towards its insertion.

The pisiform articulates with the palmar cup on the triquetral by a large dorsally located and convex articular facet and the bone bears a second confluent narrow facet on the true proximal surface for the periphery of the meniscus (Fig. 10.1D). This extensive reorganization of the pisotriquetral joint reorientates the pisiform from the heel-like monkey attitude, bringing it into a position where it projects well distally into the palm, almost reaching the hook of the hamate. Even gibbons had initiated this reorganization but with only an insignificant dorsally located articulation on the pisiform and a greater contact with the meniscus; siamangs rather more closely approximate the chimpanzee morphology.

The hamate is more narrowly wedge-shaped than monkeys, or even gibbons, and this restricts

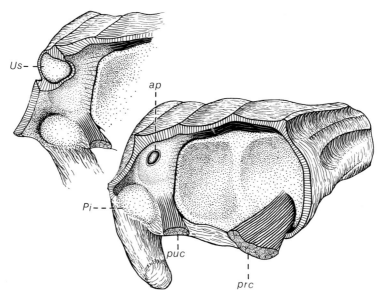

Fig. 5.9

The proximal articular surface of the right wrist joint of a specimen of *Pan troglodytes* in which the meniscus is well merged with the triangular articular disc, the two structures contributing to the concave ulnar component of the articular surface. The aperture (*ap*) leading to the proximal joint compartment is relatively restricted, and through this opening a small portion of the cartilage-clothed ulnar styloid process is exposed. The elongated pisiform (*Pi*) with its articular surface for the triquetral is apparent, demonstrating again continuity between pisotriquetral and wrist joint cavities. The thick palmar radiocarpal ligament (*prc*) has been cut from its attachment to the lunate and capitate and the palmar ulnocarpal ligament (*puc*) from the lunate. In the detail shown above the proximal synovial compartment has been opened (by incising that part of the articular surface derived from the meniscus), thus revealing the cartilage-clothed ulnar styloid process (*Us*). (Redrawn after Lewis 1969.)

its articulation with the lunate whose midcarpal surface has largely been appropriated by the capitate. The lunate is a broader bone than in monkeys or gibbons, contributing more to the radiocarpal joint, and with prominent beaks, anteriorly and posteriorly.

The scaphoid, incorporating the centrale bearing the trapezoid facet, is deeply grooved for the bifascicular palmar radiocarpal ligament. Its radiocarpal articular surface may show some residual indication of the dorsal concavity found characteristically in monkeys and prosimians, but conspicuous dorsal ridges, which could plausibly be considered as effective articular stops associated with knuckle-walking (Tuttle 1967), have not been found to be constant or even usual features of the bone. The bone is deeply notched behind at the junction of the centrale portion with the main mass, and here the centrale portion has a ligamentous union across to the triquetral; this is the homologue of the posterior centrale–triquetral ligament found in those primates where the centrale retains its independence.

The midcarpal joint surfaces of capitate and hamate are considerably elaborated from the gibbon condition (Fig. 5.10). The triquetral rides along a spiral, concave trough winding up around the convex head of the hamate from the root of its hook, and this furrow is bounded posteriorly by a raised lip; the articulation is therefore subtly different from the twisted concavoconvex surface which is found in some monkeys, and function is radically different.

The medial surface of the adjoining os capitatum presents a large articular area for the hamate, occupying the whole flattened medial surface of the bone above and prolonged as a dorsally located articular strip as far as the distal margin of the bone. In front of this articular area the two bones are united by a massive interosseous ligament. The

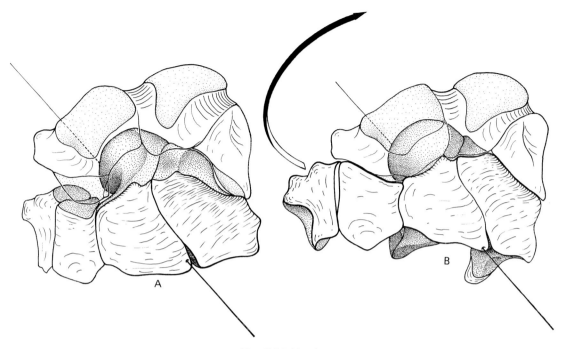

Fig. 5.10 (A, B)

A dorsal view of the left carpus of *Pan troglodytes* with a steel rod transfixing the two carpal rows along the midcarpal axis. A, the distal carpal row has been moved into flexion, ulnar deviation, and pronation about this axis. B, the distal carpal row has been moved into extension, radial deviation and supination about the axis whilst the position of the proximal row is maintained unchanged; the direction of this movement is shown by the arrow above. (From Lewis 1985b.)

head of the capitate is markedly enlarged laterally, especially in its anterior portion where there is a deeply undercut waist below the bulbous head (Figs 5.10A, 5.11); here the scaphoid (centrale component) may in some specimens be loosely tethered by a rather tenuous synovial-covered, flat fibrous band the centrale–capitate ligament. The articular surface of the head is prolonged down behind to form an articular neck on the bone (less constricted than the anterior excavation) and below this is a posteriorly located articulation for the trapezoid. Anterior to this facet trapezoid and capitate are united by a massive ligament attaching to the capitate below its hollowed waist. This posterior location of the capitate/trapezoid articulation, and the ligamentous union in front of it, are plesiomorphic features found in prosimians and monkeys (Figs 6.6, 6.7) and even *Didelphys*. The descriptive epithet 'waisted' as applied to the chimpanzee (and other) capitate bones was introduced by Clark (1967) who assumed (with, in fact,

no justification) that the deep depression on the lateral aspect represented the site of attachment of 'a strong interosseous ligament binding it to the adjacent trapezoid bone'; as noted above this ligament is located distal to the waist. The waist then is a lateral indentation most marked anteriorly (and therefore conspicuous in anterior views) but also less noticeable posteriorly in the region of the articular neck (as seen in posterior views). Unfortunately Clark (1967) incorrectly captioned the anterior view as 'posterior' giving rise to considerable subsequent confusion. 'Waisting' has even been interpreted as any indentation of the anterior surface, which, in fact, is almost flat and thus demonstrated in profile outline of the lateral aspect.

The composite articular surfaces formed by the capitate and hamate are strikingly twisted (Fig. 5.10A) and form the hub of a specialized pattern of midcarpal movement, which was already established but not completely perfected in

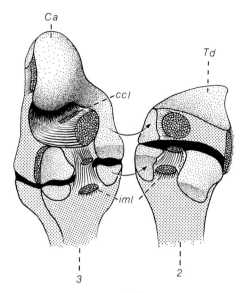

Fig. 5.11

The anatomical arrangements in the left wrist of a wet specimen of *Pan troglodytes*. The trapezoid (*Td*) with the attached index metacarpal (*2*) has been disarticulated and, as indicated by the arrows, rolled away from the capitate (*Ca*), bearing the third metacarpal (*3*). The upper arrow indicates the site of articulation between capitate and trapezoid; immediately anterior to these facets lie the cut attachments of the massive interosseous ligament. The lower arrow indicates the posterior articulation of the second metacarpal with the third metacarpal and capitate; an interosseous carpometacarpal ligament separates this posterior joint from a comparable anterior one. Articular areas are indicated by shaded stippling and non-articular areas by mechanical stippling. *iml*, intermetacarpal ligament. (From Lewis 1973.)

gibbons. Initially (Lewis 1972) this movement was figured as a whole and depicted as motion of proximal upon distal row, without any attempt to break it down into components nor to establish an axis. It was then pointed out that extension at the midcarpal joint is accompanied by conjunct rotation screwing the carpal rows together into the closed-packed maximally congruent position, with the scaphoid impacting firmly onto the articular neck of the capitate, and its incorporated centrale element becoming firmly embedded beneath the bulbous capitate head, whilst the triquetral fitted snugly upon the posterior and highest part of its spiral hamate facet. More extensive analysis of the

movement can be deduced from the experimental determination of a compromise axis (Fig. 5.10). This midcarpal axis runs from dorsally on the ulnar side of the wrist, passing through the head of the capitate with a proximal deviation towards the volar aspect on the radial side and relative movement occurs between the two rows about this axis. It could be depicted as movement of the proximal carpal row about a fixed distal row, as was initially done, but it is probably more realistic to consider the movement of the distal row, carrying with it the rest of the hand, relative to a fixed proximal row (Figs 5.10A, B). Movement between the rows follows a quite circumscribed obligatory path and the definition of a single axis is not too simplistically mechanistic. Extension of the distal row (Fig. 5.10B) is accompanied by radial deviation but also of necessity includes rotatory motion in the sense of supination; conversely flexion (Fig. 5.10A) includes ulnar deviation and pronation. Extension produces maximum congruence of the joint surfaces. The centrale portion of the scaphoid is then received snugly into the hollowed waist underlying the swollen volar part of the capitate head and the cartilage-clothed notch on the scaphoid at the rear of the centrale component locks firmly onto the articular posterior part of the capitate waist. The triquetral is then firmly lodged in the posterior part of the furrow spiralling around the ulnar aspect of the hamate and the dorsal beak on the lunate fits into confluent articular indentations shared by hamate and capitate. This whole complex movement, and the axis defining it, are exact mirror images of those found in prosimians and monkeys. Clearly such a dramatic apomorphic reconstruction, already apparent in gibbons, and in fact typifying the hominoids as a group must have significant functional connotations.

The idea that conjunct rotation accompanies midcarpal extension has been challenged in a study based partly upon cineradiography, but largely upon still radiographs of manipulated chimpanzee hands (Jenkins and Fleagle 1975). These authors claimed that they could not verify a conjunct rotation at the midcarpal joint upon wrist extension; it is clear, however, that they had no appreciation of the complex compound nature of the motions about an oblique axis for, in fact, their published radiographs seem to amply confirm the interpretation of the sequence of movement given above (Lewis 1985*b*).

The gorilla joints

In chimpanzees the meniscus at the ulnar side of the joint shows a tendency to lose its separate identity by merging with the triangular articular disc. In gorillas incorporation of the meniscus into a smoothly concave proximal articular surface is more complete and the ulnar styloid process becomes entirely excluded from direct participation in the antebrachiocarpal joint (Fig. 5.12). The opening bounded by the homologue of the meniscus is invariably small and irregular and lies adjacent to the apex of the triangular disc. This restricted opening leads into a capacious synovial cavity usually packed with a luxuriant growth of yellow fat-filled synovial villi. The shortened cartilage-tipped ulnar styloid process protrudes into this cavity and effectively articulates with the upper aspect of the thick meniscus-homologue which in passing to its anterior attachment to the lunate skirts the articular margin of the pisiform attaching to it; here again the radiocarpal cavity is in free communication with the pisotriquetral cavity. The palmar ulnocarpal ligament is again closely associated with the triangular disc and bulges into the joint cavity as a thick mass, passing to an attachment on the lunate. As in the other apes the radiocarpal ligament subdivides into two bands attaching to lunate and capitate respectively. This basic arrangement, easily derived from the chimpanzee pattern may be found in both lowland and highland gorillas. Occasionally, at least in *Gorilla gorilla gorilla* the meniscus homologue may contain one or more small ossicles or lunulae. A more derived morphology has, however, been observed in a specimen of *Gorilla gorilla beringei*. Here the neck of the proximal synovial compartment lodging the ulnar styloid process had, as it were, become pinched off, producing a closed synovial sac, located proximal to the meniscus-homologue, into which protruded the cartilage-clothed ulnar styloid process. A small opening, similar in site to that in other gorillas, persisted within the concavity of the meniscus-homologue but led only into a short blind synovial cul-de-sac; the continuity of this recess with the larger synovial cavity containing the ulnar styloid

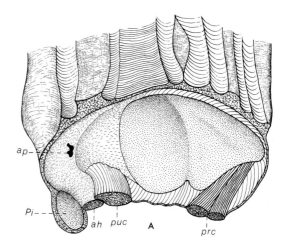

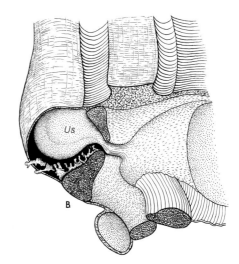

Fig. 5.12 (A, B)
A, the upper articular surface of the right wrist joint of *Gorilla gorilla beringei*. B, a detail of the same specimen with the meniscus-homologue incised along the leader line from *ap*. *Us*, ulnar styloid process; *ap*, opening into the proximal compartment of the joint cavity; *ah*, anterior horn of the meniscus-homologue; other labelling as in Fig. 5.1. (From Lewis 1969).

process had been interrupted. Thus, *three* closed synovial joint cavities were present at the wrist in this specimen: a radiocarpal cavity, an inferior radio-ulnar cavity, and a cavity related to the (still articular) ulnar styloid process. This specimen provided the key to the interpretation of the derived condition found in the remaining great ape, the orang-utan.

The carpal apomorphic specializations noted for *Pan* are quite faithfully reproduced in *Gorilla*, with the accentuation of certain features and the blurring of others. Not surprisingly the ulnar styloid process is notably short and truncated, being articular only at its blunt termination (Fig. 5.4D). The capitate is usually less markedly 'waisted' than in *Pan* but this is not the consequence of a lesser degree of enlargement of the head (as in *Pongo* or *Hylobates*) but seems likely to be derived from a chimpanzee-like morphology by bony filling of a deep lateral excavation. Unlike *Pan* there is usually no trace in dry bones of a diarthrodial contact between the capitate and trapezoid. However, in one wet specimen I have observed a posterior articulation comparable to that in *Pan*. In another I have seen a new diarthrosis between trapezoid and capitate, lying anterior to the thick interosseous ligament uniting the bones but without any trace of the primitive posterior articulation. In this specimen there was therefore a continuous articular strip located anteriorly on the lateral aspect of the capitate running from head to anterior facet for the second metacarpal, in striking contrast to the posteriorly located strip in *Pan* (Fig. 5.11). This anterior articular strip was of quite different significance from that noted above in *Hylobates*. In both this derived anterior location of the trapezoid facet, and in the masking of the waisting of the bone, this particular *Gorilla* specimen paralleled the human morphology. Suitable wet specimens have not been available to analyse the midcarpal mechanisms in *Gorilla* but the osteology leaves little doubt that motion is quite comparable to that in *Pan*.

The orang-utan joints

The specialized structure of the orang-utan antebrachiocarpal joint represents the culmination of trends similar to those noted above, which have been observed in a single mountain gorilla specimen, where the ulnar styloid process was totally excluded from the wrist joint and isolated within its own closed proximal synovial compartment. The proximal articular surface of the wrist joint is especially deeply concave but is clearly fashioned from the usual components: radius, triangular articular disc and meniscus homologue. In one

specimen which I have described (Fig. 5.13) the cartilage-covered tip of the ulnar styloid process was located within a closed synovial compartment which demarcated the upper surface of what was clearly the homologue of the typical hominoid meniscus. Within the antebrachiocarpal joint itself, a blind pit was located at the apex of the triangular articular disc doubtless representative of a developmental continuity with the proximal compartment. In this specimen, also, a synovial-lined channel led from the radiocarpal cavity to the distally located pisotriquetral joint. In the only other specimen which I have examined the proximal synovial cavity was absent, or obliterated, leaving the non-articular ulnar styloid process embedded in a thick wedge of fibrous tissue, the meniscus homologue, at the ulnar side of the joint. Moreover the homologue of the meniscus had apparently become fused to the surface of the triquetral thus sealing off the communication down to the pisotriquetral joint. The triquetral, therefore, constituted no part of the highly convex distal articular surface of the wrist joint whose dome was formed by the protuberant lunate, flanked laterally by the scaphoid. The ligamentous apparatus was like the other apes. The bones bear the clear imprint of this specialized morphology (Figs 5.14A, 6.9). The ulnar styloid process is typically very short and conical but on occasion may be longer. This may presumably be correlated with the varying conditions found in the intact state where the process may be non-articular and embedded in the meniscus homologue, or may be cartilage-covered and invaginate an independent synovial pocket.

Some suggestion of the oblique alignment of the hamate, present in monkeys and gibbons, is retained and the bone usually has a rather more extensive articulation for the lunate than that found in the African great apes and man. Its facet for the triquetral is less markedly spiral than in the other great apes.

The reduced triquetral bears a small convex facet at the distal extremity of its palmar surface for articulation with the reduced pisiform, which in some orang-utans is displaced so far distally that it has acquired an additional articulation with the hook of the hamate (Fig. 5.14A). In accord with observations in the intact state, the triquetral bears no apparent meniscal facet.

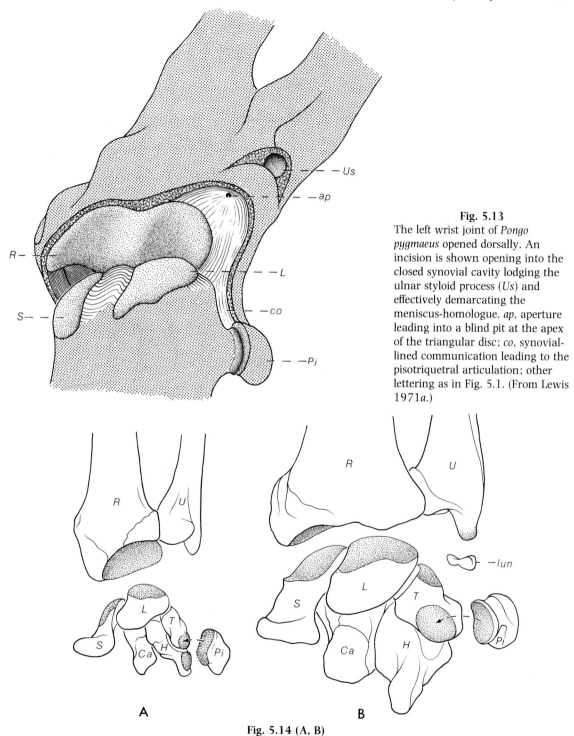

Fig. 5.13
The left wrist joint of *Pongo pygmaeus* opened dorsally. An incision is shown opening into the closed synovial cavity lodging the ulnar styloid process (*Us*) and effectively demarcating the meniscus-homologue. *ap*, aperture leading into a blind pit at the apex of the triangular disc; *co*, synovial-lined communication leading to the pisotriquetral articulation; other lettering as in Fig. 5.1. (From Lewis 1971*a*.)

Fig. 5.14 (A, B)
The palmar aspects of the bones entering into the right antebrachiocarpal joint (with the articular surfaces stippled) together with capitate and hamate in *Pongo pygmaeus* (A) and *Homo sapiens* (B). The human specimen possessed a lunula (*lun*). Other lettering is as in Fig. 5.1. (From Lewis 1972*b*.)

The scaphoid presents the usual deep groove for the palmar radiocarpal ligament and the centrale is independent except in aged specimens. Even in young individuals, however, the centrale is firmly bound to the scaphoid and moves in concert with it, and as in some monkeys it is joined to the waist of the capitate by a flimsy centrale–capitate ligament. The 'waisting' of the capitate is less obtrusive than in *Pan*, but more emphasized than in *Hylobates*, and the lateral aspect of the head of the bone is not excessively swollen. The articular surface on the head lacks continuity below with the posteriorly located trapezoid facet.

As with *Gorilla* the bones strongly indicate that the pattern of midcarpal movement, and its axis, are of the derived hominoid type.

The joints of man

The antebrachiocarpal joint

Although traditionally known as the radiocarpal joint this term is sometimes not strictly appropriate.

Unravelling of the progressive series of apomorphic changes, preserved as a scale of modifications of increasing complexity in the living apes, has brought into sharper focus a number of features of human topographical anatomy which have been either imprecisely described or even completely overlooked. Moreover, this comparative approach has provided a logical framework for interpreting the wide spectrum of variation in the human joint.

Traditional textbooks described the receptive proximal articular surface of the ellipsoid human radiocarpal joint as a smoothly concave area whose transverse extent is limited by attachments to the articular margins of the scaphoid and triquetral bones. Only two components of this surface are generally recognized, the distal articular surface of the radius and the triangular articular disc, the apex of the latter described as having an attachment to a pit adjacent to the base of the ulnar styloid process. Clearly, the medial most part of this surface, that part related to the triquetral facet, is not then satisfactorily accounted for; it obviously cannot be the triangular disc itself, and moreover illustrations of coronal sections frequently depict a disc whose thickness apparently increases medially, even encompassing the entire length of the ulnar styloid process.

Henle (1856), however, described the articular disc as presenting an apical cleavage into upper and lower laminae, the former attaching to the root of the ulnar styloid process, but with the lower continuing as part of the curved articular surface; an apparent aggregation of blood vessels between the two ligamentous lamellae prompted him to coin the term 'ligamentum subcruentum'. In contrast to this French anatomists have described a synovial cul-de-sac (the prestyloid recess) extending from the wrist joint cavity proper into relationship with the ulnar styloid process and occupying the site of the aggregation of blood vessels noted by the German author. The existence of such a synovial diverticulum has been confirmed as a normal feature of human arthrograms (Kessler and Silberman 1961). Moreover, as noted in Chapter 4 it is well recognized that the human embryo from the second to fourth months exhibits a cartilaginous nodule adjacent to the tip of the ulnar styloid process and that adults may in $\frac{1}{2}$–1 per cent of cases show a radiographic opacity (os triangulare or os intermedium) at a comparable site. The implication that this structure represents one of the primeval carpal elements, suggested by the latter term, is far from convincing but nevertheless has had a confusing knock-on effect influencing the interpretation of homologies in the whole carpus.

The prestyloid recess (Fig. 5.15) is a constant feature of the human wrist and bears a striking resemblance to the proximal compartment of the chimpanzee antebrachiocarpal joint (Fig. 5.8). This in itself is strongly suggestive that these diverticula are homologous and that the human joint therefore must be derived from one containing a meniscus, as in apes. The entrance to the human prestyloid recess is commonly an unobtrusive crenated opening located adjacent to the apex of the triangular articular disc. This opening is often masked by protruding synovial villi (Fig. 5.16) but freely admits a blunt probe. The cavity beyond it varies considerably in size but it always approaches the anterior aspect of the ulnar styloid process which itself varies greatly in length. When the ulnar styloid process is long and protrudes into the recess it is invariably clothed by articular cartilage (30 per cent); when the ulnar

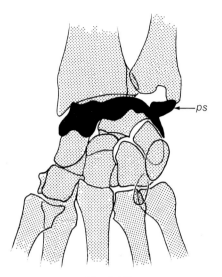

Fig. 5.15
A tracing of an arthrogram of the left wrist joint of a
human cadaver. The radiocarpal is filled with radio-
opaque medium (black) which extends into a prominent
prestyloid recess (*ps*), but not in this specimen into the
pisotriquetral joint. (Redrawn after Lewis 1969.)

styloid is even longer it may contact the triquetral
(8 per cent) via the opening into the prestyloid
recess, and then the morphology is not unlike that
of the chimpanzee shown in Fig. 5.9. Uncommonly (2 per cent) a free and independent meniscus may be found in *Homo* (Fig. 5.17) and this is of
course the common arrangement in *Pan*. It has
already been noted that in gibbons and African
anthropoid apes the pisotriquetral and antebrachiocarpal joint cavities freely communicate
around the indented anterior margin of the
meniscus, where it skirts the pisiform. A similar
arrangement may occur in the human joint and
such a communication is found in 34 per cent of
human wrists. Not surprisingly a particularly large
communication characterizes those wrists where
the ulna articulates with the triquetral. Usually
however the pisotriquetral articular cavity is
separated from the radiocarpal one by attachment
of the meniscus homologue to the triquetral,
subdividing the articular surfaces on that bone; an
analogous arrangment has been noted in some
orang-utan specimens.

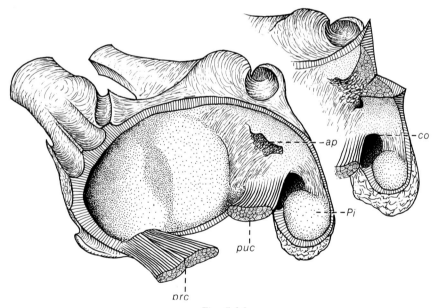

Fig. 5.16
The upper articular surface of the left wrist joint (together with the pisiform) of a human cadaver. In this specimen the
prestyloid recess fails to extend up as far as the ulnar styloid, which is not therefore cartilage-clothed. The opening (*ap*)
into the prestyloid recess is masked by a profusion of synovial villi, and the meniscus has lost its separate identity
becoming the most ulnar part of the proximal articular surface. In this specimen there was communication (*co*)
between the radiocarpal and pisotriquetral joint cavities. In the detail shown above the prestyloid recess has been
opened up. Labelling is as in Fig. 5.1. (Redrawn after Lewis *et al.* 1970.)

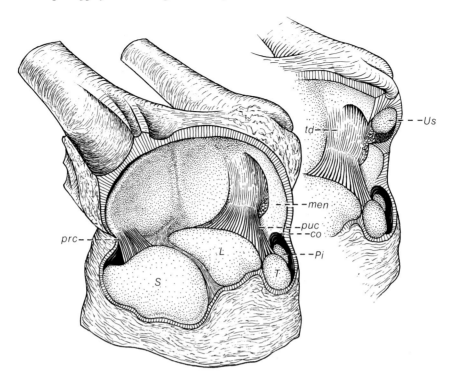

Fig. 5.17

The left wrist joint of a human cadaver opened dorsally and flexed. This specimen was unusual in retaining a virtually free semilunar meniscus (*men*), an arrangement very comparable to that shown in the chimpanzee illustrated in Fig. 5.7. The radiocarpal cavity communicates above the free margin of the meniscus with the prestyloid recess; in the detail shown to the right this recess has been opened by incising the meniscus to reveal a cartilage-clothed ulnar styloid process (*Us*) protruding into it. The radiocarpal cavity also has a communication (*co*) with the pisotriquetral cavity. Other labelling is as in Fig. 5.1. (Redrawn after Lewis *et al.* 1970.)

Radiographic opacities located adjacent to the tip of the ulnar styloid process and having an incidence of $\frac{1}{2}$–1 per cent, have long been known to radiologists and anatomists. The diversity of terms applied to these ossicles—ulnare antebrachii, triquetrum secondarium, os intermedium antebrachii, os triangulare, os styloides—reflects the uncertainty about their true nature. Some authors have suggested that all really represent un-united fractures of the styloid process, whilst others have maintained that they represent true carpal morphological elements (for example, os intermedium). Still others have suggested that they must represent detached supernumerary ossific centres for the styloid process, despite the fact that such additional epiphyseal ossifications have apparently never been observed. However, with the persuasive indications that the human joint is a

derivative of a meniscus-containing one, as in other hominoids, there can be little doubt that these ossicles are of the same nature as the os Daubentonii of the gibbon, that is they are lunulae and represent ossifications of the normally transient cartilaginous nodule found at the same site in the human embryo.

I have observed one example of such an ossicle in a wet specimen, and for the first time its relationships to the soft parts could be observed. The bone was embedded within that part of the proximal articular surface derived from the meniscus, the ulnar portion contacting the triquetral. The macerated bones of this specimen are shown in Fig. 5.14B. Thus, the ulnar part of the human joint includes a meniscus homologue although this is usually largely merged with the triangular articular disc, just as it is in some of the apes; it is becoming fashionable to call this composite mass 'the triangular fibrocartilage complex (TFCC)' (Palmer and Werner 1981; Palmer 1984).

Textbook descriptions of the ligamentous apparatus of the joint mention an array of membranous bands of parallel fibres, derived from the fibrous capsule of the joint, but lay particular emphasis on none. In fact, as is to be expected from the evolutionary history of the joint, the really significant accessory ligaments are located anteriorly, are essentially intracapsular (and therefore are best seen by opening the joint dorsally), and converge from radius and ulna onto the centrally situated bones of the carpus.

The palmar radiocarpal ligament is a massive structure, and as in apes is subdivided into two bands on its internal aspect. It is by no means a mere flat membranous sheet, as most textbooks suggest; this is an erroneous impression arising from viewing the joint only from the exterior. It arises from a large smooth impression on the anterior aspect of the styloid process of the radius and the two bands pass to attachments on lunate and capitate, grooving the scaphoid proximal to its tubercle (Figs 5.16, 5.17). The palmar ulnocarpal ligament, together with the lunate band of the palmar radiocarpal ligament form a 'proximal V', with the apex directed distally. The distal band of the palmar radiocarpal ligament is prolonged back from the capitate to the triquetral as a capitate-triquetral ligament; these together form a 'distal V'. The clinical significance of the 'proximal V' and the 'distal V' in wrist injuries is only now being appreciated (Sennwald 1987). As described above all of these ligaments have an ancient phylogenetic history. As in the other hominoids the palmar ulnocarpal ligament is merged with the triangular articular disc and forms a prominent synovial-covered protusion into the joint, misleadingly giving the impression that the anterolateral angle of the disc has a considerable attachment to the lunate.

Traditional descriptions usually include radial and ulnar collateral ligaments, a notion that seems somewhat justified by the appearance of coronal sections, but contradicted by the free movements of radial and ulnar deviation. In fact, man (along with the other members of the order Primates) possesses no clear-cut entities justifying such a description. It is true that, where the massive palmar radiocarpal ligament grooves the scaphoid bone, the investing fibrous capsule is attached to the scaphoid tubercle, and this may give a misleading impression that there is a strong collateral ligament in this situation. Similarly, there is little justification for a descriptive ulnar collateral ligament, supposedly arising from the tip of the ulnar styloid process, which as has been shown, is commonly free of all attachments, clothed with articular cartilage, and protrudes into the prestyloid synovial recess. When, however, the pisotriquetral joint is sealed off from the radiocarpal joint, such separation has clearly been effected by adherence between the meniscus homologue and the triquetral; in such cases there is a strengthened attachment between proximal and distal articular surfaces which perhaps justifies description as at least the ulnotriquetral band of an ulnar collateral ligament.

The dorsal fibrous capsule is relatively thin but presents a thickened band, the dorsal radiotriquetral ligament, part of the primitive mammalian ligamentous inheritance, and of course typically present in other primates. Also present posteriorly is a ligament passing from the scaphoid (centrale component) to the triquetral, deepening the posterior rim of the cup formed by the proximal carpal row (Fig. 5.19). It also has an ancient history. Together these two ligaments form a 'dorsal V' (Sennwald 1987).

The articular surfaces of the dry bones entering into the joint mirror the morphology described (Fig. 5.14B). The triquetral bears a convex facet proximally for the meniscal contribution to the proximal articular surface of the radiocarpal joint and a distally located pisiform facet, but these two facets on the triquetral may be confluent as in the African great apes; more often they are separated by a non-articular area representing the site of obliteration of the communication between radiocarpal and pisotriquetral joints. The pisiform itself is, of course, a squat, nubbin of bone.

There can thus be little doubt that *Homo sapiens* shares fundamental synapomorphies at the wrist with other hominoids, involving liberation of the ulnar styloid process from articulation with triquetral and pisiform. Although this evolutionary history is somewhat obscured in most specimens, nevertheless those specializations which contribute to this in man can be found to be paralleled in some great ape specimens. Indeed, in *Pongo* the morphology is even more extremely derived than in *Homo*. Embryology reinforces this interpretation of the derivation of the human joint.

Development of the human radiocarpal joint　In early embryonic stages the ulna is widely in contact with the carpus. Soon, however, its inferior extremity is remodelled into a head and styloid process, with the latter contacting both triquetral and pisiform. By the end of the embryonic period (60 days), however, a considerable mesenchyme filled interval appears between the styloid process and the pisiform and during the early part of the fetal period joint cavities are formed. Radioscaphoid and radiolunate cavities develop, separated by a thin mesenchymal septum, which is quite dense and cellular anteriorly. A more distant ulnotriquetral cavity, where the cartilaginous anlagen of the ulnar styloid process and triquetral articulate, is separated from the radiolunate cavity by a considerable extent of loose mesenchyme, located distal to the condensed tissue of the developing triangular disc, and which commonly contains a cartilaginous nodule, the so-called 'intermedium antebrachii'. As development proceeds the ulnar styloid process becomes separated from the triquetral by a mass of condensed mesenchyme, triangular in section, which is the human homologue of the pongid meniscus (Lewis 1970); when the cartilaginous 'intermedium antebrachii' persists into this stage, as it sometimes does, it lies in the interior of this mass. The ulnar styloid process is initially embedded in the upper surface of this meniscus-homologue whose lower surface is separated from the triquetral by the persistent ulnocarpal cavity. As development proceeds cavitation extends from this cavity into relationship with the styloid process thus forming the prestyloid recess, with a prominent aggregation of blood vessels in its synovial walls, and delaminating the meniscus as a free entity, triangular in section, projecting into the joint. The initially separate radioscaphoid and radiolunate cavities soon communicate and persistent vestiges of the intervening septum, especially its anterior denser part, give rise to the synovial folds (large anterior and small posterior) found here (Fig. 5.17). Indeed this anterior fold is found in all hominoids (Fig. 5.7), indicative of a similar mode

of development. There is absolutely no justification for the descriptive of the anterior fold as a ligament, radioscaphoid or radioscapholunate ligament, as has been done (Mayfield *et al.* 1976; Taleisnik 1976).

The ulnotriquetral cavity soon joins the common radiocarpal cavity by dissolution of the intervening mesemchyme, which represents the primitive septum found isolating the ulnar part of the antebrachiocarpal joint in prosimians and occasional higher Primates. A persistent septum has even been found in the malformed limb of a human fetus (Hesser 1926).

The pisotriquetral synovial cavity appears late but when it is formed it intervenes between the fetal pisiform on the one hand and the triquetral and meniscus homologue on the other, being largely related to the periphery of the latter reflecting the more proximal location of the fetal pisiform. It commonly communicates with the radiocarpal cavity. In later fetal life the pisiform becomes distally displaced and the communication between radiocarpal and pisotriquetral cavities may be obliterated. In some fetuses separation of the ulnar styloid from the carpus may be considerably retarded, and this is doubtless dependent upon variations in the relative growth rate of this process, which may result in the long articular process persisting in some adults.

Persistence and ossification of the cartilaginous nodule ('intermedium antebrachii') are almost certainly the source of the occasional adult radiographic opacities adjacent to the ulnar styloid process, and there can be little doubt that these structures are homologous with the gibbon os Daubentonii.

Thus in a quite striking way human development recapitulates phylogenetic stages comparable to the morphology found in the living apes, even gibbons, obscured somewhat by altered time relationships or heterochrony, as for example in the precocious development of the cartilaginous lunula even before a meniscus is clearly differentiated.

The human midcarpal joint

As in *Pan* the pattern of movement is determined by the form of the conjoined heads of the capitate and hamate, the two bones articulating as usual by a plane synovial joint adjacent to their posterior surfaces and united in front of this by a massive and unyielding interosseous ligament.

The lateral aspect of the human capitate and its associated articulations, are, however, strikingly remodelled (Fig. 5.18) and as in *Gorilla* 'waisting' is not usually an obvious feature. However, there can be little doubt that the human capitate was derived from a bone quite similar to that of *Pan*, with a laterally expanded head and a deep lateral excavation which has usually become even more completely obliterated than that of *Gorilla*. Some residual trace of a constricted waist is, however, usually apparent far anteriorly and into this minor depression the scaphoid becomes impacted during

wrist extension. In some specimens, however, 'waisting' is much more conspicuous in this situation and quite reminiscent of the chimpanzee condition.

Invariably the human trapezoid articulates with the capitate anteriorly, in front of the massive interosseous ligament uniting the bones. A similar remarkable transposition of this synovial joint has been noted above in one specimen of *Gorilla gorilla gorilla* and although this was certainly atypical, it paralleled the condition which is normal in man. Generally this trapezoid articulation is confluent above with the articular surface for the head (Fig. 5.18A). Sometimes, however, a non-articular area intervenes and receives the attachment of

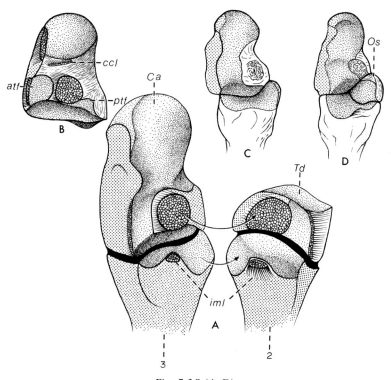

Fig. 5.18 (A–D)
A, a view comparable to that shown in Fig. 5.11 of the left capitate and associated bones of a wet specimen of *Homo sapiens*, the upper arrow indicates the severed and separated attachments of the massive interosseous capitate–trapezoid ligament, in front of which lie the articular facets on the two bones; at its distal margin the capitate exhibits a hollowed linear strip for articulation with the index metacarpal. B, a variant form of a wet specimen in which a centrale–capitate ligament (*ccl*) joined the centrale component of the scaphoid to the waist of the capitate thus isolating the anterior trapezoid facet (*atf*); in this specimen there was a persistent residual portion of the primitive posterior trapezoid facet (*ptf*). C, a macerated left-human capitate and third metacarpal; in this specimen the styloid process of the metacarpal was of more usual dimensions than in A. D, a macerated capitate and third metacarpal showing a separated os styloideum (*Os*). (A, from Lewis 1973.)

a flat fibrous band arising from the centrale component of the scaphoid (Fig. 5.18B)—the homologue of a similar structure which can occur in *Pan*; this flimsy ligament is well documented in man but there is absolutely no support for the contention, which is sometimes encountered in textbooks, that it is the adult representative of the os centrale.

Uncommonly, the trapezoid, although having its usual major articulation anteriorly located on the capitate, also retains a small articulation dorsal to the interosseous trapezoid–capitate ligament (Fig. 5.18B); this atavistic relic emphasizes the derived nature of the anterior transposition of the articulation of the human os trapezoideum. It is of interest that Fick (1904), in a purely descriptive

account of the human joints and without benefit of comparative morphological insight, accurately described the anterior location of the major articulation, the large ligamentum interosseum trapezoideo–capitatum, and the occasional small dorsal facet but this has rarely been appreciated.

The hamate, much as in *Pan*, possesses a spiral furrow for the triquetral, limited dorsally by a projecting lip, and winding up from the root of the hamulus around the convexity of the head (Fig. 5.19A). The human midcarpal joint is an important contributor to movement at the wrist, but its mobility is more restricted than that of the chimpanzee where it contributes almost three quarters of the movement of flexion–extension and two thirds of the total of radial–ulnar deviational

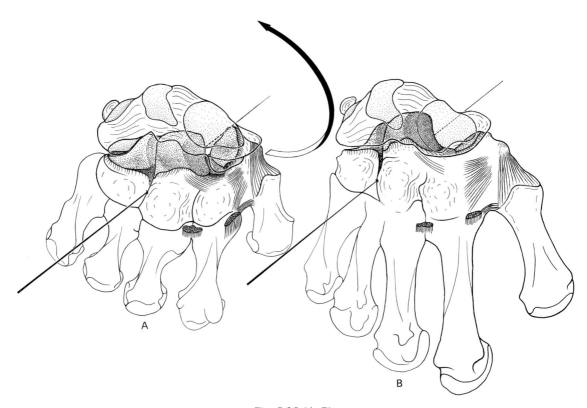

Fig. 5.19 (A, B)

A dorsal view of a preparation of the right carpus and metacarpus of *Homo sapiens* with a steel rod transfixing the two carpal rows along the midcarpal axis, as experimentally determined. A, the distal carpal row, carrying with it the metacarpus, has been moved into the position of flexion, ulnar deviation and pronation about this axis. B, the distal carpal row (and metacarpus) have been moved into extension, radial deviation and supination about the axis whilst the proximal row is retained in a constant position; the direction of this movement is shown by the arrow above. (From Lewis 1985*b*.)

movement (Jenkins and Fleagle 1975), the remainder occurring at the antebrachiocarpal joint.

In man also a stereotyped pattern of midcarpal movement is combined with compensatory motion at the radiocarpal joint, which either supplements or cancels out certain components, to produce the resultants classically defined as ulnar and radial deviation, and flexion and extension. Although the relative contributions attributed to the two joint complexes have been weighted varyingly in different studies, and apparently also show individual variation, it is usually recognized (e.g. Wright 1935) that the midcarpal joint contributes rather more than half the total excursion of about 120° in flexion–extension, virtually all the movement in radial deviation and part of ulnar deviation—a minor part according to some studies, a major part according to others.

Movements at the midcarpal joint are strikingly similar to those already described in *Pan*, and occur along a quite circumscribed path, about a comparable axis which runs from dorsally on the ulnar side of the wrist and traverses the head of the capitate with a proximal deviation, towards the volar aspect on the radial site (Fig. 5.19). Extension of the distal row (Fig. 5.19B) is then accompanied by radial deviation but also of necessity includes a rotatory motion in the sense of supination; at the other extreme flexion (Fig. 5.19A) includes ulnar deviation and pronation. As in *Pan* these stereotyped midcarpal movements are commonly associated with compensatory motion at the radiocarpal joint, reinforcing or suppressing certain components to produce the whole repertoire of movement usually described at the wrist as a whole: ulnar and radial deviation, and flexion/extension.

In moving into extension the distal row radially deviates, supinates and extends and the midcarpal joint becomes maximally close-packed (Fig. 5.19B). But on to this is superimposed further extension at the radiocarpal joint which also becomes close packed and this includes corrective components of ulnar deviation and some pronation, leaving the final cumulative result as pure extension.

In moving then into flexion the distal carpal row flexes, pronates and ulnar deviates at the midcarpal joint, but now the proximal row also flexes at

the radiocarpal joint whilst radially deviating and slightly supinating. Flexion at the two joints is additive, whilst the other components largely cancel out.

In simple ulnar deviation at the wrist, as the distal carpal row flexes, ulnar deviates and pronates about the midcarpal axis in relation to the proximal carpal row, this latter row is coincidentally undergoing not flexion but extension, together with ulnar deviation and apparently some pronation, relative to the radius. Effectively, the proximal carpal row is carrying out its own independent movement of adjustment between forearm and distal row so that the net result is ulnar deviation of the hand.

In radial deviation the converse movement occurs with the distal row extending, supinating and radially deviating about the midcarpal axis whilst compensatory movement at the radiocarpal joint causes the proximal row to flex, radially deviate and perhaps slightly supinate. Again, the proximal row effectively moves between forearm and distal row. It follows inexorably, that in moving from maximum ulnar deviation to the full extent of radial deviation the hand must change its attitude from one of pronation to supination and this can be confirmed by simple personal experiment, as noted by Bryce (1897).

It is apparent that the proximal row is moored semi-independently between the distal carpal row and the forearm bones, allowing it to carry out reciprocal movements of flexion–extension during radial–ulnar deviation largely cancelling out the opposing components which are inherent in the isolated midcarpal movement. On the contrary, in flexion–extension of the wrist the proximal row moves with the distal row into a corresponding attitude whilst carrying out the reciprocal screwing movements annulling the components of pronation–supination and ulnar–radial deviation accompanying these midcarpal movements. Aptly Bryce (1897) likened the functions of the proximal row to that of a deformable intra-articular meniscus, an idea which has been recently reaffirmed by Legrand (1983).

When the midcarpal joint is screwed into the close-packed position in either extension or radial deviation the scaphoid locks snugly about the lower back part of the head of the capitate which may here bear a clear impression of this contact

(Fig. 5.18A). The scaphoid as in *Pan*, usually possesses here a deep cartilage clothed notch between the body of the bone and the centrale component. Scaphoid and lunate then become splayed apart anteriorly, springing back together in a scissor-like action during flexion, a movement which is accommodated by a conspicuous laxity of the anterior part of the interosseous scapholunate ligament. This anterior separation of scaphoid and lunate, which is not a feature of the movement in *Pan*, appears to result from the filling out of the waist of the human capitate. This springing apart of scaphoid and lunate was noted by Bryce (1897), and recently confirmed by Kauer (1974). On the ulnar side of the wrist maximum congruence is achieved by the triquetral screwing around to become firmly impacted into the upper part of the spiral channel on the hamate (Fig. 5.19B).

A fairly strictly defined axis for the midcarpal joint seems confirmed but for an ellipsoid joint like the radiocarpal one, no anatomical constraints limiting it to one degree of freedom are to be expected. However, it is clear that when the proximal row is flexed at the radiocarpal joint it slips easily into a position also involving supination and radial deviation; conversely, when it is extended it also becomes pronated and ulnar deviated and is then close-packed. This is the habitual movement and this was appreciated by Bryce (1897) and is also implicit in an illustration by Kauer (1974). If a single axis could be defined, and it might be valid for at least the latter part of the movement of extension, it would run from dorsally on the radial side then volarly towards the ulnar side with some slight proximal angulation. Interestingly, a similar pattern of movement at the antebrachiocarpal joint occurs in other Primates, not only apes but also monkeys and prosimians. Indeed, it seems likely that this is the primitive mammalian pattern, for the notional antebrachiocarpal axis then rather parallels the primitive midcarpal axis, as retained in monkeys and prosimians, and congruent movements then occur at both midcarpal and antebrachiocarpal joints close-packing both into an attitude of extension, pronation and ulnar deviation. Retention of much of this primitive pattern of radiocarpal motion in man fulfils a radically different function by providing compensatory movements modifying the radically restructured midcarpal movement.

Synthesizing movement at the midcarpal joint by specifying a single resultant axis is supported by data culled from the extensive literature of human anatomy. Bryce (1897), in one of the earliest radiographic studies of the wrist, was probably the first to appreciate the significance of the anterior laxity of the scapholunate ligament and the role of pronation–supination movements at the midcarpal joint. MacConaill (1941) also had some awareness that a screwing action occurred at the human midcarpal joint during extension. Neither, however, attempted an overall synthesis by defining an axis. Johnston (1907a,b) provided invaluable data by meticulously describing the attitudes of the individual carpal bones in differing wrist positions by dissecting the wrists with the exposed bones stabilized by fixation in plaster. Although no attempt at an overall synthesis was made the results confirm the interpretation of midcarpal motion described above (Fig. 5.19) together with the suggested complementary radiocarpal movements. Kauer (1974) also analysed the motions of the individual bones of the proximal carpal row during flexion and extension of the wrist and his results again fit in with the model proposed here, including the scissor-like action between scaphoid and lunate. Kauer (1986) also analysed the changing attitudes of the lunate in radial and ulnar deviation and again his results, but not his interpretation, broadly fit in with the synthesis developed in this chapter. It is becoming increasingly accepted amongst clinicians that the proximal carpal row, operating as an intercalated segment, paradoxically flexes during radial deviation and extends during ulnar deviation. Linscheid (1986) has perceptively analysed how the architecture of the two bands of the massive palmar radiocarpal ligament are geared to this motion. During radial deviation, as the lunate flexes, the radiolunate band is slackened allowing the lunate to slide ulnarly effecting radial deviation of the proximal row. At the same time laxity of the radiocapitate band crossing the scaphoid waist, allows that bone to flex as well, so that its long axis swings almost to a position perpendicular to the forearm. In ulnar deviation as the lunate displaces radially the radiolunate band becomes lax allowing extension of the proximal row.

Certain previous authors have attempted to rationalize the complex movements at the wrist by describing axes. Henke (1859) deduced that independent axes operate for the movements at the radiocarpal and midcarpal joints, each axis diverging a little from the strictly transverse towards the sagittal, that for the radiocarpal joint with its ulnar end somewhat forward, that for the midcarpal with its radial extremity forward. In flexion–extension he envisaged that the greater rotational movements about the transverse component, as resolved from the resultant axis, would summate whilst the smaller movements of radial–ulnar deviation about the sagittal component would cancel out. Conversely in radial–ulnar deviation the greater parts would cancel out and the smaller summate. He did not recognize the occurrence of any pronation–supination movement between the rows and thus in the text described no angulation of the axes towards the long axis of the forearm (that is, no inclination towards the vertical in the anatomical position); unfortunately, the oblique viewpoint chosen by his artist for a key illustration does unintentionally convey an impression of obliquity. Kapandji (1966) in illustrating the 'mechanism of Henke' has reproduced this obliquity but it is noteworthy that his supposed midcarpal axis is shown as passing from the dorsal side radially, volarly, and distally, instead of with a proximal obliquity as does the true midcarpal axis. Kapandji however, follows Henke precisely in describing the movements of flexion–extension, making no mention of the supination–pronation which would be the inevitable consequence of the obliquity of the axis which he illustrated. Oddly enough, Kapandji does describe rotatory movement between the rows for radial–ulnar deviation with the distal row supinating during radial deviation, as it in reality does, yet the obliquity shown for this axis would in fact result in the opposite motion.

Fick (1911), did define a midcarpal axis corresponding to that which I have described above. He also described a radiocarpal axis paralleling this one in obliquity but intersecting it so as to pass from dorsally on the radial side distally and volarly. Rather illogically Fick believed that these axes were operative only in radial–ulnar deviation whereas in flexion–extension simple transverse hinge axes determined the movement. Fick, however, seems to have come closest to appreciating the essential features of midcarpal movement.

The definition of a midcarpal axis seems reasonably established. Whether it is advisable to utilize this descriptive device also for the radiocarpal joint is equivocal, but an axis does usefully epitomize the overall character of a movement. The axis as defined by Henke, passing from dorsally on the radial side and deviating volarly as the ulnar side is approached, does seem to summarize the main overall character of radiocarpal movements. For the pronatory–supinatory components of the movement, reflected in a proximal (my axis) or distal (Fick's) deviation of the axis in its volar passage, the position is more controversial. Indeed, there is some suggestion that this angulation fluctuates, having an orientation as I have defined it for flexion–extension, but perhaps as Fick defined it for radial–ulnar deviation.

Function in the hominoid wrist articulations

The complex remodelling of the hominoid antebrachiocarpal joint strongly suggests that it is a synapomorphy of the apes and man—a derived character complex, shared by them owing to a common phyletic history—for a high degree of integrated complexity in an innovative character makes parallelism improbable. Indeed the mode of development of the joint in man suggests that a gibbon-like stage was featured at an early phase in human evolutionary history. The reconstruction of the joint has certainly provided a considerably enhanced range of supination, by freeing the eccentrically situated ulnar styloid process from its restrictive articulation with triquetral and pisiform, and allowing it to rotate in relation to the carpus to take up a dorsal position. The meniscus provides for this rotatory transposition whilst angular movements at the joint are retained. By this means the 90° of pronation–supination found in monkeys, is increased to about 180°. It is, of course, equally apparent that other hominoid movements are also modified and in particular ulnar deviation is clearly facilitated.

Perceptive anatomists have long appreciated the locomotor advantages conferred by a free capacity for supination. Thus Gregory (1916) stated: 'the habit of brachiating, or swinging from branch to branch with the arms, trained the arms in the all-important power of supination . . .'. Avis (1962) further showed in a detailed functional analysis that the essence of brachiation involves arm-swinging in which the body swings forward, suspended from one hand, rotating as it goes, towards a grip with the other hand. This rotation is achieved by progressive supination of the grasping limb; in effect, there is rotation of the ulna (carrying with it the whole body by virtue of its non-rotatory articulation at the elbow) about the fixed radius and grasping hand.

It was therefore suggested that the hominoid antebrachiocarpal synapomorphy was a 'brachiating' adaptation (Lewis 1971b). A decade ago it was common practice to use this term in a less restrictive sense than is the current fashion, merely implying hand-over-hand suspensory locomotion. It was not intended to imply that the hominoid heritage included gibbon-like ricochetal arm-swinging (the usual connotation of the term now) but rather that the capacity for bodily rotation during forelimb suspension is diagnostic of a whole range of versatile arboreal activities such as climbing, feeding, and suspensory posturing in the small terminal branches (Lewis 1969, 1971b, 1972a, 1974).

Furthermore, it seemed clear that the hominoid midcarpal synapomorphy was a complementary adaptation for forelimb suspension, screwing the carpal rows together into a close-packed position apparently well adapted to cope with tensional stresses. Moreover, the midcarpal supination accompanying extension would complement the forearm supination and also the external rotation of the shoulder joint which accompanies abduction: all are congruent movements, acting in an additive manner to swing the body around from a grip with one hand to the other. Thus, hard morphological fact seems to support the hypothesis, which has formed the mainstream of informed opinion for a century, that man is especially closely related to the African apes and that 'brachiation' was the significant factor that moulded much of their shared postcranial morphology and provided the apprenticeship and many of the pre-adaptations needed for the assumption of a habitual erect bipedal posture. These synapomorphies at the wrist leave their clear imprint upon the bones, and thus provide useful tools for interpreting the fossil record. In an order noted for its overall conservative morphology such characters are a real boon.

The possibility of parallel acquisition of synapomorphies such as those at the hominoid wrist can certainly not be completely dismissed, for comparably complex suites of characters have certainly emerged in parallel during evolution. Cartmill and MIlton (1974) have claimed that certain of the lorisine prosimians, in particular *Nycticebus*, show specializations of the antebrachiocarpal joint parallel to, and seemingly as radical as, the Hominoidea including the presence of an intra-articular meniscus; the demonstration of this last feature is of course, indispensable for the credibility of the argument. Since lorisines are extremely cautious slow-climbing arboreal quadrupeds the suggestion was made that the supposedly comparable modifications at the wrist in both these prosimians and in ancestral hominoids must have similar functional meanings. Despite the fact, that when the detailed publication by these authors (Cartmill and Milton 1977) appeared the claim for a lorisine meniscus had been unobtrusively withdrawn, and thus the warrant for any persuasive claim for resemblance to hominoids, the so-called 'slow-climbing hypothesis' of hominoid evolution had firmly taken root rivalling the traditional 'brachiationist' theory. Indeed the supposed hominoid-like adaptations of the wrists of lorisines continue to be cited as a cornerstone (Fleagle *et al.* 1981; Aiello and Day 1982) of the idea that quadrumanous climbing provided the pre-adaptive training for human bipedalism. It followed as a logical consequence of this reasoning that the ancestral hominoids of the early Miocene must have been unspecialized monkey-like quadrupeds giving rise to parallel to the gibbons as acrobatic arm swingers, and to the great apes on the other hand as initially cautious quadrupedal climbers. This notions plays down the likely role of suspensory bimanual adaptations in the emergence of the latter group, while failing to explain why a meniscus-containing wrist joint should have appeared in parallel in two groups having such dissimilar locomotor habits. The evolutionary issues were rather fudged, however, by acknowledging an increasing role for suspensory orthograde posture in the great ape line as a consequence of increasing body size. In fact, the whole of this argument was based on a fallacy for as shown in the early part of this chapter the lorisine wrist is quite unlike the hominoid one, and these slow-climbing quadrupeds are therefore irrelevant as plausible locomotor models for the emergent hominoids.

The 'slow-climbing scenario', however, has become firmly established and there is now a tendency to play down the role of bimanual suspension as a significant factor in human ancestry. This trend has gained impetus from other recent work. Jenkins and Fleagle (1975) in their cineradiographic study of the ape wrist (mainly the chimpanzee) and Jenkins (1981) on the ateline and hylobatid wrist proposed that suspensory locomotion ('brachiation') entailed specializations in the midcarpal joint, and that these characteristic morphological trademarks are present in a semibrachiating monkey (*Ateles*) and in gibbons, but lacking in the lineage leading to the great apes and man. Again, on the face of it, this seemed to point to an essentially quadrupedal history for this particular clade.

As noted above Jenkins and Fleagle (1975), because of an apparent misunderstanding of the true mechanics of the movements, rejected the notion of any supinatory movement of the distal carpals which might stabilize the chimpanzee wrist in extension, and reached the extraordinary conclusion that the mechanism in this ape was instead similar to the action in cercopithecoid monkeys, and as in them was an adaptation to terrestrial quadrupedal locomotion. Their chain of reasoning, resting upon surprising misinterpretations of published data is as follows. I (Lewis 1972b, 1974) had described how in cerocopithecoid monkeys extension of the wrist impacts the centrale dorsally into the cleft between scaphoid, capitate and trapezoid. The mechanism in *Pan* (Lewis 1972b) is of course quite different and has been described in some detail in this chapter. The lateral aspect of the chimpanzee capitate possesses a constricted waist forming a deeply undercut non-articular groove anteriorly, but consisting of an articular surface posteriorly which is confluent below with a facet for the trapezoid (Fig. 5.11). As the scaphoid is screwed into extension its incorporated centrale element, bearing the articulation for the upper aspect of the trapezoid, is firmly embedded into the non-articular groove and the notched region of the scaphoid at the back of the centrale component fits onto the articular part of the waist (Fig. 5.10B).

Remarkably enough Jenkins and Fleagle seemingly misinterpreted my description, believing that I had said that there was dorsal orientation of the centrale facet on the chimpanzee capitate. Thus they introduced the odd notion that *Pan* and *Macaca* possessed similar morphology here, and since in the latter the structure was clearly adaptive for quadrupedalism, so it must be in the former. Other misinterpretations compounded the confusion.

Thus, on the basis of a supposed similarity of their carpal structure to that of cercopithecoid monkeys, the quadrupedal adaptations of extant African apes have come to be emphasized whilst the role of suspensory behaviour in their evolution has been minimized. As with the lorisine studies the conclusion was that an essentially quadrupedal locomotor repertoire must have characterized the ancestral apes and from this they graduated to a slow-climbing role. The extant African apes, of course, favour a form of quadruped locomotion and this was envisaged as a relatively unchanged retention of that of the ancestral apes reflected in a carpal structure comparable to cercopithecoid monkeys. In contrast, the traditional view backed by much persuasive evidence, is that the knuckle-walking proclivities of *Pan* and *Gorilla* are consequences of a secondary reversion from forelimb suspension to terrestrialism.

Surprisingly, while attention has been focused on the midcarpal joint, the much more striking apomorphic characters found in the ape antebrachiocarpal joints have come to be virtually disregarded. There seems to be a strong case that these reflect a history of suspensory posturing, and that the midcarpal reconstruction, when correctly interpreted, complements this. In contrast, the midcarpal joints of semibrachiating monkeys seem to retain all the essential hallmarks of the conservative monkey type of joint. The result of this perhaps unwarranted debate has been considerable confusion about the likely nature of human origins.

Stern (1975) in a widely circulated review publication has attempted to interpret current thinking on how the locomotor behaviour of man's ancestors might have preadapted them to bipedality. In his previous work this author has often tended to favour distancing modern or fossil man from close relationship with the African apes, but wrist morphology has clearly presented him with a dilemma. In this particular survey he is influenced by the findings of Jenkins and Fleagle (1975) and repeats the fallacy introduced by these authors that the African apes have a 'centrale' facet on the dorsum of the capitate and that this might be considered an adaptation to pronograde quadrupedalism. Again attention is focused on the midcarpal joint, and the much more spectacular specializations of the antebrachiocarpal joint receive scant attention, for there are clearly considerable difficulties in accommodating the antebrachiocarpal synapomorphies in this scenario.

Although much of the factual basis for the slow-climbing hypothesis is flawed it is not without merits and even under the often misrepresented 'brachiationist' view, climbing (and by that vertical climbing is implied) was envisaged as part of the locomotor repertoire of early hominoids (Lewis 1969, 1971b, 1972a, 1974). Studies on the myology of hip and thigh indicate that vertical climbing must have had a significant ancestral role for in this anatomical region, man seems to be most similar to cautious vertical climbers which engage in some suspension from the hindlimbs, *Alouatta* in particular possessing the morphology which could readily be conceived as pre-adaptive to the condition in erect man (Stern 1971). As regards the forelimbs Stern (1971) theorized that our hypothetical ancestor must have been characterized by a greater propensity for arm-swinging than the howler and with a more orthograde posture. Therefore, 'brachiation' is still in the picture. Developing this theme Stern (1975) depicted the likely locomotor behaviour of man's ancestors as largely involving use of both upper or lower limbs in tension during climbing, feeding, or suspended locomotion but the term brachiation (with its embarrassing history) was replaced by the new coinage 'antipronograde'. The ancestral form was then envisaged as an ape employing its forelimbs much as does the living orang-utan in climbing, suspension, and other tensile activities, but using its hindlimbs in a somewhat more pronograde quadrupedal manner than does *Pongo*. This scenario is, in fact, an acceptable statement of the more enlightened and recent developments of the 'brachiationist' view. It would follow that the hominoid wrist synapomorphies have enhanced the range of supination during manual suspension, allowing rotation of the whole trunk about the support.

Despite suggestions to the contrary, the meagre behavioural data about extant apes provide some support for the idea that forelimb suspension was probably a major part of the locomotor repertoire of ancestral hominoids. For instance, the pygmy chimpanzee, or bonobo, is said to employ a significant amount of bimanual suspension (Susman *et al.* 1980), in fact only slightly less than it does quadrumanous climbing and scrambling. The ancestral forms may have been even more committed to this behaviour which as Grand (1972) has stressed confers a considerable ecological advantage in the fruit-bearing terminal branch niche. Moreover, supinatory ability must confer a functional bonus in seeking out and grasping supports during vertical climbing and there seems little doubt that this behaviour also must have played a significant part in the ancestral locomotor repertoire. In fact, similar patterns of muscle use are found electromyographically in climbing and in suspensory locomotion (Fleagle *et al.* 1981) which merely stresses the overlapping character of these activities.

The studies cited above which played down the likely role of suspensory posturing in hominoid evolution, although based upon inadequate data and flawed interpretations, have had the unfortunate consequence that they forcibly imply that the ancestral apes must have been essentially 'dental apes'—ape-like in dentition but essentially quadrupedal and monkey-like postcranially. This issue will be taken up in Chapter 10.

6

The joints of the hand

It is commonly believed by anatomists that the human hand is essentially primitive (Sonntag 1924). Jones (1949a), who was a staunch protagonist of this view, considered that the supreme functional effectiveness of the human hand was dependent upon the elaboration of especially refined nervous mechanisms; he stated 'when we look for remarkable specializations, or wonderful human adaptations as distinguishing our hands from the hands of monkeys and anthropoid apes, our search is rather a vain one'. Certainly in overall aspect the human hand shows no gross evolutionary deformations, such as loss or reduction of digits; yet, as will be seen, the hand has its full quota of apomorphic features contained within its complex assembly of joints, which are finely attuned to its specialized role as a delicate manipulative organ.

It is tempting to see these specializations as the result of selection associated with millennia of tool use and manufacture. However, a number of the features are present, in at least emergent form, in the apes. Of course it is well known that chimpanzees indulge in occasional tool use, and even crude tool manufacture, but it is difficult to see this as the central selective mechanism in apes. Rather, in the emergent hominoids, there must have been progress towards enhancing the overall grasping repertoire of the hand.

The joints of the distal carpal row

Many of the major features here have been dealt with under discussion of the midcarpal joint in Chapter 5, but elaboration of certain points is now necessary. In general, with only a few notable exceptions the primate distal carpal row preserves the primitive mammalian architecture (Fig. 6.1). In apes, monkeys, and man (Figs 6.6, 6.7, 6.9, 10.8B) the capitate and hamate are firmly united by a massive interosseous ligament behind which is a linear plane synovial joint extending the length of the posterior part of the adjoining surfaces. Only in gibbons is this significantly changed for here the synovial articulation is subdivided into larger proximal and smaller distal parts.

With the notable exception of man and *Gorilla* the trapezoid articulates with the lateral aspect of the capitate by a posteriorly situated synovial joint. This joint is confluent below with the posterior of the two joints between the capitate and the medial side of the second metacarpal base and above with the articular head of the capitate, thus giving rise to a continuous articular strip down the posterior part of the lateral side of the capitate. Only in the occasional monkey and orang-utan (Fig. 6.9) is the trapezoid facet separated from the articular capitate head by a non-articular interruption. In front of the synovial articulation of the trapezoid is an exceedingly massive ligament uniting trapezoid and capitate. This arrangement, consisting of a posterior diathrosis, located behind an interosseous ligament is clearly the primitive mammalian condition and is found in even marsupials (*Didelphys*; *Pseudochirus*). It is generally stated that in *Gorilla* there is no synovial articular contact between capitate and trapezoid (Broom and Schepers 1946; Clark 1947) and this is certainly the conclusion which would generally be reached from the examination of dry bones. However, as noted in Chapter 5, there may occasionally be in *Gorilla* the usual posterior articulation between the bones (as in *Pan*) or even an anterior articulation (as in *Homo sapiens*).

Invariably almost the whole of the lateral aspect

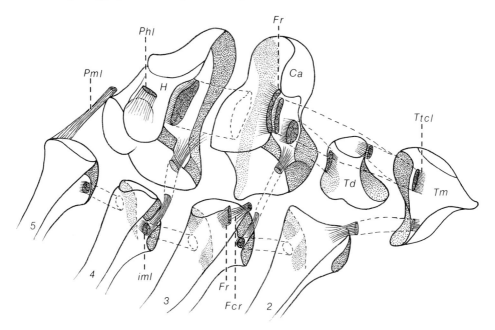

Fig. 6.1

The inferred pattern of articulations and the associated ligaments between the bones of the distal carpal row, and of the carpometacarpal joints, in a primitive therian mammal. *H*, hamate; *Ca*, capitate; *Td*, trapezoid; *Tm*, trapezium; *2–5*, the second to fifth metacarpals; *Pml*, pisometacarpal ligament; *Phl*, pisohamate ligament; *Fcr*, flexor carpi radialis tendon; *Fr*, recurrent fibres of flexor retinaculum attaching to capitate and third metacarpal; *Ttcl*, trapezio–trapezoid–capitate ligament; *iml*, intermetacarpal ligament.

of the trapezoid (except its most anterior part) is articular for the trapezium. In front of this articulation a ligamentous sheet leaves the trapezium dorsal to the groove for flexor carpi radialis, and attaches variably to the non-articular front of the trapezoid, but with its major part continued onto the capitate. As will be seen below this sheet is augmented by a reflected part of the flexor retinaculum. This arrangement is perpetuated from marsupial to man.

In hominoids, of course, the front of the capitate receives the attachment of the lower band of the palmar radiocarpal ligament and throughout the primates the capitate is attached by an anterior ligament to the triquetral (Figs 6.6, 6.7, 6.9). The capitate then is the centre of a radiating ligamentous mass on the front of the carpus.

The carpometacarpal joint of the thumb

It is instructive first to consider this joint in man,

for then as the phylogenetic history is unravelled, it will become apparent which characters are merely symplesiomorphic and which are apomorphic, either unique to man, or synapomorphic with other hominoids.

A question of terminology must first be resolved since this has been the source of not a little confusion. Due to the general concavity of the carpus, and in particular to the 'set' of the trapezium (Fig. 6.8B) the designation of the surfaces of this bone can be ambiguous. Thus the true morphological lateral surface faces forwards and outwards and the palmar (or volar) surface in an ulnar direction; the extent of this reorientation, of course, shows considerable species variation. Here the true descriptive terms—lateral, medial, palmar, dorsal—will be retained!

Homo sapiens

The key role played by this joint in manual dexterity has stimulated much detailed investigation and form and function are fairly well understood. The joint surfaces are said to be sellar (saddle-shaped) or toroidal. The latter term implies that geometrically their shape is comparable to segments cut from the inner or axial surface of a torus (a doughnut-shaped structure) and thus are curved in two directions: one curve is concave towards the centre of the ring and the other, at right angles to it, is convex. The curvatures are not, however, symmetrically arranged and as Kuczynski (1974) has pointed out, the medial part of the trapezial surface, that portion adjacent to the second metacarpal, is essentially cylindrical and the concave element converting the surface as a whole into a sellar one, is a consequence of the downward deflection of the broad and rather flattened anterolateral part of the surface, which thus creates a curved trough, passing from the radial to the ulnar side.

The fibrous capsule shows a trio of intrinsic ligamentous thickenings. The anterior oblique carpometacarpal ligament (Fig. 6.2) passes from a smooth, facet-like impression of compact bone just distal to the base of the tubercle (crest) of the trapezium to reach an impression on the volar tongue-like projection from the base of the first metacarpal. A thick lateral carpometacarpal ligament passes from a second smooth impression on the lateral aspect of the trapezium to the lateral aspect of the base of the first metacarpal; this ligament is here covered by the major insertion of the tendon of abductor pollicis longus to the metacarpal base and a bursa intervenes.

Dorsally (Fig. 6.3), a posterior oblique ligament arises from a prominent bony tubercle laterally situated on the trapezium and which should not be confused with the tuberculum ossis trapezii (PNA) which is the crest overhanging the groove for the tendon of flexor carpi radialis. The distal smooth aspect of this dorsal tubercle is confluent around the lateral border of the trapezium with the impression for the lateral ligament. From a concentrated origin on this posterior tubercle the posterior ligament fans out to become inserted

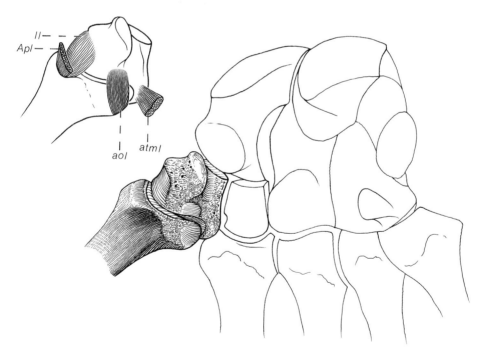

Fig. 6.2
A ventral view of the human carpus showing the osteological features described in the text for the trapezium and base of the first metacarpal. The inset shows the relationships of the attachments of the anterior oblique carpometacarpal ligament (*aol*), the lateral carpometacarpal ligament (*ll*), the anterior trapeziometacarpal ligament (*atml*), and the insertion of the tendon of abductor pollicis longus (*Apl*), to this bony conformation. (From Lewis 1977.)

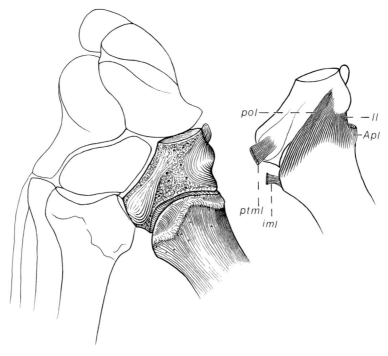

Fig. 6.3
A dorsal view of the human trapezium and base of the first metacarpal showing the osteological features described in the text. The inset shows the relationships of the attachments of the posterior oblique carpometacarpal ligament (*pol*), the lateral carpometacarpal ligament (*ll*), the posterior trapeziometacarpal ligament (*ptml*), the insertion of the tendon of abductor pollicis longus (*Apl*) and the intermetacarpal ligament (*iml*) to this bony conformation. (From Lewis 1977.)

along the whole dorsal margin of the metacarpal base, and the adjoining borders of this ligament and the lateral ligament are confluent.

Fick (1904) described the ligamentous apparatus of this joint and recognized a ligamentum trapezio–metacarpeum pollicis dorsalis, arising from a dorsoradial tubercle on the trapezium, and a ligamentum trapezio–metacarpeum pollicis volaris. His account is far from clear and the illustrations seem to bear little resemblance to the true morphology.

Haines (1944) recorded the presence of the three ligaments noted above but oddly enough he figured the posterior oblique ligament as radiating from a restricted origin on the metacarpal to a broad insertion on the trapezium, instead of vice versa; he was not concerned with the underlying osteology.

Napier (1955) gave a description in good accord with that presented here but without particularizing osteological features. Quite accurate illus-

trations of these ligaments have, in fact, long been available (Weitbrecht 1742).

To this trio of well recognized ligaments has recently been added another, the intermetacarpal ligament (Bojsen-Møller 1976). This ligament is Y-shaped with a stem attached to the base of the first metacarpal (Fig. 6.3) and a dorsal and palmar crus attached to the base of the second. The dorsal crus, the major band, has an attachment adjacent to the insertion of extensor carpi radialis longus, and as will be seen in Chapter 8 (Fig. 8.5) seems to be a derivative of that tendon. Not infrequently this particular ligament may undergo partial ossification forming a bony spur on the second metacarpal which may be visible radiographically (Koebke *et al.* 1982).

Function in the joint is well understood. Being a sellar joint it is generally considered to have two axes and two degrees of freedom (Kapandji 1966). That for flexion–extension traverses the trapezium through the centres of its convex curvatures; that for abduction–adduction traverses the metacarpal through the centres of its particular convex curvatures and is approximately at right angles to the other. However, it is well recognized that usually the thumb moves into opposition by circumducting and tracing out a cone of movement, and that during this motion the metacarpal undergoes axial rotation medially. Bojsen-Møller (1976) stressed the role of the intermetacarpal ligament as establishing a centre for this circumduction movement but it seems doubtful whether it has any particularly major role to play in this. Such a rotatory movement (about a third axis) in a sellar joint has been called conjunct rotation by MacConaill and Basmajian (1969). These authors point out that the arcuate movement of opposition can be decomposed into the diadochal sequence of extension followed by abduction, with the conjunct rotation accompanying the latter movement; flexion of the thumb is of course then required to press its pulp against one of the fingertips.

Pan troglodytes

The joint surfaces are saddle-shaped as in man but that on the trapezium is less broad, the whole bone being relatively flatter (Figs 6.8A, 6.4B). Due to the orientation of the trapezium within the carpal arch, the homologue of the human palmar surface of the bone faces even more directly medially.

The joint possesses an anterior oblique ligament, disposed similarly to that of man, although of flimsier texture, and also as in man, a strong posterior oblique ligament radiates from a prominent dorsal tubercle on the trapezium to a broad attachment on the metacarpal base. No identifiable lateral ligament exists and where it might have been expected the fibrous capsule is thin and covered by the insertion of the tendon of abductor pollicis longus to the first metacarpal. The prepollex, receiving the other larger insertion of the abductor pollicis, is sited on the trapezium proximal to its dorsal tubercle and has a sizeable synovial articulation with the tubercle of the scaphoid.

Gorilla gorilla gorilla

The joint surfaces are again saddle-shaped, that on the trapezium being broader than is usual in the chimpanzee, and thus resembles more closely the condition seen in man.

The ligamentous apparatus is similar to that of the chimpanzee. A strong posterior oblique ligament, covering the unspecialized dorsal fibrous capsule, arises from the base of a dorsal bony tubercle of massive dimensions. The trapezium has apparently sprouted a massive bony apophysis (Fig. 6.4C) lying behind the tubercle proper (or crest) of the trapezium and mimicking it and, as noted in Chapter 4, this has been achieved by amalgamation with the prepollex. This process extends far enough proximally to establish a supernumerary articulation with the tubercle of the scaphoid. In the one wet specimen where this feature has been examined there was here a tough fibrous union containing a small synovial cavity. In juvenile specimens this apophysis exists as a separate ossicle, the prepollex, which is often apparently lost in macerated preparations. The same situation occurs in the mountain gorilla. Therefore, a massive bony buttress has been elaborated at the lateral margin of the carpus in these heavy knuckle-walking apes. It is tempting to correlate its presence with their particular locomotor pattern. Because of its homology with the prepollex, not surprisingly this buttress receives the major insertion of abductor pollicis longus, and this attachment is overlaid by the second of the dual tendons which passes to the metacarpal base. In the only wet specimen examined the 'prepollex' insertion sent a prolongation across the joint line (Fig. 6.4C) to also attach to the metacarpal; could this be the way in which the human lateral ligament has been derived?

Pongo pygmaeus

The joint surfaces are generally unequivocally saddle-shaped as in the other great apes. In some specimens, however, the lateral overhang on the trapezium may be considerably reduced and the articular surface as a whole is protuberant, thus representing an intermediate stage in a trend which, if carried somewhat further, could result in the condition seen in gibbons, where the trapezium presents a convex, ball-like articular surface. A posterior oblique ligament radiates from a dorsal tubercle on the trapezium and an anterior oblique

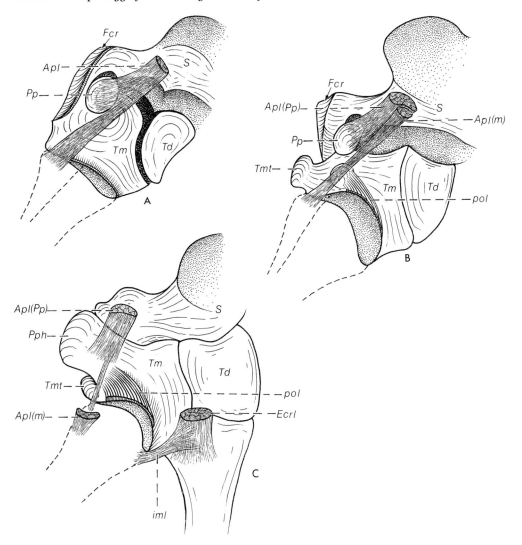

Fig. 6.4 (A–C)

Dorsal views of the region of the left carpometacarpal joint of the thumb in *Cercopithecus nictitans* (A), *Pan troglodytes* (B) and *Gorilla gorilla* (C). *S*, scaphoid; *Tm*, trapezium; *Tmt*, tubercle (crest) of trapezium; *Td*, trapezoid; *Pp*, prepollex; *Pph*, prepollex homologue in *Gorilla*; *Fcr*, groove for flexor carpi radialis tendon; *Apl*, abductor pollicis longus tendon; *Apl(Pp)*, abductor pollicis longus tendon to prepollex; *Apl(m)*, abductor pollicis longus tendon to metacarpal; *pol*, posterior oblique ligament; *Ecrl*, extensor carpi radialis longus tendon; *iml*, intermetacarpal ligament.

ligament is present. No specialized lateral ligament is demonstrable, and a prepollex sits upon the lateral surface of the trapezium, and receives part of the abductor pollicis longus insertion.

Hylobates lar

Gibbons uniquely possess a ball and socket joint at the base of the thumb metacarpal with the convex (male) surface being located on the trapezium. This is doubtless a secondary specialization and the joint surface on the first metacarpal may show suggestive indications of remodelling from a basic sellar shape. Indeed, as noted above, some specimens of *Pongo pygmaeus* show hints of a transition towards the same sort of morphology as is seen in hylobatids.

As in the apes already described, an anterior

oblique ligament is present and a strong posterior oblique ligament radiates from a bony dorsal tubercle on the trapezium.

The phylogenetic background

In marsupials (*Pseudochirus, Caluromys, Didelphys*) the trapezium tapers laterally to form a process bearing the articulation of the prepollex, which is here sited between it and the tubercle of the scaphoid (Fig. 5.2B). This projection bears on its distal aspect part of the articular surface for the first metacarpal which is deflected downward producing an overhang which imposes a concavity on the otherwise convex surface. This concavo–convex surface has more than a hint of the refined sellar surface found in man. The fibrous capsule even shows ligamentous thickenings, which may be designated anterior and posterior ligaments (Fig. 6.5) because of their apparent homology with similar structures found in Primates. The former is quite slender but the latter is a broad sheet of parallel fibres and it is not concentrated onto a bony tubercular origin on the

trapezium and is not therefore oblique and fan-shaped. In quite generalized insectivores (*Tenrec ecaudatus*) a similar joint surface is found. Haines (1958) has further demonstrated that the metacarpal surface of the trapezium is saddle-shaped in the carnivore *Herpestes ichneumon*, and in the treeshrew *Ptilocercus lowii*, and has suggested that some degree of opposability of the pollex is more common in mammals than is generally admitted, despite its usually very limited degree of independence and divarication from the other digits.

The form of the trapezium in marsupials provides the key to rational interpretation of the osetology of the primate bone. It appears that the primitive eutherian mammals which gave rise to the primates must have possessed a trapezium not too dissimilar from that in extant marsupials of generalized aspect.

In prosimians (Figs 5.1A, 5.2A) there are still clear indications of a lateral process bearing the prepollex on the trapezium, and the distal aspect of the process participates in the carpometacarpal joint. Again, as in marsupials this joint has a sellar form; Jouffroy and Lessertisseur (1959) have also

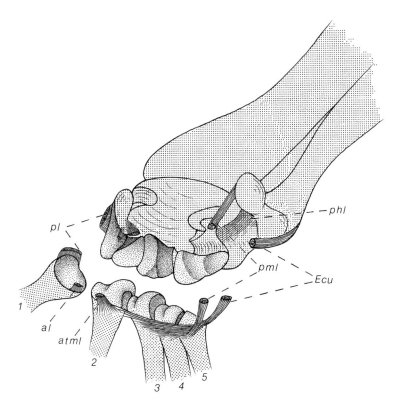

Fig. 6.5
The right hand of *Pseudochirus laniginosus*, viewed from ventrally, with the metacarpals (1–5) disarticulated from the carpus and displaced dorsally. *al*, anterior ligament of the first carpometacarpal joint; *pl*, posterior ligament of the first carpometacarpal joint; *atml*, anterior trapeziometacarpal ligament; *phl*, pisohamate ligament; *pml*, pisometacarpal ligament; *Ecu*, tendon of extensor carpi ulnaris. (From Lewis 1977.)

commented upon the saddle-shaped form of the joint in lemuroids.

The columnar process on the trapezium bearing the prepollex, which is insinuated between it and the scaphoid tubercle, persists in a particular prominent form in some New World monkeys such as *Aotus* and *Cebus* (Fig. 5.1B). This projection may almost mimic the apophysis noted in *Gorilla*, although they are not, in fact, strictly homologous. The process carries on its distal surface part of the carpometacarpal joint surface, but here the New World monkeys typically show a derived morphology. Definite indications of the presumably primitive concavo-convex form, with similar anterior and posterior ligaments, may persist (*Aotus, Lagothrix, Cebus, Saimiri*), although there is a tendency, often quite pronounced, towards suppression of the ventral concave element of the trapezial surface. In the Callitrichidae (*Callithrix, Leontideus*), the concave component of the trape-

zial surface is typically even further suppressed leaving the joint surface essentially cylindrical.

In Old World monkeys that lateral columnar extension of the trapezium has regressed to little more than a barely perceptible mound, and the prepollex articulates here with trapezium and scaphoid tubercle (Figs 6.4A, 6.6, 6.7). Only apparently in *Papio* is there any obvious bony prominence associated with the prepollex. The carpometacarpal joint surfaces in Old World monkeys are usually unequivocally concavo–convex with anterior and posterior ligaments, the latter again being a flat sheet of parallel fibres. In some (e.g. *Presbytis*) however, the concave element of the trapezial surface may be diminished, paralleling the form observed in *Cebus*.

The joint surfaces, thus, seem not to have undergone any particularly dramatic change during the evolution of the therian mammals. The commonly held notion (Napier 1961), therefore,

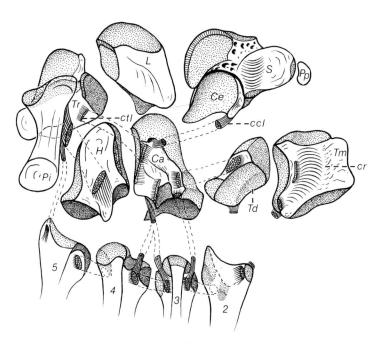

Fig. 6.6

The pattern of articulations and associated ligaments between the carpal bones and at the carpometacarpal joints in a left hand of *Colobus polykomos*. All, the basic ligamentous apparatus shown in Fig. 6.1 is here present and is unlabelled, but the reflected attachments of the flexor retinaculum are not shown, nor is its attachment to the rudimentary crest (*cr*) or tubercle on the trapezium. Flexor carpi radialis is not shown but the groove for it is indicated on scaphoid and trapezium. The proximal carpal row is shown: *S*, scaphoid; *Ce*, centrale; *Pp*, prepollex; *L*, lunate; *Tr*, triquetral; *Pi*, pisiform. The centrale–capitate ligament (*ccl*) and the anterior capitate–triquetral ligament (*ctl*) are indicated. Other labelling is as in Fig. 6.1.

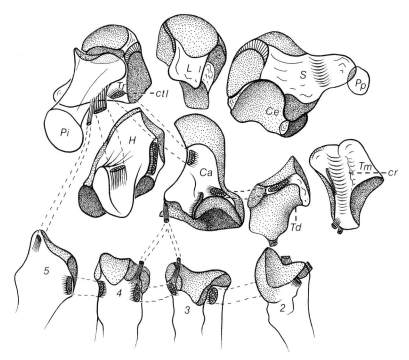

Fig. 6.7

The pattern of articulations and associated ligaments between the carpal bones and at the carpometacarpal joints in the left hand of *Cercopithecus nictitans*. Conventions and labelling are as in Figs 6.1 and 6.6. In this specimen the carpometacarpal ligament on the lateral side of the capitate was lacking, allowing confluence of the facets anterior and posterior to it. Also, there was no capitate–centrale ligament.

that a sudden change in direction of evolution produced a saddle joint which has been evolved only in Old World monkeys, apes and man, leaving other generalized mammals with a mere hinge joint, or a 'modified hinge' in New World monkeys, is misleading. The essence of the concavo–convex form seems to be a very ancient mammalian attribute, although in diverse groups some rounding off of the ventral concave element of the trapezial surface has occurred, resulting in an almost cylindrical, or even spherical, shape. Only in the Hominoidea, however, has the posterior ligament become truly oblique and fan-shaped, with a carpal origin concentrated onto a bony tubercle on the trapezium. Broadening of the lateral surface of the trapezium in man might perhaps have been accompanied by the sequestration of a part of this ligament to form the lateral ligament. Alternatively this ligament might have been refashioned from part of the insertion of

abductor pollicis longus, as is suggested by the situation in *Gorilla* (Fig. 6.4C).

The second to fifth carpometacarpal joints

A basic strategy throughout this book has been to try to infer in various situations a primitive emergent mammalian morphological pattern, which, by various apomorphic specializations, gives rise to the diverse morphologies found in extant mammals, and particularly primates. This can usually be done with reasonable assurance by applying the principles by which bones, joints or muscles are known to be modified during ontogeny or phylogeny. It is a bonus, when this inferred primeval pattern is found to persist in living mammals of generalized aspect, either eutheria or metatheria, and the whole evolutionary picture then is clarified. In the case of this set of joints, however, living marsupials and insectivores present what seems to be a considerably modified pattern, yet with clearly persistent hints of the precursor morphology. On the other hand Primates in many cases conserve what seems to be a relatively unmodified arrangement.

By backtracking in reasoning a reasonable stab at reconstructing a model of the primitive structure in the emergent therian mammals can be made (Fig. 6.1). The joint line across the articulations shows a stepped configuration, for the second metacarpal is deeply indented into the carpus, articulating not only basally with the trapezoid, but also with the adjacent sides of the capitate and trapezium. Moreover, the third metacarpal is indented medially so that this side of it abuts against the lateral surface of the hamate. The third metacarpal is braced on either side by interosseous carpometacarpal ligaments arising from the lateral side of the capitate and medially from the adjoining surfaces of capitate and hamate. Anterior and posterior to these ligaments the bases of the second and fourth metacarpals articulate with the third and the adjoining sides of these metacarpal bases thus possess dual facets. Because of the stepped configuration of the joint line the dual facets on the second metacarpal articulate with the side of the capitate by proximal extensions on these facets; the dual facets on the medial aspect of the third metacarpal similarly articulate proximally with the lateral aspect of the hamate. The second to fifth metacarpals are joined one to the other by intermetacarpal ligaments. Significantly the first metacarpal is not included in this linkage again emphasizing its individuality and special role. As noted above, however, an intermetacarpal ligament between the first and second metacarpals is found in man but it is of quite different constitution and derivation; it may be found in emergent form in the great apes (Fig. 8.5). The recessed second metacarpal bears a facet on its lateral aspect for the trapezium and a strong anterior trapeziometacarpal ligament radiates into the palm in front of this articulation attaching to the second metacarpal base. Paralleling this ligament, and in intimate association with it, is the tendon of flexor carpi radialis passing to its presumptive primitive insertion on the third metacarpal. A posterior trapeziometacarpal ligament lies behind this joint. The hamate overhangs the bases of metacarpals four

and five, articulating with them; this volar extension of the hamate is the clear precursor of the hook-like process typifying the Hominoidea, although in mammals other than this primate superfamily the process is usually an integral part of the body of the hamate. To this bony projection attaches a massive pisohamate ligament and the process is effectively embedded within its substance. Delaminated from this last named ligament is a pisometacarpal ligament which skirts the prominence on the hamate and passes to attach to the fifth metacarpal, effectively sealing the carpometacarpal joint medially.

It seems a reasonable working hypothesis to suggest that such an arrangement (Fig. 6.1) was the basic therian pattern, yet apparently all living insectivores, tree shrews and marsupials show an overlay of derived features, often comparable ones, which could suggest that the morphocline polarity was in the opposite direction. Given knowledge of how joints develop and evolve, however, it seems easier to contemplate disappearance of interosseous carpometacarpal ligaments, for instance, than their emergence *de novo*.

In the insectivore *Tenrec ecaudatus* many of the fundamentals of such a pattern persist and the stepped joint line is similar. The articulation between third and fourth metacarpals is dual and there is a thin and rudimentary interosseous carpometacarpal ligament passing distally from the capitate. The second metacarpal, however, has a single linear articulation with the capitate and third metacarpal, with no interosseous carpometacarpal ligament present; the intermetacarpal ligament, however, persists between the two metacarpals distal to their articulation. In this species flexor carpi radialis attaches not only to the third metacarpal, but also to the second, and has clearly achieved this by appropriating part of the anterior trapeziometacarpal ligament. As will be seen this is a recurring theme in primate evolution and its converse also occurs: the ligament may incorporate part of the tendon and thus project its attachment to include the third metacarpal. In other insectivores (e.g. *Suncus caeruleus*) the basic topography of articulations may be similar, but without any trace of carpometacarpal ligaments flanking the third metacarpal, and with the primitive dual facets confluent so that second metacarpal articulates by a long convex facet with the capitate, and the third similarly with the hamate. The same thing is found in tree-shrews (*Tupaia sp.*).

The marsupials *Pseudochirus laniginosus* (Fig. 6.5) and *Caluromys lanatus* have a similar stepped configuration of the joints, and there is not even a residual carpometacarpal ligament where the third metacarpal intrudes into the carpus and contacts the hamate. Both of these marsupials, however, show a striking specialization which turns out to be convergent to a modification found in the human hand, and by analogy elucidates the significance of this latter. The extensor carpi ulnaris tendon has extended from its primitive attachment to the fifth metacarpal and traverses the palm across the bases of the metacarpals reaching as far as the second (Fig. 6.5). In taking this course it runs in a groove on the distal aspect of the 'hook' of the hamate, thus separating the bases of the fourth and fifth metacarpals from the embrace of the overhanging hamate; indeed it here forms a labrum-like structure across the bases of the metacarpals. The arrangement strikingly mimics the course of the peroneus longus tendon in the mammalian foot. This changed morphology must have enhanced the mobility of the postaxial metacarpals. The medial side of the base of the second metacarpal has a long convex articulation with the capitate and is free of any interosseous carpometacarpal ligament, and mobility of the second metacarpal is also thus enhanced. These specializations are perhaps functionally correlated with the grasping mode, schizodactyly or zygodactyly, utilized by these creatures, (at least in *Pseudochirus*) in which the mobile second digit acts together with the thumb as one arm of a pincer, the other arm, also quite mobile, being formed by digits three to five.

A striking similarity in derived morphology of these carpometacarpal articulations thus characterizes such diverse groups as insectivores, tree-shrews, and marsupials (except for the specialization involving extensor carpi ulnaris peculiar to the latter group). It might be tempting to perceive this as primitive. The evidence, however, seems to favour convergent specialization imposed upon a primitive pattern such as that shown in Fig. 6.1, which must be as ancient as the emergence of the therian mammals themselves. Indeed the primitive pattern (Fig. 6.1) appears to have been transmitted

to the ancestral Primates, for certain prosimians retain it virtually unchanged.

The prosimian joints

The profile of the joint line in these primitive primates retains the character described above, with its sequence of carpal and metacarpal interdigitations. In *Lemur fulvus* and *Lemur catta* the various joint surfaces, with their attendant ligaments are all comparable in position and pattern to those in the hypothetical primeval mammal shown in Fig. 6.1. The articulation of the third metacarpal with the side of the hamate is, however, quite reduced in extent and may be restricted to the posterior parts of the bones. Moreover, in some specimens of *Lemur* the palmar and dorsal dual facets on the second metacarpal may show quite marked confluence, yet despite this the interosseous carpometacarpal ligament persists. Again a ventral beak on the hamate, embedded in the substance of the pisohamate ligament, overhangs the bases of metacarpals four and five. In other prosimian Primates this basic plan may be modified: in *Galago crassicaudatus* only the ulnar one of the pair of carpometacarpal ligaments is found, and in *Loris tardigradus* and *Perodicticus potto* both of these carpometacarpal ligaments are lacking. Their presence in *Lemur*, however, supports the idea that carpometacarpal ligaments running from either side of the capitate are part of the ancestral primate morphology. Despite the facet that the carpometacarpal ligaments may be lost and the dual articular facets become confluent in various species, typical intermetacarpal ligaments persist between the second to fifth metacarpals.

The joints in Old World monkeys

The complete array of joint surfaces and ligaments characterizing the prosimian and the primitive condition (Fig. 6.1) is clearly part of the higher Primate heritage but with one significant apomorphic advance: the third metacarpal has relinquished all contact with the hamate, and this has been exchanged for articulation of the fourth metacarpal with the capitate (Figs 6.6, 6.7). The anterior facet on the fourth metacarpal has a small edge to edge contact with a variable facet on the capitate. Posterior to the carpometacarpal ligament, however, the metacarpal has an appreciable end to end contact with the chamfered facet on the posteromedial corner of the capitate. In colobines (Fig. 6.6) the usual carpometacarpal ligament and dual facets are retained on the lateral aspect of the capitate. In cercopithecines, however, the morphology is variable. In some (*Papio*) the carpometacarpal ligament on the radial side of the third metacarpal may be reduced to a slender thread, allowing confluence of the articular facets in front and behind it on both capitate and index metacarpal. In others, such as some specimens of *Cercopithecus* (Fig. 6.7) the ligament is totally lacking and a linear hollowed facet on the capitate articulates with the lateral border of the base of the second metacarpal. Parallel acquisition of this apomorphic character has already been mentioned in certain insectivores, treeshrews, marsupials, and prosimians. Regardless of the status of the carpometacarpal ligaments typical intermetacarpal ligaments are found between the second to fifth metacarpals. A little ligament joins the trapezoid to the oddly crenated posterior part of the base of the second metcarpals (Figs 6.6, 6.7). It is also found in New World monkeys (Fig. 5.1B), apes (Fig. 6.9), and man. The ligament, but not the metacarpal notch are found in *Lemur* (Fig. 5.1A). As in the primitive condition the beak-like process—precursor of the hook—on the hamate overlies the fourth and fifth metacarpals and is embedded in the core of the massive pisohamate ligament. This is flanked by the pisometacarpal ligament and often a diverticulum of the carpometacarpal joint cavity intervenes here.

The joints in New World monkeys

The basic articular pattern here is clearly derived from a common ancestor of the higher Primates, which already possessed the major features seen in extant Old World monkeys including the establishment of dual contact between the fourth metacarpal and the capitate. It is modified, however, in various species by regression of one or other of the carpometacarpal ligaments flanking the base of the third metacarpal. Only that on the radial side is present in the marmoset *Callithrix geoffroyi* whilst in the tamarin *Leontideus rosalia* only that on the ulnar side is found. The latter condition, where the

radial of these two carpometacarpal ligaments is suppressed is, in fact, common in New World monkeys and represents a parallel to the trend noted above in certain cercopithecines. Thus, in *Aotus trivirgatus, Lagothrix lagotricha, Cebus albifrons,* and *Saimiri sciureus* the second metacarpal articulates by a long convex, linear anteroposterior facet with the capitate, no intervening carpometacarpal ligament being present and consequently no separation of this articulation into anterior and posterior parts. In *Aotus* and *Lagothrix* the ulnar carpometacarpal ligament is also lacking. Typical intermetacarpal ligaments are found. There thus seem ample grounds for believing that the basic precursor articular and ligamentous pattern in both platyrrhine and catarrhine monkeys was similar, and that there was a clear pattern of common inheritance, albeit with easily interpretable modifications in certain species, characterizing all the higher Primates.

The joints in apes

The hominoids obviously also shared this pattern of common inheritance, and indeed (with the exception of man) they have been more conservative in retaining it unchanged than the monkeys.

Pan troglodytes   The third metacarpal articulates with the complex distal surface of the capitate and is braced on either side by interosseous carpometacarpal ligaments (Fig. 6.8A); anterior and posterior to these ligaments the bases of the second and fourth metacarpals articulate with the third, the adjoining sides of these metacarpal bases thereby possessing dual facets. The second metacarpal is deeply indented into the carpus, articulating medially with the side of the capitate by proximal extensions of the dual intermetacarpal facets. The second metacarpal also bears a facet on the posterolateral aspect of its base (just anterior to the impression where inserts the tendon of extensor carpi radialis longus) for the trapezium. A strong ligament radiates towards the palm from the trapezium in front of this articulation and attaches to the second and third metacarpals, and is accompanied by flexor carpi radialis also attaching to these two metacarpals; this advance on the primitive condition has obviously been acquired by interchange of substance between tendon and

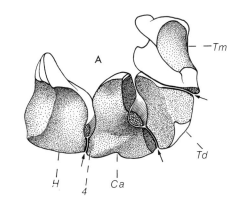

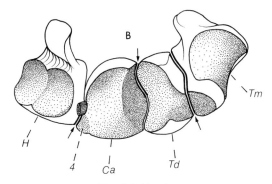

Fig. 6.8 (A, B)
The distal aspects of the left carpus of a wet specimen of
Pan troglodytes (A) and *Homo sapiens* (B). Articular
surfaces are shown in heavy line and are indicated by
arrows. *4*, articulation of fourth metacarpal with capitate; other labelling as in Fig. 6.1. (From Lewis 1977.)

ligament. A similar ligament lies behind this
articulation, joining trapezium to second metacarpal. Both the fourth and fifth metacarpals articulate with the distal aspect of the hamate which
here presents dual convex facets ballooning distally and articulating with concavities in the
metacarpal bases, which additionally have a single
intermetacarpal joint between them. The fourth
metacarpal extends beyond the hamate posteriorly
to achieve a small contact with a bevelled facet on
the corner of the capitate. There is no contact of
fourth metacarpal with capitate anterior to the
interosseous carpometacarpal ligament. The small
facet found here in monkeys has been lost in *Pan*;
this is, in fact, a synapomorphy of all the
hominoids. The ventral beak of the hamate is
remodelled into a hook which overhangs the bases

of the fourth and fifth metacarpals and here
articulates with them. The facets on the metacarpals for the hamate hook are convex and are
sharply angled to those concave ones on the
proximal aspects of their bases for the body of the
hamate. A massive pisohamate ligament anchors
the pisiform to the hook of the hamate. Lateral to it
is a pisometacarpal ligament attaching to the fifth
metacarpal. Interestingly, in one chimpanzee
which I have examined the third metacarpal was
invaginated into the carpus on its ulnar side and
articulated with the side of the hamate behind the
interosseous carpometacarpal ligament, a quite
remarkable reversion to the primitive mammalian
pattern. Typical intermetacarpal ligaments link
the second to fifth metacarpal bases.

Gorilla gorilla The whole pattern of articular
surfaces is similar to that of *Pan*, and the ligamentous apparatus is also identical with all the same
clearly definable bands: interosseous carpometacarpal ligaments flank either side of the third
metacarpal; an anterior trapeziometacarpal ligament (attaching to the second and third metacarpals) and a posterior trapeziometacarpal ligament
(passing to the second metacarpal); a pisometacarpal ligament passing to the fifth metacarpal. In one
specimen the flexor carpi radialis tendon associated with the anterior trapeziometacarpal ligament attached only to the second metacarpal and
the appearance strikingly illustrated how the
closely associated tendon and ligament, by inter-
changing material can vary their insertions to
either or both of the second and third metacarpals;
these variations in attachment occur repeatedly in
various primates.

Pongo pygmaeus Again the same articular and
ligamentous pattern is found (Fig. 6.9).

Hylobates lar The pattern is again similar but at
least some specimens show a slight modification,
which is perhaps of minor interest in itself, but
significant for the way in which it demonstrates
the sort of evolutionary trends which may affect
these joints. The interosseous carpometacarpal
ligament passing from the capitate between the
third and fourth metacarpals may be a rudimentary flattened band and with its regression the
anterior and posterior facets on each of the

Fig. 6.9

The pattern of articulations and associated ligaments between the carpal bones and at the carpometacarpal joints in the left hand of *Pongo pygmaeus*. Conventions and labelling are as in Figs 6.1 and 6.6.

adjacent metacarpals become confluent. A similar process has been noted above in various primates and other mammals affecting one or other of the interosseous carpometacarpal ligaments. Dry bones then cannot always give unequivocal evidence as to the former presence or absence of such a ligament.

The joints in man

Striking modifications have been achieved in the morphology of this set of joints in the human hand, and these are demonstrably of considerable significance in the evolution of hand function.

The articulation of the index metacarpal with the trapezoid has become saddle-shaped, rather than wedge-shaped as in other hominids (Fig. 6.8B), and the usual posterior ligament joins the trapezoid to the deep posterior notch in the second metacarpal. The interosseous carpometacarpal ligament between the second and third metacarpals is lacking and the dual facets on the second metacarpal have been refashioned into a convex articular surface for the capitate and second metacarpal (Fig. 5.18B). The intermetacarpal ligament persists. It could be inferred that this change permits a small degree of flexion into the palm coupled with conjunct rotation, in the sense of pronation. This would presumably enhance the effectiveness of the similar movements which occur during flexion at the metacarpophalangeal joint and will be described later. Although the metacarpals do seem to lack any capacity for independent rotation with the hand held flat (Landsmeer and Ansingh 1957) there are radiological indications (Van Dam 1934) that in the clenched hand the index metacarpal is in fact pronated (and that of the minimus supinated). Such movement is clearly facilitated by a change in orientation of the facet on the trapezium for the second metacarpal from the sagittal (Fig. 6.8A) to a more transverse disposition (Fig. 6.8B). Comparative analogies lend convincing support for the interpretation that loss of the interosseous carpometacarpal ligament proceeding from the radial side of the capitate, and refashioning of the articulation between that bone and the index metacarpal, are associated with increased mobility of that metacarpal. A partial development of this specialization has been noted above in *Papio papio*, and significantly baboons are noted for their quite high degree of independent control of the index finger in association with a precision grip (Bishop 1964). Further, in mammals practising a schizodactylous, or zygodactylous, grip the index finger has acquired considerable mobility and moves in concert with the thumb. Similar specializations affect the medial side of the base of the second metacarpal here. This is seen in the marsupial *Pseudochirus laniginosus* (Fig. 6.5) which is known to grip thus (Pocock 1921) and in the New World monkeys *Lagothrix lagotricha* and *Aotus trivirgatus* which also practise this gripping posture (Pocock 1925; Erikson 1963). It has been noted above how a similar apomorphic change to man has affected this articulation in the cebid New World monkeys. Examination of dry bones (Singh 1959) does, however, indicate that in rare cases dual facets persist on the ulnar side of human index metacarpals, indicating the atavistic retention of a carpometacarpal ligament in this situation: an almost ape-like condition.

A further remarkable advance on ape morphology has occurred in the articulations between the hamate and fourth and fifth metacarpals. The main joint surfaces have been remodelled so that the direction of curvature has become reversed: for the fourth metacarpal the convexity on the hamate has been replaced by a dorsoventral concavity (Fig. 6.8B) (although there is here variation and the hamate sometimes retains its convex surface, as in Fig. 6.10); for the fifth metacarpal the facet on the hamate has become markedly concave dorsoventrally but retains a lateral convexity—it has become a sellar (saddle) surface. A much more significant change, how-

ever, has involved the hook of the hamate which has become retracted from its restrictive grip on the metacarpal bases. When the joint is opened from the dorsal aspect the consequences of this separation are revealed (Fig. 6.10). A ligamentous band, part of the pisometacarpal ligament, has evolved arching across the distal aspect of the hook of the hamate and effectively excluding it from articulation, as it passes to attach to the palmar surfaces of the bases of the fifth, fourth, and third metacarpals, where it forms a labrum-like projection along their articular margins. It frequently directly forms the ventral wall of the joints and the groove on the hamate hook which lodges it may be synovial lined although in other cases a thin capsular layer separates it from the interior of the joint. The pisometacarpal ligament is invariably inaccurately described (at least in English language textbooks) as passing from the pisiform to the base of the fifth metacarpal, and the reason for this misunderstanding is not difficult to appreciate. When viewed from ventrally the arched course of the ligament is obscured by a fibrous band passing from the hook of the hamate to the fifth metacarpal. This was named the 'Haken-Mittelhandband' by Fick (1904) but, in fact, it appears to be no more than the deepest, fibrosed fibres of that part of opponens minimi digiti lying deeply to the deep branch of the ulnar nerve, and indeed, it may replace the muscle bundle. A further source of confusion has been the repeated statement that the deep groove on the distal and medical aspect of the hook of the hamate lodges the deep branch of the ulnar nerve; this is a common statement in textbooks (e.g. Frazer 1946). Lack of precision in the descriptions of the human anatomy is surprising in view of the fact that a suggestive, if not fully explanatory, illustration of the ligament was given by Weitbrecht (1742) who showed a 'reflected band' of the ligament arching around the hook of the hamate to reach the third metacarpal. This is well shown also by Poirier and Charpy (1911). Fick (1904) similarly noted the 'pars reflexa' of the ligament and that this bundle was concealed by a ligamentous band passing from the hook of hamate to the fifth metacarpal—the 'Haken-Mittelhandband'.

These joint modifications seem to play a significant role in conferring unique functional capabilities upon the human hand. The fourth and,

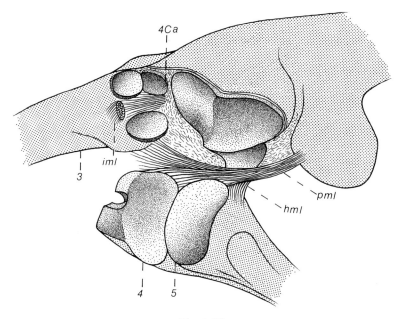

Fig. 6.10
The ulnar part of the carpometacarpal joint of a left hand of *Homo sapiens* opened from dorsally. The fifth metacarpal (5) is displaced ventrally from its articulation with the hamate; the fourth metacarpal (4) is similarly displaced from its articulations with the hamate, a small adjoining facet on the capitate (*4Ca*) and dual facets (separated by an interosseous carpometacarpal ligament) on the third metacarpal (3). The pisometacarpal ligament (*pml*) clothed on its palmar surface by the hamate–metacarpal ligament (*hml*) forms the ventral wall of the joint and is displaced from its synovial-lined groove on the hook of the hamate. (From Lewis 1977.)

particularly, the fifth metacarpals, freed from the buttressing influence of the hook of the hamate, are thus endowed with increased mobility and the sellar surface for the fifth metacarpal introduces an element of conjunct rotation into flexion which, combined with the metacarpophalangeal movements to be described, produced what might reasonably be called opposition of the little finger.

The remaining basic elements of the ligamentous apparatus of the carpometacarpal junction persist in the human hand in a form virtually unchanged from that shown in pongids. In textbook accounts there is, however, a tendency for these obvious and very phylogenetically stable structures to be obscured in a welter of description of mere adventitious bands which are simply morphological irrelevancies. From the prominent bony tuberosities on the trapezium (Figs 6.2, 6.3) in front of and behind the articulation for the second metacarpal anterior and posterior trapeziometacarpal ligaments arise. The former, lying deep to the insertion of flexor carpi radialis, attaches in

part to the second metacarpal but its major component passes in the deep and sometimes synovial-lined groove in the base of this metacarpal reaching its major attachment on the third. It was well illustrated by Weitbrecht (1742) and described as the L. trapeziometacarpeum volare bifurcatum by Fick (1904). The posterior ligament attaches to the index metacarpal at the radial margin of the impression for extensor carpi radialis longus and was similarly illustrated by Weitbrecht (1742) and described by Fick (1904).

The flexor retinaculum

The flexor retinaculum is applied as an overlay to the articulations of the carpus, and the carpometacarpal joints. Although presumably a derivative of the deep fascial investment of the limb it has come to contribute significantly to the ligamentous apparatus of the volar aspect of wrist and hand, and its true attachments, even in man, seem never to have been fully appreciated.

In marsupials it has a sling-like form with a substantial origin from the 'hook' of the hamate, from which it loops radially around the long flexor tendons and then sweeps back deep to them to attach to the capitate and to a minor extent on the base of the third metacarpal (Figs 7.1A, 9.7A, 9.8A). The recurrent portion in these mammals is greatly thickened and includes a bony or cartilaginous nodule which is related to the tubercle of the scaphoid. Outside the confines of this sleeve, with its contained ossicle, lies a synovial-lined channel conveying the tendon of flexor carpi radialis to its primitive insertion on the base of the third metacarpal; the retinaculum does, however, have some minor attachment to the tubercle of the scaphoid, sealing this conduit ventrally (Fig. 9.8A). This arrangement is a quite plausible precursor to the morphology found in more advanced mammals.

In New World monkeys (Fig. 9.5A) the primary ulnar attachment of the retinaculum is to the 'hook' of the hamate but this attachment is prolonged proximally along the pisohamate ligament to reach the pisiform and distally from the hamate as a fibrous flange. On the radial side much of the sling-like arrangement is retained. However, the retinaculum has a strong attachment to the scaphoid tubercle deep to the groove for the flexor carpi radialis tendon and its synovial sheath, and a flimsy attachment to the tubercle superficial to the tendon. More distally there is some attachment to the trapezium (and adjacent prepollex) superficial to the tendon, but no crest or tubercle, only a slight ridge, is found on the trapezium. Deep to the tendon the remaining substantial part of the retinaculum is truly recurrent, and joins the trapezio–capitate ligament (Fig. 6.1) to attach to the capitate. The lower portion of this recurrent part attaches to the base of the third metacarpal and again excludes the flexor carpi radialis tendon from the carpal tunnel. This distal 'free' radial part of the retinaculum together with an expansion from the hamate on the ulnar side, prolong the walls of the carpal tunnel as a tough fibrous sleeve into the palm. The outer walls of this sleeve afford origin to the evolving thenar and hypothenar muscles.

In Old World monkeys (Fig. 9.5B) the retinaculum retains the same essential character and although the attachments to scaphoid and tra-

pezium superficial to the flexor carpi radialis tendon are more substantial, the trapezium is merely faintly grooved by the tendon and bears no real crest or tubercle (Figs 6.6, 6.7), such as is found with a massive retinaculum attachment in hominoids. The elongated hand of cercopithecines includes a more extensive ulnar and radial prolongation of the distal margin of the sleeve-like carpal tunnel.

This latter feature is grotesquely exaggerated in the elongated hands of gibbons (Fig. 9.6A). Moreover, the distal 'free' part of the radial sleeve attaches, not to the third metacarpal, but is largely arrested on the second, which bears a prominent apophysis on its base for this purpose with only a little of its substance continuing on to the third metacarpal. Unlike monkeys, more proximally the retinaculum has a massive attachment to the trapezium superficial to the buried flexor carpi radialis tendon and here the bone presents a prominent crest or tubercle for this insertion. In *Pan* the main features of the heritage of the retinaculum which has been traced above are found, (Fig. 9.6B) with again, as in gibbons, a sturdy attachment of the retinaculum to a prominent crest on the trapezium, but the distal recurrent 'free' portion retains its primitive attachment to the third metacarpal. In *Pongo* the crest on the trapezium is little more elaborated than in monkeys (Fig. 6.9).

Remarkably enough the detailed topography of the human flexor retinaculum has been, at best, sketchily described. Yet all the aspects to be expected from the phylogenetic history detailed above are clearly apparent. On the ulnar side the retinaculum arises from the hook of the hamate with a substantial proximal extension along the pisohamate ligament and from the base of the pisiform itself; distally there is a minor fibrous prolongation of the wall of the tunnel projecting from the hook of the hamate (Fig. 9.7B). On the radial side there is a strong attachment to the scaphoid tubercle deep to the flexor carpi radialis tendon (Fig. 9.8B), contained within its sheath of synovium, and a thinner sheet passes superficial to the tendon; the scaphoid tubercle is grooved as in other primates for the tendon although in macerated bones this is scarcely noticeable. Below this the retinaculum is massively attached, superficial to the flexor carpi radialis tendon, to the crest or tubercle of the trapezium but deeply it splits so forming a recurrent lamina. This loops back medially to attach to the capitate, joining the touch ligament passing from the trapezium to trapezoid and capitate. Most distally the retinaculum is essentially free of bony attachments and forms a tough fibrous sleeve-like prolongation of the lateral wall of the tunnel into the palm. The substance of this portion of the sleeve, grooved on its internal surface by the flexor pollicis longus tendon, and related externally to the flexor carpi radialis tendon, converges onto a strong ligament attaching to the base of the third metacarpal (Fig. 9.8B). This prominent structure is unrecognized in modern textbooks but it was figured by Weitbrecht in 1742 as the 'sublime ligament of the middle metacarpal'. As will be seen in Chapter 9 important components of the musculature of the thumb arise from the external surface of this fibrous sleeve, and its ligamentous attachment to the third metacarpal acts as a peduncle tethering the deeper parts of these origins. The underlying sling-like character of the retinaculum, with its deep recurrent portion, was noted as an obvious feature in the human fetus by Lucien (1907).

The metacarpophalangeal joints

The phylogenetic background

The basic mammalian morphology and function of these joints can be seen in the generalized hands of primitive mammals such as the marsupials *Pseudochirus laniginosus* or *Caluromys lanatus* or the insectivore *Tenrec ecaudatus*. The distal articular surfaces are, of course, formed in part by the concave bases of the proximal phalanges, but hinged onto the ventral margins of these are the thick fibrocartilaginous palmar ligaments, which form a major component of the distal (female) articular surfaces. These may be conveniently called glenoid plates which is the term used in French anatomical literature for their human homologues. The palmar surfaces of the glenoid plates are deeply grooved for the long flexor tendons and form channelled guides controlling the trajectory of these tendons at their entrance to the fibrous flexor sheaths of the digits (Figs 9.7A, 9.8A). The thickened raised lateral margins of the

glenoid plates are typically reinforced by ossifications, the so-called digital sesamoids. These ossific nodules are analogous to lunulae, which have been shown by Barnett (1954) to be similar structural modifications strengthening and stabilizing the thickest parts of intra-articular fibrocartilaginous menisci. A similar notion about the role of digital sesamoids as guides for the long flexor tendons has been expressed, at least for the interphalangeal sesamoids, by Wirtschafter and Tsujimura (1961), who likened their function to that of the grooved bridge of a violin, which directs and controls the course of the strings.

The nature of these digital sesamoids has been commonly misinterpreted. It is commonly stated, at least in British textbooks (for example, Jones 1949), that these ossicles are contained within the substance of the tendons of insertion of intrinsic hand muscles, and are thus comparable to other intratendinous sesamoids, but it is difficult to see how this notion could be reconciled with the occasional occurrence of sesamoids at the interphalangeal joints. In contrast, French anatomists (Testut 1904) correctly distinguish between 'sésamoides peri-articulaires' and 'sésamoides intratendineux', the metacarpophalangeal sesamoids being of the former type. Despite this it is true that, phylogenetically, certain of the intrinsic muscles may secondarily become attached to the surface of these sesamoids as for example in the case of the human thenar muscles (see Chapter 10).

The glenoid plates seem to play a determinant role in the pattern of movement at the metacarpophalangeal joints, for as the digit is flexed, the glenoid plate moves up over the metacarpal head in a sled-like fashion, trailing the phalanx in its wake. The thickened lateral margins, with their embedded sesamoids, act as runners and travel in deep grooves in the metacarpal head, which thus presents a typically fluted appearance. Between the grooves is a prominent palmar crest or beak. The grooves are flanked by raised margins which are prolonged proximally as small tongue-like extensions of the volar surface. The articular surfaces of these rims extend onto the lateral surfaces of the head as flat cartilage-clothed areas over which collateral ligaments, which brace the joint laterally, rub during flexion.

Motion at these joints is fundamentally a hinge action in those primitive hands, with some abduction–adduction permitted in extension. The heads of the second and fifth metacarpals are typically asymmetrical dorsally, that of the second digit being somewhat bevelled off on the radial side and that of the fifth on the ulnar side. Thus, as the second digit moves into extension, it deviates to the ulnar side due to the twist of the metacarpal head and consequent slackening of the radial collateral ligament. This is associated with conjunct rotation in the sense of supination (exorotation). Movements of the fifth digit are the mirror-image of these. Even though the thumb may not be opposable in the true sense in these mammals, it does have some independence of movement. It would seem that the functional import of the modified movement in the marginal members of the remaining four digits is as follows: the digits in these quadruped animals will in this way be aligned together, with the rotation causing their volar surfaces to be flatly applied to the substrate, cancelling out the effect of the transversely arched topography of the palm. The head of the first metacarpal typically shows asymmetry similar to that of the index, with a bulbous enlargement of the radial part of its palmar surface, again facilitating application of its terminal pad directly to the ground.

This basic morphology is retained in prosimians but the crested and grooved form of the metacarpal heads is largely suppressed, despite the presence of sesamoids.

In Old World monkeys the essence of the primitive arrangement is retained and the metacarpal heads are deeply grooved and crested with dorsoradial bevelling of that of the second digit, and apparent paring away of the dorso-ulnar side of the head of the fifth metacarpal. Baboon metacarpophalangeal joints show certain apomorphic features which enable them to cope with their digitigrade mode of locomotion. The bases of the proximal phalanges are scooped out dorsally allowing them to reach a close-packed position at about 90° hyperextension on the metacarpal heads, whose distal weight-bearing ends are then covered by the sesamoid-containing glenoid plates. The heads of the metacarpals of the index and minimus are dorsally bevelled in the usual fashion, and, perhaps even more importantly than in plantigrade animals, this aligns the digits and correctly orientates their volar surfaces so that all

are applied to the substrate below; this is additionally helped by the fact that the distal extremities of the four postaxial metacarpals are almost in line (Etter 1973).

In New World monkeys such as *Cebus*, however, the fluting of the metacarpal heads is again suppressed, despite the presence of sesamoids, as is the asymmetry of the metacarpal heads of the index and minimus. The thumb metacarpal, however, shows an obvious dorsoradial bevelling and the radial margin of the lateral sesamoid groove is swollen into a cam-like formation, which by tightening the overriding collateral ligament during flexion imparts an element of abduction and endorotation to the phalanx. As noted above this is foreshadowed in more primitive mammals. This would seem to be the biomechanical basis of what Napier (1961) called pseudo-opposability, although he seems to have misinterpreted the mechanics of the joint by suggesting that the asymmetry of the head provided for abduction and endorotation in extension.

The ape joints

Pan troglodytes The heads of the metacarpals (Fig. 6.11) (except the first) in this knuckle-walking ape possess, as is well known, dorsal ridges which act as articular stops for the proximal phalanx in hyperextension, but these are variable in development and may be virtually non-existent, especially in young animals. The remainder of the metacarpal heads form relatively uncomplicated ellipsoidal surfaces. The distal articular surface is formed not only by the phanageal base but also by thick glenoid plates. As in the primitive condition these structures are deeply grooved ventrally for the flexor tendons, and are flexibly hinged at the phalangeal bases. In immature specimens sesamoid bones are absent but in adults a sesamoid may be found at the radial side of the joint of the thumb. Retterer and Neuville (1918) similarly noted the usual absence of sesamoid bones in the chimpanzee. The joint is flanked on either side by collateral ligaments having an eccentric origin adjacent to the lateral tubercles on the metacarpal heads, and with bipartite insertions onto both phalanx and glenoid plate. Effectively, the structure and function of the joints are comparable to the situation usually held to obtain in the human

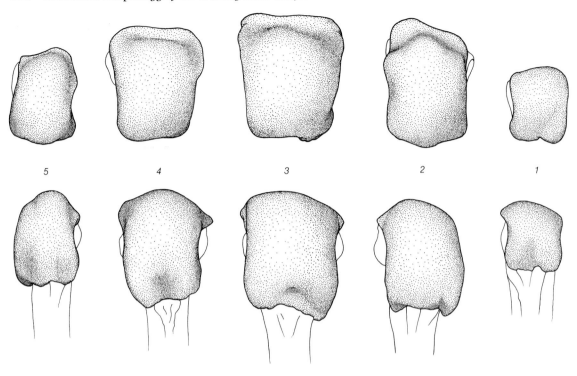

5   4   3   2   1

Fig. 6.11
The metacarpal heads (numbered) from the right hand of a specimen of *Pan troglodytes* viewed from distally (above) and from an angle of 45° distoventrally (below). (From Lewis 1977.)

joints (see, for example, Kapandji 1966): the joints have two degrees of freedom of which flexion–extension is the major one and is essentially a hinge motion; some degree of abduction–adduction is also permitted in extension. The heads of the second and fifth metacarpals show some asymmetry, that of the second being slightly bevelled off dorsoradially whilst that of the fifth gives the impression of having been pared away on the dorso-ulnar aspect. These same metacarpals show slight grooving ventrally on the heads where the margins of the glenoid plates engage, these grooves partially delimiting small marginal elevations. Well developed knuckle-walking ridges obscure the bevelling, and the asymmetry is more obvious when these ridges are poorly developed. The resultant asymmetrical form of the metacarpal heads modifies movement somewhat for extension of the joint of the second digit is accompanied by some ulnar deviation which is facilitated by relaxation of the radial collateral ligament over the

bevelled off part of the head; the converse movement occurs in the fifth digit.

The head of the thumb metacarpal is slightly bevelled off dorsally on the radial side and volarly presents a radial elevation, whose functional import will be discussed later, in the discussion of the human joint.

Gorilla gorilla   The morphology is essentially similar to the chimpanzee.

Pongo pygmaeus   No knuckle-walking ridges are, of course, found on the metacarpals of the orangutan and the form of the metacarpal heads is quite comparable to those immature specimens of *Pan* or *Gorilla* in which knuckle-walking ridges are scarcely apparent. Again the head of the second metacarpal shows dorsoradial bevelling, and that of the fifth shows dorso-ulnar bevelling with similar functional implications. The head of the thumb metacarpal is also dorsoradially bevelled and presents a radial elevation volarly.

Hylobates lar The heads of the metacarpals in this species have a morphology more in keeping with simple hinge joints and asymmetry is scarcely detectable on the head of the second metacarpal although it is rather more evident on that of the fifth. The thumb metacarpal, as in the previous species, shows some asymmetry and a radial elevation volarly.

The human joints

Textbook accounts invariably treat the human metacarpophalangeal joints as uncomplicated and simple ellipsoidal joints with two degrees of freedom. Following Fick (1911) it is usually added that radial and ulnar deviation are restricted in flexion due to the obliquity of the collateral ligaments, which pass from their dorsal attach-ments at the lateral tubercles (and depressions in front of these) to divergent attachments onto the phalangeal base and the glenoid plate.

Landsmeer (1955) refined these functional ideas and introduced the idea that rotation plays a significant part in normal movements. He suggested that ulnar deviation (his 'ulnar abduction') is freer than radial deviation (his 'radial abduction') and that the former movement is accompanied by supination (exorotation) of the finger and the latter by pronation (endorotation). According to him this differential freedom resulted from contrasting obliquity of the radial and ulnar collateral ligaments: the radial ligaments being more oblique (or in the case of ring and little finger, having a more distal metacarpal attachment) are rendered more lax by rotation than are those on the ulnar side.

The heads of human metacarpals, particularly the second and fifth (Figs 6.12, 6.13), are markedly asymmetrical. Oddly enough, Landsmeer

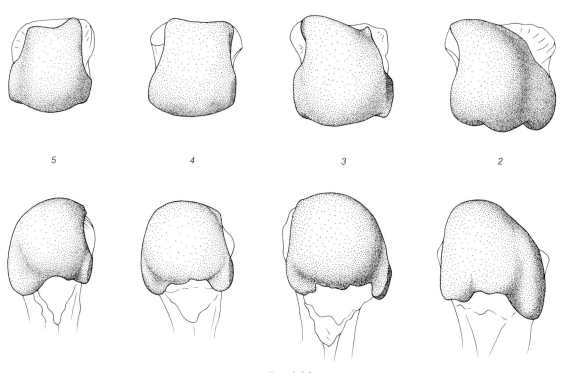

5   4   3   2

Fig. 6.12
The heads of the metacarpals of the index (2), medius (3), annularis (4) and minimus (5) of the right hand of *Homo sapiens*, all viewed from standardized positions: above, from directly distally; below, from an angle of 45° distoventrally (the flattened distal part of the dorsal surface of the shaft being used as a plane of reference). (From Lewis 1977.)

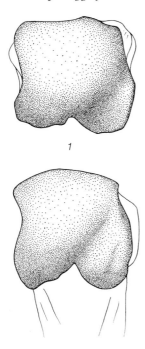

Fig. 6.13
The head of the metacarpal of the thumb (*1*) from the right hand of *Homo sapiens* viewed from distally (above) and from an angle of 45° distoventrally (below). (From Lewis 1977.)

(1955) failed to appreciate the fact that the fifth is virtually a mirror-image of the second, a feature which is clearly not consistent with his suggestion of freer ulnar deviation in all digits. However, his accurate anatomical study served the purpose of focussing attention on rotation in these joints; it appears, however, that this plays a wider role than that of a mere ancillary factor increasing deviation.

The distal (female) articular surface of each joint consists not only of the oval, concave area on the phalangeal base but an integral and functionally very important part is the so-called palmar ligament, the fibro-cartilaginous glenoid plate (Fig. 6.14). This glenoid plate is flexibly hinged to the phalangeal base and only loosely attached proximally to the metacarpal where there is a ballooning synovial pouch (McMaster 1972). Laterally also synovial-lined recesses protrude from each side of the joint between the collateral ligaments and the sides of the metacarpal head (Wise 1975). Dorsally, a small marginal meniscus

may protrude into the joint from the thin dorsal part of the fibrous capsule, which is separated from the overlying extensor apparatus by a bursa, which itself may communicate here with the joint cavity.

The proximal (male) articular surface shows considerable difference in form in the various digits (Figs 6.12, 6.13). At one extreme is the index metacarpal where the articular surface consists of a central segment which is convex and follows a spiral course dorsally, deviating to the ulnar side so that the head is effectively bevelled off dorsally and radially. Confluent with either side of the volar end of this segment, but delimited by shallow grooves, are two additional articular components: a small convexity at the ulnar side and a large bulbous protuberance radially (Fig. 6.14C).

As the index finger is extended (Fig. 6.14A) the proximal phalanx inevitably deviates to the ulnar side on the twisted metacarpal head and becomes exorotated (supinated). The form of the articular surface naturally guides it into this position, and the movement is facilitated by the dorsolateral excavation of the head which affords free passage, and ensuing laxity, to the radial collateral ligament. The glenoid plate takes up its position on the distal and volar part of the central area of the articular surface. The ulnar deviation of the phalanx determines that the groove for the flexor tendons in the glenoid plate becomes so orientated that the tendons must enter the fibrous flexor sheaths of the digit along a markedly angled course.

When the finger moves into full flexion (Fig. 6.14B) the base of the phalanx rides up the spiral articular area, preceded by the glenoid plate which eventually considerably overrides the proximal margin of the articular surface, due to the laxity here of its capsular attachment (Gad 1967; McMaster 1972). The thickened margins of the glenoid plate, which may be reinforced by digital sesamoids, fit into the grooves at the posterolateral corners of the central area. As this motion progresses the broad radial collateral ligament becomes progressively tensed over the swollen lateral articular cam-like projection and this has the effect of rotating the phalanx in the sense of pronation (endorotation) and of pulling it from an attitude of ulnar deviation towards radial deviation. The ulnar collateral ligament is similarly

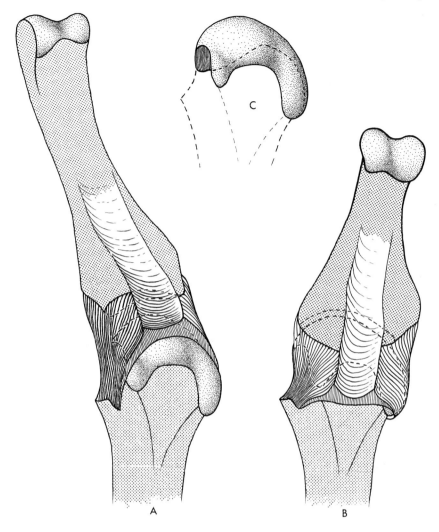

Fig. 6.14
A diagram illustrating the mechanics of the movements at the metacarpophalangeal joint of the right index finger of *Homo sapiens* reconstructed from radiographs of a ligamentous preparation in which salient features—the articular margin of the metacarpal head and the flexor groove in the glenoid plate—were outlined with metal markers. A, the joint in full extension; B, in full flexion; C, the form of the 'isolated' articular surface of the metacarpal head. (From Lewis 1977.)

stretched over the comparatively small medial articular cam. The entry of the long flexor tendons into the fibrous sheath is now no longer angulated and having entered the palm from the carpal tunnel they pursue a straight course into the fibrous flexor sheath of the finger. In executing this movement the mechanical axis of the proximal phalanx has described an arcuate course over the metacarpal head and this is associated with conjunct rotation, as one would expect from the basic principles of joint mechanics enunciated by MacConaill (1953) and MacConaill and Basmajian (1969). Hand surgeons appreciate, at least empirically, the nature of this movement, noting that as the phalanx flexes the rotatory component of the movement, about a narrowing cone, increases (Brand 1985).

The head of the fifth metacarpal (Fig. 6.12) has a

form which is the mirror-image of that of the index finger. In all essentials the movements are similarly the opposite of those occurring in the index: at full extension the phalanx is in a neutral position or slightly radially deviated and pronated; in moving into full flexion (Fig. 6.16) it becomes supinated and abducted.

The joint of the ring finger lacks the asymmetrical type of metacarpal head found in the index and minimus and flexion and extension are relatively simple hinge movements. In the middle finger the metacarpal head shows a minor degree of modelling comparable to that of the index, and clearly flexion–extension movements are accompanied by reduced components of the rotation and deviation observable in that digit.

The metacarpophalangeal joint of the thumb shows variable architecture and in at least 10 per cent of cases the metacarpal head may be surprisingly flat distally (Joseph 1951). Usually, however (Fig. 6.13), the metacarpal head shows some dorsoradial bevelling. The articular surface shows two volar prolongations, a flattened area medially and a convex cam laterally, for articulation with the appropriately shaped deep surfaces of the radial and ulnar sesamoids, designated rather aptly by Testut (1904) as 'scaphoide du pouce' and 'pisiforme du pouce' respectively. It is clear that some degree of endorotation and abduction accompany flexion, just as occurs in the index finger. This then is the biomechanical basis for the contribution made by the metacarpophalangeal joint to opposition of the thumb, noted by Napier (1956). As mentioned above this articular arrangement is apparently an ancient mammalian attribute.

The anatomy of grasping in ape and man

The anatomy and function described provide the basis for interpreting the characteristic gripping postures adopted by apes and by man, and this is a necessary prerequisite for assessing likely hand function in fossil hominids.

Ape hands

The attitude assumed by the chimpanzee hand in grasping a cylindrical object (Fig. 6.15) results from the relatively uncomplicated hinge action

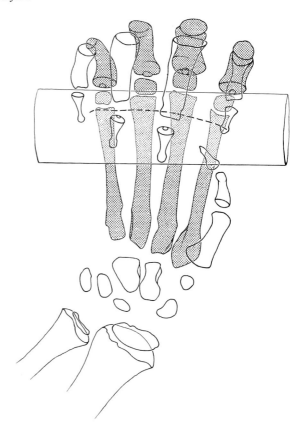

Fig. 6.15
The right hand of a two-year old chimpanzee (*Pan troglodytes*) showing the positions assumed by the metacarpals and phalanges during the power grip. Drawn from a radiograph in which the distal transverse palmar crease (broken line) was indicated by a metal marker. (From Lewis 1977.)

occurring at the metacarpophalangeal joints, and the transverse folding of the palm accompanying this produces a transverse pattern of the skin flexion creases (Biegert 1971). The minor degree of rotation at the joints of the marginal fingers on the slightly asymmetrical heads of the second and fifth metacarpals presumably correctly aligns the fingers. The shortened thumb may, on occasion, play a subsidiary role by applying some counterpressure and it is advantageously disposed to do this by the way in which the trapezium is offset from the remainder of the distal carpal row (Fig. 6.8A). A major factor contributing to this orientation of the trapezium is the acutely wedge-shaped form of the trapezoid which barely attains the anterior surface of the carpus and has its articulation with the capitate, as in all Primates other than man, posteriorly located behind a thick, rounded interosseous ligament uniting the bones.

The gripping mechanism in other apes appears to be similar in essentials. In hylobatids the grasp is even more hook-like (Tuttle 1972). Lorenz (1974) called this gibbon grip the 'comb grip' and noted how the thumb may be 'melted' into the side of the palm during this activity. The flat form of the gibbon palm, a consequence of its flat carpal arch, explains why the simplest of hinge movements at the metacarpophalangeal joints meets the needs of the gibbon grasping action.

Such grips, involving the fingers mainly, function as power grips in apes, and are generally termed a 'hook grip'. As a variant of this, apes may hold objects transversely across the palm using the fingers flexed at the metacarpophalangeal joints—this is the so-called 'digitopalmar grip' (Marzke and Shackley 1986). These authors also note that they may hold cylindrical objects laid obliquely across the palm in what is a 'cradle grip'. Apes, and for that matter other primates, may also handle objects with some precision. To do this they grasp the object between the pad of the thumb and the side of the index finger; this has been called the 'pad-to-side' or 'type I precision grip' by Shrewsbury and Sonek (1986). This may be the explanation for the common primate specialization of elimination of the carpometacarpal ligament on the lateral side of the capitate and the fashioning of a single confluent articulation between the side of the base of the second metacarpal and the capitate. Presumably this would confer the capability for some subtle resilient lateral mobility of the index digit in opposing the pressure of the thumb. Indeed, it is known that baboons, which have at least largely dispensed with this ligament, show a quite high degree of independent control of the index finger in precision handling (Bishop 1964).

Homo sapiens

The gripping posture utilized by man in taking hold of a cylindrical object (Fig. 6.16) illustrates well the role of the joint specializations described above. This attitude is the power grip as described by Napier (1956), and it contrasts markedly with the ape attitude in the way in which the object is

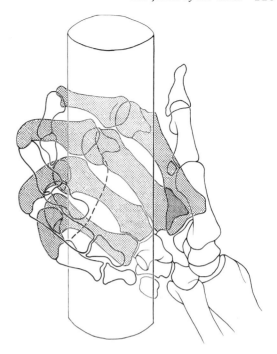

Fig. 6.16
The right hand of *Homo sapiens* showing the positions assumed by the metacarpals and phalanges during the power grip. Drawn from a radiograph in which the distal transverse palmar crease (broken line) was indicated by a metal marker. (From Lewis 1977.)

held obliquely across the palm. This is a consequence firstly of the attitude taken up by the second to fourth metacarpals which are progressively flexed in passing from the radial to the ulnar side. The immobile metacarpal of the medius is in the netural position; the metacarpal of the index is apparently slightly extended and exorotated; the metacarpals of the annularis and minimus (especially the latter) are flexed towards the palm and exorotated. These movements are undoubtedly consequent upon the apomorphic features evolved at the human second and fourth and fifth carpometacarpal joints.

The metacarpophalangeal joints make their unique contribution to the overall posture. The proximal phalanx of the index finger is extended, ulnar deviated, and exorotated, as in Fig. 6.14A; the remaining proximal phalanges are progressively flexed so that that of the minimus takes up the position of flexion, ulnar deviation, and exorotation. An oblique gutter is thus formed across the palm and this is reflected in the manner in which the palmar skin is folded with the creation of a (usually) obliquely disposed distal 'transverse' skin crease (the palmist's 'line of the heart'). Some counter-pressure is exerted on the object at the ulnar side of the hand by the hypothenar fat pad, the skin here being stabilized by the palmaris brevis muscle (Kirk 1924).

On the radial side the thumb exerts counter-pressure. It is here that the changes in the human carpal arch seem to assume particular significance, for, with the trapezium laid back towards the plane of the remainder of the distal carpal row (Fig. 6.5B), the extended, adducted and laterally rotated thumb is well disposed to exert counter-pressure against the cylinder which has the index finger curled around it. In this position the lateral ligament of the carpometacarpal joint must be taut, a fact amply attested to by radiographs which show the wide separation between its attachments; the joint is then in one of its two close-packed positions, as noted by Napier (1955) and Kuczynski (1974). The significance of the elaboration of this uniquely human lateral ligament then becomes apparent.

It seems possible also to suggest a tentative hypothesis explaining the unique human pattern of the articulations between the capitate and trapezoid. The primitive ape disposition of a posterior articulation between these bones (Fig. 6.8A) has been replaced in the human carpus by a completely new diarthrodial articulation anterior to the thick, rounded interosseous ligament between them (Fig. 6.8B). In the power grip (Fig. 6.16), I suggest that considerable compressive stresses may be transmitted from the first metacarpal base to the trapezium and thence across the expanded anterior part of the trapezoid to the capitate. The enlargement of the volar aspect of the trapezoid, readjusting the 'set' of the trapezium, and establishment of a new anterior diarthrosis with the capitate can then perhaps be interpreted as morphological markers of the human power grip.

In the power grip the wrist is typically ulnar deviated and in a neutral position. It has been shown by Wright (1935) that during ulnar deviation about 6° of the movement occurs at the midcarpal joint and the scaphoid is then swung into the long axis of the forearm; it is likely that this further moves the trapezium, and thus the base of the thumb, back towards the plane of the forearm.

The other grip described by Napier (1956) is the precision grip. Here the thumb is opposed to the tips of one or more of the other fingers, the number depending on the size of the object being handled. In contrast to the power grip, in the precision grip the ulnar deviation is corrected and the wrist is extended. The thumb moves at its carpometacarpal joint into opposition by the diadochal sequence of extension and abduction combined with conjunct rotation in the sense of pronation. Clearly then both anterior and posterior oblique carpometacarpal ligaments are then tightened and the joint is in its second close-packed position.

When the more ulnar digits are involved in this grip they occupy positions similar to those attained in the power grip and if the index finger is involved with them (as in grasping a small spherical object with the fingers) it moves toward a position of flexion, combined with pronation and abduction, as in Fig. 6.14B. In grasping a small object between index finger and thumb, the index often is disposed with its metacarpophalangeal joint rather as in the power grip (Fig. 6.14A), and with the interphalangeal joints flexed.

The idea that the power grip is relatively crude and that the precision grip represents a higher pinnacle of locomotor achievement is misleading for the same apomorphic features, although in different combinations, are necessary for the two stylized attitudes. The concept of two distinct grips then is too simplistic. Marzke and Shackley (1986) have developed the following more elaborate scheme for classifying handling repertoires which is especially useful in the evolutionary context (Chapter 10).

The grips used by apes—'pad-to-side', 'hook grip', 'digitopalmar grip'—may all be used by man. The power grip is rechristened as the 'squeeze grip'. The grips formerly lumped together as precision grips are then reclassified. A 'two-jaw chuck' involves thumb and index; it may be 'tip-to-tip', or 'pad-to-pad', or 'pad-to-side' (as in apes). A 'three-jaw chuck' involves thumb and index and middle fingers; it may be 'pad-to-pad' or 'thumb-to-finger' in which the full palmar surfaces of the index and middle fingers are used (in the evolutionary context this is a most significant posture). A 'four or five-jaw chuck' is when the other digits are called into play.

The underlying joint modifications providing for the 'squeeze grip' are just as described above for the power grip. For the 'two-jaw chuck' (with the exception of 'pad-to-side') and the 'three-jaw chuck' the metacarpophalangeal joint of the index takes up the position shown in Fig. 6.14B and this is augmented by congruent movement at the carpometacarpal joint. In the 'three-jaw chuck' the fourth and fifth digits are usually folded out of the way into the palm adopting the same attitude as they do in the power grip (Fig. 6.16).

7

The extrinsic muscles of the hand—the long flexors

The reptilian–mammalian transition

As described in Chapter 3 McMurrich (1903) established that the mammalian long flexors had been derived from a precursor pattern, found in amphibians and reptiles, in which the flexor mass was subdivided into three layers with the superficial one in turn longitudinally subdivided into radial and ulnar marginal sectors and an intermediate portion.

Subsequent authors, with access to more extensive material, have significantly refined the understanding of this basic premammalian groundplan. Thus, Straus (1942) identified the radiale of the carpus as the primitive terminal insertion of the radial sector of the superficial layer, but stressed that the muscle mass in transit also had a robust antebrachial insertion to the radius (Fig. 3.5B). Likewise, he noted that there was a comparable antebrachial insertion (ulna) of the ulnar sector of the superficial layer, even in amphibians, and that the carpal insertion in reptiles was to the pisiform. He believed, however, that the palmaris longus was a mammalian neomorph and that its tendon and the palmar aponeurosis were secondarily derived from subcutaneous mesenchyme. He thus denied that there was any persistent relic of the superficial layer of the amphibian palmar aponeurosis in mammals.

Haines (1950), however, showed that a palmaris longus was in fact differentiated from the intermediate segment of the superficial stratum in some lizards and inserted into a palmar aponeurosis, thereby vindicating McMurrich's supposition that loss of the superficial layer of the palmar aponeurosis in most extant reptiles (Fig. 3.5B) is a secondarily derived condition. Palmaris longus seems therefore to have been part of the muscular

equipment of the primitive reptilian–mammalian stem. Moreover, this same author noted that the intermediate segment in some lizards showed differentiation of a centralis from the other two parts condyloradialis and condyloulnaris, again foreshadowing the mammalian condition.

In this way, it became clear that it is only in the possession of a flexor digitorum superficialis that mammals show any really significant advance. Straus accepted the McMurrich view of derivation of the bellies of this emergent muscle by recruitment of a variable part of the intermediate sector of the superficial stratum, but he balked at the idea of derivation of its tendons from the slips of flexor brevis superficialis in the palm, conceding only that the terminations of the tendons of the one were derived from the comparable tendons of the other. He suggested that the flexor digitorum superficialis had developed new tendons from mesenchyme in the forearm and proximal hand. He reasoned that since certain mammals can have a flexor brevis manus coexisting with a flexor digitorum superficialis and even sharing the same tendon of insertion, then the flexores breves superficiales could not be totally incorporated in the flexor digitorum superficialis, since the two components can coexist.

Haines (1950) distinguished muscle slips which he designated as 'paratendinous intravaginal flexors' from the flexores breves superficiales, with which, however, they inserted; these former slips arise from the flexor tendon plate just as it splits up, but their distinction from the latter muscle slips seems to be scarcely warranted. However, Haines believed that the flexor digitorum superficialis was derived solely from these paratendinous intravaginal flexors which, by some unspecified mechanism, had migrated proximally into the forearm and even to the humerus. Again this was prompted by the belief that a mammalian flexor brevis manus, when it occurs, must be the equivalent of the superficial short flexors of reptiles. One very significant point shown by Haines was that in the lizard *Varanus* the slips of the four short superficial flexors (together with the paratendinous intravaginal flexors if those be separately considered) have a restricted insertion on the penultimate phalanges of digits two to five. This particular reptile shows an incipient reduction in the phalangeal formula from the reptilian (2,3,4,5,3) towards the mammalian one (2,3,3,3,3), as had occurred in parallel in the therapsid ancestors of mammals (Fig. 3.1); significantly the penultimate phalanges, bearing the restricted insertion of the muscle slips, are those resisting reduction and homologous with the middle phalanges of mammals. A persuasive case for identity between the insertions of flexores breves superficiales and the tendons of flexor digitorum superficialis is thereby established.

In fact there seems to be no basis for rejecting McMurrich's broad principle that flexor digitorum superficialis from its first emergence was a forearm muscle and that its progressive evolution involved increasing commitment of the condyloradialis, condyloulnaris and centralis masses to the enlarging new muscle. These particular primordial muscle masses were so named by Windle (1889), as were the radialis (proprius) and ulnaris (proprius). He catalogued their representation in a wide range of mammals and provided the groundwork for McMurrich, although the latter certainly refined ideas about the underlying evolutionary mechanism.

There also can be little doubt that the terminations of the flexor digitorum superficialis tendons are homologous with the split terminations of the reptilian flexores breves superficiales. The mode of elaboration of the intervening palmar tendons is perhaps more controversial. They could be new mesenchymal derivatives or they might be derived from the palmar part of the reptilian flexor brevis superficialis. This latter transformation of muscle into tendon would imply regression of the myoblasts and reconstruction of the connective tissue framework into tendons; in any case, this is not too different from the former alternative. Regardless of which view is correct, and the latter seems to me to be the more likely, this has little relevance to the

charting of the phylogenetic history of the muscle in mammals. The mammalian flexor digitorum superficialis muscle mass then was primitively located in the forearm and phylogenetically increased in importance by recruitment of more and more of the intermediate sector of the superficial layer at the expense of the contribution to the flexor communis digitorum. Where a flexor brevis manus is found, as it is in *Ptilocercus* (Clark 1926), some Rodentia (Parsons 1894*b*), some Carnivora (Windle and Parsons 1897) and even occasionally in man, it must then represent a secondary descent of part of this muscle to the palm, rather than persistence of a primitive condition.

McMurrich made his deductions almost entirely from transverse serial sections, in some cases of fetal material, and thereby overlooked striking features of the three-dimensional organization of the individual bellies of the muscles.

In fact, as will be seen, a basic template seems to have been established for the muscle from its earliest elaboration. Variation on this basic plan has produced a remarkable spectrum of different patterns which, seen in isolation, may appear quite unique, but when the phylogenetic progression is appreciated become readily understandable. Nowhere is this more true than when applied to the highly modified human condition in all its variant forms.

The long digital flexors

These muscles, the flexor digitorum superficialis (sublimis) and the flexor digitorum profundus together with the flexor longus pollicis are the main derivatives of the intermediate sector of the palmaris superficialis and the palmaris profundus of reptiles. The former becomes subdivided into condyloradialis, condyloulnaris and centralis (together with palmaris longus) and the latter into radialis and ulnaris (arising from the indicated forearm bones). The evolutionary history of these two groups of muscles is so inextricably interwoven that they must be considered together, along with the palmaris longus.

The primitive (marsupial) arrangement

The possum *Pseudochirus laniginosus* exemplifies

this arrangement (Figs 7.1A,B). Palmaris longus has a slender muscle belly arising from the medial epicondyle with a long thin tendon inserting partly on a cartilaginous nodule (or even ossicle) developed in the radial part of the flexor retinaculum, and articulating with the scaphoid tubercle. The rest of the tendon radiates into the palm superficial to the retinaculum as a rather flimsy palmar aponeurosis sending prolongations down into all five digits.

At its origin the palmaris longus is flanked on either side by the much more substantial condyloradialis and condyloulnaris. These muscles diverge and the greater part of their substance joins the muscles clothing the forearm bones: the radialis and ulnaris, derivatives of the premammalian palmaris profundus. Hidden between condyloradialis and condyloulnaris is the remaining major derivative of the intermediate sector, the fusiform and quite independent belly of the centralis, similarly arising from the medial epicondyle and similarly terminating by a tendon inserting on the underlying tendinous mass. In effect all five components—condyloradialis, condyloulnaris, centralis, radialis and ulnaris—combine to form a flexor communis digitorum tendon plate, which in turn subdivides into the individual five deep flexor tendons. Ulnaris arises from the upper half of the flexor surface of the ulna, including the olecranon, whilst radialis has a more restricted origin from the middle third of the radius, embracing the insertion of supinator above. The component fibres of the common tendon plate derived from these five muscle bellies are complexly interwoven. Entering the palm the plate divides into the deep flexor tendons for the ulnar four digits and the long flexor of the pollex. The interlacing character of the common tendon makes allocation of a particular tendon to a particular muscle inappropriate. However, the tendon to the first digit (the pollex) is slender and arises from the raised lateral margin of the common plate, and is clearly derived mainly from radialis but so also is the index tendon. All of the profundus tendons are broad and flattened as they traverse the palm and show a strangely twisted appearance. The initially parallel fibres diverge from the midline and pass obliquely dorsally around either margin curving onto the dorsal aspect; this creates a median longitudinal groove and here deeply lying fibres, derived as

above, emerge again to become superficial and diverge. This is just the arrangement described for the human profundus tendons by Martin (1958b), but in man the twisted character is restricted to the more distal part of the tendons where they enter the tunnels formed by the superficialis tendons.

Parting the apposed margins of condyloradialis and condyloulnaris exposes four fleshy muscle bellies, two derived from the inner aspects of each of the two condylar muscles. These comprise the bellies of flexor digitorum superficialis. The upper and lower from condyloradialis provide the perforated tendons to digits three and two respectively; the upper and lower bellies from condyloulnaris are for the tendons of digits four and five respectively. In each case the lower bellies (for digits two and five) have extended their origin slightly to the surface of the underlying emergent common flexor tendon plate. The slender superficialis tendons derived from these bellies pass down in the grooved common tendon plate, those for digits three and four being most superficial, and all enter the palm deep to the flexor retinaculum.

Each pair of tendons for a digit, profundus and superficialis, sharing a synovial sheath, enter the respective digit by traversing the deeply grooved palmar or glenoid plates of the metacarpophalangeal joints, with their supporting sesamoids. Here they are bound down by a strong, glistening fibrocartilaginous loop within the commencement of the fibrous flexor sheaths. These sheaths are prolonged down to the distal phalanges.

Within the sheath the superficialis tendon splits into two halves and each flattened half passes backwards and downwards to encompass the profundus tendon and in doing so undergoes a twist so that the fibre bundles which were originally lateral become medial. These fibres decussate, reinforcing the ventral capsule of the proximal interphalangeal joint and insert together with the uncrossed fibres on the intermediate phalanges. The profundus tendons, whose spiralling fibres reflect the twist of the superficialis tendons, traverse the tunnel and gutter thus created, to reach their insertion on the distal phalanges.

Within the fibrous tunnel, and independent of it, the tendons are tied to the proximal phalanx by two very strong fibrocartilaginous loops or pulleys, a narrow one proximally and a broader one more

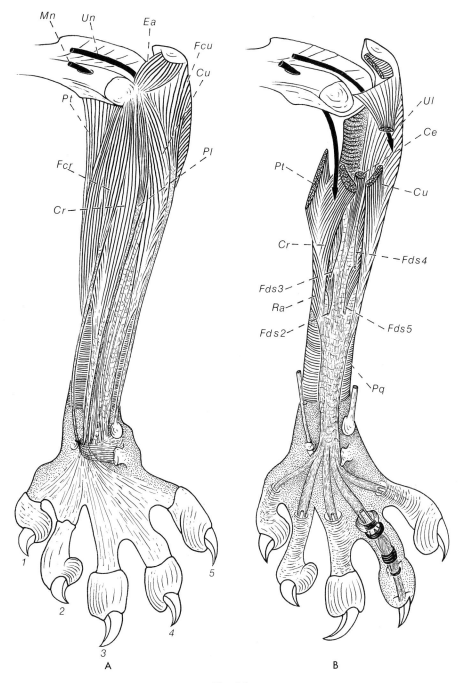

Fig. 7.1

A superficial dissection (A) of the right forearm musculature of *Pseudochirus laniginosus* and the same specimen with the superficial layer removed (B). *Un*, ulnar nerve; *Ea*, epitrochleo-anconeus; *Fcu*, flexor carpi ulnaris; *Cu*, condylo-ulnaris; *Pl*, palmaris longus; *Fcr*, flexor carpi radialis; *Cr*, condylo-radialis; *Pt*, pronator teres; *Mn*, median nerve; *Ul*, ulnaris; *Ce*, centralis; *Pq*, pronator quadratus, *Fds* 2–5, flexor digitorum superficialis tendons to digits 2–5; *Ra*, radialis.

distally (Fig. 7.1B). It is to be noted that these loops are intrasynovial and very substantial and are revealed as shining tough independent structures when the fibrous sheath is opened and peeled away. Opposite the middle phalanx there is a similar small loop. There are no loops worthy of mention at the interphalangeal joints. A single loop ties the long flexor of the pollex to its proximal phalanx and there is a small loop at the glenoid plate (Fig. 9.8A).

The course of the median nerve into the forearm provides a useful guide to the muscle planes and will be used in later discussion. Here it enters the forearm deep to pronator teres after traversing a supracondylar foramen on the humerus, and then courses down the forearm passing lateral to the contribution of condyloradialis to radialis, and between this muscle belly and pronator teres (Fig. 7.1B).

The arrangement of the long flexors in this little marsupial provide an excellent model representing the basic mammalian plan, and the diverse patterns to be described in primates could only satisfactorily be derived from such a precursor arrangement. There are sound reasons for believing that such a pattern was therefore the basic therian arrangement. It would be expected that it would be found in other marsupials, and in particular in the American didelphids. Here adequate data are lacking. The monographic literature on these marsupials (Sidebotham 1885; Coues 1872; Stein 1981; Barbour 1963) is simply not helpful; adequate detail is not provided and the insights to be derived from a wider comparative study are lacking.

In the subsequent descriptions the model provided by *Pseudochirus laniginosus* (Figs 7.1A, B) will be used as the point of departure, aspects of modifications or advance being emphasized.

The prosimian arrangement

This is typified by *Lemur catta* (Fig. 7.2A). There are tantalizing similarities to the marsupial condition, leaving little doubt that the derivation was from a very similar precursor. There is also an overlay of derived features and these will be seen to be transitional towards the condition in the higher primates.

The contribution of condyloulnaris to ulnaris, and so to the common flexor tendon plate is reduced to a slender tendon, derived from a relatively small part of the muscle. This tendon joins the ulnaris just as it becomes continuous with the tendon plate, paralleling here the similar rather thicker tendon of centralis which is joining the radialis. There is a complementary greater contribution of the condyloulnaris to the flexor digitorum superficialis. Two bellies separate out from it to provide the tendons for digits four and five, the belly for the former lying higher. The portion providing the tendon for the minimus may be at least partially digastric, with an intermediate tendon, and its lower belly is then located well down the forearm.

The contribution of condyloradialis to radialis remains substantial, much as it was in *Pseudochirus*. The remainder of this muscle is drawn down into the two flexor digitorum superficialis bellies providing tendons to digits three and two, the former again lying higher.

The common tendon plate again has its fibres interwoven so that the contribution of individual muscle bellies to individual tendons is obscured. The slender long flexor of the pollex again arises from the raised lateral margin of the grooved tendon plate and thus includes part of the substance of radialis, but certainly not all; this particular muscle mass provides in addition much of the substance of the deep flexor tendon of the second digit and probably some of that of the third.

The flexor digitorum superficialis tendons split as in marsupials and the flexor digitorum profundus tendons show a comparable spiral fibre arrangement. Each pair of tendons enters its digital fibrous flexor sheath. Within the fibrous tunnel are tough, shining retaining loops, exactly comparable in position to those featured in the marsupial: a small one at the glenoid plate, two at the proximal phalanx, and one at the middle phalanx. In *Lemur*, however, the one at the middle phalanx is larger, as is the proximal one at the proximal phalanx. Moreover, a small additional relatively insubstantial loop is developed opposite the glenoid plate of the proximal interphalangeal joint, but there is no loop worthy of the name at the distal interphalangeal joint. There is a loop at the glenoid plate of the pollex and one at the proximal phalanx. These loops do not immediately stand free on opening the fibrous sheath as in *Pseudochirus*, since the sheath is somewhat adherent to their surface; however it readily peels away revealing

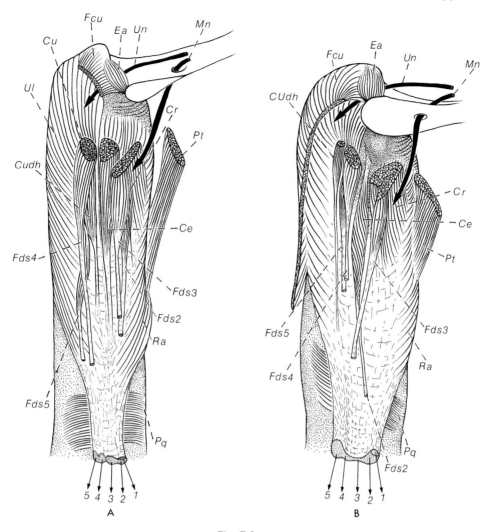

Fig. 7.2
Deep dissections, comparable to that in Fig. 7.1B, of the left forearm flexor musculature in *Lemur catta* (A) and *Cebus capucinus* (B). The common deep flexor tendon plate showing contributions to the individual digital deep flexor tendons 1–5. *Cudh*, contribution of condylo-ulnaris joining ulnaris; other labelling as in Fig. 7.1.

them as largely independent structures with sharply defined proximal and distal margins. A fundamental primate pattern for these loops or pulleys has been established in *Lemur*.

Again the median nerve enters the forearm through a supracondylar foramen and its major branch skirts the lateral surface of the condyloradialis contribution to radialis, lying between it and pronator teres. In *Lemur*, however, the median nerve also gives a small branch passing the other side of condyloradialis (and deep to all the other

condylar origins) to communicate with the ulnar nerve.

The New World monkey pattern

This arrangement, exemplified by that in *Cebus capucinus* (Fig 7.2B) shows advances in pattern from that seen in *Lemur*, but the derived features are easily interpretable as progressive elaborations of the prosimian pattern.

The contribution of condyloulnaris to the ulnaris, and so to the deep common flexor tendon plate, is even further reduced. This portion of the muscle is reduced to a slender spindle shaped belly arising from the medial epicondyle and terminating as a thread-like tendon upon the aponeurotic surface of ulnaris. The remainder of the muscle is employed in providing the tendons of flexor digitorum superficialis to digits four and five, the belly for the latter again lying lower and more deeply.

The contribution of condyloradialis to radialis, in contrast, remains as massive as in *Lemur*. Again the median nerve after traversing the supracondylar foramen of the humerus enters the forearm deep to pronator teres and its main trunk continues down between this muscle and the lateral surface of the condyloradialis attachment to radialis. Again, as in *Lemur* a communication to the ulnar nerve passes on the other (deep) side of condyloradialis. The remaining part of condyloradialis, at its margin adjoining condyloulnaris, is continued as a single flexor digitorum superficialis tendon, that descends to digit three.

Between condyloradialis and condyloulnaris lies an independent muscle which is clearly centralis. It arises from the medial epicondyle as a flattened tendon continuous below with the muscle substance; another tendon arises from the other surface and this is continued as the flexor digitorum superficialis tendon of the second digit. Clearly the condyloradialis portion devoted to this tendon has become amalgamated with the centralis which thus assumes a quite new role. The deep common flexor tendon plate, grooved on its surface for the superficialis tendons, enters the palm and subdivides into the five deep flexor tendons. The slender flexor pollicis longus tendon arises from the raised lateral margin of the gutter and is thus largely derived from radialis; this same muscle mass, however, clearly contributes much of the substance of the tendons for the second and third digits.

The architecture of the profundus tendons, and the specialized retaining loops tying them to the phalanges, follows the pattern established in marsupials, and perpetuated and somewhat enhanced in prosimians, including the new primate acquisition of a loop opposite the proximal interphalangeal joint (Fig. 9.5A).

The Old World monkey pattern

Cercopithecus nictitans (Fig. 7.3A) shows an especially illuminating advance on the basic pattern. Condyloulnaris has relinquished any deep attachment to ulnaris, and the whole of its residual belly is effectively part of flexor digitorum superficialis, but only providing the tendon for the fourth digit.

As in *Cebus*, condyloradialis persists in making a very substantial contribution to radialis, and has similar relationships to the median nerve and its sizeable communicating branch to the ulnar nerve. Also as in *Cebus*, its remaining part contributes the flexor digitorum superficialis tendon to the third digit.

The other two tendons of flexor digitorum superficialis, those for the second and fifth digits, are derived from the terminal two bellies of a trigastric muscle lying more deeply, between the condylar muscles. The upper belly doubtless represents the centralis but below it has apparently appropriated the superficialis bellies proper to the tendons of both the second and the fifth digits. This represents a progression beyond the derived situation in *Cebus*, but here the intermediate tendon clearly signals the tripartite nature of the composite muscle. This arrangement, in fact, as will be seen, remarkably parallels the human condition. It is not suggested that this indicates any ancestral relationship, but it strikingly shows the sort of potentialities for modification inherent in the basic (marsupial) plan.

The common flexor tendon plate gives rise to the long flexor tendons of all five digits. The slender tendon for the pollex arises superficially but the trajectory of most of its constituent fibres indicates derivation largely from the ulnaris; in no way can it be seen as particularly the property of the radialis muscle.

The architecture of the perforated and perforating digital tendons is as before and the retaining loops in the digits (Fig. 9.5B) are disposed as in *Cebus*.

The ape pattern

In the apes fundamental innovations are grafted onto the basic pattern and these have clear implications for the interpretation of the human

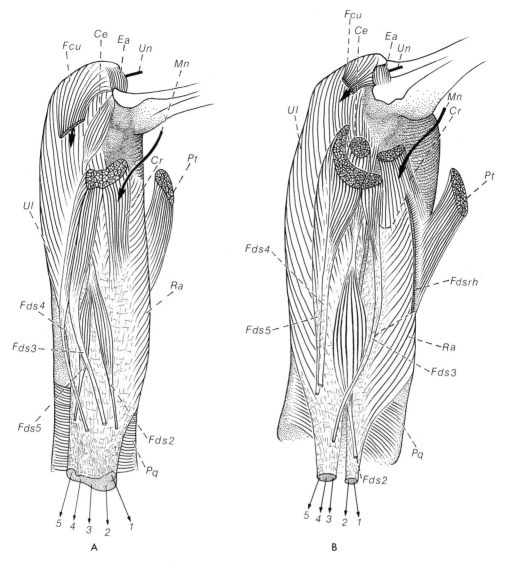

Fig. 7.3
Deep dissections comparable to those in Fig. 7.2, of the left forearm flexor musculature in *Cercopithecus nictitans* (A) and *Pan troglodytes* (B). *Fdsrh*, additional radial origin of flexor digitorum superficialis (condyloradialis portion); other labelling as in Figs 7.1, 7.2.

condition. *Pan troglodytes* (Fig. 7.3B) clearly shows these transitional derived features.

As in Old World monkeys condyloulnaris makes no contribution to the deep common flexor mass, but is totally assigned to flexor digitorum superficialis, ending in the tendons to digits four and five; as would be expected the belly for the former is located higher and more superficially.

A persistent considerable contribution from

condyloradialis to radialis (and so to the common deep flexor mass) is retained. The uppermost of these fibres, however, attach directly to the radius, which is exposed above by some retreat of the origin of radialis down the shaft; these fibres insert with pronator teres and their presence has a wider significance, as will emerge, when the human condition is considered. The remainder of condyloradialis furnishes the tendon of flexor digitorum superficialis to the middle digit, but this part of the muscle is no longer solely condylar for it has acquired an additional new attachment to the exposed part of the radius although apparently in some specimens this new origin may be lacking (Le Double 1897). When present an arch is formed between the two origins bridging across the median nerve, which as usual insinuates itself between the pronator teres and the deep contribution of condyloradialis. In *Pan*, however, the communicator to the ulnar nerve is given off more distally than in monkeys, and the deep 'head' of condyloradialis is no longer embraced by the two subdivisions of the median nerve; the latter seems to be the basic primate arrangement and the former a derived hominoid specialization.

The tendon of flexor digitorum superficialis to the second digit is the termination of a deeply lying, quite independent digastric muscle. By analogy with the monkey arrangements there can be little doubt that the upper belly, having a condylar origin, is the centralis, whilst the lower belly represents the appropriated muscle belly of the second digit tendon, primitively part of condyloradialis. Both centralis, and to some extent condyloradialis, have extended their origins across the elbow joint to include the upper and outer corner of the coronoid process. This ape modification apparently results from the way in which the head of the radius is rather laid back in order to facilitate supination, bringing the muscles emanating from the medial epicondyle into more intimate contact with the coronoid process.

The architecture of the superficialis and profundus tendons is as usual, twisting of the constituent fibres of the profundus tendons starting only in the digits as in man.

In contradistinction to monkeys the radial and ulnar parts of the deep common flexor tendon are fairly independent at the wrist. The former provides the tiny flexor pollicis longus tendon and also the profundus tendon of the second digit; the latter part splits into the remaining three tendons.

The restraining loops within the fibrous flexor sheaths of the digits (Fig. 9.6B) retain the typical primate pattern.

The human pattern

The Primates already described provide the groundwork for an enhanced understanding of the complex human forearm flexors and the frequent variations encountered. Several sources of confusion pervade textbook accounts and the literature.

The deep head of pronator teres has always seemed to be a muscular oddity. Following Macalister (1869), it has been almost universally accepted that it represents a persistent proximal part of the deep pronator sheet of muscle uniting radius with ulna in amphibians and reptiles and thus its morphological affinities are with the pronator quadratus derived unequivocally from the distal part of this sheet. As such it has been considered (Taylor and Bonney 1905) as the serial homologue of popliteus in the hindlimb. The tentative suggestion has been made, however, that it might be part of the radial segment of the superficial layer which has migrated distally over the elbow joint (Straus 1942), that is, of the same derivation as the superficial head. It has been said to be restricted to the apes and man (Hartman and Straus 1933), an odd circumstance indeed if it is truly a representative of a reptilian muscle. Reports of the presence of the deep head in the chimpanzee (for example, Swindler and Wood 1973), or orangutan and chimpanzee (Sonntag 1924) seem to rest entirely on some sort of perceived resemblance to the human condition and do nothing to elucidate the true significance of the muscle.

Even more obscure is the true nature of the so-called 'accessory' or 'occasional' head of the flexor pollicis longus. this is an accessory muscle belly, commonly arising from the medial humeral epicondyle, or anywhere down the medial border of the coronoid process of the ulna, often in close association with the deep head of pronator teres. It has been well illustrated by Dykes and Anson (1944) and occurred in over half the specimens examined by them. Long before that it was clearly figured by Albinus (1734). The 'occasional' head of flexor pollicis longus is not only inappropriately named but, it should be noted (as will soon emerge) that it has no phylogenetic affinity with the so-called 'additional' head (Martin 1958*a*); this is a slip of origin of flexor pollicis longus, which has been well illustrated by Frazer (1946), arising from the *lateral* side of the tuberosity of the ulna and running down the anterior border of supinator.

Both the 'occasional' head of flexor pollicis longus and the deep head of pronator teres, in their various manifestations, are in fact predictable components of human anatomy and their presence does much to emphasize the basic similarity in structure between man and the apes.

In man the deep common flexor tendon plate is completely split longitudinally to form a quite independent flexor pollicis longus, separated from the flexor digitorum profundus. Moreover, the index finger portion is variably split from the latter muscle mass. The flexor pollicis longus cannot simply be equated with the radialis, nor can the flexor digitorum profundus be merely homologized with the ulnaris. It is true that sometimes the whole radial origin may be dedicated to the flexor pollicis longus, as in the specimen shown in Fig. 7.4. More commonly the flexor digitorum profundus (index finger portion) extends it origin from the interosseous membrane onto the radius, and thus includes in its composition some substance which should be designated radialis. This should occasion no surprise for after all the radialis in the chimpanzee also furnishes the tendon to the index. The converse may occur, and part of the ulnaris—that part embracing the radial side of the ulnar tuberosity which receives the insertion of brachialis—may be included with flexor pollicis longus; thus is explained the common so-called 'additional' head of flexor pollicis longus. As noted by Testut (1884) the flexor pollicis longus and flexor digitorum profundus may even be inseparably united. The upshot of all this is that it is clear that the subdivision ending in individual tendons of the muscle mass comprising radialis and ulnaris has been variable, not only between primate species, but even among individuals of the same species. Sonntag (1924) put it rather well by saying that 'the deep flexor mass undergoes vertical cleavage, the line of splitting moving mediolaterally in zoological order'.

The flexor digitorum superficialis of man has reached the ultimate stage in a progressive sequence of primate evolution (Fig. 7.4). The terminal tendon of the centralis is captured not only by the muscle belly of the tendon for the index, as in the chimpanzee, but also by that for the fifth digit. It has already been seen that this situation has been paralleled in at least some of the Old World monkeys (Fig. 7.3A). Moreover, again as in *Pan* the belly of centralis has become adherent to the upper medial corner of the coronoid process

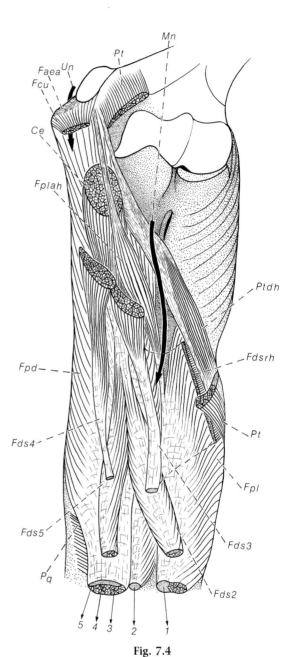

Fig. 7.4
Deep dissection of the left forearm of *Homo sapiens*, comparable to those in Figs 7.2, 7.3. *Ptdh*, deep head of pronator teres; *Fpl*, flexor pollicis longus; *Fpd*, flexor digitorum profundus; *Fplah*, accessor head of flexor pollicis longus; *Faea*, fibrous arch representing epitrochleo-anconeus; other labelling as in Figs. 7.1, 7.2, 7.3.

and thus the flexor digitorum superficialis (sublimis) has acquired some origin from this bony prominence, which has therefore been dubbed the sublime tubercle in human anatomy. The deep trigastric part of the muscle is usually well defined and separate (Le Double 1897) and has been excellently illustrated by Jones (1949a).

The condyloulnaris and condyloradialis have become even more completely committed to continuity with the superficialis tendons of the fourth and third digits. Just as in the chimpanzee the condyloradialis belly for the third digit (medius) has almost always secured a linear body origin from the radius, thus creating the muscular arch spanning the passage of the median nerve down the forearm. It has been seen that the increased dedication of the condyloulnaris to the flexor digitorum superficialis in the higher primates has been matched by a reciprocal reduction in its contribution to the deep common flexor. Only in certain of the New World monkeys does a rudimentary belly regularly retain this termination. In man a similar slip may occasionally be found detaching itself from the belly of superficialis for the fourth digit (condyloulnaris); this is the muscle of Gantzer (accessorius ad flexorem profundum digitorum) which has also been noted as a variant in various New and Old World monkeys and even in *Gorilla* and *Pongo* (Le Double 1897; Testut 1884).

The most striking and uniquely human apomorphic features affect the deep head of the condyloradialis. In the chimpanzee it has been noted that the uppermost fibres of this muscle mass, which may have some attachment to the coronoid process at passage into the forearm, peel away to gain some tendinous attachment to the exposed bone of the radius, in quite close association with the insertion of pronator teres. The more massive part of the muscle joins the flexor pollicis longus, on the other side of the radial origin of flexor digitorum superficialis. As in the whole primate series, the median nerve crosses the radial or superficial side of this muscle mass. In man this deep part of the condyloradialis has been disrupted into a variable number of residual shreds (Fig. 7.4), arising in the tangled tendinous mass extending from the medial humeral epicondyle down to the medial border of the ulnar coronoid process. The commonest of these is the canonical deep head of pronator teres.

Although traditionally described as arising from the medial border of the coronoid, it may possess some humeral origin (Fig. 7.4) and the whole muscle belly may also be rudimentary, fibrous, or default entirely. Having regard to its derivation, it not uncommonly sends a slip to the remaining part of condyloradialis, the belly for the superficialis tendon of the third digit (medius). Often associated with it is the accessory head of flexor pollicis longus (also a 'muscle of Gantzer') which may have a humeral origin or have this arrested on the coronoid, and which may also send a slip to flexor digitorum superficialis. Coexisting with, or replacing this, a further slip may peel away from the condyloradialis part of flexor digitorum superficialis to join the flexor pollicis longus; this is the well known connection between the superficial and deep muscle layers. In accord with their derivation from the deep head of condyloradialis, all of these muscle slips are crossed by the median nerve (Fig. 7.4). A quite simple understanding of the phylogeny of this muscle mass thus unites into one coherent scheme a whole spectrum of muscular variations.

The anecdotal accounts (Sonntag 1924; Hartman and Straus 1933; Swindler and Wood 1973) which attribute a deep head of pronator teres to certain of the apes (chimpanzee and orang-utan and perhaps sometimes in gibbons and gorillas), and which are usually not illuminated by any satisfactory illustrations, have been most misleading. The impression is given that this muscle has made a unique appearance in the hominid clade. Doubtless these authors were influenced by the resemblance of the upper part of the deep head of condyloradialis to the human condition but this takes no account of the ancient and interesting phylogenetic history of the muscle mass.

In some cases the belly for the flexor digitorum superficialis tendon of the fifth digit may have descended further into the palm (Le Double 1897; Testut 1884) arising from the flexor profundus tendons, palmar aponeurosis or ligaments of the carpus. It is then a flexor brevis digitorum paralleling the condition which is normal in some mammals. For reasons given above it is misleading to consider it merely as an atavistic reappearance of the reptilian condition.

It is well known that the other representative of the intermediate part of the superficial muscle stratum, the palmaris longus, may not uncommonly be absent in man, sometimes in the chimpanzee, and usually in the gorilla. In these cases its muscle substance has clearly been subsumed into the flexor digitorum superficialis, which unsurprisingly may provide a tendon of insertion into the palmar aponeurosis.

The spiralling architecture of the human perforating and perforated tendons and their manner of insertion resembles that of *Pan* and other primates; it was well described by Martin (1958*b*). The system of restraining loops for the human long flexor tendons, lying within the fibrous flexor sheaths of the digits, have to some extent lost their individuality, nevertheless, clear hints of the ancestral pattern remain. Both loops related to the proximal phalanx, and that at the metacarpophalangeal glenoid plate, have merged to a variable degree (Fig. 9.7B) and also blended in with the investing fibrous sheath; but clearcut free proximal and distal margins of this whole complex are often apparent on the interior of the sheath. The loop related to the middle phalanx retains a high degree of independence within the fibrous sheath proper. The loop at the proximal interphalangeal joint is little more than an insubstantial thickening of the fibrous sheath itself and that at the distal interphalangeal joint is even more flimsy. In the thumb (Fig. 9.8B) the ancient primate pattern is retained but with minor modification. The metacarpophalangeal loop persists, as does that related to the proximal phalanx but the latter has its basic character obscured by an oblique overlay derived from the adductor pollicis insertion. A flimsy loop lies at the glenoid plate of the interphalangeal joint.

The traditional descriptions in anatomical textbooks of the fibrous sheaths for the flexor tendons as being thick and transverse opposite the proximal and middle phalanges but thinner and cruciate opposite the joints is clearly quite inadequate. Among hand surgeons knowledge of the true anatomy has fared rather better, prompted by the practical problems posed by reconstructive surgery of the flexor tendons, and unrelated to the insights provided by comparative anatomy. All of the retinacular loops or annuli (usually inappropriately called 'pulleys') described above, and traceable through a long mammalian history, are recognized. The nomenclature used for the main annuli

(A1–A5) after Schneider and Hunter (1982) is shown appended to Fig. 9.7B. Rather surprisingly these authors failed to appreciate the essentially dual character of the A1 pulley; this is strange in view of the fact that its midpoint (the weak junction of its two parts) is a common site for the protrusion of synovial derived ganglia in the hand (Angelides 1982). The standard description recognizes the prime significance of pulleys A2 and A4 in reconstructive surgery for their repair is essential to prevent bowstringing of the tendons (Froimson 1982)—no surprises here for the comparative anatomist. The insignificance of pulleys A3 and A5, and even the inconstant nature of the latter, is also on record. The pulley opposite the proximal phalanx of the thumb (Fig. 9.8B), a structure as ancient as the mammals themselves, is named as the 'oblique pulley'; this merely acknowledges the fact that in man it has acquired a masking overlay derived from the adductor pollicis. The varying degree of coalescence, but often with intervening gaps, of the three pulleys opposite the metacarpophalangeal joint and the proximal phalanx (the phylogenetically dual A1 and the strong A2) has been recognized in a further study using a different terminology in which this whole annular pulley complex is designated AP1 (Strauch and De Moura 1985). With comparative insight what would otherwise be a dauntingly complex subject becomes relatively simple and logical.

The radial sector muscles

The radial sector of the superficial stratum has remained relatively conservative, showing only a minor elaboration of the prototypal mammalian pattern, which was already established in reptiles, where this sector inserted terminally on the radiale although with a considerable part of its mass peeling off to an antebrachial insertion. This latter is the forerunner of the mammalian pronator teres (and its superficial head only in man); the remainder is designated as the mammalian flexor carpi radialis.

Pronator teres has shown little change throughout mammalian evolution, although the extent and precise disposition of its radial insertion varies, as documented by Macalister (1869). The muscle arises from the supracondylar ridge on the humerus, and when in the more primitive mammals a supracondylar foramen (Figs 7.1, 7.2) is present, this canal traverses the bone deep to the origin, so that the contained median nerve leaves the foramen deep to pronator teres. In most of the higher primates (including man) where the foramen is lost the median nerve passes around the medial humeral border to pass deep to pronator teres (Fig. 7.3). Not uncommonly of course in man remains of the disrupted medial wall of the foramen are to be found in the shape of a supracondyloid process and a ligament (of Struthers); the pronator teres then arises from these structures, and the median nerve traverses the fibro-osseous canal.

As noted above the origin of the flexor carpi radialis from the medial epicondyle is unremarkable given its derivation from the superficial muscle stratum. In *Pan*, however, in skirting over the radius it has acquired a secondary attachment to that bone via the sheath of pronator teres, in a manner analogous to that affecting the flexor digitorum superficialis (condyloradius portion) in the same species; this has also been noted in the other apes (Le Double 1897). It is not surprising therefore that such an origin should occur as an anomaly in man: the so-called flexor carpi radialis brevis (Effendy *et al.* 1985). The flexor carpi radialis tendon in a generalized hand such as that of the marsupial *Pseudochirus* (Fig. 9.8A) inserts by a small tendon to the scaphoid (which of course here incorporates the radiale) and by a further tendon prolonged down to the base of the third metacarpal. This latter extension has doubtless been acquired by a mechanism similar to that shown in Fig. 3.7A, and traverses an independent synovial lines canal related to the scaphoid and trapezium and excluded from the carpal tunnel by the deep reflected sheet of the flexor retinaculum. As it passes to its third metacarpal insertion it is brought into intimate contact with the anterior trapezio-metacarpal ligament attaching to the second metacarpal. This close relationship between tendon and ligament persists even in man, where the two lie in a deep sometimes synovial lined groove on the base of the second metacarpal. By the interchange here of component fibres between ligament and tendon, the wide spectrum of variation of attachments encountered in man and other primates is readily explained; either the

ligament or the tendon may swap attachments between second and third metacarpals or obtain dual attachment to each of these metacarpals. Similar variations occur in man, and also the tendon may be arrested at the scaphoid or trapezium (Le Double 1897).

The ulnar sector muscles

The ulnar sector muscles have, if anything, been even more resistant to change. The major part of the muscle mass inserts on the pisiform, just as it did in reptiles, as the mammalian flexor carpi ulnaris. The antebrachial insertion of the sector is simply renamed as the mammalian epitrochleo-anconeus. This muscle bridges the entry of the ulnar nerve into the forearm and has been illustrated here in the phylogenetic sequence from marsupial to chimpanzee (Figs 7.1, 7.2, 7.3). Some published accounts imply that it is inconstant in the apes, but it is equivocal whether this is a fact, or a consequence of the vagaries of observation and description. Moreover, it sometimes occurs in man and as described in detail by Le Double (1897) there can be little doubt that when absent it is represented by the fibrous arcade spanning the interval between the epicondylar and olecranon heads of the flexor carpi ulnaris (Fig. 7.4). A variable extension of the origin of the muscle down the posterior border of the ulna is seen in different primate species (Fig. 7.2B) and the aponeurotic origin here in man is well known.

Comparative anatomy gives no support for the notion, which is part of the folklore of human anatomy, that the pisiform is a mere sesamoid bone developed within the substance of the flexor carpi ulnaris tendon. On the contrary, there is every indication that the pisiform is a primordial carpal element, and indeed it has consistently functioned in mammalian evolution as a sizeable weight-bearing element in the 'heel' of the hand. From marsupial to man it has been distally tethered by the massive pisohamate and pisometa-carpal ligaments (Figs 6.5, 6.6, 6.7, 6.9). Its reduction to a mere bony nodule in man, means that it has secondarily come to be engulfed to some extent by the flexor carpi ulnaris tendon and the distal ligamentous bands, and so mimics a true sesamoid.

The deep pronator layer

In mammals this layer is restricted to the more distal part of the forearm, uniting radius and ulna, as the pronator quadratus. Its extent up the forearm shows considerable species variations (Macalister 1869) but otherwise its evolution has been unremarkable.

8

The extrinsic muscles of the hand—the extensors

The reptilian–mammalian transition

The basic groundplan of the extensor musculature of the forearm and hand was established by Brooks (1889) and refined in points of detail, following further investigation, by Howell (1936) and Haines (1939). In its essentials this muscle group has a primitive bilaminar arrangement with the superficial layer subdivided longitudinally into three sectors and the deeper layer running obliquely downwards from the postaxial borders to enter the dorsum of the hand. This pattern is broadly similar to the opposing flexor musculature on the reverse aspect of the limb (although there is no extensor counterpart to the deepest pronator layer of the flexor surface); moreover, as will be seen in Chapter 15, the pattern is strikingly similar to that of the primitive cruropedal extensors in the hindlimb.

The primitive extant reptile *Sphenodon punctatum*, the tuatara, conserves a pattern in this muscle group (Fig. 8.1) which cannot be too far removed from that which characterized the reptilian precursor of the mammals. Other extant reptiles, and also amphibia, retain many of the hallmarks of this same basic arrangement. The authors referred to above have used different terminology for the various muscular components. Brooks (1889) directly applied the nomenclature of human anatomy, but as noted in Chapter 3, an approach which assumes precise homology of muscles across wide taxonomic boundaries is unwise, and can even be self defeating; it may obscure the real evolutionary processes which have been involved. Again Haines (1939) excessively subdivided certain of the major components in the belief that the partitioning which he described foreshadowed the mammalian condition. Howell (1936), however, followed by

Straus (1941*a*), used non-committal descriptive terms for the pre-mammalian subdivisions of this muscle group; the same approach is followed here, although the terms used are different (Howell's are given in parentheses).

The superficial stratum (brachio-antebrachial) arises from the lateral epicondyle of the humerus. Its radial sector (extensor humero-radialis) has an extensive insertion to the radius, but its most superficial part reaches the carpus and terminates on the radiale. The ulnar sector (extensor humero-ulnaris) has a comparable extensive forearm insertion onto the ulna, but the distal prolongation reaches the pisiform and base of the fifth metacarpal. The intermediate sector (extensor humero-dorsalis) inserts by four slips into the bases of a pair of adjacent metacarpal bones, the first to the first and second metacarpals, and so on. The whole of the superficial stratum (including the ulnar insertion of the ulnar sector) is quite separate from the muscle triceps.

The deep stratum (the antebrachio-manual or ulnocarpal extensors) takes origin from the ulna and below that from the intermedium and ulnare in the carpus. From the uppermost part of this origin a muscle belly, called by some the supinator manus, descends obliquely to insert at the radial border of the hand into the base of the first metacarpal. Five bellies radiate from the remainder of the ulnar and carpal origin to furnish extensor tendons to all five digits, inserting on the phalanges as far as the distal ones. The tendons of the intermediate sector reach the metacarpal bases between these five short extensor bellies and send aponeurotic slips onto the surface of the bellies.

The five short or deep extensors of the stratum profundum are joined by metacarpal heads (manual-metacarpal extensor series or dorsometacarpales) arising from the dorsum and sides of the

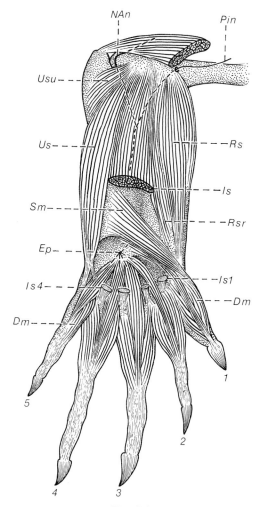

Fig. 8.1

The extensor aspect of the right forelimb of *Sphenodon punctatum*, redrawn after Brooks (1889) and Haines (1939). Superficial muscle stratum: *Rs*, radial sector; *Rsr*, deep radial insertion of radial sector; *Us*, ulnar sector; *Usu*, ulnar insertion of ulnar sector; *Is*, intermediate sector (cut) and its metacarpal insertions (first, *Is1* and fourth, *Is4*). Deep muscle stratum: *Sm*, the so-called supinator manus; *Ep*, the extensores profundi or short extensors of the first to fifth digits; *Dm*, the first and last of the series of dorsometacarpales. *Pin*, posterior interosseous nerve (n. extensorius cranialis); *NAn*, nerve to anconeus (n. extensorius caudalis).

metacarpal bones. These are arranged in pairs, one either side of each short extensor, except in the fifth digit where the ulnar head is lacking.

Some very similar arrangement of this muscle group must have been present in the reptilian precursors of the mammals.

The primitive marsupial arrangement

Pseudochirus laniginosus (Fig. 8.2) can be taken as a specimen characterizing this pattern. Essentially similar arrangements, differing only in minor details, have been described in the marsupials *Didelphys virginiana* (Haines 1939; Stein 1981) and *Trichosurus vulpecula* (Barbour 1963).

Superficial layer, radial sector

Unlike the condition in the primitive tetrapod limb this sector shows clearcut subdivision into four individual muscles with independent insertions: brachioradialis, the extensores carpi radialis longus et brevis, and supinator.

The brachioradialis has an extensive origin from a crest on the humerus above the lateral epicondyle and inserts below to the scaphoid; in effect then the terminal carpal insertion of the radial sector to the radiale in primitive tetrapods is retained, for the radiale is incorporated within the marsupial scaphoid.

Closely associated with this muscle and arising from the lateral humeral epicondyle is the belly of extensor carpi radialis subdividing below into two tendons, longus and brevis. The former inserts to the base of the second metacarpal and the latter to the third. This part of the radial sector has thus had its terminal insertion prolonged more distally in mammals.

All of these three muscles are clearly derived from that part of the primitive radial sector with a carpal insertion. The part with a forearm insertion becomes the supinator. It has a tendinous origin from the epicondyle and immediately below this invests a sizeable sesamoid bone which forms an integral part of the annular ligament of the superior radioulnar joint, and thus articulates directly with the head of the radius. Below this the muscle expands to insert into the upper third of the shaft of the radius. There is no ulnar origin.

The interpretation of the derivation of the mammalian radial sector muscles given above is essentially in accord with the views of Brooks (1889), but other conflicting opinions have been expressed. Haines (1939) described three parts to the radial sector in primitive tetrapods—superficialis, profundus and intermedius—the first characterized by an insertion on the radiale and the other two onto the radius. He homologized the first

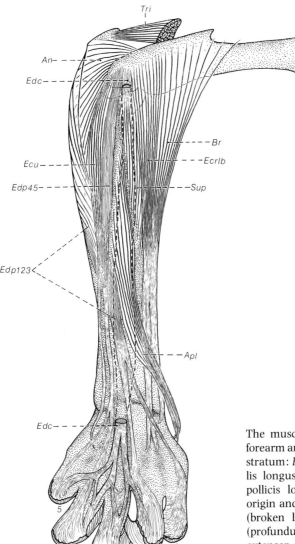

Fig. 8.2
The musculature of the extensor aspect of the right forearm and hand of *Pseudochirus laniginosus*. Superficial stratum: *Br*, brachioradialis; *Ecrlb*, extensor carpi radialis longus and brevis; *Sup*, supinator; *Apl*, abductor pollicis longus; *Edc*, extensor digitorum communis, origin and insertion, intervening muscle belly removed (broken line); *Edp123*, extensor digitorum proprius (profundus) of first, second and third digits; *Edp45*, extensor digitorum proprius of fourth and fifth digits; *Ecu*, extensor carpi ulnaris; *An*, anconeus; *Tri*, triceps.

with the mammalian extensor carpi radialis longus et brevis, the second with supinator and the third with brachioradialis. The subdivision seems artificial and the scheme of homologies rests largely upon the probably erroneous belief that an insertion into the radius for brachioradialis was primitive.

Another discordant opinion has had a much more confusing effect. Howell (1936) believed that the supinator represented the uppermost fibres of the deep lamina (part of supinator manus) and thus had phylogenetic affinities with the abductor

pollicis longus. The appearance in the human forearm may make this a tempting proposition but the close association of the two muscles in man is secondary to the acquisition of a large ulnar origin by supinator; the theory gains little support from the broader comparative perspective.

Superficial layer, ulnar sector

The anconeus is derived from the part of this sector having a forearm insertion, and the extensor carpi ulnaris from the part inserting on the manus.

In marsupials the anconeus (also known as anconeus lateralis to distinguish it from its flexor counterpart, the epitrochleo-anconeus) arises from the lateral epicondyle and has an extensive ulnar insertion, reaching halfway down the shaft. Its uppermost fibres inserting on the olecranon are contiguous with the triceps muscle terminating there. This is clearly a secondary feature because the counterparts of the two muscles are quite independent in *Sphenodon* (Fig. 8.1). Those familiar with the anatomy of lower tetrapods (Haines 1939; Brooks 1889) have stressed the primitive separate identities of the two muscles. Howell, (1936), however, believed that anconeus was a subdivision of triceps, as did Straus (1941*a*), and this has been an almost universal belief among human anatomists, who have clearly been influenced by the peculiar way in which it shares its nerve supply with the triceps, by the nerve to anconeus emerging from that muscle. This 'evidence' crumbles, however, when it is realized that the nerve to anconeus of mammals is merely the attenuated remnant of a branch (sometimes called n. extensorius caudalis) which enters the forearm to join the posterior interosseous nerve (n. extensorius cranialis) in more primitive tetrapods (Figs 3.1, 8.1) and participate in the nerve supply of the forearm extensor musculature.

The remainder of the ulnar sector forms the extensor carpi ulnaris which typically in mammals retains its primitive tetrapod insertion to the base of the fifth metacarpal. In at least some marsupials, however, a derived feature is added to this basic mammalian (and reptilian) insertion: a prolongation from the tendon traverses the palm as far as the second metacarpal. This specialized arrangement occurs in *Pseudochirus* and *Caluromys* as described in Chapter 6 (Fig. 6.5) and an apparently similar arrangement has been recorded in *Trichosurus* (Barbour 1963).

Superficial layer, intermediate segment

This segment becomes the mammalian extensor digitorum communis, but the insertion has been moved from the metacarpal bases to the dorsal expansions of the ulnar four digits. It has usually been assumed that this distal migration of insertion was achieved by incorporation of the tendons of the short extensors (Haines 1939; Straus

1941*a*), since it was noted that even in *Sphendon* aponeurotic slips of the metacarpal tendons of insertion of the intermediate sector (humerodorsalis) were prolonged onto the bellies of the short extensor (Brooks 1889). This theory, however, fails to explain adequately the actual anatomical findings. Howell (1936) proposed, as an alternative, the idea that the humerodorsalis tendons relinquished their metacarpal attachments and reached the digits by seizing on, or modifying, the dorsal fascia of the manus, this process being independent of any merger with the short extensors. There is little doubt that in broad principle Howell had reached the correct conclusion.

In *Pseudochirus* (Fig. 8.2) the tendons of extensor digitorum communis are incorporated in the deep fascia on the dorsum of the hand and together with it form a clearly defined layer overlying the subjacent tendons of the short (deep) extensors. Indeed it is almost impracticable to characterize individual communis tendons, for they form an anastomosing web of fibrous tracts traversing the deep fascia. Such an evolutionary mechanism for transfer of a tendinous insertion, by utilizing fascia, is quite in accord with the principles of myological evolution (Fig. 3.7A). As will be seen (Chapter 15) a comparable mechanism has operated in the hindlimb and here Bardeen (1906) has shown in an embryological study that the extensor digitorum longus tendons are initially attached to the metatarsals but later extend towards the digits as an initially undivided tendinous sheet.

Deep layer

Few would quibble with Brooks' (1889) contention that in mammals this stratum of short or deep extensors has shown a proximal migration onto the forearm, forsaking all carpal origin. From it is derived the abductor pollicis longus and the deep (proprius) extensor tendons of the digits. One consequence of this ascent is that the radially emerging tendons of the deep stratum—abductor pollicis longus and the deep extensor of the pollex—are interleaved between the muscles of the radial and the intermediate sectors.

The origin of the abductor has climbed high onto the radius, to become contiguous with the insertion of the supinator, and it retains only a small origin below from the ulna, being excluded from the bone above by the insertion of anconeus. Its tendon divides into two, one inserting into the prepollex, with some spread onto the adjoining trapezium, and the other onto the base of the first metacarpal (Fig. 9.8A).

The next part of the deep extensor, providing the proprius tendons for the first, second and third digits has an origin prolonged deep to extensor carpi ulnaris and high up the posterior border of the ulna on the other side of anconeus.

The final part of the deep extensor, providing the proprius tendons to the fifth and fourth (and a slip to the third) digits, which in reptiles arose low from the carpus, has an origin which has ascended to the humerus. This identification of the muscle follows the interpretation of Brooks (1889). In effect the more ulnar part of the deep extensor has ascended as dual prolongations lying either side of the ulnar segment muscles, extensor carpi ulnaris and anconeus, which remain located in the interval between them. Of course, as described in Chapter 3, the actual mechanism of ascent has presumably entailed recruitment of myoblastic material from the superficial stratum, and most likely its ulnar sector. The realistic approach to homology, however, dictates that these upwardly prolonged muscle bellies should still be considered as ascended deep extensors, and indeed the disposition of their tendons confirms their true nature, for these tendons occupy the deep plane on the dorsum of the manus.

The true nature of the deep extensor muscle for the fourth and fifth digits (the extensor digiti quinti et quarti proprius), has been a subject of controversy, and the resulting confusion has greatly clouded understanding of the whole extensor muscle group. The muscle has been considered to be a mammalian neomorph and as such has been called either extensor digitorum lateralis or extensor digitorum ulnaris. As such it has been considered to be a derivative of the superficial muscle stratum, split off either from the intermediate (humerodorsal) sector (Ribbing 1907; Howell 1936) or from the ulnar (humero-ulnar) sector (Haines 1939; Straus 1941a); indeed, Lewis (1910) described its development from the anlage of the superficial layer. This misunderstanding seems largely to result from misapplication of the principles determining muscle homologies.

There are no detectable remnants of the dorso-metacarpales in marsupials.

The monotreme arrangement

Somewhat surprisingly, the arrangement in the monotremes *Echidna* (Haines 1939) and *Ornithorhynchus*, (Brooks 1889) is strikingly similar to the marsupial plan, with only a few trivial modifications. The extensor digitorum communis is similarly prolonged down the digits, but unlike the generalized marsupials also furnishes a tendon to the pollex. Appreciation of the mechanism which has transferred insertion from the reptilian metacarpal bases to the digital extensor expansions and so to the phalanges in mammals, makes the occasional elaboration of a pollicial tendon not unlikely; indeed, it occurs normally in the koala and occasionally as an anomaly in man (Le Double 1897). The monotreme extensor carpi ulnaris has acquired some additional ulnar origin and sends a tendinous slip down the fifth digit, but both of these derived features are also paralleled in some primates and the latter may even be found in man (Fig. 8.6D). In other features, the monotreme muscle group shares with marsupials a derived pattern of segmentation and migration which exhibits a significant advance in organization over the reptilian condition. This pattern must clearly have emerged early in mammalian evolution and is obviously relevant to current views on the origin of the monotremes discussed in Chapter 2.

The primate arrangement

Prosimian primates and monkeys possess forearm extensors which are essentially arranged according to the basic therian plan, which in its turn is readily derived from the marsupial pattern (Fig. 8.2); there are, however, a few relatively minor apomorphic features grafted onto this pattern. In contrast to this conservatism, the hominoids show an overlay of significant derived features, which are of particular relevance to the interpretation of the human condition. Important studies by Straus (1941a,b) have provided invaluable data, even if his interpretations are sometimes unacceptable.

The monkey pattern

The basic primate, and thus therian, morphology is well shown by a New World monkey such as *Cebus* (Fig. 8.3A) although most Old World monkeys might also have been chosen as a type specimen. Some individual species show particular specializations but the essence of the plan seems to be universal.

Superficial stratum, radial sector   The brachioradialis shows an altered insertion. In marsupials its flattened tendon, passing to an insertion onto the scaphoid, is bound to the rounded lower extremity of the radius by the abductor pollicis longus (Fig. 8.2). There is little doubt that a carpal insertion is primitive, it is merely a persistence of the reptilian terminal insertion of the sector to the radiale, and is also found in monotremes. Typically in Therians the abductor pollicis longus tendon is deeply embedded in a groove in the radius, which seems to have cut short the transit of the brachioradialis tendon to the carpus: its insertion is left arrested at the ridge forming the proximal margin of the groove. On general principles one might expect that the distal 'severed' part of the tendon would persist as a ligament. In *Lemur*, in fact, such is the case and a sizeable lateral ligament of the radiocarpal joint, connecting radial styloid to scaphoid (radiale), apparently represents the terminal part of the tendon; in higher Primates, however, this ligament has withered to a scarcely distinguishable capsular thickening. The extensor carpi radialis longus and brevis, lying more medially, are not subject to the same pressure from abductor pollicis longus and pass to unmodified insertions on the second and third metacarpal bones. As will be seen later, very occasionally in monkeys the long carpal extensor has gained an additional attachment to the base of the first metacarpal bone.

As in *Pseudochirus* the supinator has its main origin from the humerus and these tendinous fibres merge with the surface of the annular ligament of the superior radio-ulnar joint. At this concentrated confluence of tendinous and ligamentous tissue, in at least some specimens of *Cebus*, the annular ligament shows a conspicuous cartilaginous or fibrocartilaginous thickening (in effect, a hemisesamoid), recalling the sesamoid

bone found in this situation in *Pseudochirus*, which was probably a primitive mammalian feature. A sesamoid 'included in the tendon of origin of supinator' has been recorded in certain mammals (pangolins) and even as an anomaly in man (Le Double 1897) and a comparable ossicle to that of *Pseudochirus* exists in certain rodents (Kesner 1986). The higher primates, however, show as an apomorphic feature, the emergence of an ulnar head of origin of supinator, although even in *Lemur* a little bundle of the posterior fibres of the muscle may arise from the ulna. In monkeys this origin is more sizeable, but still quite restricted (Fig. 8.3A), and this new origin brings the muscle into close contiguity with the abductor pollicis longus.

Superficial stratum, ulnar sector   Extensor carpi ulnaris is unremarkable, although it quite often has a secondary ulnar head of origin, gained in its passage down the forearm. Anconeus, however, has a markedly restricted insertion which has retreated up the ulna occupying merely the olecranon. This has had repercussions on the arrangement of the deep muscle layer.

Superficial stratum, intermediate sector   The extensor digitorum communis arises from the lateral epicondyle and its tendons, as would be expected from their mode of derivation, tend to be united into a matted aponeurotic mass on the back of the hand, which in turn is embedded within the investing deep fascia. As the digits are approached, however, the individual tendons passing to the ulnar four digits separate out and are here quite intimately associated with the fascia covering the dorsal interossei, but no dorsometacarpals are discernible. However, an extensor brevis manus (the occasional mammalian representative of the reptilian muscles) has been reported in the spider monkey, *Ateles* (Straus 1941a).

Deep stratum   Monkeys retain a full complement of deep extensors, and just as in *Pseudochirus*, their origins have migrated far up the forearm. Abductor pollicis longus has extended its origin above from the radius across the interosseous membrane onto the ulna, where it abuts against the lower border of supinator. This creeping expansion of the upper origin of abductor pollicis longus has

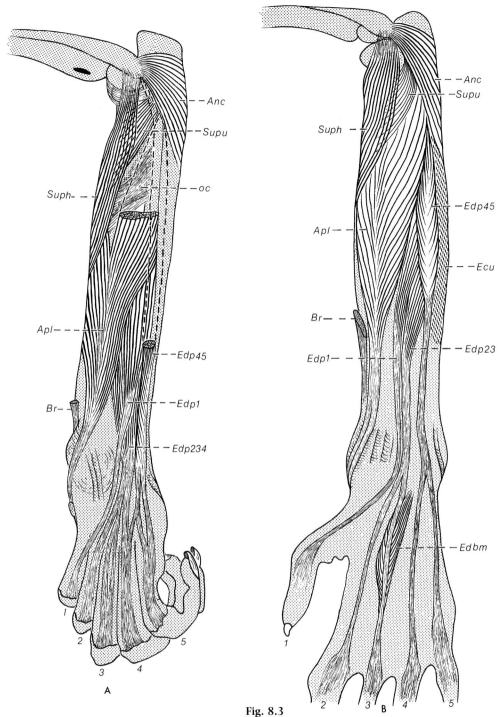

Fig. 8.3

The deep stratum musculature, together with supinator and anconeus, of the left forearm and hand in *Cebus capucinus* (A) and *Pongo pygmaeus* (B). *Supu*, ulnar head of supinator; *Suph*, humeral head of supinator; *oc*, oblique cord; *Edbm*, extensor digitorum brevis manus. The muscle belly of *Edp4*5 has been removed and is shown by the broken line in A. The origin of extensor carpi ulnaris from the ulna is shown in dark stipple in B. Other labelling is as in Fig. 8.2.

presumably occurred deep to the posterior interosseous nerve. This would explain why this nerve in higher primates lies superficial to abductor pollicis longus (and extensor pollicis brevis in man) but retains its primitive situation deep to the other derivatives of the deep stratum; in *Sphenodon* and monotremes the corresponding nerve (n. extensorius cranialis) lies deep to all of the deep stratum including its uppermost part, the supinator manus (Brooks 1889; Haines 1939). Lying under cover of the uppermost part of abductor pollicis longus, and just below supinator, is an extremely strong ligamentous band, running obliquely downwards and laterally from ulna to radius. It is especially prominent in some specimens of *Cebus* (Fig. 8.3A) but also occurs in other New and Old World monkeys. I have also observed a comparable structure in *Didelphys*, but not in other marsupials. It gives every indication of representing a fibrous transformation of the deepest part of the abductor pollicis longus muscle, and with little doubt is the homologue of the oblique cord (ligament of Weitbrecht) of human anatomy; it will be discussed later in that context. The abductor pollicis longus tendon splits at its termination into two slips, one inserting into the prepollex (with perhaps spread onto the trapezium), and the other to the base of the first metacarpal.

The remainder of the deep stratum in monkeys typically provides the deep or proprius extensor tendons to all five digits. As in marsupials and monotremes, the belly terminating as tendons for digits four and five has 'ascended' so that it arises from the lateral humeral epicondyle, where its belly is in close association with the extensor carpi ulnaris; as noted for the marsupials this extensor digiti quinti et quarti proprius is commonly called the extensor digitorum lateralis or extensor digitorum ulnaris, names reflecting the authors' uncertainty as to its real nature. Retreat of the anconeus insertion up the ulna has restored the remainder of the extensor digitorum profundus to its rightful place adjoining the other most radial part of the deep stratum, the abductor pollicis longus, and arising from the ulna below it. The more modest size of the anconeus means that no longer, as in marsupials, are the extensor digitorum proprius bellies for digits one, two and three left isolated along the ulnar border of the forearm by the anconeus. The extensor digitorum proprius

(profundus) for digit one, the homologue of the human extensor pollicis longus, has a belly quite separate from the remainder of the muscle which somewhat underlies it, and inserts into an extensor expansion on the thumb, reaching as far as its distal phalanx. There is no extensor pollicis brevis. The proprius tendons to the remaining four digits typically pass to insertions into the ulnar margins of the extensor expansions of the fingers, but, as noted by Straus (1941b) there is considerable variation, and sometimes a deep tendon gives an offshoot to the radial side of the adjoining digit; in view of the arrangement shown in the marsupial in Fig. 8.2 this is not surprising. Thus in the monkey shown in Fig. 8.3A the extensor digiti quinti et quarti proprius provides two tendons, one exclusively to the fifth digit and one primarily to the fourth digit but also furnishing a slip to the fifth. Moreover, the remaining part of the deep extensor with an ulnar origin also contributes a tendon to the fourth digit; the coexistence of deep tendons to this digit from both humeral and ulnar bellies corroborates the belief that both of these muscle bellies are primitively derived from the deep stratum.

The hominoid pattern

Whilst monkeys have retained a fundamentally primitive pattern in this muscle group the hominoids show major remodelling, affecting particularly the deep stratum. This quite major reconstruction shows considerable intra-species variation, although this is inadequately documented, and likely to remain so given the scarcity of available specimens. However, the main evolutionary trends can be fairly easily defined. The specimens illustrated in Figs 8.3B and 8.4 do not necessarily show the commonest pattern but they do clarify the underlying processes of change.

Superficial stratum, radial sector   In all hominoids brachioradialis possesses the typical eutherian insertion into the radius. It may have its primary attachment rather high up the bone, as in *Pongo* (Fig. 8.3B), but then an attenuated tendinous continuation prolongs the attachment further down to the groove for abductor pollicis longus. Extensor carpi radialis brevis is unremarkable and inserts as usual to the dorsoradial side of the third

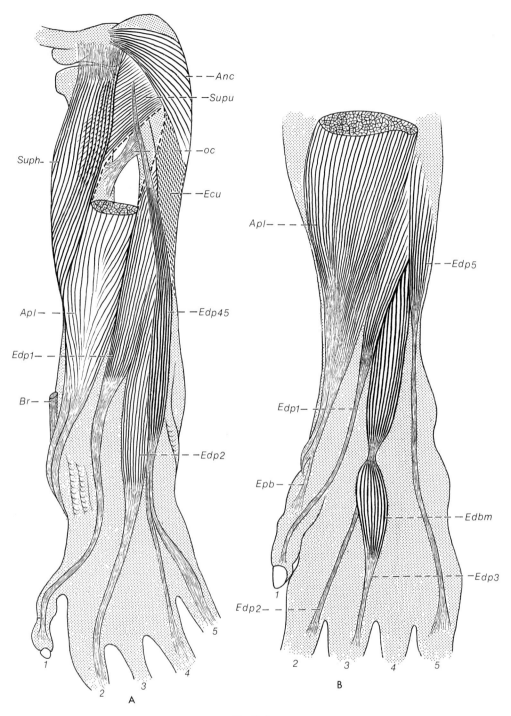

Fig. 8.4

Deep dissections, comparable to those in Fig. 8.3, of the left forearm and hand, in *Pan troglodytes* (A) and *Gorilla gorilla* (B). The ulnar origin of extensor carpi ulnaris (*Ecu*) in *Pan* is shown in dark stipple. *Epb*, extensor pollicis brevis. Other labelling is as in Figs 8.2, 8.3.

metacarpal. Extensor carpi radialis longus, however, shows a progressive tendency to span the first intermetacarpal space at its insertion by attaining an additional slip of attachment to the first metacarpal. Bojsen-Møller (1978) has described four types of insertion (Fig. 8.5). Type A is the typical mammalian insertion, restricted to metacarpal two, and is found in the great majority of monkeys, at least many gibbons, and occasional specimens of *Pan* and *Gorilla*. Type B is a rare occurrence in monkeys, but seems to be reasonably common in *Pan* and *Pongo*, whereas type C (a natural advance on Type B) seems to be the prevailing arrangement in the great apes. Type D is the exclusive and usual property of man, where the tissue spanning the root of the first intermetacarpal space has been converted into an intermetacarpal ligament; type C, however, occasionally

occurs in man, and when it does so it provides clear testimony to the true ultimate source of the intermetacarpal ligament. The functional role of this ligament has been dealt with in Chapter 6. The supinator has expanded its ulnar attachment in the great apes whilst retaining a ligamentous humeral origin which merges with the surface of the annular ligament of the superior radio-ulnar joint. In at least some specimens of *Pan* there is a substantial tendinous oblique cord (Fig. 8.4A) lying below the lower border of supinator, and deep to the ulnar origin of abductor pollicis longus, from the deeper fibres of which it has presumably been derived. In man the ulnar origin of supinator is even more expansive, so that its lower part comes to cover the oblique cord. This close association led Martin (1958a) to propose that the cord was a degenerate fibrous part of supinator. This might have been plausible from the restricted perspective of human anatomy but this suggestion is not sustained by comparative findings. The tendinous epicondylar origin of the muscle in man is again well merged with the annular ligament and some partially segregated fibres may be relayed on anterior and posterior to the superior radioulnar joint, to attach to radius and ulna respectively; these have long been known (Le Double 1897) and have recently been described and figured in some detail as the medial and lateral tensor muscles of the annular ligament (Hast and Perkins 1986).

Superficial stratum, ulnar sector Anconeus is quite restricted in size as in monkeys. Extensor carpi ulnaris in the apes may have a considerable ulnar origin. Of course, a muscle crossing exposed bone may secondarily adhere, and 'take root' so to speak. In the chimpanzee specimen illustrated in Fig. 8.4A this ulnar origin was considerable and the primitive humeral one, although present, was quite minor. In the orang-utan shown in Fig. 8.3B the sole origin was from the ulna; in effect then, the muscle had shown phylogenetic descent. In man, of course, there is merely a fascial attachment of the muscle to the posterior border of the ulna, but this may well be a persistent relic of a once more substantial attachment. At its insertion in man the tendon may be continuous with a prolongation distally over the surface of opponens digiti minimi which, as will be described in Chapter 9, appar-

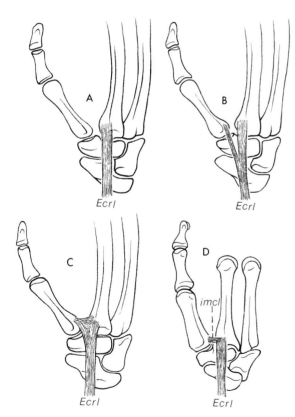

Fig. 8.5

The varying types of insertion of extensor carpi radialis longus (*Ecrl*) in higher primates showing the derivation of the intermetacarpal ligament (*imcl*). Redrawn from Bojsen-Møller (1978).

ently represents a tendinous transformation of a segregated deep belly of the abductor digiti minimi; this variation has not been recorded in other primates. Also in man, and sometimes *Pan*, the tendon may provide an accompanying slip to that of the extensor digiti minimi and so contribute to the extensor expansion of the fifth digit (Fig. 8.6D). This slip has long been known as the m. ulnaris quinti digiti (Le Double 1897). This name incorrectly suggests serial homology with the m. peroneus quinti digiti of the hind limb which is, in fact, of quite different nature (see Chapter 15).

Superficial stratum, intermediate sector   In the apes and man the extensor digitorum communis retains its humeral origin unchanged and it inserts into the extensor expansions of the ulnar four digits by tendons which are incorporated in the deep fascia investing the dorsum of the hand. Anson *et al.* (1945) have described how in the human hand the tendons are interleaved between superficial and deep layers of the deep fascia, and are themselves connected together within this compartment by another thin layer of fascia. Lucien (1947) described how this last layer in the embryo gives rise to both the tendons and connective tissue between them. This, of course, accords excellently with the view given above of the phylogenetic derivation of the tendons and also explains the considerable variability of the arrangement of the tendons on the dorsum of the hand. Little information on these variations has been recorded for the apes but they are well documented for man. The arrangements recorded by Schenck (1964) are shown in Fig. 8.6 and the approximate percentages of the different arrangements are given below. The tendon to the fifth digit may be large and separates proximally from the remainder (Fig. 8.6A, 10.5 per cent); commonly when large, however, it only separates from that for the annularis quite far distally, even near to the metacarpal heads (Fig. 8.6B, 12 per cent); the tendon for the fifth digit may be quite tiny, but there is then a substantial junctura tendinii passing from ring to little finger (Fig. 8.6C, 21.5 per cent); at the extreme, the contribution to the fifth digit is represented only by a juncture tendinii leaving the annularis tendon adjacent to the metacarpal heads (Fig. 8.6D, 56 per cent) and it is in these cases where the communis tendon to the little finger is said to be 'absent'. There is a

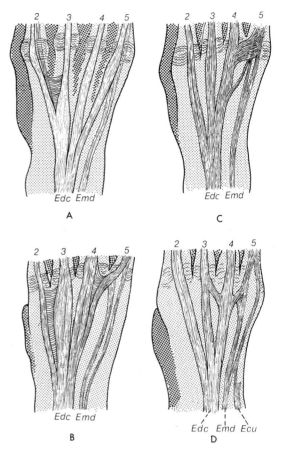

Fig. 8.6

The four main arrangements of the tendons of extensor digitorum communis (*Edc*) and extensor digiti minimi (*Emd*) in right human hands, redrawn from specimens illustrated by Schenck (1964). From A to D there is a progressive trend, described in the text, for a more distal separation of the tendinous contribution of extensor digitorum communis to the little finger. In D the extensor digiti minimi has dual tendons—the radial one has a double insertion into ring and little fingers, and the ulnar one receives a slip from the extensor carpi ulnaris (*Ecu*) tendon.

reciprocal relationship between the size of a juncture tendinii to the little finger and the size of an independent tendon to that digit; in effect, there seems to be a progressive trend towards more distal separation of the tendinous contribution to the little finger. So-called 'absence' of the tendon to the fifth digit has also been recorded in *Pan* (Le Double 1897). Connections may also be found between the extensor communis tendons in the second and third intermetacarpal spaces (Mestdagh *et al.* 1985): an oblique fascial or tendinous connection may be found in the third (Fig. 8.6D) and transverse fascial connections joining the index and middle finger tendons are commonly found in the second intermetacarpal space even occupying its whole extent (Figs 8.6A, B). All of these findings are quite predictable from what is known about the phylogeny and ontogeny of the tendons.

Deep stratum   It is here that the most extreme modifications of the extensor musculature have occurred, and where the most variability is found. The abductor pollicis longus in apes and man arises high up from radius and ulna, just as it did in monkeys (Figs 8.3B, 8.4A, B). In *Hylobates, Pan* (Fig. 6.4B) and *Pongo*, it terminates as dual tendons, one inserting on the prepollex (with some overspill onto the trapezium), and the other onto the first metacarpal base. In *Gorilla*, however, as described in Chapter 6, the prepollex is amalgamated with the trapezium and the upper tendon inserts to that composite bone (Fig. 6.4C). In contrast, in man the prepollex has undergone dissolution; nevertheless, the tendon of the abductor pollicis longus is double more often than not (Giles 1960), one of these tendons retaining a metacarpal insertion, and the other, in the absence of the prepollex passing to attach to the trapezium or alternatively blend with the surface of abductor pollicis brevis (Fig. 9.8B).

The extensor pollicis brevis differentiates as an offshoot of the metacarpal tendon of the abductor pollicis longus, and reaches the base of the proximal phalanx presumably by the mechanism of fascial transformation shown in Fig. 3.7A. It is usually said (Straus 1941*a*) that apart from man it is found only in *Gorilla* (Fig. 8.4B), and not invariably present even in that species (Le Double 1897). However, Giles (1960) has reported a rudimentary version, not quite reaching the proximal phalanx, in *Pan*. In man the tendon and muscle belly are usually completely split off from the abductor; in occasional cases, however, the extensor brevis tendon retains the emergent form and exists as a mere offshoot of the abductor tendon (Stein 1951). Not uncommonly the tendon in man may even be prolonged, at least partially, as far as the distal phalanx (Straus 1941*a*).

The extensor pollicis longus, both tendon and belly, is independent of the remainder of the deep or profundus extensor sheet in apes and man (Figs 8.3B, 8.4A, B), but this can occur, despite contrary reports, even in platyrrhine monkeys (Fig. 8.3A). Apart from its usual insertion to the distal phalanx of the thumb, it can also have a subsidiary attachment, arrested at the proximal phalanx, in both apes and man.

Part of the common inheritance of apes and man is an extensor digitorum profundus, or proprius, of digits four and five (extensor digitorum ulnaris or lateralis, or extensor digiti minimi) which has 'ascended' to the lateral epicondyle, and is not unlike that in *Pseudochirus* (Fig. 8.2) or *Cebus* (Fig. 8.3A). The tendon to the fourth digit is, however, usually (but not always) lacking in the African apes and man but it is usually present in orang-utans (Figs 8.3B, 8.4A, B). Despite the common absence of any insertion to the ring finger in man, the tendon is usually double (Mestdagh *et al.* 1985); this apparent inconsistency is made intelligible by reference to Fig. 8.3A; here, loss of attachment to the fourth digit would still leave a dual tendon serving the fifth digit. This interpretation is supported by a study of the ontogeny of the insertion of extensor digitorum lateralis (extensor minimi digiti) in man (Kaneff and Cihak 1970): for a time during development the tendon is expanded radially to attach to the fourth, or even the third digit, and only later is it retracted to serve only the fifth digit, although uncommonly the attachment to the ring finger may persist (Fig. 8.6D). In the apes the belly of the extensor digitorum proprius of the fifth digit (or fourth and fifth) shows a clear-cut tendency to descend and gain a forearm origin thus re-establishing its primitive association with the remainder of the deep extensor sheet. In the chimpanzee shown in Fig. 8.4A no humeral origin persisted, but the path of this 'descent' was revealed by an aponeurotic slip attached to the surface of the supinator; such descent is apparently unusual in *Pan*. In the *Gorilla* shown in Fig. 8.4B, origin of the muscle was entirely from the ulna, and although this is probably an extreme case, it is on record that in *Gorilla* the muscle rarely has a direct bony origin from the humerus but springs from the intermuscular septa between it and the adjoining muscles (Straus 1941*a*). In the orang-utan shown in Fig. 8.3B the origin was entirely from the ulna, and this is acknowledged as the usual arrangement in this species. It is not surprising then that occasionally the muscle should have an ulnar attachment in man.

The remainder of the extensor digitorum profundus sheet has typically been markedly reduced in the African apes and man, usually only a tendon to the second digit being present (Fig. 8.4A), leaving the third and fourth digits without proprius tendons, but this is not always so and there is considerable variation. In the orang-utan (Fig. 8.3B), however, the full complement of proprius tendons is typically retained. The tendons usually lost in African apes and man may atavistically reappear as variants in some specimens. That to the middle digit sometimes occurs in *Pan* (Le Double 1897), sometimes in *Gorilla* (Fig. 8.4B), and rarely in man; at the other extreme even the human extensor indicis proprius may be lacking (Mestdagh *et al.* 1985).

The specimens of *Pongo* (Fig. 8.3B) and *Gorilla* (Fig. 8.4B) illustrated here are especially interesting since each shows an extensor brevis manus associated with the proprius tendon to the middle digit. Reports of the extensor brevis manus in man also crop up sporadically in the clinical literature (for example, Bingold 1964). There is little doubt that these muscles represent an atavistic reawakening of the ancient reptilian dorsometacarpales (Fig. 8.1). In fact, it seems that the developmental specification for these muscles must have lain dormant throughout the long span of primate evolution. Cihak (1972) has described how in the early human embryo quite separate muscular primordia appear dorsal to the anlagen of the dorsal abductor components of the dorsal interossei in the second, third and fourth intermetacarpal spaces. As development proceeds these show some degree of amalgamation with the underlying dorsal interossei but commonly or even usually in the adult, particularly in the second and third intermetacarpal spaces, relatively independent muscle strips (variably showing fibrous transformation) persist. Cihak noted how in reconstructions they strikingly mimic the human muscular variation known as the extensor brevis manus. Yet he rejected this homology, apparently solely on the basis of their nerve supply from the deep branch of the ulnar nerve, and so gave them the enigmatic name of 'interossei dorsales accessorii'. As noted in Chapter 3 such faith in the inviolability of the principle of nerve–muscle specificity is no longer tenable and there is little doubt that, in fact, these rudiments represent the reptilian dorsometacarpales. The muscle strips usually variably merge and insert with the dorsal interossei, but this does not seriously conflict with the interpretation of the homologies of these muscles given in Chapter 9; regressing contrahentes muscles may also make some myoblastic contribution to the interossei. Persistence and progressive development of certain of these dorsal slips almost certainly gives rise to the anomalous extensor brevis manus muscles. No data are available for this developmental sequence in other Primates but it would be surprising if it were not similar; doubtless the short extensor muscles shown in Fig. 8.3B and 8.4B were derived in this way. These statements certainly apply to the short type of extensor brevis originating close to the intermetacarpal spaces or from the dorsum of the distal carpal row. A long variant of the extensor brevis is, however, also found on occasion, originating from the lower part of the radius; almost certainly these merely represent some secondary descent of the deep muscular stratum. Kadanoff (1958) also distinguished long and short varieties of the extensor brevis manus, and noted fusion of the latter type with the dorsal interossei.

9

Intrinsic muscles of the hand

The reptilian–mammalian transition

The groundwork for understanding the architecture of the intrinsic hand muscles was provided by Cunningham's (1882) studies on the marsupial hand, which established the primitive mammalian pattern of the intrinsic musculature (Fig. 9.1A). Cunningham demonstrated the presence of a basic trilaminar arrangement of the digital muscles consisting of a dorsal layer abducting from the middle digit, an intermediate layer of bicipital flexores breves with metacarpal origins and insertions either side of the corresponding digit, and a superficial layer of four adductors (Mm. contrahentes). This basic mammalian pattern was realized to be a relatively simple elaboration of that characteristic of reptiles, as described by McMurrich (1903) in a study supplementary to his work on the forearm flexors, which was summarized in Chapters 3 and 7. It was noted there that the palmar aponeurosis of amphibia and reptiles splits and sandwiches between its layers the slips of the flexor brevis superficialis, which by tendinous conversion or replacement, provides the insertions of the flexor digitorum superficialis of mammals. Deep to the amphibian palmar aponeurosis are additional muscle layers: flexor brevis medius, the flexor brevis profundus and the intermetacarpals. In reptiles the flexor brevis medius becomes subdivided into two layers: a superficial stratum retains attachment to the dorsal surface of the palmar aponeurosis, just where it divides to form the flexor profundus tendons, and furnishes four slender slips to the four ulnar digits. These are the precursors of the mammalian lumbricals; the deep stratum is the anlage for the superficial adductor layer of Cunningham's primeval trilaminar mammalian palmar musculature, and immediately

deep to it lies the plane of the palmar nerves. Deep to this the reptilian flexor brevis profundus is the source of Cunningham's intermediate layer of short bicipital flexors whilst most dorsally of all a set of intermetacarpal muscles (often converted into tendinous bands in extant reptiles) provided Cunningham's dorsal abducting layer. Strangely, in an earlier study, Cunningham (1878) had confused the fate of his superficial (contrahentes) and intermediate (flexores breves) layers, in the mistaken belief that the intermediate layer was unrepresented in eutherian mammals, except in the first and fifth digits; although corrected in 1882 this early error has caused considerable confusion in subsequent work.

Since the flexor brevis superficialis and medius of reptiles are redeployed in mammals, and the changed function of their derivatives is reflected in a changed nomenclature, this leaves only one group of muscles designated as short flexors in mammals: Cunningham's intermediate layer (homologous with the reptilian flexor brevis profundus). In the remainder of this chapter this group of bicipital short flexors will be referred to simply as the flexores breves.

The midpalmar musculature

The contrahentes

The contrahentes comprise the superficial adductor layer in Cunningham's scheme for the primeval palmar musculature, and, has been seen, are derived from the deep stratum of the reptilian flexor brevis medius. Topographically this layer is demarcated from the deeper ones (contributing to the interossei) by the deep branch of the ulnar nerve.

In marsupials there are typically four of these muscles adducting digits one, two, four, and five towards the middle one, by insertions to the ulnar side of the proximal phalanges of the first and second digits, and the radial side of the proximal phalanges of the fourth and fifth digits. Primitively these muscles shared a common origin from the floor of the carpal tunnel and the bases of the metacarpals and fanned out symmetrically to their appropriate insertions. There has been a phylogenetic trend occurring in parallel at least several times for distal expansion of the origin of the contrahens of the pollex, or that of the minimus, or both, distally along a raphe located ventral to the midline of the palm—the third metacarpal. A corresponding trend has affected the homotypical muscles in the foot—the pedal contrahentes (Figs 16.1A, C). In the manus of an Australian marsupial such as *Trichosurus* this specialization is largely restricted to the contrahens of the fifth digit, and to a minor extent that of the first; thus the contrahens of the fourth digit is overlapped by the extension of that of the fifth and indeed may be quite tiny or even absent (Barbour 1963). In *Didelphys* both the muscles of the first and fifth digit have enlarged thus, with the fan-shaped muscles meeting in a midline raphe (Stein 1981) and largely covering the reduced muscles of the second and fourth digits. In Primates a similar specialization is largely confined to the contrahens of the pollex which shows just such a progressive fan-shaped expansion, and becomes the adductor pollicis, variably subdivided into oblique and transverse parts.

Monkeys generally retain the full complement of four muscles, although that to the index is commonly diminutive, may be largely covered by the adductor transversus, and may sometimes be absent. In New World monkeys (Fig. 9.5A) a quite simple primitive arrangement may be retained, with only some distal extension of the origin of the first contrahens. In Old World monkeys (Fig. 9.5B) this trend is exaggerated and may also affect to some degree the contrahens of the fifth digit. Invariably, the insertions of these muscles are to the sides of the glenoid plates of the metacarpophalangeal joints (with contained sesamoids if present) and to the adjacent bases of the proximal phalanges.

In hominoids the contrahentes of digits two, four and five show a progressive tendency to undergo involution and replacement by fibrous tissue, and in that order. Generally the three are present in gibbons but they may be poorly developed; in the specimen shown in Fig. 9.6A the contrahens of digit two was rudimentary. It is generally said that the contrahentes of digits four and five are usually present in chimpanzees although poorly developed (Jouffroy 1971; Hartman and Straus 1933); in the specimen shown in Fig. 9.6B the contrahentes of digits four and five had lost much of their individuality, at least at their insertions, by blending with the flexor brevis muscles (numbers 7 and 9 in Fig. 9.1A) on the radial sides of those digits. In orang-utan and man the contrahentes of digits two, four, and five are generally lacking.

Interestingly the complete contrahentes layer makes a transient appearance during human development (Cihak 1972, 1977). The ulnar part shows only an abortive transformation of myoblasts into myotubes and progressively this part of the layer is converted into or replaced by fibrous tissue (Figs 9.2A,B) and forms the anterior interosseous fascia. As Cihak suggests, some of these myoblasts may be amalgamated with the underlying flexores breves, and so contribute to the interossei. This would conform with the situation noted in the chimpanzee shown in Fig. 9.6B. The radial part of the contrahentes sheet becomes the adductor pollicis; again, the fact that this may include embryologically material referable to contrahens two, does not disturb the rational approach of homologizing the adductor pollicis with the contrahens of the pollex.

It is well known that in the human hand the anterior interosseous fascia when traced towards the metacarpal heads becomes especially thickened as the deep transverse metacarpal ligaments joining together the palmar or glenoid plates of the metacarpophalangeal joints of the ulnar four digits. It has become traditional, following Forster (1916), to consider these ligaments as derivatives of the contrahens layer, but embryology (Cihak 1972) lends no support to this idea. In fact, primates retaining a full complement of contrahentes also possess these ligaments, lying in a plane ventral to the tendons of insertion of the contrahentes. In New World monkeys (Fig. 9.5A) they are quite flimsy but are much more substantial structures in Old World monkeys (Fig. 9.5B). In *Pan* (Fig. 9.6B) and man they are strong fibrous bands. In *Hylobates* (Fig. 9.6A), however, they are rudimentary and it seems likely that this represents a secondary regression. Doubtless they are derived merely as a fibrous condensation.

The interossei

The interosseous muscles, dorsal and palmar, have presented generations of morphologists with a perplexing puzzle. These palmar adductors and dorsal abductors are characteristically grouped about the middle digit, but the functional axis of the primate hand can vary between species and even intraspecifically. It may be transferred to the second or even the fourth digit. Any such shift apparently requires a drastic reshuffling of the muscle insertions, contrary to the usual principles of myological evolution enunciated in Chapter 3. Moreover, the muscles seemingly vary numerically, some Primates being described as having as many as six, or even seven, in the palmar series. The supernumerary ones, however, although part of the same palmar sheet as the remainder, are so disposed as to act as abductors, closely related to and reinforcing the normal dorsal abductors. Any satisfactory phylogenetic explanation must account both for shifts in axis and variations in number. Various explanatory hypothesis have been proposed and these are shown in Fig. 9.1.

Cunningham (1882) proposed that the human condition could be derived from his primitive mammalian model by assuming that the dorsal abductors became the dorsal interossei while certain of the flexores breves became the palmar interossei, with the remainder disappearing (Fig. 9.1B); in his earlier admittedly mistaken view (1878), he had derived the palmar interossei from the contrahentes layer. This view gained a wide following and was incorporated in the standard British textbook of the day (Thane 1892).

A second hypothesis which also gained a wide following was derived from the work of Ruge (1878*b*) on the development of the pedal interossei. Ruge claimed to have shown the development of both plantar and dorsal interossei from paired muscles lying initially on the plantar aspects of the metatarsals (in the position of Cunningham's short

flexors), those bellies destined to become dorsal interossei later reaching their definitive position by migration into the intermetatarsal spaces. Windle (1883) followed this up and reported similar findings for the human hand (Fig. 9.1C). This hypothesis recognizes no persistent derivatives of Cunningham's dorsal abductor layer in the human hand, and was supported by Campbell (1939), Keith (1948), and Haines (1950). It provides an attractively facile mechanism for accounting for shifts in the palmar or plantar axis, the axis becoming dependent solely upon which individual muscles undertake the dorsal migration. This view, based purely upon the interpretation of serial sections of staged embryos came at a time when inflexible views on ontogenetic recapitulation prevailed. Ruge (1878*b*) and Windle (1883) were thus seeking evidence of an ontogenetic migration of the future dorsal interossei, which they believed they had observed. Re-examination of their illustrations, however, strongly suggests that no actual migration occurs, but instead there is a relative delay in the full development of the intermetacarpal parts of the dorsal interossei, a delay associated with the late widening of the intermetacarpal spaces. This is but one more example of the disruptive effect of heterochrony on any stylized pattern of recapitulation during development.

A third hypothesis portrays the dorsal interossei as composite muscles, resulting from the phylogenetic amalgamation of a dorsal abductor with a short flexor, in each case one of those short flexors which Cunningham supposed to have disappeared (Fig. 9.1D). Forster (1916) in a detailed comparative analysis supported this proposal and Bunnell (1942) and Eyler and Markee (1954) applied it to the interpretation of the human morphology. Earlier a somewhat similar view had been proposed by McMurrich (1903), modified, however, by his rather unorthodox views upon the phylogeny of the thenar and hypothenar musculature, which had a knock-on effect in his attempts to account for the appropriate number of flexores breves in the midpalmar interosseous musculature. He postulated (Fig. 9.1E) that the first and tenth flexores breves became the first and fourth human palmar interossei, and that the first dorsal interosseous muscle, like the fourth, incorporated not one, but two flexores breves. This involves not

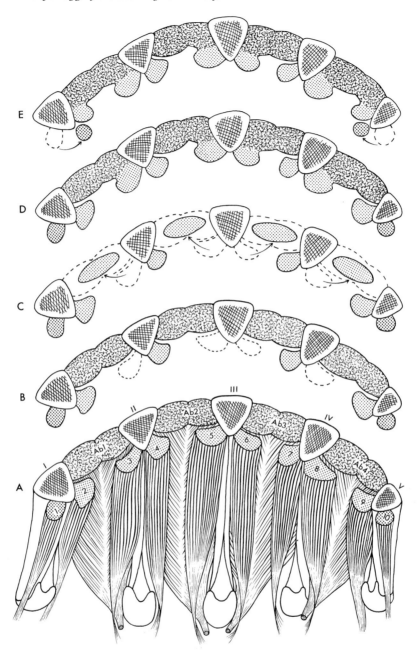

Fig. 9.1

A diagrammatic representation of the various suggested theories concerning the evolution of the interosseous muscles. A, the arrangement of the dorsal and intermediate muscle layers in the primitive mammalian hand (the contrahentes layer is not shown); the four bipennate dorsal abductor muscles are labelled Ab1, Ab2, Ab3 and Ab4; the ten flexores breves are stippled and indicated by arabic numerals (the marginal members of this series, the first and the tenth, are shown in darker stippling); the metacarpals are indiated by roman numerals. The shading conventions shown in A are retained in all other parts of the figure and a muscular component which is believed to disappear is represented by a broken line, as is the primitive site of a muscle which has undergone a supposed migration; the arrows indicate the direction of such postulated movement. B, shows the view of Cunningham (1882); C, the view of Windle (1883); D, the view of Forster (1916); E, the view of McMurrich (1903). (From Lewis 1965*b*.)

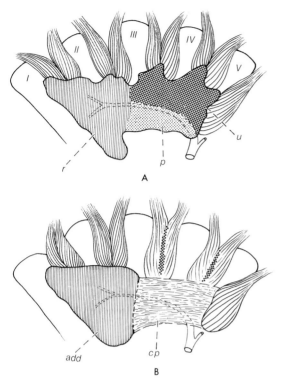

Fig. 9.2

Stages in the development of the human contrahentes layer (redrawn after Cihak 1972). A, at crown–rump length 20 mm; B, at crown rump length 32 mm. *I–V*, the metacarpals; *r*, radial part of the contrahentes layer; *u*, ulnar part of the contrahentes layer; *p*, proximal part of the contrahentes layer; *add*, the differentiated adductor pollicis primordium; *cp*, the regressed contrahentes plate, forming the interosseous fascia; the flexores breves profundi are shown deep to the contrahentes layer.

only the unlikely translocation of the first and tenth flexores breves from one side of their digit to the other, but also the transfer of the insertions of the second and ninth across an interdigital cleft. This view that two flexores breves are incorporated with each of the first and fourth dorsal interossei, as will be seen, derives no support from human or comparative anatomy. Indeed, McMurrich's figure of a section from the hand of a 6 cm human embryo appears instead to confirm the view of Forster (Fig. 9.1D), namely, that each human dorsal interosseous results from the partial incorporation of a *single* flexor brevis with a dorsal abductor muscle, and that each palmar interos-

seous is a virtually unmodified flexor brevis. From what follows, this will be seen to be a virtually inescapable conclusion.

Marsupial hands Those Australian marsupials which retain fairly generalized hands (e.g. *Trichosurus*, *Pseudochirus*) provide ample endorsement for Cunningham's findings, but certain details of muscular insertion (Fig. 9.1A), which did not particularly concern him, became of critical importance in assessing subsequent apomorphic changes. The four bipennate abductor muscles of the dorsal layer are orientated about the middle digit. These are overlaid ventrally by the sheet of flexores breves; the marginal members of this layer, being components of the thenar and hypothenar groups, will be dealt with later in this chapter. Thus eight muscles (numbers two to nine in Fig. 9.1A) remain for consideration in the midpalmar region, the first of these being related to the ulnar side of the first metacarpal and the last to the radial side of the fifth metacarpal. The remainder are grouped in three pairs with confluent tendinous origins from the bases of the central three metacarpals and the capsules of the carpometacarpal joints. It is preferable to speak of pairs of individual muscles rather than to adopt the more traditional designation of single bicipital muscles, since the individual members of each pair are, in fact, fairly separate entities and diverge to separate insertions. The classical literature provides little detail about the insertions of the muscles of the two layers merely implying that both attach to the proximal phalanges and adjacent metacarpophalangeal sesamoids. More detail here is a necessary requirement for charting the phylogenetic modifications of these muscles.

The tendons of the dorsal abductors merge with the surfaces of the metacarpo–phalangeal joint capsules and certainly insert into the bases of the proximal phalanges. The tendons of the flexores breves, however, pass to a more dorsal insertion by crossing the abductor tendons and are totally attached to the margins of the corresponding extensor tendons (not the base of the phalanx as hitherto described) here forming 'wing tendons'. American marsupials such as the opossum *Didelphys marsupialis*, which are generally considered closer to the marsupial stem pattern, here clearly present what seems to be an amalgam of derived and primitive features in the tendinous insertions. The flexores breves are only partially prolonged as 'wings' to the extensor tendons; as they skirt the metacarpophalangeal sesamoids, they gain an attachment thereto which is doubtless secondary. The marsupial fourth dorsal abductor muscle generally has a double tendon of insertion, with slips passing to both sides of the fourth interdigital cleft where they find insertion into the contiguous sides of the proximal phalanges of the fourth and fifth digits. This is a common, if not constant, insertion for this muscle in marsupials, and was noted earlier by Brooks (1886) in *Didelphys* and *Trichosurus* (although not observed in my own specimens of *Trichosurus vulpecula*); it is probably primitive.

Monkey hands   Again, as in marsupials, the tendons of the derivatives of the flexores breves cross those of the abductors to insert as wing tendons joining the expansion of the extensor tendons, whilst the abductors basically insert on the proximal phalanges. The individuality of these tendons and their differing mode of insertion is emphasized by a variably developed sheet of fibrous tissue passing from the extensor tendon around the sides of the metacarpophalangeal joints to reach the palmar plates of the joints; together with the extensor tendon this tissue forms a mobile sleeve about the joint. It is much less obvious than the comparable structure in hominoids, where it is known as the transverse lamina (see below). However, when reasonably well developed it intervenes between the phalangeal tendon of the abductor which runs deep to it and the wing tendons which are superficial.

In monkeys of the New or Old World, there are typically four dorsal abductor muscles, similar in disposition to those in marsupials, each arising from a pair of adjoining metacarpals, and abducting digits two, three and four from an axis formed by the middle digit; these muscles are usually designated as dorsal interossei. Situated on their palmar surface is a layer of fusiform muscles, each arising from a single metacarpal, and clearly homologous with the flexores breves. Only seven such may be present, corresponding to numbers three to nine inclusive in the primitive palm (Fig. 9.1A); these are the muscles commonly described as palmar interossei. In some monkeys,

however, there is an additional member of the flexor brevis series present on the ulnar side of the first metacarpal (corresponding to number two in Fig. 9.1A), making eight palmar interossei in all (Figs 9.5A, B). The presence of the full quota of eight has also been recorded by Brooks (1887) in *Cercopithecus, Macacus, Pithecia,* and *Hapale.*

Derived states may obscure this basic simplicity of the dorsal and palmar interossei in some monkeys and these are of interest mainly for the way in which they provide models elucidating comparable specializations in man. For instance, in a specimen of *Cebus nigrivittatus*, those four members of the flexor brevis series which are homologous with the human palmar interossei (corresponding to numbers two, four, seven, and nine in Fig. 9.1A) all exhibited a typical insertion as wings of the extensor aponeurosis, but the bellies of the remaining flexores breves (numbers three, five, six, and eight) showed some fusion with the subjacent dorsal interossei. The tendons of insertion of these amalgamated flexores breves and dorsal interossei on the radial sides of digits two and three, however, retained the typical mutual crossings as they passed to their respective insertions into the wings of the extensor aponeurosis and the proximal phalangeal bases. The insertions on the ulnar sides of the third and fourth digits were, however, modified. In each instance the partly united bellies of the flexor brevis and the dorsal interosseous gave rise to their own tendons, crossing as usual alongside the metacarpophalangeal joints, but both tendons inserted into the wing of the aponeurosis, there being no phalangeal insertion. The tendon of the dorsal interosseous had, in effect, relinquished its phalangeal attachment and merged with the deep aspect of the overlying flexor brevis tendon, which thus prolonged its insertion distally into the aponeurosis wing. This accords well with the principles of myological evolution described in Chapter 3. In a specimen of *Colobus polykomos* the bellies of flexores breves, three, five, six, and eight were even more intimately merged with the related dorsal interossei than in *Cebus*. Despite this, in each instance the tendons of the two morphological components retained their separate identities and crossed each other as usual. The primitive simplicity of insertion was not, however, here completely retained: on the radial side of the second digit the flexor brevis wrapped over the tendon of the dorsal interosseous (as it does in *Pan* and sometimes in *Homo*) but did not attain the wing, the whole mass inserting on the phalangeal base. In this specimen on the ulnar side of the middle digit some of the dorsal component was appropriated by the flexor brevis tendon, only a very slender phalangeal tendon being retained. On the radial side of the middle digit and the ulnar side of the fourth the insertions were uncomplicated, with tendons crossing, as usual, as they passed to their respective insertions into the proximal phalangeal bases and the extensor aponeurosis wings.

The potentiality for amalgamation of a flexor brevis with a dorsal abductor (dorsal interosseous) is established by this natural experiment, as is the capability of one of the tendons of the composite muscle taking on the dominant or even sole role of insertion.

The chimpanzee interossei Four dorsal interossei, corresponding to those of monkeys, and to the dorsal abductors of marsupials, are invariably present. On their palmar surfaces lie the fusiform bellies of the palmar interossei, homologues of the flexores breves. There may be seven of these (Fig. 9.6B), corresponding to numbers three to nine in the primitive palm (Fig. 9.1A).

Most authors (Hepburn 1892; Sonntag 1923) have recognized only six of these muscles, failing to observe the most radial member of the series (number three). Forster (1916), however, recognized the full complement of seven in a specimen of *Troglodytes niger*. In specimens which I have examined muscle number three was an entity clearly separated from the underlying dorsal interosseous but perhaps, as will be shown for the homologous muscle in man, there is here individual variation. Of these muscles, numbers four, seven, and nine correspond in position to the ulnar three human palmar interossei, and will be shown to be homologous therewith: muscles three, five, six, and eight are, however, intimately related, both at origin and insertion, to the subjacent dorsal interossei. The various muscular components are, however, quite independent and the adjacent muscle bellies show little evidence of that fusion which has occurred in some monkeys.

The pattern of insertion may best be considered by first describing the situation where there are

tendons of both a dorsal interosseous and a palmar interosseous (flexor brevis); for example, the radial side of the middle digit (Fig. 9.3). There is a well differentiated fibrous sheet passing from each side of the extensor tendon to invest the metacarpo–phalangeal joint and attach to the palmar ligament. This is clearly homologous with the human transverse lamina. It constitutes a useful morphological landmark and is much more clearly defined than in monkeys. The tendon of the dorsal

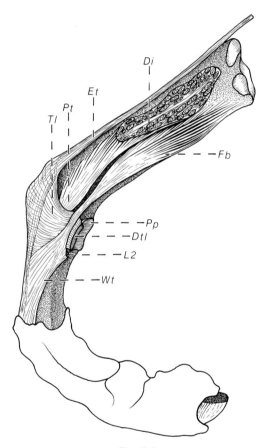

Fig. 9.3
The muscles and extensor apparatus on the radial side of the middle finger of the right hand of the chimpanzee. The radial head of the second dorsal interosseous has been cut from its origin on the second metacarpal; the flexor tendons and the second lumbrical muscle are removed. *Di*, second dorsal interosseous; *Fb*, flexor brevis; *Et*, extensor tendon; *Pt*, phalangeal tendon; *Tl*, transverse lamina; *Pp*, palmar pad (glenoid plate); *Dtl*, deep transverse ligament of palm; *L2*, tendon of second lumbrical; *Wt*, wing tendon. (From Lewis 1965*b*.)

interosseous commences on the volar surface of the muscle and the fully constituted tendon passes deep to the transverse lamina to insert into the base of the proximal phalanx. In contrast, the tendon of the palmar interosseous (flexor brevis) passes superficial to the transverse lamina and is inserted entirely into the extensor aponeurosis, forming its wing tendon, and is joined here by the second lumbrical tendon.

This description is also applicable to the insertions on the ulnar side of the middle digit, the ulnar side of the fourth and the radial side of the second. In this last digit the separate identity of the flexor brevis is usually emphasized by the well-marked aponeurotic surface of the first dorsal interosseous; the flexor brevis is represented proximally by a slender tendon of origin only, but its muscular belly extends far distally onto the dorsal interosseous tendon, before giving rise to a rather feeble tendon which inserts in the usual way as a wing tendon. This foreshadows the human condition.

On those sides of the digits which lack a dorsal interosseous insertion—the ulnar side of the second digit and the radial sides of the fourth and fifth—the arrangements are comparable to those shown in Fig. 9.3, if the dorsal interosseous is omitted: the flexor brevis (i.e. a palmar interosseous) passes superficial to the transverse lamina to insert exclusively as the wing tendon of the extensor apparatus. Usually in *Pan* there is no muscle corresponding to the human first palmar interosseous (number two of the flexor brevis series) and its absence has already been noted in many monkeys. However, in this situation Brooks (1887) described a fibrous band in a chimpanzee and a muscular belly in the orang-utan and gibbon. When absent, as in Fig. 9.6B, it is presumably inseparably merged with the oblique part of adductor pollicis. All the flexor brevis tendons, where they lie superficial to the transverse lamina, are invested by a thinner more superficial layer of the transverse lamina or hood. A similar thin superficial lamina of this hood ligament, as will be seen, occurs in man.

The palmar interossei (flexores breves) have a fusiform appearance often with a small proximal tendon of origin attached adjacent to the metacarpal base. The deep aspect of the fusiform belly, however, gains appreciable origin from that ventral part of the metacarpal shaft which is free of

dorsal interosseous attachment and an intervening fibrous septum usually emphasizes the separate identity of the two muscles.

The insertion of the fourth dorsal interosseous may show a departure from the general plan elaborated above, recalling the common double insertion of the homologous marsupial muscle, for in addition to the usual abducting phalangeal tendon passing deep to the transverse lamina on the ulnar side of the fourth digit, a second slender tendon may arise from the aponeurotic palmar surface of the muscle, and disappear beneath the transverse lamina on the radial side of the fifth.

Homo sapiens   It is clear that the seven or eight so-called palmar interossei of other primates are inappropriately named since this term, borrowed from human anatomy carries with it functional implications which are not fulfilled by all. For those corresponding to flexores breves two, four, seven, and nine (Fig. 9.1A) this designation is appropriate since part of their role (even if it is a minor one) is adduction of their digits towards the middle one; the human palmar interossei are homologous with these. The remaining ones (numbers three, five, six, and eight) accompanying the subjacent abductors towards their insertions have been shown to have a tendency towards fusion with their companion muscles in some monkeys; this specialization is more advanced in man, so that the canonical dorsal interossei of human anatomy are double-bellied, each belly, dorsal or ventral, having a characteristic mode of insertion, the flexor brevis component into a wing tendon and the dorsal abductor component by a phalangeal tendon. These insertions may, however, be modified, according to the quite limited rules of myological evolution, and in ways paralleled in certain of the digits in monkeys and apes. Thus the prolongation of the flexor brevis into the wing tendon may be attenuated, and terminate by insertion into the transverse lamina; a dorsal abductor phalangeal tendon, beyond its passage deep to the transverse lamina, may become incorporated with the deep aspect of the related wing tendon, leading to diminution (or even disappearance) of its primary phalangeal insertion; the dorsal abductor muscle may even merge completely with its companion flexor brevis and the composite whole give rise to a single tendon passing superficial to the transverse lamina into the extensor aponeurosis wing.

Surgical interest in the detailed anatomy and function of the digital extensor apparatus of man has refined knowledge about its complex architecture. Figure 9.4 summarizes current knowledge and is taken from a dissection performed with the hindsight provided by the work of Bunnell (1942), Braithwaite *et al.* (1948), Landsmeer (1949, 1955), and Haines (1951). Certain minor details have been omitted from this diagram: thus, Haines (1951) described an additional thin superficial layer to the hood (already noted above in *Pan*), and Landsmeer (1949) described the retinacular ligament as having an additional transverse band. It is apparent that, typically a dorsal interosseous muscle comprises two bellies, the more dorsal belly inserting on the proximal phalanx and the more ventral terminating as a tendon which crosses

more superficially to a dorsal insertion as the wing of the extensor expansion.

These two bellies, respectively the homologues of the dorsal abductor component and the ventral flexor brevis component have been confusingly designated in the literature as 'superficial' and 'deep'; Landsmeer (1955) and Haines (1951) used these terms in opposite senses, the one author considering arrangements from a dorsal viewpoint, the other from the palmar aspect. The individual identities of the tendons of insertion of the two bellies are emphasized, not only by their typical mode of crossing one another, but also by their passage either side of the tough, clearly defined transverse lamina or hood ligament. This hood is not part of the metacarpophalangeal joint but, together with the common extensor tendon, constitutes an entirely separate mobile sleeve investing the dorsum and sides of the joint. Its

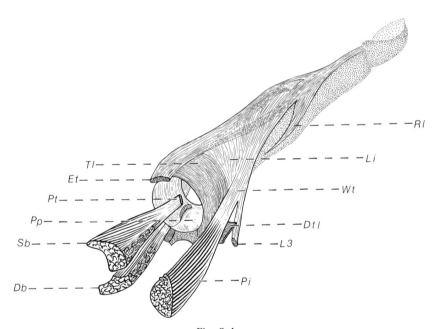

Fig. 9.4

The extensor apparatus of the human left ring finger. The terminology used is that of Landsmeer (1949, 1955); alternative terms used by Haines (1951) are shown in brackets. *Db*, deep belly (superficial belly) of fourth dorsal interosseous; *Dtl*, deep transverse ligament of palm; *Et*, extensor digitorum tendon; *Li*, lamina intertendinea; *L3*, tendon of third lumbrical; *Pi*, third palmar interosseous; *Pp*, palmar pad (with the attachment of a slip of the metacarpophalangeal collateral ligament to its margin); *Pt*, medial tendon (phalangeal tendon) cut short of its insertion to the proximal phalanx; *Rl*, oblique band of retinacular ligament (link ligament); *Sb*, superficial belly (deep belly) of fourth dorsal interosseous; *Tl*, transverse lamina (deep layer of hood ligament); *Wt*, wing tendon (side slips). (From Lewis 1965*b*.)

sliding movement about the joint is facilitated by large bursae intervening between it and the joint capsule. Besides the well-known dorsal bursa (which may communicate with the joint cavity) there is frequently a large bursa deep to the transverse lamina on each side. Where only a palmar interosseous inserts the lateral bursa separates the transverse lamina from the collateral ligament of the metacarpophalangeal joint; where a dorsal interosseous inserts the bursa intervenes between the transverse lamina and the subjacent phalangeal tendon, which in turn is usually separated from the collateral ligament of the joint by a second deep bursa. The presence of these lateral bursae incidentally facilitates the clear and unequivocal demonstration of anatomical arrangements in this region. The bursae have been described in some detail by Wise (1975) emphasizing their likely role in the causation of the typical deformities of rheumatoid arthritis.

As Salsbury (1937) first pointed out, the different dorsal interossei vary in the relative size of the insertions of the two bellies: the first muscle, according to him, inserts solely into the proximal phalanx; the other three insert additionally into the extensor expansion, the phalangeal insertion being greatest for the second, least for the third, and intermediate for the fourth. Landsmeer (1955), however, showed that the first dorsal interosseous sometimes may also have a separate ventral belly inserting into the transverse lamina or even into the wing of the extensor expansion. Various authors (Forster 1916; Salsbury 1937; Landsmeer 1955) have stressed that the palmar interossei (like their homologues the ventral bellies of the dorsal interossei) are inserted entirely into the extensor expansion, reaching it by passing superficial to the transverse lamina; yet nevertheless many textbooks still describe a partial bony phalangeal insertion as the normal arrangement. The interesting generalization which emerges from these considerations is that the extensor apparatus of each finger displays symmetrically disposed wing tendons, derived either from a palmar interosseous or from the ventral belly of a dorsal interosseous (and reinforced on the radial side by a lumbrical). The subdivision of the dorsal interossei into dorsal and ventral bellies has been interpreted (Salsbury 1937; Haines 1951; Landsmeer 1955) as a functional modification related to

the different insertions and actions of the two bellies—the phalangeal tendons acting on the metacarpophalangeal joints, the wing tendons having their actions projected to the interphalangeal joints, as described by Stack (1962, 1963). Now, however, it is apparent that this arrangement, with its undeniable functional importance, is not a mere human specialization but really owes its existence to the phylogenetic history of the muscles.

The insertions of the two components of the dorsal muscles show variations, particularly characterizing certain digits. In the following account the approximate percentage frequency (taken from a small series) of any particular variant is shown in brackets. In the hand of man a first palmar interosseous (flexor brevis two) is commonly separately identifiable. The eight flexores breves components present a characteristic gross appearance, that is, they are fusiform, often have a slender proximal tendon of origin from the metacarpal base but also arise from the metacarpal shaft, and certain of them show a marked tendency towards merging with a related dorsal abductor component.

Flexores breves two, four, seven, and nine, however, remain independent as the palmar interossei. Many textbooks perpetuate the fallacious idea that the palmar interossei possess a dual insertion, including both the extensor aponeurosis and the proximal phalangeal base despite the fact that this has repeatedly been refuted. It is here confirmed that, with certain reservations, the normal insertion of a palmar interosseous is by a wing tendon into the extensor aponeurosis. To reach this attachment the tendon passes superficial to the transverse lamina, to which it is firmly attached, an attachment which has perhaps been mistaken for a phalangeal insertion. In the case of the fourth palmar interosseous, however, (as previously noted by Eyler and Markee 1954), the deep aspect of the wing tendon, just about the point where it is joined by the fourth lumbrical, often (44 per cent) becomes adherent to the underlying phalangeal base. This is clearly a secondary attachment and wholly different from the direct primary insertion of the phalangeal tendon of a dorsal interosseous. (Occasionally there is a true phalangeal tendon on the radial side of the minimus, which will be considered with the fourth dorsal interroseous.) Theoretically, it might be expected that the first palmar interosseous would also retain an insertion as a wing tendon and, in fact, its tendon does so often (44 per cent) and may in its entirety pass superficial to the transverse lamina of the thumb (here lying immediately dorsal to the metacarpophalangeal sesamoid) to reach the ulnar wing of the pollicial extensor apparatus (Fig. 9.8B). Such an insertion has been noted before by Brooks (1886) and by Poirier and Charpy (1901). Most textbooks, however, give the insertion as the sesamoid and base of the proximal phalanx; in this present series the muscle was completely arrested at the sesamoid in only 16 per cent but more often (32 per cent) was attached to sesamoid and wing jointly. Sometimes (8 per cent) the muscle is not distinguishable from the M. adductor obliquus (Fig. 9.8C) which may then have appropriated the wing tendon, as its partial insertion.

The descriptive human dorsal interossei are clearly composite muscles derived each from a dorsal abductor (comparable to the whole descriptive dorsal interossei of, for example, *Pan* or monkeys) together with a flexor brevis (numbers three, five, six and eight). The second dorsal interosseous exhibits this derivation in almost unmodified form, arrangements being virtually identical with those described for this region of the hand in *Pan* (Fig. 9.3). The muscle bellies of the flexor brevis and the dorsal abductor components may fuse to a variable degree, or may remain quite independent, but their tendons always retain complete individuality as they pass respectively superficial and deep to the transverse lamina to reach their appropriate insertions into the wing of the aponeurosis and the proximal phalangeal base. Similar features are usually shown by the fourth dorsal interosseous, where the component bellies are again variably fused and the flexor brevis is occasionally almost independent. The tendons show the usual insertions but part of the substance of the dorsal abductor may sometimes (16 per cent) accompany the flexor brevis tendon superficial to the transverse lamina and into the wing of the extensor aponeurosis, resulting in a diminution in the size of the phalangeal tendon. Rarely (4 per cent) the whole of the dorsal abductor may be thus inserted, there being no phalangeal tendon. Also rarely (4 per cent) the dorsal abduc-

tor tendon may pass as usual deep to the transverse lamina, but lack a phalangeal insertion and reach instead an attachment to the deep aspect of the wing tendon. Such modifications, exceptional for the fourth dorsal interosseous, become the standard pattern for the third. The component bellies of this third dorsal interosseous may occasionally be separately identifiable but are most commonly inseparably fused although the tendons again clearly indicate the muscle's dual phylogenetic derivation. Invariably a well-defined tendon passes superficial to the transverse lamina to join the wing of the aponeurosis. Rarely (8 per cent) a slender phalangeal tendon passes deep to the transverse lamina to insert upon the proximal phalanx. Sometimes (24 per cent) a tendon similarly passes deep to the transverse lamina but then joins the deep aspect of the wing tendon and less commonly (4 per cent) such a tendon also retains a slender phalangeal insertion, signalling its derivation from the dorsal abductor component. Generally (68 per cent), however, the whole of the third dorsal interosseous inserts into the extensor aponeurosis by passing superficial to the transverse lamina.

The constitution of the first dorsal interosseous is especially informative. It is commonly observed in the dissecting room that a volar portion of this muscle, arising from the second metacarpal, wraps over the thick, phalangeal tendon of the remainder, *en route* to a more dorsal insertion. The more dorsal main mass of the muscle, arising bipennately from the adjoining first and second metacarpals, continues as the phalangeal tendon, and has long been known as the M. abductor indicis of Albinus (1734); the volar portion has been named as the M. interosseous prior indicis of Albinus (or M. flexor brevis indicis or M. extensor tertii internodii indicis of Douglas 1750) (Henle 1855; Wood 1866, 1867, 1868; Le Double 1897; Jones 1949). This arrangement clearly recalls that described above in the comparable situation in *Pan* and in *Colobus polykomos*. Again the dorsal bipennate portion clearly represents the dorsal abductor component, and terminates as the usual thick phalangeal tendon reaching its insertion by passing deep to the transverse lamina, where the typical superficial and deep bursae may invest it. The volar portion (flexor brevis three) is variably separate and crosses the phalangeal tendon with a variant degree of obliquity, to reach the surface of the transverse lamina with which it merges. It may be prolonged (4 per cent) into the wing tendon, which on the index finger is however mainly formed by the large first lumbrical, or it may terminate as an abortive wing tendon adherent to the shaft of the proximal phalanx (16 per cent). Most frequently, however, it simply blends into the substance of the transverse lamina (80 per cent). The common description that the first dorsal interosseous inserts solely into bone is clearly an over-simplification.

Occasionally in the human hand (28 per cent) there is a tendon on the radial side of the metacarpophalangeal joint of the minimus, passing deep to the transverse lamina, and whose muscle belly of origin attaches to the fifth metacarpal shaft between the origins of the ulnar head of the fourth dorsal interosseous and the fourth palmar interosseous; it is effectively a delamination of the former. The tendon has the typical topographical relationships of the phalangeal tendon of a dorsal abductor and usually (20 per cent) inserts into the base of the proximal phalanx but may, however, fail to reach this, and so terminate by merging into the metacarpophalangeal joint capsule (8 per cent). It is doubtless homologous with the fibrous band noted above in the same situation in some specimens of *Pan*, and with the common additional insertion into the fifth digit shown by the marsupial fourth dorsal abductor. It seems likely then that the bifid marsupial insertion of this muscle represents the primitive condition, and its reappearance in some other Primates including man is an atavism.

The specialized modification of the attachments of the four dorsal interossei of the human hand is beautifully attuned to the remodelling of the metacarpophalangeal joints and allows the attitude of the hand to be adjusted to the requirements of the power grip as has been described in Chapter 6. The almost complete loss of the phalangeal insertion of the third dorsal interosseous causes the hand to divaricate about the interval between the middle and ring fingers (Fig. 6.16)—abduction is not simply about an axis formed by the medius! The peculiar architecture of the human first dorsal interosseus muscle complements the remodelled structure of the metacarpophalangeal joint of the index finger. In flexion of the proximal phalanx (Fig. 6.14), so important in variants of the precision grip, the angle of attack of the flexor brevis component is admirably suited to imposing a component of endorotation upon flexion. Indeed electromyographical evidence confirms the pre-eminence of this part of the muscle in such grips (Masquelet *et al.* 1986).

Developmental studies could be expected to provide confirmatory data for the idea of a dual derivation of the human dorsal interossei. Fairly recent studies have, in fact, been done (Cihak 1972, 1977) but their interpretation seems only partially to vindicate this notion. The dual origin (from intermetacarpal abductor plus flexor brevis) of the second, third, and fourth interossei is undisputed and clearly confirmed; in contrast Cihak suggests that the first dorsal interosseous differs and is instead derived from two flexores breves, corresponding to numbers two and three in Fig. 9.1A. How can this be, when it is so at variance with the comparative findings? It should be noted that Cihak (1972) was little concerned with phylogenetic data, and even mistakenly believed that the insertion of the intermetacarpal component of a dorsal interosseous was to the extensor expansion, and the flexor brevis to the proximal phalanx. Moreover, a topographical duality of the human first dorsal interosseous, recognized in a specific nomenclature for its component parts, has been appreciated for 250 years.

Cihak's view of derivation of the first dorsal interosseous recalls that proposed by Windle (1883) for all these muscles (Fig. 9.1C) and is apparently based on similar misconceptions about the relationship between ontogeny and phylogeny. Static histological sections, frozen in time, certainly give the appearance of myoblasts belatedly streaming into the intermetacarpal region between the first and second metacarpals from the region of the anlagen of flexores breves two and three; myoblasts from this source may actually be migrating to populate the intermetacarpal region, or a process of myoblastic differentiation may be spreading, after some delay, to this site. But the lack of a separate blastema appearing at this site in no way provides evidence for an actual phylogenetic translocation of the flexores breves into the intermetacarpal region, so as to become the dorsal interosseous. The process is analogous to the way in which osteoblasts derived from the radiale element of the human scaphoid usually invade and ossify the centrale portion; this in no way negates the idea that a centrale homologue is a component of the scaphoid. Both represent the interplay of the processes of heterochrony. As noted earlier the interossei quite likely receive a myoblastic contribution from the embryonic contrahentes. In a comparable way the interossei seem to receive a contribution dorsally from rudiments representing the reptilian dorsometacarpales; this was discussed in Chapter 8.

The marginal musculature

The cupped form of the mammalian palm distorts the basal simplicity of the layered musculature, for the exposed muscles at the margins are rolled inwards, particularly at their origins, so that the dorsal layer takes up a more volar position, corresponding to the plane of the most superficial midpalmar layer in amphibians and reptiles—the flexor brevis superficialis. This positional change, more apparent than real, has unduly influenced the determination of homologies. Cunningham (1882) was well aware of this distortion and decisively recognized the abductor pollicis brevis and the abductor minimi digiti as derivatives of his dorsal abductor layer. Moreover, he designated those dual slips of muscle belonging to his intermediate layer (the flexores breves) and related to the first metacarpal (1 and 2 in Fig. 9.1A) as flexor pollicis brevis; and those related to the fifth metacarpal (9 and 10 in Fig. 9.1A) as flexor brevis minimi digiti. Despite this terminology he explicitly stated that only the radial slip of the muscle related to the first metacarpal was homologous with the flexor pollicis brevis of human anatomy, the other slip becoming the diminutive first palmar interosseous. He implied similar conclusions for the fifth digit: only the ulnar slip corresponded to the flexor brevis minimi digiti of human anatomy, whilst the other was homologous with the fourth palmar interosseous (Fig. 9.1B). The opponens pollicis and opponens minimi digiti were clearly seen to be derived from partial metacarpal insertions of the proper short flexors of the marginal digits. This simple and elegant solution to the problem of homologies in more advanced mammals was

destined to become remarkably confused by subsequent work, which was often based upon inadequate material and rested upon faulty theoretical assumptions.

The hypothenar muscles

McMurrich (1903) derived the whole of the hypothenar musculature, abductor, flexor brevis, and opponens from his primitive tetrapod flexor brevis superficialis sheet. This deduction was based solely upon the circumstance noted above: the superficial location of these muscles, as seen in sections of the arched palm of embryonic hands. Only partial agreement with this was reached by Cihak (1972) in a later embryological study. He held that only the abductor was derived from the flexor brevis superficialis and the remaining hypothenar muscles were developments from the most ulnar of the flexores breves profundi. This assessment of the homology of the abductor was again based on the appearance of the anlage of this muscle, apparently in the same plane as the developing flexor brevis superficialis—the future palmar part of the flexor digitorum superficialis. Of course, in reality the flexor brevis superficialis furnishes the perforated tendon of the fifth digit, and for it simultaneously to provide the source of the short abductor seems improbable in the extreme.

Easily the most influential interpretation of the morphology of the hypothenar muscles was devised by Brooks (1886); his ideas were quickly accepted in British textbooks (Thane 1892) and have been widely followed to the present day. Whilst Brooks concurred with Cunningham that the abductor minimi digiti was a derivative of the dorsal abductor layer he maintained that the most ulnar of the flexor brevis series (10 in Fig. 9.1A) was represented in the human hand only by that small part of the opponens minimi digiti which is situated deep to the deep branch of the ulnar nerve; all of the remainder of the opponens, together with the flexor brevis minimi digiti, were in Brooks' view, derivatives of the adductor layer (the contrahens of the fifth digit). Strangely, nothing was made of the fact that this would entail displacement of the insertion of this adductor from the radial to the ulnar side of the digit, with migration of its origin in an ulnar direction before following the periphery of the carpal tunnel to reach the flexor reticulum. This theory relied for its validity solely on the notion that any part of the palmar musculature superficial to the deep branch of the ulnar nerve must indisputably belong to the contrahens sheet. In fact, Brooks' theory of hypothenar homologies seems to have been effectively demolished in a little appreciated embryological paper by Frazer (1908), and moreover a more plausible scheme of homologies was there tentatively proposed. Frazer showed (as did Cihak 1972, later) that the contrahens digiti quinti has a transient existence in the fetal human hand and this anlage has no connection with the developing opponens; moreover, it was known that in other adult primates the contrahens can be coexistent with a well developed opponens. Frazer, in fact, showed that the opponens minimi digiti first appears at the site appropriate for the most ulnar of the flexor brevis series, and then enlarges so that its origin engulfs the deep branch of the ulnar nerve, and then extends up around the margin of the carpal tunnel, leaving only a small part of its substance deep to the nerve. Rather surprisingly Frazer denied that the flexor brevis minimi digiti develops from this same source, but rather he suggested that it emerges as a late delamination of the abductor minimi digiti. He agreed with Cunningham that this last named muscle belonged to the dorsal abductor layer. Noting its origin from the pisiform, or post minimus, he was not surprised to find the muscle with a transient embryonic attachment to the adjacent metacarpal, the fifth, behind the opponens.

In the following account it will be seen that comparative anatomy largely corroborates Frazer's view, but with the qualification that the flexor brevis minimi digiti represents part of the ulnar flexor brevis profundus (10 in Fig. 9.1A), just as does the opponens.

The marsupial muscles   The basic marsupial pattern, representing also a quite generalized mammalian arrangement is well shown in *Pseudochirus* (Fig. 9.7A). The abductor digiti minimi has its sole origin from the pisiform and inserts by an independent phalangeal tendon into the base of the proximal phalanx; its homology with other dorsal abductors is strikingly apparent. The ulnar flexor brevis of the fifth digit, the tenth of this series of

muscles (Fig. 9.1A), however, shows some progressive advance from the presumed protomammalian condition. The muscle has extended its origin around the margin of the carpal tunnel, here formed by the hamate, to overlie the deep branch of the ulnar nerve and none of the muscle retains a persistent metacarpal origin on a deeper plane. No developmental studies have been performed, but this result would obviously be achieved by a migration similar to that shown by Frazer (1908) in the human hand. Part of the muscle substance, the deepest and most ulnar part, has been arrested on its passage to an insertion by attachment to the metacarpal shaft; thus a small opponens minimi digiti is formed as an offshoot of the flexor and lying entirely superficial to the deep branch of the ulnar nerve. The prime insertion of the muscle (i.e. the flexor insertion) is by a wing tendon coursing dorsally behind the sesamoid-containing glenoid plate to join the extensor expansion. This insertion is the symmetrical counterpart of the insertion of the flexor brevis, (number nine) on the other side of the digit. Despite the modified origin and insertion of the tenth flexor brevis, its comparability with the remainder of the series is clear. Essentially similar arrangements are found in *Trichosurus vulpecula* and in the little American opossum *Caluromys lanatus*. In some specimens the insertions of the abductor and the flexor may not be entirely independent, for the abductor may make a flimsy contribution to the wing tendon and the flexor may contribute a few fibres to the phalangeal tendon. In *Didelphys* a residual part of the flexor brevis minimi digiti retains some of its primitive metacarpal origin, deep to the plane of the deep branch of the ulnar nerve (Brooks 1886; Brandell 1965); the last named author used a complex classification, which has no morphological validity, for the shredded out slips of the various muscles but his observations are in accord with the interpretation given above. The recognition of a persistent deep slip of the flexor brevis is a significant argument in favour of the theory that the muscle has shown a territorial expansion of its origin around the margin of the carpal tunnel.

New World monkeys   *Cebus* (Fig. 9.5A) and other fairly generalized platyrrhine monkeys such as *Pithecia*, give an informative insight into how the

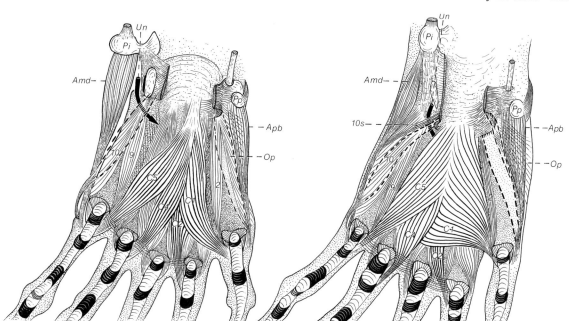

Fig. 9.5

The complete intrinsic musculature in the left hands of *Cebus capucinus* (A) and *Cercopithecus nictitans* (B). *Pi*, pisiform; *Pp*, praepollex; *C1–5*, the contrahentes; *2*, flexor brevis 2 (the first palmar interosseous); *9*, the flexor brevis 9 (the palmar interosseous of the fifth digit); *Un*, deep branch of ulnar nerve; *10*, flexor brevis 10, arising in common with flexor brevis 9, but transformed into the opponens minimi digiti; *Amd*, abductor minimi digiti; *10s*, portion of *10* whose origin has spread across superficial to the deep branch of the ulnar nerve and constitutes the so-called 'radial head' of *Amd* in B; the outline of the flexor digiti minimi brevis is represented by the broken line superficial to *10*; *Op*, opponens pollicis overlaid by abductor pollicis brevis, *Apb* (shown as semi-transparent) and flexor pollicis brevis (broken line). The toughened restraining loops for the long flexor tendons, which are incorporated in the fibrous flexor sheaths and were described in Chapter 7, are shown. The deep transverse ligaments of the palm are also shown.

higher primate morphology must have been derived from the simple primitive mammalian pattern, such as can be deduced from living marsupials. Moreover, the morphology revealed here provides an excellent introduction to the more extremely derived condition seen in other monkeys and apes, and also in man. Abductor digiti minimi, as in the marsupials, arises from the pisiform and inserts by a phalangeal tendon, which skirts the metacarpophalangeal glenoid plate to reach its attachment on the phalangeal base. The flexor brevis minimi digiti, at its origin from the hook of the hamate and the adjoining flexor retinaculum, lies entirely superficial to the deep branch of the ulnar nerve and inserts by a wing tendon wrapping over the phalangeal tendon to attach to the extensor expansion. All of this is quite in keeping with the marsupial condition and accords with the derivation of the muscles proposed there, but in the monkey no opponens portion detaches from the flexor. instead the entire opponens, inserting by definition along the metacarpal shaft, arises deep to the deep branch of the ulnar nerve by a tendinous origin shared with the most medial of the series of palmar interossei; just as this palmar interosseous is undeniably number nine of the series of flexores breves profundi, so there can be little doubt that the opponens represents that part of number ten of this layer which has remained *in situ* and failed to migrate around the carpal arch. It has already been seen that part of this muscle remains in this situation in some marsupials. It is to be noted that the contrahens of the fifth digit is normally situated and there can be no question of any contribution from it to the hypothenar musculature.

Old World monkeys   The extant catarrhine monkeys (Fig. 9.5B) show aberrant derived features, which are fairly obviously off the main stream of progressive primate evolution, yet these same specializations are informative in their own way. All the components described in *Cebus* are present, with comparable attachments and apparently similar phylogenetic derivations. The *in situ* part of flexor brevis 10 has an insertion onto the shaft of the fifth metacarpal and so forms an opponens minimi digiti comparable to that of *Cebus*. In continuity with it are further muscular origins around the arched ulnar margin of the carpus onto the surface of the flexor retinaculum. The most anterior fibres, those from the retinaculum, constitute the flexor brevis minimi digiti and insert into the wing tendon of the fifth digit. The intervening fibres of origin, some deep and some superficial to the deep branch of the ulnar nerve, peel away and attach, not as might be expected to the deeper opponens minimi digiti, but to the tendon of abductor digiti minimi. These latter fibres have been designated as the 'radial head' of the abductor (Hartman and Straus 1933), a confusing and unfortunate choice of name. If these fibres had merged instead with the opponens they would be considered as a superficial layer of it, and as such a migrated part of the tenth flexor brevis; doubtless this is their true nature. There can be little doubt that the whole extended arc of muscular origins around the ulnar margin of the carpus are derived by migration of the origin of the last of the flexor brevis series.

The gibbon hand   The analysis given above of the cercopithecine morphology, and its likely implications, are reinforced by the hylobatid morphology (Fig. 9.6A). The migrated part of the tenth flexor brevis, associated at its origin from the flexor retinaculum with the flexor brevis minimi digiti, forms a lamina of the opponens lying superficial to the deep branch of the ulnar nerve: these fibres are clearly homologous with the cercopithecine supernumerary 'radial head of abductor minimi digiti'. The basic hominoid morphology is thus established in gibbons. The tenth flexor brevis arising in common with the last palmar interosseous (flexor

brevis nine), forms the deep part of opponens, but has also expanded its origin across the deep branch of the ulnar nerve to form a superficial lamina of opponens, whilst its most extremely migrated portion becomes the descriptive flexor brevis minimi digiti, retaining the primitive insertion by a wing tendon which wraps around the phalangeal tendon of the abductor.

The chimpanzee hand   The pattern established in *Hylobates* is further refined in *Pan* (Fig. 9.6B) so that an even more human arrangement is realized. The superficial part of the opponens is more substantial. In some specimens part of it may be segregated as an almost independent belly terminating in a tendon fading away on the surface of the remainder of the opponens. The attachments of the remaining muscles are as in the other Primates described.

The human hand   In man (Fig. 9.7B) the abductor digiti minimi arises as expected from the pisiform. As it approaches its termination the superficial part peels away to insert into the ulnar wing of the extensor expansion, but additionally a rounded and separate tendon emerges from the core of the muscle, often from a quite separate belly, to pass deep to the transverse lamina and insert on the base of the proximal phalanx. The flexor brevis minimi digiti, which is quite often absent, usually retains some of its primitive insertion to the wing tendon but its deeper substance is adherent to the glenoid plate and phalangeal base. The contribution of the abductor to the wing tendon is clearly a secondary association and represents an interchange of insertions comparable to that noted in certain of the composite human dorsal interossei. A conservative *in situ* part of flexor brevis ten forms the deep portion of opponens minimi digiti, lying deep to the deep branch of the ulnar nerve, and it arises from the hook of the hamate and the related pisohamate and pisometacarpal ligaments. Again, as in the chimpanzee the superficial part of the opponens is substantial, and again as in this ape parts of this may be loculated into virtually independent bellies applied to the surface of the rest of the muscle. Apparently of a quite different nature is another supernumerary muscular formation, associated with the surface of the opponens and apparently unique to man. It may take

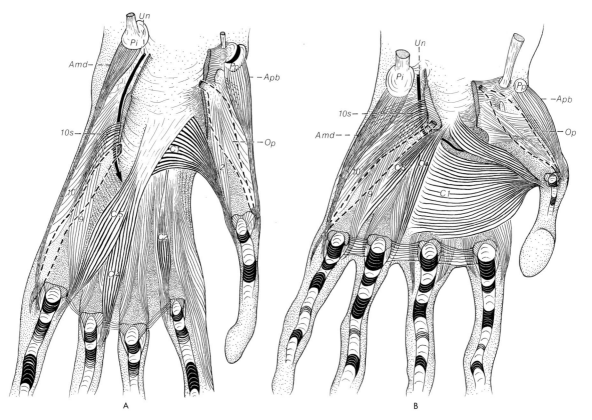

Fig. 9.6

The complete intrinsic palmar musculature of the left hands of *Hylobates lar* (A) and *Pan troglodytes* (B). Lettering is as in Fig. 9.5, but in this case *10s* has become the superficial part of opponens minimi digiti; the outlines of the short flexors of the thumb and fifth digits are again shown by broken lines.

the form of a small fleshy belly arising adjacent to the metacarpal base, often from the pisometacarpal ligament, and terminating in a flattened tendon sometimes inserting in common with the phalangeal tendon of the abductor, but often not reaching so far and fading out in the fascia on the opponens (Fig. 9.7B). It seems that this slip is not infrequently entirely converted into a tendinous band, which proximally may then establish secondary continuity with the termination of the tendon of extensor carpi ulnaris. Such tough fibrous bands, on the surface of the opponens are a common finding in the human hand (Thane 1892). There seems little doubt that the muscle slip represents a deep detached part of the abductor, arising more distally than the main mass of the muscle. It is noteworthy that Frazer (1908)

observed a transient embryonic attachment of the abductor to the fifth metacarpal.

The thenar muscles

As with the hypothenar musculature McMurrich (1903) considered the whole of the thenar group to be derived from the flexor brevis superficialis. It would not be unreasonable, however, to suppose that the thenar muscles have had a comparable history to those of the hypothenar eminence: this would entail derivation of the abductor brevis pollicis from the dorsal abductor layer and the flexor brevis pollicis and opponens pollicis from the first of the flexor brevis profundus series which has extended its origin around the margin of the carpal tunnel. This is essentially the interpretation sug-

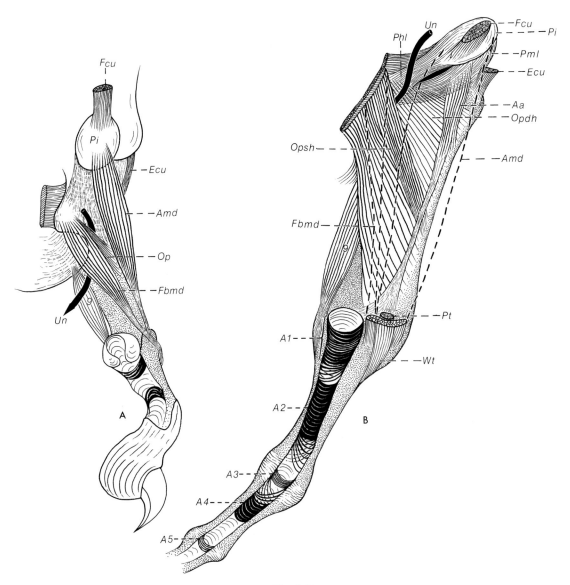

Fig. 9.7

The hypothenar musculature of the right hands of *Pseudochirus laniginosus* (A) and *Homo sapiens* (B). *Fcu*, flexor carpi ulnaris tendon; *Pi*, pisiform; *Ecu*, extensor carpi ulnaris tendon; *Amd*, abductor minimi digiti (broken line in B); *Op*, opponens minimi digiti; *Fbmd*, flexor brevis minimi digiti (broken line in B); *9*, palmar interosseous of fifth digit (flexor brevis 9); *Un*, deep branch of ulnar nerve; *Phl*, pisohamate ligament; *Pml*, pisometacarpal ligament; *Aa*, accessory belly of abductor minimi digiti; *Opdh*, deep head of opponens minimi digiti; *Pt*, phalangeal tendon; *Wt*, wing tendon; *Opsh*, superficial head of opponens minimi digiti; *A1–A5*, pulleys for the long flexor tendons as described in Chapter 7.

gested by Cunningham (1882). Expansion of the origin of the marginal flexor brevis on the thenar side is of course less impeded than at the hypothenar side, for here there is nothing comparable to the deep branch of the ulnar nerve. Just as with the hypothenar muscles Cihak (1972) only partly agreed with this on embryological grounds, preferring to derive the abductor from the flexor brevis superficialis; the same strictures apply to this interpretation as were stated in the discussion on the hypothenar group.

In fact, the comparative findings which follow indicate that the first of the flexor brevis muscles has undergone an expansion of its origin comparable to that of the tenth.

The marsupial pattern Pseudochirus (Fig. 9.8A) shows a delightfully simple basic thenar morphology, establishing a plausible groundplan for the somewhat more derived forms to be described later. The true nature of the different components is here even more convincingly demonstrated than in the hypothenar muscles for there is nothing comparable to the deep branch of the ulnar nerve to complicate issues. The abductor pollicis brevis has virtually its entire origin from the bony prepollex, which articulates with the scaphoid and trapezium and receives part of the insertion of the tendon of abductor pollicis longus. This is a striking mirror-image arrangement, reflecting that of the abductor digiti minimi from the pisiform, and accords with the deduction that both muscles represent dorsal abductors and that both pisiform and prepollex are rudiments derived from marginal digital rays. The muscle inserts by a phalangeal tendon onto the base of the proximal phalanx as would be expected for one of the dorsal abductor series. The origin of the first flexor brevis has enlarged around the margin of the carpal tunnel to reach the flexor retinaculum which here contains a cartilaginous or bony nodule articulating with the scaphoid tubercle. Almost the entire muscle terminates distally, as would be expected, in a wing tendon wrapping around the phalangeal tendon of the abductor to reach the extensor expansion. A few fibres peel away from the muscle belly to insert on the metacarpal shaft, constituting a rudimentary opponens pollicis. In other marsupials with generalized hands the opponens may be more extensive and Cunningham (1882) has noted the

variable development of the muscle in other species. At its origin the flexor brevis pollicis is closely associated with the second of the flexor brevis series, just as are the pairs of these muscles for other digits. In essence, then, the arrangements merely duplicate those in other digits, when the rolled in orientation of the first is taken into account. The true identity of the muscular components can be in little doubt.

New World monkeys Cebus (Fig. 9.5A) shows a simple transition from the marsupial pattern. The core of the abductor pollicis brevis again arises from the sizeable bony prepollex, but its origin has encroached on the neighbouring bones: the tubercle of the scaphoid and the trapezium. It inserts by the usual phalangeal tendon. The first and second of the flexor brevis series are again quite clearly recognizable for what they really are and are associated at their origins as would be expected. The second is the first palmar interosseous and terminates by the usual wing tendon. The origin of the first of the flexor brevis series has extended around the fibrous margin of the carpal tunnel (the reflected part of the flexor retinaculum) to encroach on the superficial surface of the flexor retinaculum. The fibres from the deepest part of this arched origin, in closest association therefore with the origin of the first palmar interosseous, have their insertion arrested on the shaft of the metacarpal, and thus by definition constitute the opponens pollicis. The more superficial portion, arising from the flexor retinaculum, continues to a digital insertion, as the flexor pollicis brevis. This insertion is, however, no longer into a clear-cut wing tendon. The fibres merge with the phalangeal tendon of the abductor and with the side of the sesamoid-containing glenoid plate. The main advance over the marsupial condition is the augmentation of the opponens pollicis subdivision.

Old World monkeys Cercopithecus (Fig. 9.5B) exemplifies the typical morphology. The abductor again has its core origin on the prepollex, with encroachment onto neighbouring bones, and inserts by a phalangeal tendon. In the specimen illustrated a clear-cut first palmar interosseous (flexor brevis two) occurred, and was associated with the arched origin of the derivatives of the first flexor brevis from the fibrous carpal tunnel mar-

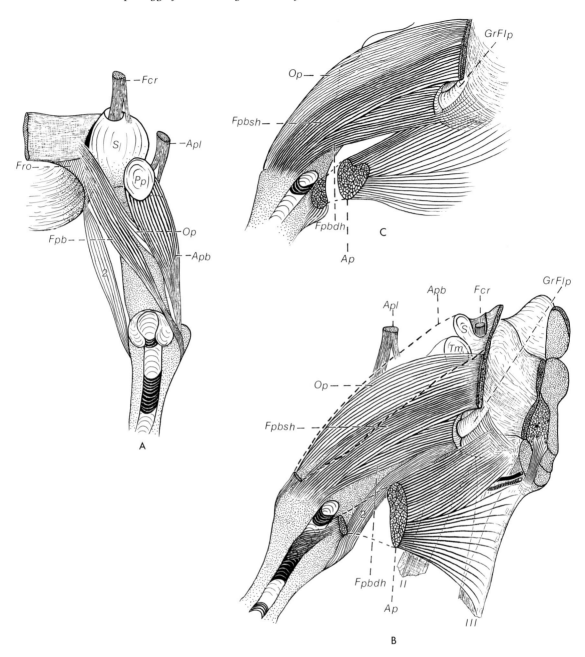

Fig. 9.8
The thenar musculature in the left hand of *Pseudochirus laniginosus* (A) and in two right hands of *Homo sapiens* (B, C). *Fcr*, flexor carpi radialis tendon; *S*, tubercle of scaphoid; *Apl*, abductor pollicis longus tendon; *Pp*, prepollex; *Op*, opponens pollicis; *Apb*, abductor pollicis brevis (broken line in B and removed in C); *Tm*, trapezium; *GrFlp*, groove for tendon of flexor pollicis longus; *II*, *III*, second and third metacarpals; *Ap*, adductor pollicis; *Fpbdh*, deep head of flexor pollicis brevis; *Fpbsh*, superficial head of flexor pollicis brevis.

gin. The opponens, of considerable size, was derived from that part of the first flexor brevis having the most external origin, excluded from the actual tunnel margin. Thus the belly of flexor brevis pollicis took origin from the whole arched tunnel margin extending from the ventral wall (flexor retinaculum) onto the dorsal wall (carpal bones and reflected part of retinaculum). This whole origin was thus folded around the entrance of the tendon of flexor pollicis longus into the palm. The insertion was mingled with the phalangeal tendon of the abductor and the attachments of both muscles were to the base of the phalanx, the sesamoid-containing glenoid plate, and the wing tendon. The latter was not the exclusive termination of either muscle.

The gibbon thenar muscles Again the abductor arises from the prepollex and terminates as a phalangeal tendon attaching to the base of the phalanx (Fig. 9.6A); before reaching this insertion, however, some of its substance peels away to be prolonged as a rudimentary wing tendon. In fact, the abductor has commandeered the wing tendon as its own partial insertion leaving the flexor brevis attachment arrested at the glenoid plate and phalangeal base. A significant, and fortuitous, specialization of the gibbon palm heightens the distinction between the various muscle groups and clarifies homologies. The elongation of the palm is accompanied by an exaggerated prolongation of the walls of the carpal tunnel as tough fibrous, or fibrocartilaginous, extensions as noted in Chapter 6. This is particularly accentuated on the radial side where the reflected part of the retinaculum forms a tough septum, with a distal free border, effectively separating the origin of the first contrahens (adductor pollicis) from those of the flexors breves (one and two) of the thumb. These two muscles share a confluent origin from the outer side of this partition. The first palmar interosseous (flexor brevis two) proceeds distally, at first separated from the adductor by the septum, to its usual wing tendon insertion. The origin of the first flexor brevis has spread around the outer tubular wall of the extension of the carpal tunnel. The most dorsally arising fibres constitute the opponens pollicis, as determined by their metacarpal insertion. The flexor brevis pollicis belly is provided by the most ventral extension of

this origin onto the flexor retinaculum. The ventrally lying flexor has its insertion secondarily arrested at the sesamoid-containing glenoid plate and ventral part of the phalangeal base. As has been seen, the wing tendon, such as it is, has largely become the property of the abductor.

The chimpanzee thenar muscles In *Pan* (Fig. 9.6B) there is only a limited degree of distal prolongation of the tubular carpal tunnel. The abductor pollicis brevis has its core origin on the prepollex, but as usual this has spread onto the adjoining scaphoid tubercle. The first flexor brevis has expanded its origin to cover the entire outer aspect of the fibrous extension of the carpal tunnel reaching ventrally onto the flexor retinaculum. The deepest and most external of the fibres derived from this origin insert on the metacarpal as the substantial opponens pollicis. The most superficial fibres, those from the flexor retinaculum, comprise the flexor pollicis brevis. In the specimen illustrated in Fig. 9.6B there was no first palmar interosseous (flexor brevis two), but in another specimen it was found, and, as would be expected under the phylogenetic proposals elaborated here, it took origin in close proximity to the most dorsally arising opponens fibres. Whether there were single or dual flexor brevis origins, in this situation they were segregated from the contrahens origin (adductor obliquus) by the distal prolongation of the margin of the carpal tunnel. The insertion of the flexor pollicis brevis terminated at the radial side of the glenoid plate and base of the proximal phalanx, in close association with the more dorsally lying phalangeal tendon of the abductor. The latter peeled off from its surface some fibres contributing to the rudimentary wing tendon of the extensor expansion.

The human thenar muscles No dramatic evolutionary change has effected the human muscles, and anyone familiar only with the primates described above would feel quite at ease dissecting the muscles of man. A critical feature affecting their interpretation, however, is the detailed topography of the radial attachment of the flexor retinaculum: as noted in Chapter 6 the distal part of the retinaculum is essentially free of bony attachments and forms a tough fibrous sleeve-like prolongation of the lateral wall of the tunnel into the palm, and converges onto a strong ligament attaching to the base of the third metacarpal (Fig. 9.8B). The derivatives of the flexores breves of the thumb (one and two of the series) largely arise from the external surface of this sleeve, which demarcates their origin from the more centrally arising contrahentes derivatives. The sleeve is conspicuously channelled by an obvious synovial-lined groove for the tendon of flexor pollicis longus and its sheath. From the lower and outer aspect of the fibrous sleeve, and to some extent from the adjacent base of the first metacarpal, arises the first palmar interosseous (flexor brevis two) when present (Fig. 9.8B). It commonly inserts partly on the sesamoid-containing glenoid plate at the metacarpophalangeal joint of the thumb, but usually retains some of its primitive continuity with the rather flimsy ulnar wing tendon of the extensor expansion of the thumb. In close association with it arise the derivatives of the first flexor brevis which have, as in other primates, extended their origin right around the outer aspect of the fibrous sleeve and often quite far across on the flexor retinaculum—the parts arising ajdacent to the margin of the carpal tunnel comprise the flexor pollicis brevis of human anatomy. The deepest part, closely associated with the origin of the first palmar interosseous, is the part now almost universally designated as the deep head, and it joins the remainder, the superficial head, by curling deeply and around the flexor pollicis longus tendon. The whole muscle has its insertion arrested at the radial sesamoid in the glenoid plate of the metacarpophalangeal joint of the thumb, little if any being continued on into the primitive insertion—the radial wing tendon. The opponens pollicis is closely associated with the external aspect of the former muscle, sharing a common derivation with it. It arises from the outer aspect of the fibrous sleeve, often far across the flexor retinaculum, and additionally has some bony origin from the trapezium. The abductor pollicis brevis has an extensive origin from bone and flexor retinaculum encircling the tunnel for the flexor carpi radialis tendon since the propollex is lacking in man, but it commonly receives a slip from the abductor pollicis longus, a reminder of its ancient virtually exclusive origin from the prepollex. The deeper and more medial portion of this muscle inserts on the glenoid plate with its sesamoid and

the proximal phalanx. The remainder is continued into the wing tendon on the radial side of the thumb. In effect the primitive insertions of abductor and flexor have been interchanged, just as has been noted to a variable degree in other primates. This may be consequent upon the changed orientation of the thumb. In the functional context, this specialization of the insertion of the short abductor means that the muscle can function not only as an abductor but also as a medial rotator at the metacarpophalangeal joint, movement then being congruent with that occurring at the carpometacarpal joint of the thumb, and potentiating the important action of opposition (Napier 1952). The adductor pollicis obliquus arises within the fibrous sleeve of the carpal tunnel, which thus demarcates it from the encircling flexor pollicis brevis and opponens complex. The adductor obliquus muscle mass, as it passes to its insertion with the transverse part into the ulnar metacarpophalangeal sesamoid, the base of the phalanx, and perhaps with a minor extension into the wing tendon, has its radial border closely approximated not only to the first palmar interosseous, but also to the deep head of flexor pollicis brevis. Indeed, it may be variably merged or inseparably fused with the former. Moreover, in some specimens this first contrahens may also make a contribution to the deep head of the flexor brevis pollicis (Fig. 9.8C) in a manner analogous to the way in which contrahentes five and four may merge with the related flexor brevis (palmar interosseous) in *Pan* (Fig. 9.6B). In the case of the thumb muscles, however, the fusion is not with the corresponding flexor brevis, but with that inserting on the opposite side of the digit; such fusion has clearly been accommodated by the rotation of the thumb.

The identity of the different components of the human thenar muscles, and their morphological significance, shown in Figs 9.8B, C can be in little doubt, but it has not always been so. The descriptive anatomy of the human thenar muscles has been a minefield of terminological confusion. When this is extended into the comparative field the chaos is compounded.

Albinus (1734) designated both the true abductor pollicis brevis and the superficial head of flexor brevis pollicis as superficial and deep heads of the former muscle; for him then the flexor pollicis brevis was the true deep head together with the adductor obliquus. This idea was essentially followed by the German school and we find Henle (1855) using the same concept of a two-headed abductor but the flexor pollicis brevis in his nomenclature was restricted to the deep head; significantly, he correctly identified the first palmar interosseous (flexor brevis profundus two).

Brooks (1887) recognized the abductor for what it really was, defined the superficial head of the short flexor as the 'radial' head but strangely considered that the first palmar interosseous was the 'ulnar' head of this same muscle. The true deep head he considered to be merely part of the adductor obliquus. The essence of this view until very recently permeated the British literature, although the terms 'superficial' and 'deep' were substituted for 'radial' and 'ulnar'. In contrast the French typically followed the convention shown in Figs 9.8B, C, recognizing the true deep head for what it really was, although this was based upon precise observation of the human condition rather than on comparative data; the French position has been summarized and excellently illustrated by Poirier and Charpy (1901). British textbooks have gradually fallen into line with this interpretation, influenced perhaps by the clear discussion of the issue given by Jones (1949*a*).

A recent morphological study on the flexor pollicis brevis by Day and Napier (1961) has unfortunately added to the confusion. The chain of reasoning developed by these authors was as follows: McMurrich (1903), on purely theoretical grounds—the exposed superficial situation of the muscle—had derived the flexor pollicis brevis from the reptilian flexor brevis superficialis; Cunningham (1878) had stated that the short flexors of the digits were derived from the superficial intrinsic palmar layer (but this was the contrahentes, or adductors, and quite different from McMurrich's superficial layer); Day and Napier therefore concluded that the deep head (as shown in Fig. 9.8B) was properly defined as the 'deep head of flexor pollicis brevis' for after all, following Brooks, British anatomists had usually considered this slip as part of the adductor obliquus. All that seemed to be involved was some semantic reshuffling.

The flaw in this reasoning was that Cunningham had committed an error in his preliminary paper in 1878, which he later realized and corrected in his detailed publication of 1882, then correctly stating that the flexores brevis profundi, the intermediate layer having no affinity with the adductors, were the true antecedents of the flexor pollicis brevis, flexor brevis minimi digiti and the palmar interossei. Thus for quite faulty morphological reasons Day and Napier had arrived at the correct conclusion: the deep head is truly part of the flexor pollicis brevis. Their method of illustrating the morphology, by depicting the various components of the thumb musculature flayed and laid apart flat is quite unnatural and erroneously suggests that the deep head is morphologically aligned with the adductor obliquus which, of course, suited their viewpoint. They were aware that if their reasoning was correct, part of the adductor sheet must have migrated from the ulnar sesamoid of the thumb to the radial one, a possibility which seems unlikely on the basis of the usual principles governing change in muscle attachments. In a sense there was a germ of truth in this belief in affinity between the deep head and the contrahens layer, for as has been seen the first contrahens can on occasion (Fig. 9.8C) make a contribution to the deep head, but merely as an adjunct to the main mass. Day and Napier (1963) attempted to read special significance into this supposed migration of part of the contrahentes layer to produce a deep head correlating the fact that it was apparently restricted to Old World monkeys, man and *Pongo*, with the acquisition in catarrhines of true opposability of the thumb; they explained its apparent absence in *Pan*, *Gorilla* and *Hylobates* as being a consequence of very deep carpal tunnels produced by an in-set orientation of the trapezium. This theory gives an impression of a discontinuity in evolution, with the appearance of an innovative apomorphic change in catarrhines, which is quite at odds with the facts. As shown in Figs 9.5, 9.6, 9.8, a common basic pattern involving migration of the first flexor brevis around the radial margin of the carpal tunnel is characteristic of all higher Primates. Differences occur in just what part of this expanded origin terminate in a metacarpal insertion and thus form the opponens. If it is the middle portion (Figs 9.5B, 9.8B) then the deepest part, occupying the original unexpanded origin, remains as a contributor to the flexor pollicis brevis—a deep head—but this so-called 'head' is very different from the unique innovation envisaged by Day and Napier.

10

Fossil hand bones

It is a fundamental tenet among anthropologists that a watershed in primate evolution was achieved by the firm establishment of tool use and manufacture. The dawn of this new era was a long time ago, primitive tools have been found at Lake Rudolf, Kenya, dated at 2.6MY ago (Brock and Isaac 1974) and in the Omo valley dated at approximately 2MY ago (Merrick *et al.* 1973).

However, it has been widely believed by anthropologists that the emergence of tool-making was unrelated to any spectacular advance in the functional anatomy of the human hand (Oakley 1972), but, as noted in previous chapters the hand of *Homo sapiens* has its full quota of apomorphic features. This chapter will not be a comprehensive review of all the fossils found but will concentrate on certain specimens where the presence or absence of those apomorphies can be reasonably determined. Thus, the evolutionary scenario deduced from comparative anatomy can be put to the crucial test, and related to the historical perspective.

The emergent primate hand

It has been noted in Chapter 2 that the fossil Adapidae are widely considered to be ancestral not only to the living prosimian Primates but also to the higher Primates. Until recently, however, their hand anatomy has been poorly known. Now, however, almost complete hand skeletons are known for an American middle Eocene notharctine (*Smilodectes gracilis*) and a European late Eocene adapine (*Adapis parisiensis*). In Chapters 5 and 6 *Lemur* was used as a model for the emergent primate hand, but with the recognition that it had its own apomorphic characters, although these could reasonably be deduced from comparative studies. These hypotheses are largely confirmed by the fossils.

The notharctine hand

The hand of *Smilodectes gracilis* the earliest, and probably most primitive, of the known adapids has recently been described by Beard and Godinot (1988). The wrist and midcarpal joints show a topography similar to that of *Lemur* (Fig. 5.1A) differing only in two respects: the centrale is not enlarged as in *Lemur*, to the extent that it extends across the back of the carpus to contact the hamate, and the embrasure receiving it between capitate and trapezoid is not cramped and sloped as in *Lemur*; the facet for the ulnar styloid process on the pisiform is not deeply excavated and very broad mediolaterally as in *Lemur*, but corresponds more in width to that of the triquetral pisiform facet as it does in higher Primates (Fig. 5.1B). The two extant lemuriform synapomorphies, which are lacking in *Smilodectes* must therefore have been established at some later stage in adapid evolution.

The carpometacarpal joints of *Smilodectes* go a considerable way to confirming the evolutionary scenario derived from comparative studies. The trapezium has a sellar shaped joint for the first metacarpal and bears an enormous palmar tubercle, as in *Lemur*, where it overlies the flexor carpi radialis tendon.

The particular stepped character of the carpometacarpal junction which characterizes the higher Primates is not established, as would be expected, and the capitate articulates only with the third metacarpal, and not the fourth also, although precise information is lacking about whether the third metacarpal contacts the hamate. Dual facets between third and fourth metacarpals are demarcated by the presence of a groove for a typical carpometacarpal ligament. On the other side of the capitate the dual facets for the second metacarpal base are confluent. As noted in Chapter 6 this is an apomorphic development which has occurred independently in some insectivores, marsupials, monkeys, and man. Yet it does not necessarily mean the total disappearance of the primitive carpometacarpal ligament occupying this site: in *Lemur* the facets are confluent, yet a persistent shred of ligament remains (Chapter 6).

The adapine hand

In the later fossil *Adapis parisiensis* (Godinot and Jouffroy 1984), at least in one version of the reconstruction it seems clear that the third metacarpal base is not restricted to the capitate but does abut slightly against the hamate. The expectation that this is the primitive primate condition, and indeed the basal mammalian condition, is fulfilled.

It would not be surprising if this later fossil had not accumulated further specializations. Indeed the ulnar articulation on the pisiform has been expanded mediolaterally as in extant lemuriforms (Fig. 5.1A). This synapomorphy linking the late Eocene Old World *Adapis* with extant lemuriforms suggests a more recent common ancestry between these two taxa than either share with *Smilodectes*. *Smilodectes* in lacking these lemuriform specializations, might therefore be nearer the ancestry of the haplorhines, but the derived form of its articulation between capitate and second metacarpal excludes it from direct ancestry.

Fossil ape and hand bones

Early Miocene apes

From the time of their discovery, it has been assumed that the ancestral apes, the dryopithecines of the African early Miocene, were basically only 'dental' apes: ape-like in dentition but essentially quadrupedal, and monkey-like postcranially. The discovery of the almost complete forelimb skeleton of the early Miocene ape designated either as *Dryopithecus (Proconsul) africanus* or alternatively as *Proconsul africanus* (the preferred taxo-

nomy here has, in fact, gone full circle and the latter name conferring separate generic status is now generally used) formed the keystone of this hypothesis. The classical first detailed description of these important fossils was by Napier and Davis (1959) who concluded that this fossil primate preserved an essentially quadrupedal type of wrist articulation not unlike that of cercopithecoid monkeys (the ulna articulating with pisiform and triquetral), but that this was accompanied by incipient brachiating features in other parts of the forelimb skeleton. These conclusions were reasonable at the time since knowledge of the fundamental differences between ape and monkey wrists was then quite rudimentary and the presence of a meniscus (which might even contain a lunula) in the joints of hominoids, accompanied by drastic bony remodelling, had not then been documented.

Subsequently I reassessed these fossils on the basis of the osteological criteria described in Chapter 5, Figs 5.4, 5.15 (Lewis 1971b, 1972a,b), in order to try to determine whether they too shared the synapomorphy of a meniscus-containing wrist with the known existence of this trait in all the extant hominoids. The key features were as follows (Fig. 10.1).

I described the ulnar styloid as having a 'hook-like' form (Fig. 10.1C). In retrospect this description was not particularly apt and has certainly been widely misinterpreted. It was meant to convey the idea that an articular surface clothes its distal and peripheral aspects, as in *Hylobates* or *Pan* (Fig. 5.4B, C), rather than on the surface facing the interior of the wrist joint as in monkeys (Fig. 5.4A). In the fossil there is some indication of subdivision of this articular area into a small facet at the tip (for the triquetral) and a much larger peripheral area (presumably for a meniscus). Thus, the appearance of the ulnar styloid process, even considered in isolation, is strongly suggestive of a meniscus-containing joint; it is not conclusive, however, for this appearance can very occasionally be mimicked in the macerated limbs of monkeys which were known to have facets in the fresh state on the usual aspect facing the interior of the joint.

The hamate is clearly vertically aligned and bears only a narrow articulation at its apex for the lunate and its facet for the triquetral is markedly spiral. In these features the hamate bears a striking

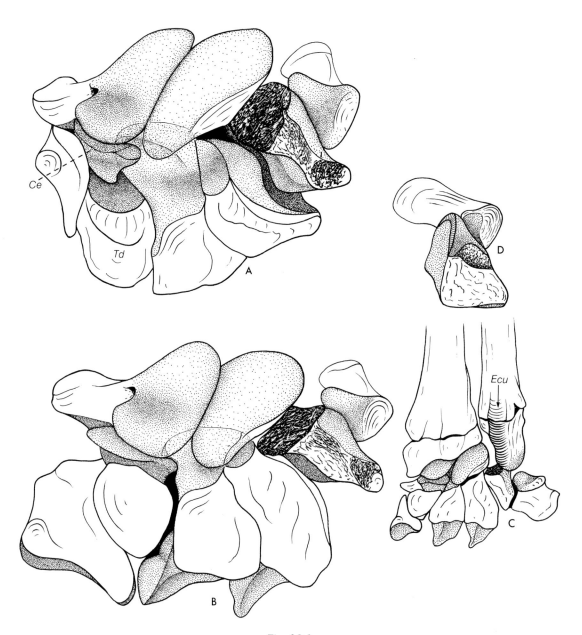

Fig. 10.1
A, the carpus of *Proconsul africanus* (KNM-RU2036) with the addition of a gibbon centrale (*ce*) and trapezoid (*Td*) in the position of midcarpal flexion, and viewed from proximally and dorsally. B, the same specimen in midcarpal extension. C, the same specimen with the addition of radius and ulna in the position of full wrist extension, viewed from posteromedially and showing the groove for extensor carpi ulnaris (*Ecu*). D, the articulated triquetral and pisiform of a specimen of *Pan troglodytes* viewed from behind.

resemblance to that of the chimpanzee. However, so-called 'spiral' facets have been noted in Chapter 5 as occurring in some dry monkey bones, and even in prosimians, although they function quite differently from those of apes.

The disposition of the pisiform and triquetral is much less equivocal and is highly suggestive of an essentially hominoid type of wrist organization. The triquetral is broken and lacks its proximal portion. The remaining part, however, has the form of a triangular pyramid with the palmar surface largely devoted to an extensive shallowly concave facet for the pisiform which is strikingly like the comparable part of the chimpanzee bone (Fig. 5.4C). The pisiform is also like that of *Pan*, and bears a large dorsal facet for the triquetral which is confluent with a smaller (meniscal) facet on the proximal aspect of the base; this latter facet was interpreted by Napier and Davis (1959) as being for the ulnar styloid but of course they took no account of the possible presence of a meniscus. The whole bone must clearly have been displaced and orientated distally into the palm as in extant hominoids.

The scaphoid is strikingly like that of gibbons, being deeply grooved between the long tubercle and body and again, as in these lesser apes, the proximal surface shows a residual dorsal concavity, impressed in extension by the locking onto it of the dorsal rim of the radius (Fig. 10.1C). There had clearly been an independent centrale, as there is in *Hylobates* but this was not recovered. The form of the capitate provides additional confirmation of the ape-like character of the wrist for the bone is quite markedly 'waisted', the neck being articular posteriorly and non-articular anteriorly, although these features are noticeably less obtrusive in the fossil than in the chimpanzee.

The outcome of these observations was that this early Miocene ape had already established those anatomical changes at the wrist, which could be envisaged as facilitating supination and thus favouring suspensory locomotion. This suggestion triggered a rather remarkable series of papers which almost invariably defended the original proposition that *Proconsul* possessed a cercopithecoid type of morphology at the wrist, but usually they arrived at this conclusion by a very dubious chain of reasoning which often betrayed a less than clear understanding of the features to be looked for as evidence for a meniscus-containing joint. Some of the publications, however, equivocated to some extent. None were based on new date and indeed they often demonstrated a lamentable lack of understanding of basic points of morphology, but despite this they did have the effect of totally confusing what was a very important issue: the likely locomotor specializations of the emergent apes.

Schön and Ziemer (1973) postulated that the fossil wrist would have become closed-packed during dorsiflexion and that the triquetral would then be moved down the spiral hamate facet and impacted against the hamate hook, rather as it does in *Alouatta* or *Ateles* (Fig. 5.5). These platyrrhine monkeys were therefore suggested as satisfactory models for the pattern of palmigrade quadrupedal locomotion which they attributed to *Proconsul africanus*. It follows from what has been described in Chapter 5 that this would only be valid if the fossil had retained a monkey-like midcarpal axis. As will be seen later, the evidence, in fact, is in favour of a hominoid type of axis which would produce quite the opposite functional result. These authors did not consider the question of the presence or absence of a meniscus, but the idea had again been introduced that the fossil possessed a monkey-like hand and wrist, albeit a ceboid-like one.

Morbeck (1975, 1977) accepted this idea that the spiral facet of the fossil hamate indicated a habitual palmigrade posture but cited other features which indicated an essentially monkey-like morphotype for the carpus. She asserted, that the *Proconsul* ulnar styloid process articulated not only with the triquetral but also with the pisiform and that the form and relationships of the articular surfaces here, as well as the size, shape, and placement of the articular facets on the pisiform, were comparable to those in palmigrade quadrupedal cercopithecoids. In fact, a triquetral facet alone, adjacent to the tip of the ulnar styloid, would also be quite compatible with hominoid affinities: such a facet is regularly present in the meniscus-containing joint of *Hylobates*, and commonly also found in *Pan* as has been indicated in Chapter 5. In the later paper (Morbeck 1977) she, however, equivocated about pisiform articulation and suggested that it might be for a meniscus; if true, this would radically change the conclusion.

As regards the capitate she reached the conclusion that 'waisting' of this bone 'approaches the condition present in *Papio* and *Macaca*'. This statement was apparently the result of the way in which this term had been used in this particular context, as is evidenced by her Figure 2, a profile outline of the lateral aspect of this carpal bone, in which 'waisting' is interpreted as an anterior indentation (a definition of the term has been given in Chapter 5). Clark (1967) first applied the term to hominoid capitates but his figures of anterior views were incorrectly captioned as being 'posterior' and confusion has persisted ever since. These studies by Morbeck have, in my opinion, no secure logical basis, but are often quoted as evidence favouring a quadrupedal palmigrade locomotor pattern for *Proconsul africanus*.

O'Connor (1975) studied the wrists of cercopithecoid monkeys and with this background proceeded to reconsider the *Proconsul africanus* material (O'Connor 1976). His suggestion that the head of the fossil ulna is cercopithecoid in form is misleading, since a comparable appearance can occur in gibbons, as has also been appreciated by Corruccini *et al.* (1975), especially when the immature condition of the fossil is taken into account. O'Connor's contention that the fossil hamate lacks a hamulus and is cercopithecoid-like is also indefensible. The *Proconsul* hamate hook may not be quite as clearly demarcated as that in *Pan* but it is broadly similar. This author did rightly note that some cercopithecoids possess an apparently spiral facet on the hamate for the triquetral but as described in Chapter 5 the monkey facet contrasts in form and function with that of apes, and *Proconsul africanus* had the hominoid morphotype. In likening the 'waisted' form of the fossil capitate to that of cercopithecoids, he demonstrated the widespread confusion over the use of this term, and even more surprising was his belief that the *Proconsul* pisiform was similar to these monkeys. O'Connor's overall conclusion, like that of Morbeck, was that *Proconsul africanus* possessed the wrist anatomy of a quadruped of cercopithecoid type.

Multivariate statistical methods have also been enlisted to solve the problem but with conflicting results. Corruccini *et al.* (1975) using such an approach, incorporating thirteen measurements relating to wrist osteology, concluded that *Proconsul africanus* had a 'monkey-like' structure. Their basic metrical data, however, quite inadequately reflected the subtleties of the underlying wrist morphology for only one of their measurements (ulnar styloid length) could conceivably discriminate between a meniscus-containing antebrachiocarpal joint of hominoid type and one of monkey morphology and their capitate measurements were mistakenly based upon a lateral profile view in order to assess 'waisting'. These same authors in a later publication (1976) incorporated other data from elbow, shoulder and the foot of *Proconsul africanus* and again stressed the resemblances to cercopithecoid monkeys. In contrast to these studies Conroy and Fleagle (1972) and Zwell and Conroy (1973) in earlier multivariate analyses had aligned the *Proconsul africanus* forelimb with those of the African apes and attributed a knuckle-walking locomotor habitus to it. That such a supposedly objective technique should lead to such divergent results might be thought surprising; what these results really demonstrate are the limitations of multivariate methods particularly when applied to a complex area such as the wrist where there are formidable difficulties in formulating metrical data which adequately reflect the morphology.

The 'dental ape' school was given new encouragement by the cineradiographic studies of Jenkins and Fleagle (1975) and Jenkins (1981) which have been discussed in Chapter 5 and which seemed (mistakenly) to show that the wrist articulations of African apes were essentially quadrupedally adapted and that the real morphological hallmark of 'brachiation' was a midcarpal joint of enhanced mobility as found in *Ateles* or *Hylobates*. Indeed Jenkins (1981) speculated that the carpus of *Proconsul africanus* was suggestive of a fundamentally quadrupedal habitus.

With an implicit acceptance of Jenkins' work Corruccini (1978) again applied comparative morphometrics to the hominoid wrist, using thirty measurements mainly reflecting those interpretations. Not surprisingly he concluded that the unique specializations of the African apes, and even man, are related to knuckle-walking. These measurements were used by McHenry and Corruccini (1983), in a study widened to include *Proconsul africanus* and *Macaca*, which produced the proposition that the *Proconsul africanus* wrist in essence resembled that of the macaque, and indicated that the fossil practised the generalized quadrupedalism typical of monkeys. In line with their earlier study they went to the surprising length of asserting that the whole postcranium was equally plesiomorphic. The pendulum had swung firmly the way of the dental ape school.

The situation was transformed by the discovery of many additional parts of the skeleton of this same *Proconsul africanus* specimen (KNM-RU2036) and its description by Walker and Pickford (1983). As these authors say: 'It has been the consensus that the postcranial skeleton of *Proconsul* was, in many features, reminiscent of those of certain New and Old World monkeys. It now seems that in other respects there are many features that are also found in living apes, particularly *Pan* . . . it appears that in the forelimb there is a gradient of features along the limb such that features that are more hominoid-like are found more proximally. In the hindlimb the gradient is reversed . . .'. They noted for instance that the scapula looks like a diminutive version of a chimpanzee bone, the brachial index is very close to *Pan* and the upper part of the radius cannot be distinguished from *Pan*.

Mosaic evolution is, of course, a well documented occurrence, but to have it in such starkly contrasting form that the upper end of the radius, so concerned in supination, is hominoid in morphology whilst the wrist is cercopithecoid, surely stretches credibility. These authors' reconstruction of the forearm and hand bones, is reminiscent of that by Napier and Davis (1959), but whereas these latter correctly show the pisiform as distally directed, Walker and Pickford (1983) have contrived to show it proximally directed, in the monkey 'heel-like' attitude, an orientation quite impossible if any anatomically plausible articulation of the bones is to be realized. Clearly the time seems to be ripe, for a fresh look at the wrist of this fossil. Much of such a reassessment, or defence, has already been published (Lewis 1985*b*) but now a number of additional isolated carpal bones of *Proconsul africanus*, and also the larger *Proconsul nyanzae*, have been recovered and described (Beard *et al.* 1986) which should provide a critical check on those interpretations. Regrettably the illustrations provided by Beard *et al.* (1986) are rather fuzzily uninformative on key anatomical points of

functional importance (this even applies to the well known bones of apes and monkeys which were used for comparison) and certainly the authors' conclusions cannot be unreservedly accepted, but some provisional corroboration of previously doubtful points can be obtained. Interestingly these authors themselves, while paying lip service to the fashionable view of a monkey-like morphology, state 'the structure of the ulnar side of the wrist, particularly the ulnocarpal joint, is significantly different from that of extant monkeys and suggests some functional affinities with extant hominoids.'

Functional morphology of the wrist of P. africanus

This discussion will concentrate particularly on the form of the ulnar styloid process and its relationship to the triquetral and pisiform, and the nature of the probable midcarpal mechanism. Just those parts which Beard *et al.* (1986) were clearly unhappy about. My initial coining of the term 'hook-like' to describe the form of the ulnar styloid was in retrospect unfortunate. Several authors have seized upon this and countered by stating that it is straight or columnar (Morbeck 1977; Corruccini *et al.* 1975) and it is, but then so are those of most primates. Indeed when portrayed in anteromedial view, as is the usual practice (Walker and Pickford 1983; Corruccini *et al.* 1975) the fossil and monkey styloid processes may appear virtually indistinguishable. When viewed from posterolaterally, however, a very different picture is revealed. In monkeys the extensor carpi ulnaris tendon enters the hand between the ulnar styloid process and the head of the bone, and then winds around medially to spiral down (Fig. 5.1B) across the exposed non-articular surface, sometimes burnishing it smooth so that in anterior view it may mimic that of the fossil. In hominoids, including gibbons, this same tendon is effectively moved ulnarly, and courses straight down the back of the ulnar styloid process, grooving it, before contacting the periphery of the meniscus which then separates the tendon from the triquetral (Fig. 5.7). *Proconsul africanus* has the hominoid type of morphology (Fig. 10.1C).

Any accurate understanding of the anatomical relationships at the ulnar side of the wrist, depends critically upon the correct articulation of radius and ulna at the inferior radio-ulnar joint. Almost invariably in reconstructions of the *Proconsul africanus* material the ulna is shown too distally located and indeed, it is only then that it is feasible for the ulnar styloid process to contract the pisiform directly. In fact the immature condition of the fossil, with persistent growth plates at the lower ends of both bones, facilitates accurate positioning. The ulnar head should be quite proximally located with respect to the radius, largely articulating with the diaphysis of that bone, and leaving room for the attachment of a substantial triangular fibrocartilaginous disc, which as in other higher primates undoubtedly separated it from the proximal carpal row. The arrangements should be comparable to those shown in the immature chimpanzee illustrated in Fig. 5.8. Thus articulated (Fig. 10.1C) the likelihood of direct articular contact between the ulnar styloid process and the pisiform is highly improbable.

The ulnar-sided articular gap is accentuated by the form of the triquetral. Even though the fossil is incomplete, missing the portion closely adjacent to articulation with the lunate, its form and position in the carpus can be determined with a reasonable degree of certainty. It is totally unlike the comparable monkey bone, and is essentially a triangular pyramid, the homologue of the anterior surface of its evolutionary precursors being reduced to a narrow border. The hollowed and sloping anterolateral surface is almost entirely occupied by the pisiform facet, as in *Pan*. This articular surface is continuous on the adjoining part of the dorsal, or dorsomedial, surface with another presumptive articular surface which takes the form of an elongated strip. In *Pan* the meniscus plays over just such a cartilage-clothed area (Fig. 10.1D). In some chimpanzee specimens the upper part of this area adjacent to the lunate makes a minor contact with the ulnar styloid within the encircling meniscus, which below this intervenes between the extensor carpi ulnaris tendon and the smooth facet. Distal to this area in *Pan* is a roughened excavation into which the fibrous capsule and periphery of the meniscus attach. All these features are reproduced in the fossil except that the missing portion adjacent to the lunate includes at least most of the part where a presumed residual contact with the ulnar styloid might have been found. With the description now of additional triquetra by Beard *et*

al. (1986) these findings can be confirmed, and it is clear that it is only the posterolateral corner of the KNM-RU-2036 triquetrum which is missing. Beard *et al.* (1986) appreciated that the bone was quite unlike that of cercopithecoids in the way in which it tapered away to a point distomedially and in the form of the mediolaterally extended articular strip on its proximomedial surface, yet they persisted in regarding this strip as a receptive area for the ulnar styloid process. There can be little doubt that it was for a meniscus as in *Pan*. In all its functionally critical features then the fossil has as its closest counterpart the chimpanzee triquetral. In contrast, both to *Pan* and *Proconsul*, the gibbon triquetral (Fig. 5.4B), despite the presence of a massive meniscus largely retains a monkey-like morphology. If many workers have interpreted the *Proconsul* triquetral as retaining a monkey-like ulnar styloid articulation, how much more likely they would have been to reach a similar conclusion about an isolated and unidentified gibbon bone; but they would have been wrong. The siamang shows an interestingly transitional triquetral morphology between gibbon and chimpanzee for a much larger hollowed articulation for the pisiform is impressed into the anterior surface of the bone than is typical for other hylobatids.

The morphology of the fossil pisiform is particularly informative and would seem to confirm that *Proconsul africanus* had a fairly advanced type of meniscus-containing hominoid wrist. The fossil pisiform articulates with the triquetral in such a way that it is pointed palmwards and distally (Fig. 10.1C), contrasting markedly with the proximal angulation which characterizes monkeys so that in the palmigrade position the pisiform acts as a 'heel'. This leaves exposed a quite large proximally facing area on the base which has usually been interpreted as an articulation for the ulnar styloid. This exposed portion shows laterally (Figs 10.1A, B) a rounded, smooth apparently articular protuberance which is separated by a depressed furrow from another medial elevation. Exactly the same features are shown in the immature chimpanzee pisiform illustrated in Fig. 10.1D. But the other wrist of this specimen was dissected in the wet state (Fig. 5.7) and the meaning of these features is there revealed. The lateral prominence is in fact articular, not for bone but for the periphery of the meniscus, and in

Fig. 5.7 is exposed through the communication between the wrist and pisotriquetral joints. The margin of the meniscus bounding the communication lodges in the groove and the periphery of the meniscus together with the fibrous capsule, then attaches to the margin of the more medial elevation. In gibbons in this situation the meniscus contains the large os Daubentonii, which is thus effectively attached to the base of the pisiform. It is within the realm of possibility that a similar lunula occurred in the fossil. In the additional pisiforms available to Beard *et al.* (1986) similar features appear to have been present but these authors focused their attention on the central concavity, interpreting it as a receptive facet for the ulnar styloid.

I suggested (Lewis 1985*b*) that there were features in the *Proconsul africanus* wrist that suggested that its midcarpal joint operated in a similar way, and about a similar axis, to that characterizing the hominoids. This, if true, would be a compelling reason for distancing the fossil morphology from that of monkeys. The clues prompting this interpretation were firstly the reception of the triquetral into a spiral concave trough of hominoid type on the hamate which is markedly aligned toward the proximodistal direction, and secondly the waisted form of the capitate (by which is meant a ballooning of the capitate head anterolaterally, the 'waisting' then being manifest in anterior view, although it is significantly less accentuated than in *Pan*). Dorsally both the fossil hamate and the capitate present confluent articular indentations (Fig. 10.2A), which clearly should lodge the dorsal beak of the lunate in extension, just as happens in *Pan* (Fig. 5.10B). All this, of course, was of necessity rather speculative in the absence of the centrale (which was assumed to be rather like that of a gibbon) and the trapezoid.

All this has been transformed now that the morphology of the centrale with its associated scaphoid, and of the trapezoid, are known (Beard *et al.* 1986). The *Proconsul* scaphoid is indeed remarkably like that of gibbons, possessing the same prolonged tubercle and the same dorsal concavity geared to a locking mechanism onto the radius in extension; the centrale also is strikingly like that of gibbons (Fig. 4.3B) and so is the trapezoid.

In fact, a gibbon centrale and trapezoid of appropriate size can be substituted for the missing

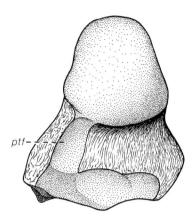

Fig. 10.2
Lateral view of the cast of the right-sided Sterkfontein capitate TM1526. *ptf*, posterior trapezoid facet.

elements in the KNM-RU-2036 reproducing closely the relationships known to occur in the new *Proconsul* material. This has been done in Fig. 10.1.

Manipulation of such a preparation from flexion (Fig. 10.1A) to extension (Fig. 10.1B) leaves little doubt that the movement had similar rotatory characteristics to those of *Pan* and occurred about a comparable axis (Fig. 5.10). Thus flexion would be accompanied by ulnar deviation and pronation; extension by radial deviation and rotation in the sense of supination. If this is truly so, the monkey-like midcarpal locking mechanism in palmigrade posture, as envisaged by Schön and Ziemer (1973), which would indicate retention of the primitive midcarpal axis, clearly could not operate.

Beard *et al.* (1986) were aware of a number of aspects of this morphology, which had nothing in common with that of monkeys, but they were clearly obsessed with the prevailing conclusion among American workers that the fossil had a monkey-like wrist retaining a direct ulnocarpal articulation. They noted the proximodistal orientation of the facet on the hamate for the triquetral (rather than a mediolateral disposition as in cercopithecoids). They noted also the relatively great proximolateral–distomedial extent of the facet on the triquetrum 'for the ulnar styloid process' and the proximal, rather than medial orientation of the 'facet for the ulnar styloid process on the pisiform'. Clearly this was an ulnocarpal articulation distinctly different from that of extant cercopithecoids, but they missed the point. They concluded that the fossil wrist must have allowed a greater degree of ulnar deviation than that of extant cercopithecoids and was thus derived for catarrhines, and 'in some ways fore-shadows' the specialized morphology of homi-noids. In fact, there can be little doubt that the wrist joint of *Proconsul africanus* was a meniscus-containing one and that the midcarpal modifica-tions of hominoid type were also under way at this evolutionary stage.

In carpometacarpal joints of *Proconsul* have figured little in discussions yet the bases of the second to fifth metacarpals of KNM-RU-2036 are known. The trapezium (Lewis 1977) has an ape-like 'set' in the carpus, largely determined by its articulation with the second metacarpal, and has a narrow sellar surface for the first metacarpal with a dorsal tubercle, presumably for a posterior oblique ligament: a hominoid characteristic. Yet it has only a faintly discernible tubercle, or crest, adjoining the identifiable groove for the flexor carpi radialis tendon; as seen in Chapters 6 and 9 this is characteristic of the monkey trapezium, reflecting a relatively flimsy attachment of flexor retinacu-lum fibres superficial to the tendon, whereas the massive attachment in hominoids has raised a substantial apophysis. How significant then is this in the fossil? Although chimpanzees have a massive tubercle or crest, in the macerated bones of *Pan* specimens of comparable immaturity no greater crest is apparent than in the fossil–ossifica-tion apparently extends into the crest late. In the other trapezium specimens available to Beard *et al.* (1986) the palmar surface was not well preserved yet it seems apparent that they possessed only a moderate trapezial tuberosity (crest). Yet as noted earlier in this chapter a prominent tuberosity is found in adapids, and extant lemurs. It seems therefore that this early excrescence was subse-quently reduced in the emergent higher Primates and is not homologous with the crest of hominoids. In monkeys the crest is negligible and they have retained the primitive anthropoid condition. In *Proconsul* it is inconspicuous and in *Pongo* it is small (Fig. 6.9). The prominent structure of Afri-can apes and man must therefore have been a relatively late acquisition. Somewhat surprising therefore is the finding that a very prominent tubercle occurs in *Hylobates*. It can only be

concluded that this was a parallel acquisition in the gibbon lineage.

For the remaining carpometacarpal joints it is clear that the fourth metacarpal had achieved contact with the capitate and that dual facets with an intervening interosseous carpometacarpal liga-ment characterized the opposed surfaces of third and fourth metacarpals: the primitive higher primate condition. On the lateral side of the capitate and third metacarpal the situation is rather different. Because of the linear facet for the second metacarpal on the capitates available to Beard *et al.* (1986) they assumed confluence of dorsal and ventral facets here with loss of the primitive interosseous carpometacarpal ligament. In KNM-RU-2036, with the metacarpals available, it is clear that the situation is just like that described in certain cercopithecoids in Chapter 6 where the interosseous ligament is in an incipient stage of regression. However, as noted there, regression of this primitive ligament has occurred in parallel in certain prosimians, certain New and Old World monkeys and *Homo sapiens*. The same thing seems to have been initiated in *Proconsul*.

The amalgam of quite primitive hominoid features (scaphoid and centrale) in *P. africanus* and more advanced features (triquetral and pisiform) would suggest that first the gibbon and then the orang-utan clades had split from the hominoid line, the gibbons at least going their own way at a stage earlier than *Proconsul*. On the other hand, the African apes and man were clearly later offshoots, although the incipiently apomorphic character of the joint between second metacarpal and capitate in *Proconsul* seems to preclude this fossil from a direct ancestral relationship to the latter clade.

Middle–Late Miocene Eurasian apes

Only one significant specimen is known—a capi-tate (GSP17119) from the Late Miocene of Pakis-tan and referred probably to *Sivapithecus indicus*. This specimen has been well described and illus-trated by Rose (1984), but the interpretations of this author were heavily influenced by the views of Jenkins and Fleagle (1975) and Jenkins (1981) on the workings of the midcarpal joint (which were critically discussed in Chapter 5) and thus should not be unreservedly accepted.

In fact, the bone has just the morphology to be expected in a derivative form evolved from an ape not too dissimilar to *Proconsul africanus* described above. As Rose (1984) noted, the head of GSP17119 is slightly more expanded than Early Miocene capitates and thus it would make a perfectly plausible ancestral form antecedent to the characteristic morphology of orang-utan capitates. This would be in accord with the scenario postulated above for the derivation of the extant apes.

Fossil hominid hands

South African Australopithecine hand bones

The merest glimmer of insight can be obtained into hand function of the emergent hominids on the basis of the scanty hand fossils which have been recovered from the South African sites of Sterkfontein and Swartkrans dated at about 3.3MY and 2MY ago respectively (Partridge 1973; Tobias 1973; Vrba 1975).

The Sterkfontein capitate TM1526 This fossil came from the Sterkfontein lower breccia, site, and was not associated with tools, although stone artefacts have been found at the more recent Sterkfontein extension site. It was originally described and figured and attributed to '*Plesianthropus transvaalensis*' by Broom and Schepers (1946); only gracile australopithecines have been obtained from this site and the fossil is now usually classified as *Australopithecus africanus*. Broom and Schepers decided, on very flimsy grounds that the bone was 'intermediate between that in the orang and that in man, but distinctly nearer to the human condition'. Clark (1947) elaborated upon this account, drawing attention to the apparent disposition of the articulation for the second metacarpal, as much on the distal as on the medial surface of the bone, and likening this to the human condition. However, he stressed that the bone contrasted with that of man in possessing a pronounced excavation on its lateral surface, giving it a 'waisted' appearance, reminiscent of that of chimpanzee or gorilla capitates. He assumed (with, in fact, no justification) that this deep depression represented the site of attachment of a strong

interosseous ligament which, in a later publication (Clark 1967), was more explicitly identified as 'a strong interosseous ligament binding it to the adjacent trapezoid bone'. Despite this, Clark (1947, 1967) concluded that the bone had a very 'human appearance' and that 'the capitate and the adjacent bones of the hand were less free in *Plesianthropus* than in Europeans, but much more free than in the hand of modern anthropoid apes'. It can now, however, be seen that these views were insecurely based, for the basic anatomical information was not then available: the significance of 'waisting' in the capitate was not appreciated nor was the true location of the interosseous ligament binding it to the trapezoid; most importantly, the striking human apomorphy of substitution of an anterior articulation between these bones for the primitive posterior one was unrecognized (Chapter 5). However, both sets of earlier authors were mistakenly under the impression that the second metacarpal in the African apes had a single small dorsally located lateral contact with the capitate and that '*Plesianthropus*' could be likened to man in the possession of an elongated, enlarged, and rather distally facing facet for the second metacarpal. This episode could be cited as a cautionary tale for those who would venture opinions on form and function in fossil bones without first acquiring the adequate detailed background of comparative experience which can underpin those opinions.

A later assessment of the fossil (Lewis 1973) led to the following revision. The fossil when viewed from laterally (Fig. 10.2) shows a swollen head, and thus a quite marked degree of 'waisting' comparable to that of the immature specimens of *Pan troglodytes*. Again the trapezoid facet as in apes is posteriorly located and continuous via a narrow articular isthmus on the neck with the articular surface of the head. Below, it is confluent with a quadrilateral posterior facet for the second metacarpal which is linked via a faint depression (the presumptive site of attachment for the typical hominoid interosseous carpometacarpal ligament) with an anterior second metacarpal facet on the prominent beak of bone forming the anterodistal corner of the lateral surface. This beak also presents a nonarticular (ligamentous) area and the process is thus of rather similar morphology to that of the other subhuman hominoids (Fig. 5.11). It

African and Ethiopian capitates have been considered above. Like TM1526 the bones are strikingly similar in fundamentals to the chimpanzee capitate although the waist may be slightly more filled out. Although the anterior and posterior facets for the second metacarpal are confluent there are fairly suggestive indications of a persistent remnant of a carpometacarpal ligament. The morphology of the bases of the second and third metacarpals reinforce this view for in certain specimens the adjoining surfaces show dual facets and in others a composite confluent one. The morphology is like that shown in certain cercopithecoid monkeys when there is variable regression of the carpometacarpal ligament at this site; but the morphology is unlike that of *Homo*, although that is how it is usually portrayed. Moreover, there is no true styloid process on the third metacarpal, but a stepping up of the dorsomedial corner of the base of the metacarpal, which is accommodated by a shallow hollow in the capitate, as occurs in *Pan*. There is thus no deep receptive cup for the second metacarpal base, fashioned out of capitate, third metacarpal and its styloid process, and no provision for the significant movement which occurs here in the human hand. Not surprisingly then the facet for the trapezium on the radial side of the second metacarpal, and its counterpart on the trapezium, are aligned more as in *Pan* than in man. The second carpometacarpal joint, could clearly not accommodate the movements characteristic of man but presumably operated more in the manner of those New and Old World monkeys which have evolved a derived morphology here. Yet, presumably the australopithecine morphology would have provided an excellent platform for subsequent remodelling into a fully human pattern.

If the base of the second metacarpal and its articulation with the capitate had been remodelled along human lines one would have expected that similar human characteristics would be found in the head of the bones. They are not. The heads of the australopithecine second metacarpals are no more asymmetrical than those of juvenile chimpanzees (which lack knuckle-walking 'stops'). This, of course, is in excellent conformity with the structure of the base. The human type of movement shown in Fig. 6.11 would clearly not have been possible. Just as the lateral part of the palm has not acquired human apomorphic characters,

so the same obtains at the ulnar side. It is obvious that the bases of the fourth and fifth metacarpals articulated with the hamate as in *Pan* (Fig. 6.5A). There was none of the derived character imposed in man (Fig. 6.7) by the intervention of the pisometacarpal ligament, releasing the metacarpals from the constricting embrace of the hamate hook.

The bases of the metacarpals, particularly the fifth, clearly show a wrapping over of the articular surface onto the ventral aspect, as for the Swartkrans fifth metacarpal (see above). It is noteworthy, however, that metacarpal facets on the hamate were essentially dorsoventrally concave, lacking the central protruberances seen in *Pan* (Fig. 6.5A), although the available metacarpal bases indicate that this conversion to a simply concave surface was variable. However, there is some progress here towards a condition which could be pre-adaptive to the human one. In no way, however, could the articulation of the fifth metacarpal have had the mobility achieved in man where it is accommodated by a sellar articulation. Again the fifth metacarpal head shows only minor asymmetry, with slight bevelling on the dorso-ulnar aspect, comparable to that of *Pan* (Fig. 6.8), if the knuckle-walking stops are disregarded. In no sense was it remodelled along human lines (Fig. 6.9).

This impression of general resemblance of the hand bones to *Pan* is greatly heightened by the morphology of the pisiform. if the disorientation shown in the published figures is appreciated it is clear that virtually exact counterparts can be found among chimpanzee bones even in respect to the detailed morphology of the articulations at the base.

The articulation of the first metacarpal with the trapezium was again like *Pan*, and there was clearly a posterior oblique ligament fanning out from a dorsal tubercle on the trapezium. The first metacarpal, however, was certainly not relatively reduced in size, as it is in *Pan*. Indeed Marzke (1983) has calculated that the thumb was longer relative to the index finger than in chimpanzees and that the proportions were very similar to those of modern humans.

The form of the proximal phalanges reinforces the impression of a basically ape-like morphology, for they are bowed like those of *Pan* and like that species show prominent marginal impressions or ridges on the middle third of the ventral surface. These have been described as for the fibrous flexor sheaths. They represent the attachment of the extremely tough 'pulleys' found on the proximal phalanx in apes (Fig. 10.6B) and monkeys, corresponding to the less clearly demarcated A2 pulley in the human hand (Fig. 10.7).

The hand bones of Homo habilis (OH7)

This collection of subadult hand bones was found at Olduvai Gorge, associated with stone tools of the Oldowan culture (Leakey 1971) and have been reliably dated at about 1.75MY. They were first described by Napier (1962). The collection, consisting of a right trapezium, right scaphoid, the base of a right second metacarpal and several phalanges, all apparently from the same hand, was subsequently reassessed (Lewis 1977) and the following account is based on that study.

The trapezium (Fig. 10.3A) is almost complete but the tuberculum (crest) is broken off. The distal articular surface for the first metacarpal is concavo-convex but broader in dorsovolar extent than that of *Pan*, although some specimens of *Gorilla* are comparable. In this feature it approaches the human condition. It possesses a dorsal tubercle and so presumably a posterior oblique ligament was present. This prominence immediately adjoins the articular margin as in *Pan* and it is not truncated and separated from the border so leaving an area for the origin of a lateral ligament, as in man; the lack of elaboration of this ligament can

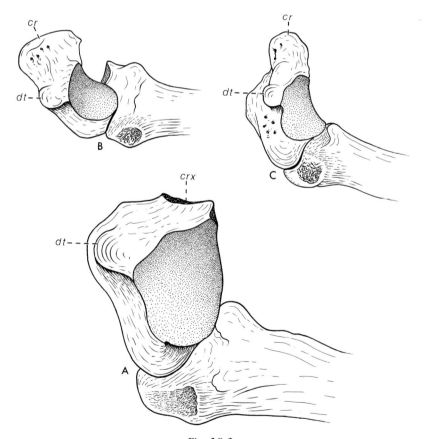

Fig. 10.3
A, casts of the trapezium and second metacarpal of *Homo habilis* (OH7) articulated together and viewed from laterally. B, the same bones of the right hand of *Homo sapiens*. C, the same bones of the right hand of *Pan troglodytes*. *dt*, dorsal tubercle of trapezium; *cr*, crest of trapezium; *crx*, broken base of crest of trapezium of OH7. (Redrawn from Lewis 1977.)

tentatively be inferred. There is also a distally located tubercle on the dorsal surface for a dorsal trapeziometacarpal ligament reinforcing the back of the articulation with the second metacarpal.

It has been noted (Fig. 6.5) that the important human specialization by which the trapezium has been re-orientated was largely achieved by remodelling of the trapezoid and capitate and the articulation between them. Neither of these bones is available in the Olduvai find but a reasonable indication of the 'set' of the trapezium can be deduced, since the damaged base of the second right metacarpal is available and this fortunately conserves the area of insertion of extensor carpi radialis longus and the adjacent articulation for the trapezium. The second metacarpal facet of the trapezium is a direct continuation of that for the trapezoid, with only slight angulation between the two, as in *Pan* (Figs 6.5A, 10.3C).

In man, with splaying apart of the carpal arch the second metacarpal facet on the trapezium has become marked offset from that for the trapezoid, presenting as a truncated end to the tapering distal part of the bone (Fig. 10.3B). Napier (1962) correctly discerned of the fossil trapezium that 'the evidence provided by the other articular surfaces indicates that its "set" in the carpus was unlike that found in modern man and similar to the condition in *Gorilla*'. He did not, however, provide any reasoned analysis supporting this statement. Its 'set' is also similar to that of *Pan* (Figs 10.3A, C). The lateral surface of the bone is expanded when compared with the ape condition, but indications are that the situation and form of the missing crest were more like *Pan* than like man.

Because of the known characteristics of midcarpal motion in extant primates the scaphoid would be expected to yield additional evidence about the orientation of the trapezium in relationship to the other distal carpal bones. Thus, in *Pan* (Fig. 5.10A) the bunched up arrangement of the distal carpal row means that facets on the scaphoid for the trapezoid and trapezium closely adjoin the margin of the capitate surface, and the tuberosity of the bone is free from articulation and presents as an upturned beak. In *Homo sapiens* in contrast (Figs 4.3A, 5.19A) the distal carpals are splayed apart radially and the facet for the trapezium extends out to the under surface of the tuberosity of the scaphoid. In the fossil the tip of the tuberosity is broken off but there are strong indications that it had a beak-like character and that the articulations for the trapezoid and trapezium bore more resemblance in their location to *Pan* than to *Homo sapiens*.

Susman and Creel (1979) have also studied the OH7 hand and essentially repeating the findings given above for the carpal bones. They noted in addition, however, that the middle phalanges are robust and strongly curved and infer from this the retention of considerable climbing and suspensory activity. On the other hand they noted that the distal phalanges show greater apical development than those of apes and are thus more human in appearance.

Hand function in fossil hominids

That the austrolopithecine hand retains an ape morphology in essentials and that there are clear indications of an arboreal heritage emerge strikingly from the descriptions given above. There is virtually none of the remodelling of the heads of the second and fifth metacarpals which plays such a significant role in the repertoire of human gripping postures. In the absence of this restructuring one would scarcely expect the apomorphic changes found at the basal joints of these bones in man—after all the index metacarpal of man seems to be tugged into an attitude of radial deviation, flexion and pronation only in response to comparable movement at the metacarpophalangeal joint, enacted by a modified first dorsal interosseous muscle.

The variable modification of the joint between the second metacarpal and the capitate in *A. afarensis* is no more than that seen in various monkeys, and surely lacked the functional characteristics of the human joint; yet it could clearly be preadaptive for that more advanced structure. Again the fifth (and fourth) metacarpals clearly lacked the unique supinatory capacity of the human bones. Yet, in the suppression of the central elevations of the hamate seen in apes, there was some hint of things to come.

The fragmentary *Homo habilis* hand shows evidence of somewhat more advanced capabilities in the structure of its thumb carpometacarpal joint. However, the clear retention of an ape-like joint between the trapezium and second metacarpal precludes significant rotatory ability of the latter bone. However, the metacarpal heads are unknown and significant changes in that of the second might already have been initiated; as noted above, they would be expected to precede modification of the basal joints of the second metacarpal.

It is impossible to establish whether *Homo habilis* was capable of a 'squeeze' (power) grip for none of the bones from the ulnar side of the fossil hand have been recovered. However, the ape-like orientation of the trapezium, which can be deduced from its articulation with the second metacarpal, is circumstantial evidence that the factors which fit the thumb for its role in the power grip, had not yet been acquired.

Considerable play has been made of the 'set' of the trapezium as a distinguishing feature between pongids and man, but the functional import of this has not been clearly specified. The implication appears to be (Napier 1962) that the human arrangement is an essentially negative feature and that the 'set' in pongids indicates a deep carpal tunnel and thus strong long finger flexors. Following this type of reasoning, Napier (1962, 1964) concluded that the Olduvai hand, with its apparent rather pongid-like 'set' of the trapezium, was capable of a strong power grip but lacked perfection of a precision grip, because the ape-like 'set' of the trapezium was said to indicate an ape-like proportionality between thumb and index finger; why this should follow is not obvious.

In an innovative piece of work Marzke and Shackley (1986) tried to identify the sort of grips which might have been used by the early hominid flint knappers. These authors found that human volunteers with no previous instruction slipped easily into the practice of holding the preforms with a 'cradle grip' and holding the hammerstone in a 'three-jaw chuck thumb-to-finger' grip. It is clear from the descriptions given above that such grips were beyond the capabilities of the australopithecines. It is true that they might have used a 'cradle grip'—chimpanzees use what has been called a relaxed power grip which is fundamentally similar—but the sophisticated three-jaw chuck certainly could not be accommodated by their morphology. However, as with chimpanzees, a 'pad-to-side' grip could have been utilized for rather cruder chopping actions.

Using essentially the data described in Chapter 6

(which had previously appeared in the papers cited therein) Marzke (1983) and Marzke and Shackley (1986) attempted an assessment of the capabilities of the *A. afarensis* hand. Surprisingly they reached very different conclusions from those just proposed, and it can only be said that their perception of the human, comparative, and fossil anatomy was very different from that described here. They concluded that a 'three-jaw chuck' grip was quite feasible for *A. afarensis* and based this upon their belief that the joint between capitate and second metacarpal, although different from that of modern humans, probably functioned similarly. They made no mention of the lack of restructuring at the metacarpophalangeal joint of the index finger, a basic necessity for the 'three-jaw chuck' grip.

Marzke (1983) had concluded that the Hadar hamate facet for the fifth metacarpal was continued onto the hook as in apes, precluding the use of a 'squeeze' (power) grip. Yet in the later joint paper it was suggested that there was likely to have been a pisometacarpal ligament intervening (as in man). Yet they stuck to the position that the only likely grip to have been beyond the powers of *A. afarensis* was the squeeze grip. implying that the unique morphological pattern of the hominid hand was first established in the region of the second and third metacarpals and only later progressed to involve the ulnar side of the hand.

In a later paper Marzke and Marzke (1987) elevated the little styloid process of the third metacarpal to the status of having a key role in the evolution of the hand. There can be no doubt that the human styloid process by its encroachment on capitate territory has come to form an important part of the receptive cup for the base of the second metacarpal (Chapter 6); doubtless these authors are also right in stating that it stabilizes the third metacarpal against palmar subluxation and hyperextension and that it might thereby be of significance in resisting stresses when the three-jaw chuck grip is used in tool use; but it is a quite variable structure in man (Fig. 5.18). The claim for special evolutionary status for this process rests on the belief that it is represented as a separate chondrifying element in the embryonic carpus, which in other primates fused with the capitate, but in man has been transferred to the third metacarpal; the authority for this statement is

Dwight (1910). In fact, Dwight showed nothing of the sort: in a brief paragraph he merely echoed the belief at that time that the process separately chondrified and the main thrust of his argument was to challenge the idea that the array of accessory ossicles in the hand, so painstakingly documented by Pfitzner, all had their genesis in chondrifying nodules scattered about the hand, and which could be traced back to elements in an ancestral flipper-like appendage and which variably fused or disappeared in later development. The idea presented by Marzke and Marzke is really a revival of this discredited notion, discussed in Chapter 3.

Other recent studies (e.g. Kootstra *et al.* 1974) continue to echo the apocryphal statement that the styloid process chondrifies separately. All these statements ultimately rest for their authority on the embryological work of Thilenius (1896). This author examined 181 embryonic hands yet described and figures a separate cartilaginous styloid (his 'carpometacarpale 4') in only one. Without the rigours of serial sectioning and reconstruction it is impossible to determine whether or not the apparent independence of the nodule was merely an artefact produced by the plane of section. Of course, this particular example might have been a truly separately chondrifying process. There is no doubt that a separately ossifying os styloideum does occur (Fig. 5.18D) and presumably in some such cases it may also chondrify separately; in this it is no different from other anomalous ossicles occurring in bony apophyses (Chapter 3). Thilenius illustrated three other cases which he presented as examples showing recent 'fusion of carpometacarpale 4' to the third metacarpal. They carry no conviction, for expanded processes on bones often have constrictions or grooves at their root and sections shaving through an appropriate plane may show some indication of peripheral demarcation of the process. Thilenius' illustrations show no more than this and his work gives no real support for the notion that the carpus regularly includes a separately chondrifying element representing the styloid process. More recent detailed studies (O'Rahilly *et al.* 1957) confirm this view.

Clearly the human styloid process is simply derived as an accentuation of the step-like elevation at the posterolateral corner of the third metacarpal base in apes, which bears the posterior facet for the second metacarpal. This elevation is even present in *Proconsul africanus*, is quite marked in *Pan* and *Pongo*, and in all is impressed into a hollow in the capitate which reduces the contribution of that bone to the posterior articulation with the second metacarpal.

Evolutionary mechanisms and the hand

The evolution of the human hand, far from showing saltatory jumps, seems to have been characterized by the cumulative acquisition of small changes, in themselves seemingly insignificant, but by their additive effect they have realized quite new dimensions in function. Thus change in form of the trapezoid has altered the 'set' of the trapezium, in turn altering the orientation of the trapezium–second metacarpal joint, which with remodelled articulations on the other side of that metacarpal changed the function of that bone. The second carpometacarpal joint, however, has probably been unduly emphasized in evolutionary studies. Its limited movement is congruent with that at the index metacarpophalangeal joint; the latter is the prime focus of movement, carpometacarpal movement merely amplifies it somewhat. Other secondary changes have affected the base of the fifth metacarpal.

Subtle changes in form of the heads of the second and fifth metacarpals paved the way for new gripping postures and appropriate changes in extrinsic and intrinsic muscles accompanied this remodelling. All this is classically Darwinian.

The evolution of the wrist joint, however, seems to belong in a quite different category. The hominoid wrist is a true evolutionary novelty and seems to have opened up new vistas for the apes, distancing them from their primitive catarrhine ancestors. The punctuated equilibrium model seems to offer a more likely mechanism for the establishment of such reorganization; palaeontological data does not seem to be at odds with this view.

11

The skeleton of the foot

Fin to limb—the embryological approach

Few would doubt that the hindlimb, like the forelimb, was derived from the fin of a lobefin ancestor. Just as with the forelimb attempts have been made to interpret the phylogeny of the skeleton from purely ontogenetic studies of human embryos; the most recent (Fig. 11.1), epitomizing

the others, is by Cihak (1972). In this scheme it is envisaged that the primordial elements for the tarsus (Fig. 11.1A) corresponded almost exactly to those for the carpus (Fig. 4.1): five distal tarsalia; four centralia; a proximal row consisting of tibiale, intermedium and fibulare; representatives of marginal rays, prehallux (in this case double) and 'pisiform homologue'; in the tarsus in addition

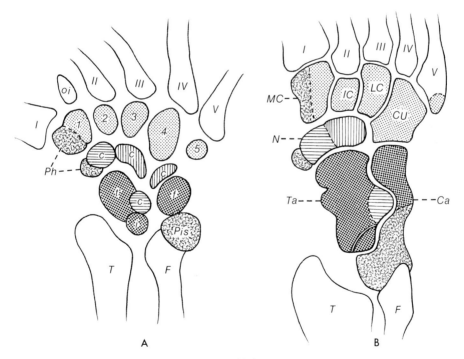

Fig. 11.1

A, the canonical elements identifiable in the embryonic human hand, according to Cihak (1972), with shading comparable to that shown for the embryonic carpus in Fig. 4.1A. *T*, tibia; *F*, fibula; *t*, tibiale; *i*, intermedium; *f*, fibulare; *ph*, prehallux; *Pis*, pisiform homologue; *c*, the four centralia; *1–5*, tarsalia distalia; *I–V*, the metacarpals. B, the identification of the same elements in the adult human foot according to Cihak, shaded as before. *Ta*, talus; *Ca*, calcaneus; *N*, navicular; *MC*, medial cuneiform; *IC*, intermediate cuneiform; *LC*, lateral cuneiform; *Cu*, cuboid.

there is said to be an 'intermetatarseum' rudiment, reputedly a separated part of the first tarsale distale. Contrary to the situation in the carpus, almost all these canonic elements so-called, are said to be represented within the adult tarsus. It is claimed that the talus originates by fusion of the tibiale, a centrale (representing the talar lateral tubercle) and the intermedium (forming the posterior process of the talus, the region of the anomalous os trigonum). The calcaneus is envisaged as the product of fusion of the fibulare and the 'pisiform homologue' (harking back to the original suggestion by Owen 1866). The naviculare is represented as the product of fusion of two centralia, and part of the prehallux, the remainder of which is supposedly amalgamated with the medial cuneiform. The cuboid is interpreted as representative only of the fourth tarsale distale, the fifth being incorporated as part of the tuberosity of the fifth metatarsal (region of the anomalous os Vesalianum). The supernumerary intermetatarsal element is said to disappear normally, but may persist as the anomalous os intermetatarseum. As critically discussed in Chapters 3 and 4, such studies are based upon identification of dubious mesodermal condensations and prochondral outcrops from major centres of chondrification; these studies carry no conviction. The identification of the various components is heavily influenced by classical comparative and embryological studies of the last century.

The classical comparative approach

Anatomists in the past have been in fairly general agreement that the primitive tarsal elements have had the following fate in man (and other mammals): the talus represents a fused tibiale and intermedium, the fibulare has become the calcaneus, the navicular is a derivative of the centrale (or centralia) and the primitive five tarsalia have given rise to the cuneiforms and cuboid. The anomalous os trigonum of the human foot has been said to represent the intermedium tarsi. This view has been perpetuated in innumerable textbooks of human anatomy. Gegenbaur (1864) initiated this interpretation, a key part of which was the notion that the mammalian talus resulted from the coalescence of tibiale and intermedium.

He noted the apparent fusion of tibiale and intermedium in certain of the Chelonia (for example, *Chelydra*) producing a tarsus resembling that of mammals, and reasoned that the same fusion must have occurred in mammals, and even suggested that *Chelydra* itself represented a survivor of the reptilian stem which had given rise to the mammals. Bardeleben (1883a,b,c), investigating this hypothesis, noted the anomalously separate posterior tubercle of the human talus, which he named 'os trigonum', and held to represent the intermedium tarsi, (homologue of the carpal lunate); furthermore, he maintained that such a free intermedium was a normal feature of the tarsus in a number of marsupials. The small marsupial bone in question was first described by Owen (1841) in *Dasyurus macrourus* and later (1874) in *Phascolomys*; Owen, however, drew no conclusions regarding its morphological identity. If such a free intermedium is really a feature of marsupial anatomy it could be seen as a telling point in favour of Gegenbaur's hypothesis. Bardeleben seemed further to strengthen his case for a dual phylogenetic origin of the talus by reporting a separate anlage for the posterior tubercle of the talus in the two-month human fetus.

Baur (1884, 1885a), however, raised a difficulty: in rodents there is an extra tarsal bone giving the appearance of a double navicular. Although this bone had long been known (even to Gegenbaur) it had been dismissed as an accessory element, either a sesamoid or the product of secondary division of the navicular. Baur, however, suggested that it was really the tibiale (as did Albrecht 1884) and that in other mammals it fused to the centrale to form the navicular. He was also unable to verify Bardeleben's finding of a dual ontogenetic origin of the talus, which bone he held to consist entirely of the intermedium; the extra element in marsupials he described as a 'Sehnenverknöcherung'. He believed that the calcaneus was a derivative of the fibulare. This alternative scheme of homologies seemed quite irreconcilable with that of Bardeleben.

Subsequently a compromise incorporating parts of both views, was reached. Bardeleben (1885a) accepted Baur's identification of the tibiale and even reported that the human navicular showed a dual developmental origin—tibiale and centrale. He continued, however, to regard the os trigonum as the intermedium, but decided that the remainder of the talus, since it could no longer be the tibiale, must represent a second centrale. Baur (1885*b,c*) agreed with this composite view.

Bardeleben (1885*c*) seemingly had second thoughts and subsequently rejected this comprise and then returned to a modified version of his original view, again including the tibiale in the talus. Baur (1886) also, after a further detailed comparative and embryological study, favoured a return to his own original view and Leboucq (1886) came to his support.

The evidence from palaeontology

Palaeontologists have agreed that the talus is the homologue of the intermedium. Broom (1901, 1904) at first supported Bardeleben's original hypothesis but later (1921, 1930) he considered the talus to consist entirely of intermedium and so suggested that the navicular was the homologue of the tibiale. Most other authorities (Schaeffer 1941*a*; Romer 1955) have agreed on this homology of the talus, but have held that the tibiale is no longer represented in the mammalian tarsus, and that the navicular is the homologue of the persistent centrale alone.

In contrast, Cope (1884, 1885) had maintained that there was justification (partly palaeontological) for regarding the mammalian navicular as a fusion of the tibiale and the centrale, rather than as the one or the other alone. He drew attention to the belief that the bone bearing the horny perforated spur in the monotreme foot was a tibiale and further suggested the presence of such an element in pelycosaurs. Baur (1885*d*, 1886) welcomed this support from the palaeontological camp for his view of a dual derivation of the navicular, which he himself had based on the morphology of the rodent tarsus. Paradoxically, Cope's views were based on a false assumption, since, as will be seen, the monotreme spur-bearing bone (os calcaris) is certainly not the tibiale. However, the monotreme foot does possess a true tibiale, additional to the os calcaris (Lewis 1963). Meckel (1826) had confused this tibiale with the os calcaris and the true tibiale virtually disappeared from subsequent accounts; figures in textbooks generally omitted it entirely: it has occasionally been figured without

comment, but only Emery (1901) seems to have clearly observed it and to have realized its significance. It thus transpires that Cope's conclusions as to the representation of the tibiale in the mammalian tarsus had some justification, though for the wrong reasons.

It is well established that the emergent tetrapod tarsus (Fig. 11.2A) had a proximal row of three bones, a distal row of five, and four centralia between. In this it is closely comparable to the corresponding carpus (Fig. 4.2A). However, where the carpus possessed both preaxial and postaxial representatives of marginal rays—prepollex and pisiform—in the hindlimb there is only a prehallux. There is no evidence that the hindlimb at any stage possessed a homotype for the pisiform (as suggested by Cihak for instance).

Progressive evolution in reptiles from this basic plan shows similar trends to those described in the carpus in Chapter 4, largely consisting of reduction in the number of centralia. In pelycosaurs, two bones, which have been identified as centralia (Fig. 11.2B), occupy the position of the mammalian navicular. It is generally supposed that the medial one later disappears, leaving the persistent one to become the navicular. However, it seems more probable, as Broom (1921) maintained, that

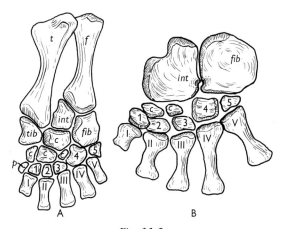

Fig. 11.2
A, the crus, tarsus and metatarsus of the primitive amphibian *Trematops milleri* (after Schaeffer 1941*a*). B, the tarsus and metatarsus of the pelycosaur *Ophiacodon retroversus* (after Romer and Price 1940). *c*, centrale; *f*, fibula; *fib*, fibulare; *int*, intermedium; *p*, prehallux; *t*, tibia; *tib*, tibiale; *1–5*, tarsalia one to five; *I–V*, metatarsals one to five. (From Lewis 1964*a*).

the medial one is in fact the tibiale. It is probable therefore that the tibiale persisted through the therapsid line, either as an independent element or in some species fused to the centrale, the composite bone then being the counterpart of the mammalian navicular. For instance, there appears a distinct possibility that a cartilaginous tibiale existed in the very mammal-like foot of the therapsid *Bauria*. The fate of the other three centralia found in the labyrinthodont foot remains problematic. They may have fused to other tarsal components, all four centralia may have united to form the single reptilian centrale, or some may have disappeared; the virtual impossibility of resolving such a problem is discussed in Chapter 3. Certainly it seems that at the time of attainment of the reptilian stage of evolution only a single free centrale persisted. In those pelycosaurs possessing a mammal-like fused centrale and tibiale, resembling a mammalian navicular, an additional small ossicle may be found which is doubtless a persistent prehallux. Palaeontologists have generally believed that the cuboid represents tarsale four alone. However, five tarsalia were present in pelycosaurs (Romer and Price 1940) and in the therapsids *Ictidosuchoides* and probably *Bauria* (Schaeffer 1941*a,b*). This supports the comparative embryological evidence, to be mentioned later, favouring the view that the cuboid is homologous with tarsalia four and five.

As in the forelimb, the primitive reptilian hindlimb phalangeal formula seems to have been 2-3-4-5-3, and was converted to the typical mammalian one by reduction and then loss of certain of the bones.

Recent morphological evidence

The manner of conversion of such a therapsid reptilian tarsus to the mammalian pattern can now be stated with considerable conviction, based on detailed morphological grounds (Lewis 1963, 1964*a*): the talus is the intermedium alone, the calcaneus is the fibulare, the usual mammalian (and human) navicular is the tibiale coalesced with the single persistent centrale, and the five distal tarsalia form the three cuneiforms and the cuboid.

The monotremes *Ornithorhynchus anatinus*

(Fig. 11.3) and *Tachyglossus aculeatus* possess a tarsus of strikingly primitive form. Both the talus and the calcaneus retain articulations with the fibula, and the tibia projects far distally to articulate with the talus alone. There is a free tibiale, greatly resembling a separated navicular tuberosity. The cuboid is a single bone (earlier accounts have sometimes incorrectly accounted it as double) bearing the articulations of the fourth and fifth metatarsals. The os calcaris is a flat bony concretion developed about the base of the horny perforated spur which conveys to the exterior the secretion of the femoral (poison) gland. It is attached to the talus by a syndesmosis and it

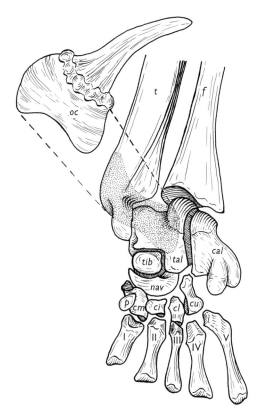

Fig. 11.3
The flexor aspect of the right tarsus and metatarsus of *Ornithorhynchus anatinus*; the os calaris bearing the horny spur is shown lifted away from its site of articulation which is indicated by stippling. *cal*, cancaneus; *ci*, intermediate cuneiform; *cl*, lateral cuneiform; *cm*, medial cuneiform; *cu*, cuboid; *f*, fibula; *nav*, navicular; *os*, os calcaris; *p*, prehallux; *t*, tibia; *tal*, talus; *tib*, tibiale; *I–V*, metatarsals one to five. (From Lewis 1964*a*.)

articulates through a small synovial joint with the tibia; between these attachments it bridges over the tibialis posterior and flexor tibialis tendons as they enter the foot. Strangely, as was seen above, the os calcaris and the tibiale have unaccountably become confused in the literature and in many accounts the true tibiale has disappeared entirely. In wet specimens (Fig. 14.19A) it is apparent that the distal terminal attachment of the tibialis posterior tendon is to the tibiale; to reach it, however, the tendon must cross the projecting talus to which a considerable part of it gains attachment and it is here deep to the os calcaris to which it has a minor insertion. It may be presumed that the tibiale was the primitive insertion for it is not unusual for tendons crossing bony prominences to gain a partial attachment thereto, as has here happened in respect of the talus.

In rodents (with the exception of one family) a comparable situation exists, for the navicular is double the medial element being traditionally called the 'tibial navicular' and the other the 'fibular navicular' (Fig. 11.4). The tibialis posterior tendon inserts into the tibial navicular, reaching this insertion by running deep to a strong ligament passing from tibial malleolus to the medial cuneiform (Fig. 14.15A)—the precursor and partial homologue of the primate Y-shaped ligament (Chapter 14). Moreover, the tibial navicular receives the attachment of a well-defined spring ligament (plantar calcaneonavicular ligament) passing from the sustentaculum tali. Hildebrand (1978) confirmed the insertion of the tibialis posterior to the tibial navicular (his 'medial tarsal' bone) in 25 families of extant rodents but failed to appreciate the true nature of the overlying ligament calling it the 'principal strand of the collateral ligament'. The insertion of the tibialis posterior tendon to the monotreme tibiale and to the rodent tibial navicular strongly suggests that these bony elements are homologous. There can be little doubt that they both represent the tibiale and this is supported by the finding (Schaeffer 1941*a*) that the amphibian tibiale receives the insertion of the precursor of the tibialis posterior, the pronator profundus muscle (plantaris profundus I of McMurrich 1904). In each case the adjacent bone—the 'navicular' of monotremes and the 'fibular navicular' in rodents—is clearly the persistent centrale. Apparently a double navicular

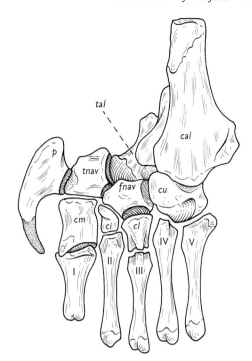

Fig. 11.4
The flexor aspect of the right tarsus and metatarsus of *Coendou prehensilis. cal*, calcaneus; *ci*, intermediate cuneiform; *cl*, lateral cuneiform; *cm*, medial cuneiform; *cu*, cuboid; *fnav*, 'fibular navicular'; *p*, prehallux; *tal*, talus; *tnav*, 'tibial navicular'; *I–V*, metatarsals one to five. (From Lewis 1964*a*.)

(free tibiale) is not confined to rodents: Cope (1884) reported a similar arrangement in certain fossil ungulates and Leche (1900) described it in *Galeopithecus, Hyrax,* and the Dinocerata.

There is little doubt that in Metatheria and the great majority of Eutheria (exceptions noted below) these two tarsal components, tibiale and centrale, have coalesced to form the descriptive navicular. The medial part of this composite bone including the tuberosity retains the attachment of the spring ligament and the tibialis posterior tendon, although this latter insertion may be modified according to certain well defined principles (Chapter 14). Embryological evidence supports this view, just as it does the view that the human carpal scaphoid is a composite of radiale and centrale. As with the carpal elements heterochrony is demonstrated. Usually the human pedal navicular chondrifies singly (O'Rahilly *et al.*

1957); sometimes, however, it originates as two (tibiale and centrale) cartilaginous anlagen (Emery 1901). Certainly in most marsupials tibiale and centrale chondrify independently, thereafter fusing to form the navicular (Emery 1897), a course of events quite comparable to the development of the human carpal scaphoid. It is of significance that in the Macropodidae, with their digital reduction and narrowed tarsus, no anlage of the tibiale appears, and that in these marsupials the tibialis posterior, lacking its insertion, is also absent (Lewis 1962*a*). Apparently also in some Eutheria (for example, dog, cat, bear, gibbon) tibiale and centrale chondrify separately (Baur 1884; Testut 1904). As will be seen in Chapter 14 the tibiale can chondrify and ossify separately in man: this is one of the anomalous bones which have been designated as the 'os tibiale externum' or 'naviculare secundarium'. Thus, a developmental trend observable in the carpus is repeated in the tarsus; the indisputable embryological evidence for a dual origin of the carpal scaphoid is applicable with equal force to the tarsal navicular.

It could be that the rodent condition represents a case of evolutionary reversal and that the composite bone was established early in therian history. A more likely scenario is that the two components were independent in the early Theria and fusion was achieved in parallel in the Metatheria and the Eutheria (or most of them). Certain points of quite fundamental significance support this proposal. Hildebrand (1978) observed that, uniquely among the rodents the family Anomaluridae lack a free 'medial tarsal bone' and in them the strong tendon of tibialis posterior inserts upon the unusually large medial cuneiform. Hildebrand concluded that it was probable that the 'medial tarsal bone' (the tibiale) had fused to the first cuneiform, carrying with it the insertion of the muscle. This seems a perfectly reasonable assumption and could of course be confirmed by a developmental study, given appropriate material.

Interestingly, I have observed a comparable morphology in the insectivores *Suncus caeruleus* and *Tenrec ecaudatus*: in each the medial cuneiform has a posteriorly directed process, flanking the reduced navicular, and receiving the insertion of the tibialis posterior, which here lies deep to the homologue of the Y-shaped ligament (tibial–medial cuneiform band). Again the obvious conclu-

sion is that in these species the tibiale has fused to the medial cuneiform, and the navicular is represented by the centrale alone. Whether this is the rule in insectivores has not been investigated. Perhaps most significantly of all in the treeshrew *Tupaia sp.* (Fig. 14.5) the same condition is found, although the tibialis posterior tendon is well merged with the Y-shaped ligament as it approaches its insertion to the backwardly elongated prong (presumptive tibiale) on the medial cuneiform. This observation serves to even further distance the treeshrews from primate relationships. Moreover, if the observation by Leche above, that '*Galeopithecus*' (= *Cynocephalus*, the colugo, the sole living representative of the Dermoptera) has an independent tibiale, it would have considerable relevance to the current debates about the relationships of the Insectivora, Dermoptera and Scandentia or treeshrews mentioned in Chapter 2. There are, therefore, persuasive grounds for postulating that in the earliest Therians, and also in the first Eutherians, the tibiale was still separated from the centrale, and while these two fused in the ancestors of most Eutheria, in another lineage (or lineages) the tibiale in contrast fused with the medial cuneiform.

With the tibiale removed from contention as a contributor to the talus, that leaves the whole bone derived from the intermedium. Indeed the talus normally chondrifies singly (Bardeen 1905) and its developmental position between the tibia and fibula in marsupials (Emery 1897) points to its homology with the intermedium. This is strongly supported by its relationship to the perforating artery of the tarsus. In the adult salamander in both carpus and tarsus there is a perforating artery between the intermedium on the one hand and the fibulare or ulnare on the other; similar vessels appear in the development of the human carpus, between lunate and triquetral, and in the tarsus between talus and calcaneus, though only the latter vessel persists (Leboucq 1886). The vessel clearly identifies intermedium and fibulare and their homologues in the mammalian tarsus. Again there is revealed a considerable fundamental unity in plan between carpus and tarsus.

How then can the so-called free 'intermedium tarsi' of the marsupial foot be explained away? The supposed existence of this element, of course, (along with the occasionally occurring os trigonum of the human foot) was a key point in prompting speculation that the talus was a composite of tibiale and intermedium. Marsupials in general possess a large intra-articular meniscus at the ankle, intervening between the fibula and the talus (Fig. 11.5). This structure is seen in what is apparently the prototypal marsupial form in the Australian phalangers and in these, and in other Australian marsupials, (Lewis 1964*a*) it contains a large pyramidal ossicle (lunula) embedded within its substance. In macerated preparations this supernumerary bone strikingly but fortuitously resembles the anomalous os trigonum of the human foot. The attachments and function of the meniscus will be more fully considered in Chapter 12. It is clear then that this ossicle is a functional modification in the marsupial foot and has nothing whatsoever to do with the intermedium or any other of the basic components of the tarsus for that matter. As for the human os trigonum it has no homology with either the intermedium or the marsupial ossicle: it results from an anomalous secondary ossification centre (having no morphological value or equivalence) in the posterior tubercle of the talus.

These morphological findings then, taken together with the evidence from palaeontological and embryological sources clearly point to the following conclusions: that the calcaneus is homologous with the fibulare, the talus with the intermedium, the navicular with the coalesced tibiale and centrale, and the three cuneiforms with tarsalia one to three. Evidence from morphology and palaeontology is equivocal as to whether the

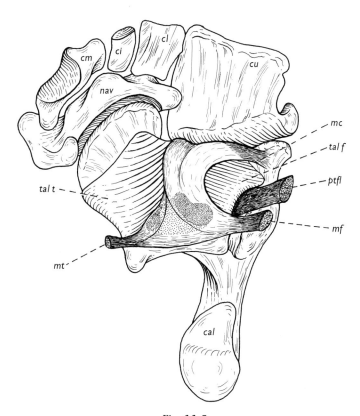

Fig. 11.5

The dorsal aspect of the right tarsus of *Phascolarctos cinereus* showing the intra-articular meniscus, the lunula of which is indicated by stippling. *cal*, calcaneus; *ci*, intermediate cuneiform; *cl*, lateral cuneiform; *cm*, medial cuneiform; *cu*, cuboid; *mc*, calcaneal attachment of meniscus; *mf*, fibular attachment of meniscus; *mt*, tibial attachment of meniscus; *nav*, navicular; *ptfl*, posterior talo-fibular ligament; *talf*, fibular facet on the talus; *talt*, tibial facet on the talus. (From Lewis 1964*a*.)

cuboid represents tarsale distale four alone, or is the product of fusion of the fourth and fifth. It is generally agreed that the human cuboid chondrifies and ossifies singly, but in marsupials (Emery 1897) there are two cartilaginous anlagen. Thus, invoking the principle of heterochrony the likelihood is that the cuboid is representative of the coalesced fourth and fifth tarsalia.

The prehallux

The ancient pre-axial ossicle, the prehallux, is a common persistent feature of the mammalian foot. In the primitive condition it presents as a sesamoid within the complex tendon of insertion of the flexor tibialis; it is found thus in most marsupials and in some of the Rodentia, Insectivora, and Edentata. The way in which this insertion is modified, affecting the precise topographical relationships of the prehallux will be described in Chapter 14. There has often been considerable confusion in the literature between the prehallux and the true tibiale of rodents and monotremes (Flower 1876; Schafer and Thane 1899). This is reflected in confusion of the anomalous human os tibiale externum (Chapter 14) with the prehallux, and presumably prompted Cihak (1972) to include a

prehallux component in the developing human navicular (Fig. 11.1). Bardeleben (1885*b*, 1894) considered the prehallux as a remnant of a true sixth digit, hence the name. There is no evidence, however, that it prepresents a true sixth tetrapod digit, for even in the primitive pentadactyl amphibian *Trematops* (Fig. 11.2A) it was merely a small bony nodule. It is probable that like the prepollex it is representative of a preaxial fin ray. These names then are not entirely appropriate, but are worthy of retention, if only to avoid confusion.

In the skeleton of some mammals—certain Rodentia and Edentata (Jones 1953), certain Insectivora (Dobson 1883*b*) and even *Elephas africanus* (Leche 1900)—the prehallux is enlarged to such an extent that it mimics an additional preaxial digit. This specialization is particularly exaggerated in the Brazilian tree porcupine (Fig. 11.4) where the prehallux is grotesquely expanded and forms the skeletal basis for a pad which may be opposed to the remainder of the sole, thus fulfilling a prehensile function. This remarkable secondary specialization for arboreal life was fully described by Jones (1953); this author, however, failed to appreciate the significance of the bipartite navicular (a normal rodent feature), which he regarded as a secondary specialization for increasing the mobility of the prehensile pad.

12

The ankle joint

The basic architecture of the ankle joint in the tetrapod ancestors of the mammals has been exhaustively studied by Schaeffer (1941a), and from this it is clear that the earliest amphibians and the mammal-like reptiles possessed an ankle joint incorporating a weight-bearing articulation between the fibula and the evolving calcaneus (fibulare). Even in the monotremes the fibula provides the major weight bearing surface at the ankle since it retains articulation with the calcaneus (Fig. 11.3). Schaeffer rightly stressed that the therapsid foot was converted into a mammalian one primarily by superposition of the talus upon the calcaneus, thereby separating the fibula from the calcaneus and resulting in a loss of weight-bearing contact between the bones. However, in a number of mammalian species a calcaneofibular contact, although of different nature, has been restored as a new apomorphic character. On the basis of the condition in cynodonts and monotremes it is usually assumed that any calcanoeofibular contact in mammals is primitive or plesiomorphic; this is not so, but the notion has significantly influenced contemporary evolutionary theory, especially speculations about the origin of the order Primates. It follows from this preamble that the human *Nomina Anatomica* term 'Articulatio talocruralis' is too restrictive and not always appropriate in the comparative context; here the simple term 'ankle joint' is preferred. The name 'upper ankle joint' used by some comparative anatomists to distinguish it from the subtalar joint complex ('lower ankle joint') is also needlessly confusing and is only really apt in certainly highly specialized feet.

It is reasonable to suppose that the anatomy of living marsupials (metatherians) gives a good indication of the structure of those advanced early Cretaceous therians from which the first placentals (eutherians) evolved (Chapter 3). There is little doubt that the primitive marsupial foot was prehensile (Bensley 1903) with a divergent opposable hallux such as that now represented in the Didelphidae; among the Australian marsupials, the Phalangeridae present a foot close to the prototypal marsupial pattern except for the specialization of syndactylism of the second and third digits. The phalangers and didelphids should, therefore, provide a good insight into the probable structure of the primitive metatherian ankle joint, and by implication, also clarify the probable morphology of this region in the ancestors of the Eutherians. Moreover, the specializations found in various marsupial species, particularly in those of the diverse Australian radiation, should indicate those potentialities for modification inherent in this basic structure, and which might reasonably be expected to have appeared convergently in eutherians.

The monotreme joint

At first sight the foot architecture of monotremes seems far removed from the therian condition, and superficially seems to exhibit much closer affinity with that of their reptilian precursors. Critical examination, however, uncovers revealing insights into the ancestry of form and function in the more advanced mammals. For this purpose the foot of the echidna is most informative: although possessing obviously aberrant specializations, these are less profound than those in the platypus.

Tibia and fibula both participate in the ankle joint, as do the calcaneus and talus (Fig. 13.1). The shallowly concave lower extremity of the fibula articulates directly with both talus and calcaneus, and no meniscus intervenes. The fibular articulation with the calcaneus is obviously a primitive inheritance from the therapsid ancestors. The tibia has a rather hook-like articular surface, spiralling downward to terminate below as a prominent knob, which is received into a concave depression, on the talus. Adjacent to this depression, which is most clearly demarcated from the remaining convex part of the talar surface in fresh specimens, the tibia and talus are united by a strong ligament, which, as will be seen, is clearly the homologue of the therian posterior tibiotalar ligament. The joint is walled posteriorly by another strong ligament passing from the crural bones to attach to the talus, predominantly into a posterior cleft in that bone. It is likely that the fibular portion of this is the precursor of the higher mammalian posterior talofibular ligament. Anteriorly there is a strong anterior tibiocalcaneal ligament passing obliquely across the joint, and laterally there is a less well-defined fibulocalcaneal ligament.

The form of the tibial articular surface, with its attached posterior tibiotalar ligament, as will be seen, is clearly ancestral to the marsupial morphology, which in turn foreshadows the architecture characteristic of the prosimian and higher primates. The articular surface on the talus similarly represents a logical precursor morphology to the marsupial pattern. The oblique anterior ligament may have provided the basis of the fibrous anterior horn of the marsupial meniscus when that structure became elaborated, but the remainder of the therian ligamentous apparatus except for the posterior tibiotalar ligament is not clearly differentiated at the monotreme grade.

The Metatherian joint

The primitive Metatherian morphology

The arboreal Australian phalangers (*Trichosurus* and *Pseudochirus*) provide a convincing picture of the likely structure of the primitive marsupial ankle joint (Fig. 12.1). The joint cavity is continuous medially around the side of the talus with the subjacent talocalcaneonavicular joint. There is here a medial wall common to these confluent joints. This is the medial ligament, which possesses

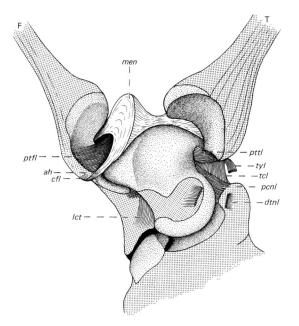

Fig. 12.1

The right ankle joint of *Trichosurus vulpecula* opened from dorsally and with the tibia (T) and fibula (F) rolled apart to demonstrate the lunula-containing meniscus (*men*) with posterior attachment to tibia and fibula and an anterior fibrous horn (*ah*) arching to an attachment on the calcaneus; part of this horn is secondarily annexed in this marsupial by the fibula to form a calcaneofibular ligament (*cfl*). Also indicated are: *ptfl*, posterior talofibular ligament; *lct*, ligamentum cervicis tali; *dtn*, dorsal talonavicular ligament (cut); *pcnl*, plantar calcaneonavicular ligament; *tcl*, tibiocalcaneal ligament; *tyl*, tibial band of Y-shaped ligament; *pttl*, posterior tibiotalar ligament. (From Lewis 1980*a*.)

two fasciculi, the homologues of similar parts of the human medial (deltoid) ligament: the posterior tibiotalar ligament arises from the knob-like termination of tibial articular surface (cf. echidna); in continuity with this is a more anterior band (tibiocalcaneal ligament) passing down to attach to the plantar calcaneonavicular (spring) ligament and so indirectly to the calcaneus. The wall of the joint is completed dorsally by the anterior ligament of the ankle joint passing to the neck of the talus which is joined at the same situation by another very tough ligament (dorsal talonavicular ligament), sweeping over from the dorsal margin of the navicular. The tibialis posterior tendon passes over the surface of the two bands of the medial ligament and is here under cover of the homologue of the tibial–medial cuneiform band of the primate Y-shaped ligament (Chapter 14) which takes no part therefore in forming the wall of the joint, being lifted clear of it by the tibialis posterior tendon.

The form of the inferior surface of the tibia is strikingly similar to that of monotremes (Figs 12.1, 13.1). There is no demarcation of a malleolar area. The broader upper part of the tibial surface articulates directly with a large convex area on the talus which is hollowed out laterally. Medially this talar area shows an ill-defined depression which lodges the terminal articular knob on the tibia in dorsiflexion. There is therefore no clear-cut subdivision of the tibial area on the talus into medial and lateral facets. The hollowed lateral part of the tibial facet is separated by a ridge from a splayed out area on the talus for the fibula. Here a prominent meniscus intervenes between the fibula and the talus and also extends up to separate the apposed cartilage clothed surfaces of the tibia and fibula. As noted in Chapter 11 this fibrocartilaginous structure has a posteriorly located body of triangular pyramidal shape containing a lunula (bony or cartilaginous) and the posterolateral and posteromedial corners have strong ligamentous attachments to fibula and tibia respectively (Fig. 11.5). Anteriorly the meniscus is prolonged forwards between the fibula and talus and terminates as a fibrous anterior horn, which takes a recurved course to attach to the calcaneus. The oblique anterior ligament of monotremes may have provided the structural basis for the evolutionary fashioning of this fibrous attachment. As the anterior horn runs off the surface of the talus, its deep synovial covered surface abuts on a lateral projection on the calcaneus which is, in fact, the exposed margin of the articular surface of that bone entering into the subtalar joint; here the subtalar and ankle joints are in continuity. Thus, both medially and laterally the ankle joint is in communication with what may be conveniently called the subtalar joint complex. In *Pseudochirus laniginosus* (Fig. 12.3A) the anterior horn of the meniscus is the only 'ligamentous' structure on the lateral aspect of the ankle joint. In *Trichosurus vulpecula* (Fig. 12.1, 12.3B), however, the fibula has annexed part of the substance of the anterior horn of the meniscus, secondarily forming a calcaneofibular ligament, strikingly similar to and

presumably homologous with that of man and other Primates.

Passing from the fibula to the talus, encircled by the meniscus, is a posterior talofibular ligament, part of the lateral ligament of the joint, which is essentially intracapsular and intervenes between the fibula and the exposed projection of the subtalar articular facet on the calcaneus. Thus, in these marsupials there is an indirect approximation of these two bones, but no actual calcaneofibular contact as in monotremes. The lateral ligament is completed by a thin anterior fasciculus passing from the fibula to the talar neck, really no more than part of the fibrous capsule.

An ankle joint meniscus appears to be of universal occurrence in marsupials, with a single exception to be noted below, but it does not always contain a bony lunula. The varying morphology shown by the meniscus and the related articular surfaces in different marsupials can be used to provide, by analogy, new interpretations of the structure of the eutherian joint.

Apomorphic derivatives in Metatheria

In the American didelphids *Caluromys* and *Didelphys* (Figs 12.2, 12.3C) the fibrous anterior horn of the meniscus is completely divorced (not partially as in *Trichosurus*) to form a calcaneofibular ligament, leaving the meniscus proper secondarily attached to the talus. Morever, in *Didelphys* the articular surface on the tibia is remodelled. Its lower knob-like end is clearly demarcated from the remaining upper part of the surface, by a non-articular groove; in effect it has become a medial malleolus of almost eutherian aspect. The tibial surface on the talus has become sharply sculpted to accommodate this remodelled tibia, in contrast to the smoothly rounded contours in the phalangers. A deepened furrow accommodates the upper (or lateral) tibial facet and an even deeper excavation, only hinted at in phalangers, accommodates the medial (or 'malleolar') tibial facet. Essentially the same description of these osteological features in *Didelphis* has been given by Jenkins and McLearn (1984). These authors were in error, however, in thinking that this type of architecture was universal in didelphids and shared by phalangerids. It is clear that the phalanger type of architecture (Fig. 12.1) is the primitive one, easily derived as it

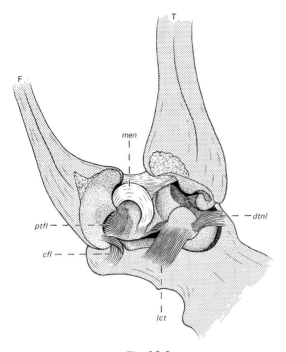

Fig. 12.2

The right ankle joint of *Didelphys marsupialis* opened from dorsally. Labelling as in Fig. 12.1. (From Lewis 1980*a*.)

is from the monotreme morphology, and indeed I have observed this same structure in a specimen of the didelphid *Caluromys*. In contrast the *Didelphys* arrangements present a suite of apomorphic changes which go some way towards paralleling the architecture typical of the Eutheria.

The broad expanded form of the talar body, with divergent weight-bearing surfaces for tibia and fibula which is seen in the arboreal phalangerids, has become remodelled into a cylindrical trochlear shape in terrestrial Australian marsupials.

The transformation may be imagined as essentially involving compression of the talar body between the evolving malleoli, converting the joint into a stable hinge type. An incomplete transitional stage in such evolutionary remodelling of the talar body is retained in the more terrestrial dasyurids such as *Sarcophilis harrisi* and *Antechinus swainsoni*. Here, however, the antecedent phalanger-like morphology is still clearly apparent, and a typical marsupial meniscus persists and often contains a lunula. The form of the tibial

articular surface and the sharper modelling of the talar body recall the situation in *Didelphis*.

In the terrestrial kangaroos and wallabies the transformation of the articular surfaces is more extreme (Fig. 12.3D) and the body of the talus is remodelled into a trochlear shape, mimicking that found in placental mammals. The splint-like fibula is expanded below into a lateral malleolus articulating with the side of the talus, but with a reduced semilunar meniscus intervening. The fibula is also in close relationship with the calcaneus where it protrudes quite far laterally from beneath the talus; here, the flattened posterior talofibular ligament intervenes and has virtually been converted into an intra-articular meniscus (Fig. 12.4A). Thus a much more substantial approximation of fibula and calcaneus is present than that seen in *Trichosurus*. The latter is doubtless the primitive arrangement and the considerable area of apposition of the two bones seen in *Macropus* is not a retention of the ancient tetrapod calcaneofibular contact but is a derived feature correlated with remodelling of the subtalar joint complex. Medially, in the kangaroos, the lower end of the tibia is fully remodelled so as to present a clearcut malleolar facet, accentuating the trend heralded in *Didelphis* or the terrestrial dasyurids. Moreover, the ankle joint is separated by a fibrous septum, in effect a deep portion of the medial ligament, from the talocalcaneonavicular joint, again converging here on advanced eutherian structure.

In the bandicoots the transition towards eutherian form (Fig. 12.3E) is even more striking, at least when the joint is viewed from laterally, for the ankle meniscus is here completely lacking. There is, however, a tough ligamentous connection—an inferior transverse tibiofibular ligament—between tibia and fibula, which doubtless represents the residual connections of the meniscus to these bones, and a strong calcaneofibular ligament remains as the homologue of the anterior fibrous horn of the meniscus. Surprisingly, the medial aspect of the bandicoot joint is relatively conservative for the talus here retains its splayed out form, and the tibia lacks a malleolar knob. The remodelling of the talus in these kangaroos and bandicoots may be seen as an evolutionary device providing increased stability at the ankle joint, but at the cost of sacrificing rotatory mobility of the leg bones.

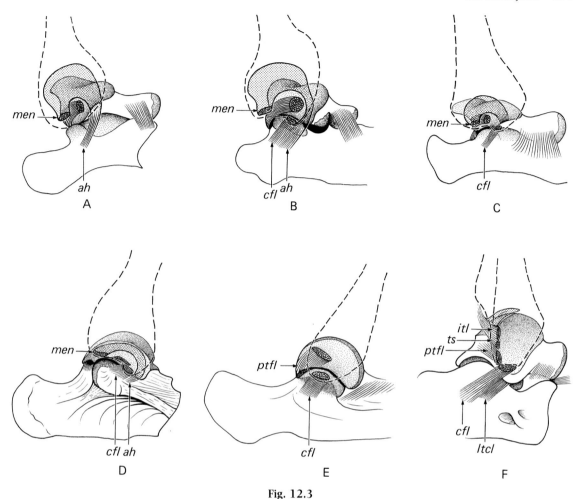

Fig. 12.3

The lateral aspects of the ankle joints of *Pseudochirus laniginosus* (A), *Trichosurus vulpecula* (B), *Didelphys marsupialis* (C), *Macropus major* (D), *Perameles gunni* (E), *Homo sapiens* (F) with the outline of the overlying fibula shown in broken line. Where a meniscus is present (A, B, C, D) it is stippled and the attachment of its posterior horn (*men*) to the fibula is indicated. The meniscus then arches forward, encircling the fibular attachment of the posterior talofibular ligament (*ptfl*), to attach by its fibrous anterior horn (*ah*) to the cancaneus; in B and D part of this horn is detached and in C the whole of it, to form a calcaneofibular ligament (*cfl*); in E, where the meniscus is lacking only this ligament (*cfl*) remains. In *Homo sapiens* (F), similarly, the cancaneofibular ligament (*cfl*) remains, but the presumptive meniscal remnant is shown in stipple projecting forward from the inferior transverse tibiofibular ligament (*itl*) which in this specimen is joined by a tibial slip (*ts*) of the posterior talofibular ligament (*ptfl*); this particular specimen also shows a lateral talocalcaneal ligament (*ltcl*) delaminated from the calcaneofibular ligament (*cfl*). (From Lewis 1980*a*.)

Function in Metatheria

Such a substantial evolutionary novelty as the marsupial ankle meniscus must have conferred some special functional advantage. The marsupial fibula then is provided with intra-articular fibro-cartilages in the joints at both upper and lower ends. Reptiles possess one only, at the upper end: the femorofibular disc (Chapter 14). This structure accommodates rotatory movement of the fibula about its long axis, which coupled with a rotation of the tibia in the same direction, allows the reptilian foot to be rotated into an abducted or adducted position (Haines 1942). The movement achieves a similar result to forearm pronation and

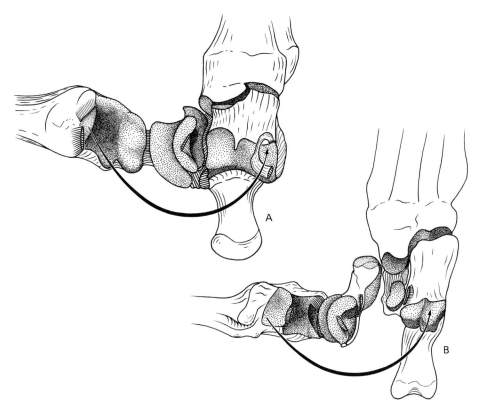

Fig. 12.4
Right feet of *Macropus major* (A) and *Oryctolagus cuniculus* (B). In each case the talus and fibula have been cut from their attachments laterally, and rolled away medially. The arrows represent how the conarticular fibular and calcaneal surfaces may be brought back into contact. (From Lewis 1983.)

supination, although the mechanical solution is different. The marsupial ankle meniscus, denied to reptiles or monotremes, adds a new dimension to this pattern of movement.

The foot can then be externally rotated on the leg bones, movement being centred on the fibula, and at the same plantarflexed; it is for such combined movements, involving rotation with angular displacement, that menisci are required in joints (Barnett 1954). At the same time the fibula externally rotates about its long axis, but there is negligible accompanying movement of marsupial tibia, unlike the situation in reptiles. The upshot of this is that the foot is reversed. As Cartmill (1974) has shown such a movement allows for placement of the plantar surface of the clawed foot against near vertical supports during controlled head first descent. The capability for such movement is

characteristic of the arboreal didelphids and phalangerids.

As Barnett and Napier (1953) noted, however, rotatory movement of the fibula is lost in terrestrial marsupials such as the macropodids (kangaroos) and peramelids (bandicoots). As we have seen above, in these families the talus is restructured into a trochlea form of almost eutherian aspect, providing for hinge movements. Moreover, in these marsupials the femoro fibular disc is remodelled to form what is, in effect, an intra-articular popliteus tendon (Fig. 14.18). The erroneous idea, however, that the ancestral mammalian hind limb has at no stage been capable of any appreciable fibular rotation (Carleton 1941), was a consequence of the failure to consider the possible relevance of marsupial structure in mammalian evolution.

Foot reversal is not confined to marsupials but also occurs in various eutherians—some sciurids, procyonids, felids, viverrids, tupaiids, and even prosimians. Jenkins and McLearn (1984) did a study, largely radiographic, of these movements. Remarkably they described the movement in marsupials without a single reference to the existence of an ankle meniscus. They focused their attention entirely upon the movement of talus upon tibia, and while this is true so far as it goes, the tibiotalar movement is in a sense subsidiary. Significantly, however, they did show that hind foot reversal in eutherians was the outcome of a quite distinct morphological solution: exaggerated subtalar inversion combined with transverse tarsal supination.

The eutherian joint

Basic eutherian morphology and apomorphic trends

As has been seen in Chapter 3 the divergence of the emergent Eutheria seems to have been correlated with an initial terrestrial commitment. One would therefore expect that the stem Eutherians would have evolved a trochlear talar body, with accompanying loss of the ankle meniscus. A good insight into the likely early eutherian morphology is provided by the treeshrews. In *Tupaia*, for instance, the body of the talus is remodelled into a trochlear shape reminiscent of that convergently acquired in the marsupials *Macropus* and *Perameles*. Significantly at the medial side of the ankle the joint cavity is continuous with that of the talocalcaneonavicular joint, just as in the unspecialized marsupial phalangerids and didelphids, and the ligamentous apparatus here is also similar. Laterally, again like those same marsupials, the ankle joint is continuous with the subtalar (posterior talocalcaneal) joint. There is no meniscus within the ankle joint and, as would therefore be expected, the ligamentous apparatus is comparable to that of *Perameles*: the posterior talofibular ligament is of the usual form and a calcaneofibular ligament passes downwards and backwards partly walling the confluent joint cavities laterally. The treeshrews, then, are apparently unique among the placental mammals in lacking a medial separation

of the ankle and subtalar joints and in this respect strikingly resemble the generalized marsupials. The most parsimonious conclusion is that this arrangement has been derived from the last common ancestor of marsupials and placentals. This is one more addition to the catalogue of morphological characters which are reminiscent of marsupials and which persist in treeshrews (Jones 1917; Clark 1926).

Lateral communication between the ankle joint and the posterior talocalcaneal joint, however, is not restricted to marsupials and treeshrews but persists as a common eutherian feature, being present even in prosimian Primates and New and Old World monkeys. Here, to a varying extent, an indirect approximation of calcaneus (posterior talar facet) to the fibula may occur, but the interposed thick posterior talofibular ligament prevents direct contact.

The lateral continuity of the joints has provided the raw material for an apomorphic character which has evolved in parallel in a number of mammalian groups: renewed calcaneofibular articulation. It must be stressed that such articulation is not a retention of a primitive character. The earliest mammals certainly possessed a calcaneofibular articulation as part of their reptilian inheritance. In mammal-like reptiles, from pelycosaurs to cynodonts, the fibula participates in the ankle joint as a significant weight-bearing component, articulating with the calcaneus and often with the talus also. The monotremes retain a similar morphology. It has therefore been assumed that any calcaneofibular contact in extant or fossil mammals merely represents a persistence of this primitive condition. In fact, it seems clear that calcaneofibular contact in various extant mammals, and by analogy in certain fossils, is an apomorphy, utilizing a different part of the calcaneus, and having arisen in parallel a number of times in response to new functional requirements. In the emergent therian mammals, in contrast to monotremes, increased superposition of the talus upon the calcaneus has resulted in the fibula retreating from direct contact with the calcaneus and a massive new ligament, the posterior talofibular ligament, which has no clear monotreme homologue, has emerged. This ligament in phalangerids for example (Fig. 12.1) is essentially intracapsular and intervenes between the lower extremity of the fibula and a lateral projection of the proximal facet on the calcaneus, which emerges from under cover of the talus, bringing subtalar and ankle joints here into continuity. There is therefore only an indirect relationship between fibula and calcaneus and no direct contact. It is to be noted that the calcaneal articular surface underlying the ligament is the homologue of part of the proximal articular surface of the calcaneus for the talus in echidna (Fig. 13.1) and is not homologous with the calcaneal facet for the monotreme fibula. It is also apparent that continuity between talocalcaneal and ankle joints is no new acquisition, it is merely retention of the monotreme condition. This basic morphology seems to have included a ready potentiality for modification in response to new functional needs. As noted above (Fig. 12.3D) the saltatory kangaroos, derived from phalanger-like ancestors, nicely demonstrate such a progressive change. The subtalar joint in the grey kangaroo, *Macropus major*, for instance, is drastically remodelled into what is effectively a 'lower ankle joint' (Fig. 12.4A) with an almost transverse axis and an overall morphology converging on that described for artiodactyls by Schaeffer (1947). The ankle joint is also remodelled into a complementary hinge, with a talus of trochlear shape and a splint-like fibula expanded below into a lateral malleolus for articulation with the side of the talus. Here lies a reduced marsupial meniscus, but significantly the fibula is in intimate relationship with the talar calcaneal facet being only partially separated from it by a compressed posterior talofibular ligament, which has been converted into something resembling an intra-articular meniscus.

This combination of conversion of the subtalar joint into a 'lower ankle joint', associated with substantial calcaneofibular articulation, and often with progressive degrees of amalgamation of the fibula with the tibia, has been realized convergently a number of times in mammalian evolution, apparently as an adaptation to saltatory or cursorial locomotion. No attempt will be made here to give a comprehensive survey of the Eutheria for evidence of this adaptation but representative examples will illustrate the significant features of the derived morphology. These examples have been chosen because of the way in which misconceptions about their nature have unduly influenced cladistic reasoning.

The rabbit *Oryctolagus cuniculus*, (Fig. 12.4B) clearly shows the manner in which direct calcaneofibular contact may be attained in eutherian mammals. In effect the talus is displaced medially, exposing the lateral part of the posterior calcaneal facet for articulation with the fibula. Furthermore, the residual surface for the talus is remodelled, the anterior slope becoming hollowed and forming with the talocalcaneonavicular joint a concave trough for the talus—a 'lower ankle joint'. To accommodate the remodelled calcaneal facet, which retains its convex posterior rim, a wedge-shaped articular notch is excavated into the lateral aspect of the talus. The large exposed lateral portion of the calcaneal facet has a broad articular contact with the distal extremity of the fibular component of the fused tibiofibula. Unlike *Macropus major* this contact is extensive and direct, since it lies lateral to the posterior talofibular ligament.

Certain of the shrews (*Suncus*) and elephant shrews, particularly those with a lengthened foot adapted to a hopping gait, possess a posterior talocalcaneal joint which has also been re-orientated into an essentially transverse disposition, similar to that of kangaroos or rabbits. Again, convergently, the calcaneus protrudes laterally from under cover of the talus to achieve a direct articulation with the fibula. All the species described here which show unequivocal direct calcaneofibular articulation are characterized by a lengthened pes, fusion of the tibia and fibula, and remodelling of the subtalar joint complex comparable to the fossil transitional stages described by Schaeffer (1947) for the emergence of the fully elaborated 'lower ankle joint' of artiodactyls. In the extant artiodactyls the talus is effectively converted into a roller bearing, interposed between leg and foot, and calcaneofibular articulation is extensive.

Such derived types of direct calcaneofibular articulation, result from redeployment of pre-existing morphological features. The primitive lateral continuity between the ankle and posterior talocalcaneal joints provides the potential for direct articular contact between fibula and calcaneus, only the posterior talofibular ligament presenting an effective barrier. Modification of the subtalar joint complex into a transversely disposed 'lower ankle joint' seems often to be associated with a bodily medial shift of the talus. The exposed lateral portion of the posterior talar surface on the calcaneus is then appropriated by the fibula which thus establishes a major direct weight-bearing articulation lateral to the posterior talofibular ligament. Since the fibula in therian mammals (and particularly Eutheria) has relinquished weight-bearing transmission from the femur it is not surprising that a correlated morphological change should involve varying degrees of amalgamation of the shaft of the fibula with that of the tibia.

The very nature of the gradual transition from mere contiguity to substantial direct articulation between fibula and calcaneus means that a study restricted to osteological material may yield equivocal or even frankly misleading results.

The primate joint

Subhuman Primates In lemuroids such as *Lemur catta* (Fig. 12.5) or *Perodicticus potto* the body of the talus is modified, as it is in *Tupaia*, into a

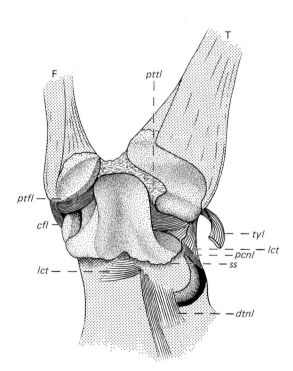

Fig. 12.5
The right ankle joint of *Lemur catta* opened from dorsally. *ss*, synovial septum separating ankle joint from talocalcaneoavicular joint. Other labels are as in Fig. 12.1. (From Lewis 1980*a*.)

characteristically eutherian trochlear shape with elevated borders, the lateral being most prominent. The medial ligamentous apparatus is quite comparable to that described above for generalized marsupial feet and consists of a posterior tibiotalar ligament and a tibiocalcaneal ligament, these two bands being the homologues of the corresponding parts of the human medial or deltoid ligament. It should be noted that the tibiocalcaneal ligament, so-called, in reality attaches to the plantar calcaneoavicular (spring) ligament, its attachment to the sustentaculum tali being largely indirect. These ligaments are related to the tibial malleolar surface of the medial aspect of the talar body. This facet is comma-shaped, encircling the attachment of the posterior tibiotalar ligament, and is expanded anteriorly into a cup-like depression which receives the knob-like tibial malleolus in dorsiflexion. The articular surface on the tibia is remodelled along the lines found in the marsupial *Macropus*, and which was at least incipiently realized in *Didelphys*. The ancestry of the knob-like malleolus can readily be traced back to the situation in generalized marsupials (Fig. 12.1) and even in echidna (Fig. 13.1). In marked contrast to generalized marsupials and *Tupaia*, the ankle joint is medially separated from the subtalar joint complex by the elaboration of a synovial septum which passes across the neck of the talus, skirting the front of the cup for the medial malleolus, and then attaching on to the deep aspect of the tibiocalcaneal band of the medial ligament. It forms a complete barrier sealing off the ankle joint synovial cavity from the subtalar joint complex. There appears to be some variation in this feature in different prosimians and in some it may be thicker and more fibrous, virtually forming a deep lamina of the medial ligament. It is to be noted that such a deep part of the medial ligament, closing off the ankle joint from the subtalar joint complex, is not an exclusively primate specialization but constitutes the usual eutherian condition, and it is found even in insectivores such as *Tenrec ecaudatus; Tupaia* is a conspicuous exception. Related to the external surface of the deltoid ligament is the synovial covered tendon of tibialis posterior, held firmly in place by the typical primate Y-shaped ligament (Chapter 14). Laterally, the talar body bears a triangular surface for the lateral malleolus. This is markedly splayed out laterally, and posteriorly it bears the imprint of the related posterior talofibular ligament. This latter ligament effectively intervenes between the fibular malleolus and the exposed posterior talar facet on the calcaneus, which here protrudes a little laterally from under cover of the talus. The ankle joint and subtalar joints are here in communication. Arching across this articular protrusion of the calcaneus is a calcaneofibular ligament, and it seems reasonable again to interpret this as the residual anterior fibrous horn of a pre-existent meniscus such as that found in marsupials. An inferior transverse tibiofibular ligament, which is also present, perhaps represents the posterior meniscal attachments. In some lemuroids, a fibrous tongue projects forward from the inferior transverse tibiofibular ligament and contacts a bevelled off triangular facet on the posterior part of the lateral margin of the trochlear surface of the talus; from its situation it seems plausible to suggest that this is a persistent meniscal remnant.

Monkeys, both New World and Old World varieties, retain the essentials of this arrangement (Fig. 12.6). Without exception the ankle joint is sealed off medially from the subtalar joint complex by a variably developed deep lamina of the medial ligament which in some species is represented by little more than a synovial septum. In contrast the joint retains communication with the subtalar joint laterally, where the calcaneus may project slightly from under cover of the talus to retain an indirect calcaneofibular relationship, and the posterior talofibular and calcaneofibular ligaments are lined by the synovial membranes of the confluent joints. In general the sidewards flaring of the malleolar facets on the talus is reduced in cercopithecines and the mortise and tenon character of the joint is enhanced. In some monkeys (Fig. 12.6) a fibrous tongue projecting forward from the inferior transverse tibiofibular ligament may represent a meniscal remnant.

In *Hylobates* the essentials of the morphology seen in *Pithecia* are retained. The malleolar facets are markedly flared out sideways (Fig. 12.8A) and the medial ligament possesses a deep lamina. However, the ankle joint is sealed off laterally from the subtalar joint, although the calcaneus protrudes somewhat here and is covered by a flattened posterior talofibular ligament which intervenes between it and the fibula. Again there is a fibrofatty

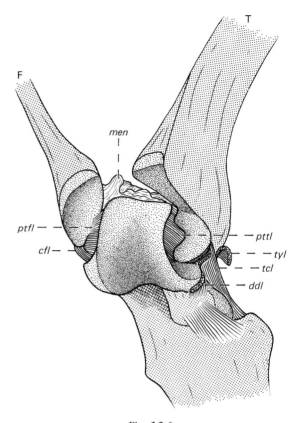

Fig. 12.6
The right ankle joint of *Pithecia monachus* opened from dorsally. *men*, possible remnant of meniscus; *ddl*, deep part of medial ligament of ankle joint. Other labels are as in Fig. 12.1. (From Lewis 1980a.)

tongue, possibly representing the marsupial meniscus, contacting the bevelled off facet on the posterolateral margin of the upper aspect of the talus. *Pan* and *Pongo* both show similar morphology and the ankle joint is also sealed off laterally from the subtalar joint. In one specimen of *Pongo*, the fibrous tongue, which possibly is an atavistic meniscal remnant, actually attached to the bevelled off facet on the talus.

The human joint The articular surface of the human talus is clearly derivative from that of the primates described above, particularly the great apes. Sewell (1904) described it in the following way. The upper surface is generally concave from side to side but in 20 per cent of cases the more lateral part may be convex. The lateral border is longer than the medial one and somewhat curved, so that the surface as a whole is broader in front than behind. The resultant 'wedging' is however, less noticeable than in the apes. The lateral malleolar facet usually drops sharply downwards with only a slight and uniform outward flare, but sometimes it turns sharply outwards. The medial malleolar facet is comma shaped and almost vertical but it may be prolonged forwards onto the neck and there curves sharply medialwards in 1 per cent of cases. Sewell noted that this represents an atavistic character reminiscent of the ape morphology and that this feature is usually more noticeable in fetal feet; he related it functionally to a habitual inverted attitude of the foot, and speculated that in the adult it might be associated with persistent use of the sartorial position. Barnett (1954) also noted the occasional forward extension of the comma shaped medial malleolar facet on the talus, observing also that it is much commoner in the fetus.

It is clear that the human talus is a somewhat specialized primate type, the main distinction being the usual diminution in the lateral and medial flaring of the malleolar articular surfaces. A similar remodelling has, however, occurred in other Primates, for example *Cercopithecus nictitans*. It is also noteworthy that an approximation to the commoner primate form, with outwardly flared malleolar surfaces, may sometimes occur in human tali.

The ligamentous apparatus (Fig. 12.7) shows the clear hallmarks of derivation from the basic primate pattern, although with certain uniquely human apomorphic characters. The deep lamina of the medial ligament is strongly developed, separates the ankle and talocalcaneonavicular joints, and commonly stands isolated and sharply demarcated by the extrasynovial fat of the two joints; this part of the medial ligament is designated in the *Nomina Anatomica* as the pars tibiotalaris anterior, and it has been well figured by Kapandji (1968) and Brizon and Castaing (1953). The pars tibiocalcanea of the medial ligament is as in other Primates, and again has its major attachment below to the plantar calcaneonavicular (spring) ligament rather than directly to the sustentaculum tali as is usually described in

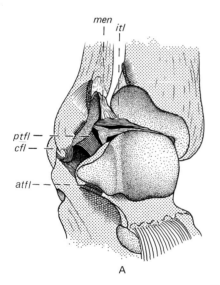 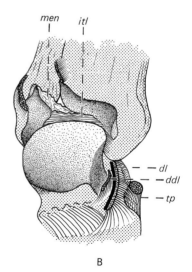

Fig. 12.7

The right ankle joint of *Homo sapiens* opened from dorsally. A, viewed from antero-laterally; B, viewed from antero-medially. In this specimen there was a tough, fibrous, presumptive remnant of the marsupial meniscus (*men*) projecting forward from the inferior transverse tibiofibular ligament (*itl*). *ddl*, deep part of medial (deltoid) ligament; *tp*, tendon of tibialis posterior; *dl*, medial (deltoid) ligament (tibionavicular and tibiocalcaneal parts) attaching to the plantar calcaneonavicular ligament and spreading to the surface of the navicular and the tibialis posterior tendon; *atfl*, anterior talofibular ligament. Other labels are as in Fig. 12.1. (From Lewis 1980*a*.)

textbooks of human anatomy. This relationship was, in fact, pointed out long ago by Smith (1896): the tibocalcaneal band is quadrilateral and descends to interlace with the plantar calcaneonavicular ligament and the interwoven pair of ligaments thence attach to the front of the sustentaculum tali. The human medial ligament has acquired an additional anterior band, unrepresented as such in other Primates although its derivation is patently obvious. In man, as described in Chapter 14, the stem and lower limb of the primate Y-shaped ligament have merged with the tendon of tibialis posterior, thereby prolonging its forward attachment from the navicular to the medial cuneiform and creating the recurrent attachment to the sustentaculum tali; the upper limb has simply amalgamated with the joint capsules of the ankle and talocalcaneonavicular joints creating the descriptive anterior band of the medial ligament (Fig. 12.7B), the pars tibionavicularis (which in reality also spreads down over the tibialis posterior tendon). The pars tibiotalaris posterior attaches to the posterior aspect of what has been called the 'anterior colliculus' of the medial malleolus by Pankovich and Shivaram (1979). This anterior colliculus is no more than the knob-like process at the lower end of the tibia whose history is clearly traceable from monotremes, through marsupials and other primates to man. Indeed the relationship of the attachment of the posterior tibiotalar ligament in the vicinity of this colliculus has been unchanged throughout this lengthy mammalian history. In one feature, however, man is unique: the articular surface on this anterior colliculus for the talus has become flat (Fig. 12.7A). It is common practice to restrict the term deltoid ligament to the more superficial tibiocalcaneal and tibionavicular parts; the anterior and posterior tibiotalar parts are then lumped together as the deep layer of the medial ligament of the ankle (Kapandji 1968; Brizon and Castaing 1953; Pankovich and Shivaram 1979).

Laterally the fibula and posterior talofibular ligament have withdrawn appreciably from the calcaneus and, as in the apes, there is no communication between the ankle and subtalar joint cavities. Sometimes part of the substance of the calcaneofibular ligament may be secondarily attached to the lateral tubercle of the talus, forming the inconstant lateral talocalcaneal liga-ment (Fig. 12.3F). As in other Primates, it seems likely that the calcaneofibular ligament itself is a persistent derivative of the fibrous anterior horn of the ancestral marsupial meniscus. Similarly, the inferior transverse tibiofibular ligament is the likely homologue of the posterior attachments of this structure; often associating itself with this ligament, and presumably of similar derivation, is the so-called tibial slip of the posterior talofibular ligament (Fig. 12.7A). It is not uncommon to find a tough fibrous tongue projecting forward from the inferior transverse tibiofibular ligament where it contacts the bevelled-off posterolateral facet (the intermediate articular facet of Sewell 1904) on the trochlea of the talus. This anterior fibrous prolongation skirts the upper margin of the articular surface of the lateral malleolus, where it partly seals off the entrance into the synovial cul-de-sac between the tibia and fibula. Although it is usually dismissed as a mere synovial fringe this structure is often tough and fibrous. Its position suggests that it is a residual homologue of the marsupial meniscus, and it is even more prominent in the human fetus, as might be expected if such is its derivation. The human ankle also presents an anterior talofibular ligament (Fig. 12.7A) which is of interest since it is the structure usually damaged in inversion sprains. Apparently only gibbons share with man the unique distinction of possessing this elaboration of the fibrous capsule (Inman 1976).

Function in Primates   Movement at the ankle joint appears to be basically similar in all subhuman primates, and this function seems to be particularly attuned to arboreal life; moreover, the fundamental character of the movement seems to be retained in those primates which have occupied an essentially terrestrial niche. The human joint, however, shows significant specialization in function.

Inspection of the superior articular surface of the talus of any arboreal primate such as *Lemur catta* (Fig. 12.5) or *Hylobates lar leuciscus* (Fig. 12.8A) makes it apparent that this enters into a sellar joint, or more accurately, a modified sellar joint as defined by MacConaill (1973), for the trochlea is obviously concave centrally and this region is flanked by convexities adjoining the lateral and medial margins. As MacConaill (1973) has pointed out such joints permit only a single

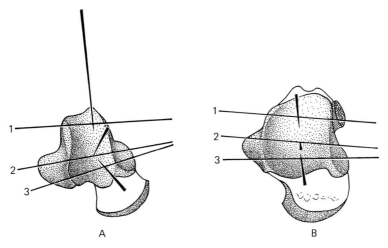

Fig. 12.8
Drawings made from superimposed radiographs of ligamentous preparations of the right ankle joints of *Hylobates lar leuciscus* (A) and *Homo sapiens* (B), taken at successive stages of movement at the joint. In each case a metal marker is inserted along the mechanical axis of the tibia to indicate swing and another transverse pointer through the tibia just above its articular surface (to indicate spin). The three positions shown in each case (1, 2, 3) represent successive stages in movement of the tibia over the stationary talus from a position of full plantarflexion (1) to full dorsiflexion (3). (From Lewis 1980*a*.)

arcuate movement. Indeed the direction of that movement can be confidently predicted by mere inspection of the arched course of the concave channel in the *Lemur* talus (Fig. 12.5), which curves medially as it passes forwards. Further, the articular surface for the medial malleolus splays medially onto the talar neck when it is traced forwards, forming here a deeply concave cup for the nodular medial malleolus. The lateral malleolar surface also flares out laterally.

The mechanics of the movement as the tibia rides forward over the trochlea from posteriorly (the position of plantar flexion) to anteriorly (the position of dorsiflexion) are shown in Fig. 12.8A, for *Hylobates lar leuciscus*, and in general terms this action is fairly typical of sub-human primates. As the tibia swings forward it becomes angulated medially. But not only that, due to the arcuate course, it also undergoes a medial rotation (spin), and close-packs on to the dorsum of the neck of the talus. In so doing, the rounded protuberance forming the articular surface of the medial malleolus impacts into the articular cup located on the talar neck. The fibula, of course, accompanies the tibia in its medial swing and its lateral malleolus retains contact with the lateral surface of the talus

because of the splayed-out character of the latter. Some anterior separation of the tibia and fibula is produced with consequent lateral rotation of the latter bone about its long axis.

Whilst it is easiest to depict the movement as that of tibia on talus, in most circumstances in real life it might be better to think of movement of the foot relative to the leg. In the final close-packed dorsiflexed position the foot is somewhat laterally rotated (or abducted) on the leg and its sole is slightly inverted. Whether the foot is then involved in grasping a horizontal support during quadrupedal locomotion, or a vertical support during vertical clinging, weight is clearly largely directed down through the medial malleolus into the receptive cup-like articular surface. In this dorsiflexed position of the foot, the trochlea is stabilized within the mortise formed by the leg bones by the sling-like action of the posterior tibiotalar and talofibular ligaments. The latter, as it passes around the back of the talus, is accommodated by an indented groove in the splayed-out lateral malleolar surface of that bone. The lateral flare of this surface which is so disposed as to close-pack on the fibula in dorsiflexion, together with its ligamentous groove, are then morphological features correlated with the total arboreal pattern. The compromise axis which would determine such a movement cannot be simply transverse, as this would only provide for a simple hinge action of plantar flexion–dorsiflexion. The axis must be directed downwards and laterally to provide for the component of external rotation of the foot during dorsiflexion and also somewhat forwards to introduce the element of inversion. Manipulation of the bones, however, suggests that the various components of the movement are selectively emphasized in different phases of the movement and definition of a single stationary axis is not feasible. Nevertheless an axis can be determined experimentally which is valid enough at least for the latter part of the movement into dorsiflexion and it is a useful aide mémoire to the main characteristics of the movement (Fig. 12.9A). As will be seen in the next chapter the lateral rotation of the foot upon the leg and the inversion of the sole accompanying

dorsiflexion are enhanced by movements of considerably greater amplitude at the subtalar and transverse tarsal joint complexes, to bring the foot into its most effective grasping attitude, for example about the side of a horizontal branch. The sum total of these movements then compensates for the loss of crural and ankle pronation–supination seen in marsupials; this loss presumably accompanied the terrestrial (or semi-terrestrial) apprenticeship of the earliest Eutherians (Chapter 2).

The human ankle still bears the hallmarks of derivation from a morphology such as that described above. The trochlea is clearly of modified sellar form and presents an arched guiding groove veering medially as it is traced forwards (Schmidt 1981). The malleolar facets are usually almost vertical. Variation is found but never is the medial facet prolonged forward as a deeply indented articular cup, nor does the lateral facet, despite being flared to some extent, ever bear a deep

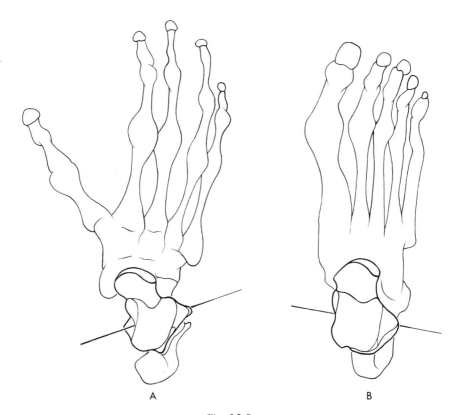

A   B

Fig. 12.9
A, a ligamentous preparation of the right foot of *Pan troglodytes* showing the ankle joint axis determined experimentally. B, a ligamentous preparation of a right foot of *Homo sapiens* showing the ankle joint axis.

impression for the posterior talofibular ligament.

As the human tibia moves forward over the talus (Fig. 12.8B) it also traces an arcuate course which includes a component of conjunct medial rotation (Schmidt 1981), although this is less accentuated than in sub-human primates. It is to be noted that in man the medial border of the trochlea has usually become relatively elevated (Elftman and Manter 1935; Schmidt 1981), and this counteracts the inward angulation of the leg bones which is seen in other Primates during dorsiflexion, so that the human foot in contrast becomes everted in full dorsiflexion at the ankle and inverted in plantar flexion (Barnett 1955; Bowden 1967).

Barnett and Napier (1952) have analysed movements at the human ankle joint in terms of two descriptive axes having different types of inclination, one for dorsiflexion and one for plantar flexion, derived not by experiment but by geometrical deduction from the form of the bones. This approach does not adequately depict the quite complex movements occurring at the joint, but the concept of dual axes did lead to interesting deductions. Thus, Barnett and Napier (1952) reasoned that as the joint is plantar flexed the body of the talus must medially rotate, thereby providing an explanation for the 'wedging' of the trochlea. Barnett (1966) later interpreted the joint as a sellar one in which movement included a component of rotation. This medial rotation of tibia upon talus during dorsiflexion has been demonstrated above (Fig. 12.8B) and it is clear that it is an ancestral primate trait, but one which seems to have nevertheless been pre-adaptive for function in the erect posture. As will be seen in the next chapter, the medial rotation at the ankle joint at the end of the stance phase of gait is associated with a similar rotatory movement at the subtalar joint complex. Further, as the tibia (with fibula) medially rotates it is obvious that the calcaneofibular ligament is tensed and that this presumably produces the downward movement of the fibula during dorsiflexion which has been noted in the living (Weinert *et al.* 1973) and which apparently stabilizes the ankle mortise during loading. Barnett and Napier (1952) reasoned that because of the deflection downwards and laterally of their postulated dorsiflexion axis this attitude would be associated with lateral rotation of the fibula. In

fact, it has been shown that such a rotation of the fibula is highly variable in man (Reimann and Anderhuber 1980).

It has been conclusively shown by Inman (1976) that the pattern of movement shown in Fig. 12.8B can be neatly expressed in most human ankle joints by a single stationary axis, experimentally determined. This axis deviates slightly downwards and posteriorly when traced from medial to lateral (Fig. 12.9B); this contrasts strikingly with the situation in subhuman primates where the deviation of the axis (or more likely axes) is downward, laterally and *forward*. A useful pointer to the changed direction of the axis is the way in which the apex of the triangular facet on the talus for the fibular malleolus has been moved relatively posteriorly in man (Fig. 12.9).

Latimer *et al.* (1987) investigated the talocrural axis in the African apes and man but they concentrated their attention on its obliquity in the coronal plane, noting correctly that it is more oblique in the pongids where it enhances the amount of axial rotation of the tibia. Strangely, they took no account of any variation of the axis in the transverse plane, which determines components of inversion or eversion, yet Inman (1976) had convincingly shown how in man inversion accompanies plantar flexion and eversion is correlated with dorsiflexion. Inman also stressed that the really telling point is the relationship between the ankle axis in the transverse planc and the midline of the foot; illustrations of the axis on isolated tali fail adequately to convey the significant difference which exists here between the apes and man (Fig. 12.9) and this subtle distinction eluded Latimer *et al.* (1987).

The contrasting osteological correlates of these dissimilar talocrural movement patterns in subhuman primates and in man are fundamental to the interpretation of fossils and are particularly clear on the talus. In subhuman Primates the lateral margin of the trochlea is elevated, especially anteriorly, thus imposing an inversion component on dorsiflexion; in man the medial margin is elevated producing eversion.

Minor osteological subtleties matching the contrasting types of movement are shown on the tibia and fibula. In subhuman Primates the anterior lip of the tibial articular surface is most prominent and the surface faces slightly backwards, facilitating close-packing at the end of dorsiflexion. In man the posterior lip projects furthest and the joint surfaces faces slightly forwards. The disposition of the proximal margin of the malleolar surface on the fibula reflects the differing slope of the tibial articular surface; in subhuman primates it slopes anterodistally.

In apes the lateral part of the tibial surface is excavated to accommodate the protuberant lateral margin of the talar trochlea, and thus this part of the surface shelves proximally and faces somewhat laterally. This is not usually the case in man although some hollowing of the lateral part of the tibial surface may sometimes recall the situation in apes. Latimer *et al.* (1987) laid considerable emphasis on the way in which the long axis of the tibia in man is perpendicular to its distal articular surface, contrasting with the situation in *Pan* and *Gorilla* where it inclines superolaterally. Turning

logic on its head, they interpreted this as a consequence of a varus position of the knee in these apes; in truth it clearly reflects the characteristically inverted attitude of the ape foot. The African apes certainly have a less valgus knee (bicondylar angle about 1° to 2°) than man (bicondylar angle about 10°) and *Pan* may show a rather spurious bowlegged appearance when walking bipedally because of the flexed attitude of the knee during the stance phase.

In apes the malleolar articular surface on the fibula shelves away laterally, and so faces somewhat distally, to accommodate the flared triangular surface on the talus. In man the facets are vertical. All these descriptive features have been applied with varying effect to the interpretation of hominid fossils; these points will be taken up in Chapter 17.

13

The intrinsic joints of the foot

The subtalar joint complex

The nomenclature of the joints subjacent to the talus in therian mammals has been complicated by conflictingly varied terminologies and the confusion is very evident when the movements involved are considered; this becomes especially troublesome in the comparative context. It is well recognized that the talus articulates with the remainder of the foot—the lamina pedis (footplate) as it has aptly been termed by MacConaill and Basmajian (1969)—by two anatomically distinct articulations consisting of a posterior talocalcaneal joint and an anterior talocalcaneonavicular joint. Movement of the lamina pedis upon the talus involves both joints simultaneously and the functional composite of the dual articulations has frequently been termed the subtalar 'joint' (Manter 1941). In deference to this common clinical usage this particular term was incorporated in the official anatomical nomenclature, but illogically it was then applied only to the posterior of the two functionally complementary joints. The term 'subtalar joint' was also used restrictively by Shephard (1951) for the posterior talocalcaneal joint together with only part (the talocalcaneal portion) of the anterior anatomical joint, the remainder of this latter articulation (the talonavicular part) being included as a component of the midtarsal joint; Shephard (1951) used the comprehensive term 'peritalar joint' for the whole of the two anatomically separate joints. The articulations underlying the talus have also commonly been described as 'the lower ankle joint' in comparative anatomy but this term is really only an accurate one in cursorial or saltatory forms where the axis of movement has become almost transverse and complements and amplifies movement of the talocrural or ankle joint (Barnett 1970); it is particularly appropriate in the highly modified subtalar joints of artiodactyls (Schaeffer 1947). In this book the term 'subtalar joint complex' is used for the articulations (or single confluent articulation in some marsupials) between the talus and the lamina pedis. In the human foot then, this complex is represented by the talocalcaneal and talocalcaneonavicular joints.

The movements at this joint complex in man occur about an oblique compromise axis and are usually defined as 'inversion' and 'eversion'. These movements can be artificially resolved into two components: a rotation of the foot about its long axis (supination–pronation) and a rotation of the whole foot about a vertical axis (adduction–abduction). Thus, inversion can be considered as a combination of adduction of the front part of the foot together with partial supination. It should be noted that Shephard (1951) and Manter (1941) used the terms in a contrary sense. For them the composite movements were supination–pronation; supination for instance, in their usage, consisted of inversion and adduction. In this book the terms inversion–eversion are used for the composite movements at the subtalar joint complex and the terms supination–pronation are restricted to the description of movements at the midtarsal joint complex.

The monotreme joints

The monotremes, and particularly the echidna, provide an informative insight into the emergent mammalian pattern of structure and function in the subtalar joint complex. The morphology here establishes a bridge for understanding the transition between the foot structure in cynodont reptiles and in the earliest mesozoic mammals on the one hand, and the more advanced architecture of therian mammals on the other. In *Tachyglossus aculeatus* can be found the best model of monotreme structure. The hemispherical talus lies alongside the calcaneus and articulates with it by dual articulations (Figs 13.1, 13.2A, B): a dorsal one, predominantly concave on the talus, and convex on the calcaneus (but here hollowed out at its two extremities) and a distal one with reversed curvatures. Between is a non-articular zone and here the bones are united by a tough interosseous talocalcaneal ligament. The distal facet on the calcaneus forms the posterolateral margin of a complex cup-shaped cavity for the talus (Fig. 13.2B). The navicular and a separate tibiale form additional components of this complex articular surface and are united by a plantar calcaneonavicular ligament to the calcaneus below its distal facet. This distal calcaneal facet is continuous with the articulation for the cuboid and here the latter bone intrudes into the cup-shaped cavity, thereby forming a component of its wall. This cup-shaped cavity in its totality, is the precursor of the therian acetabulum pedis.

Three ligaments—anterior, middle, and posterior—bind the talus to the calcaneus and all have persistent homologues in therian mammals. The anterior ligament is clearly the homologue of the ligamentum cervicis tali and is so named here, even if the term is not entirely appropriate because the talus has as yet no discernible neck. The middle one lies in the canalis tarsi, as noted above, and is the interosseous talocalcaneal ligament. The posterior talocalcaneal ligament, at its calcaneal origin, lines the fairly deep channel medial to the calcaneal tuber for the entry of the flexor fibularis tendon to the sole, and attaches on the talus into the posterior inflection or cleft on that bone. Merging onto its surface here is the posterior ligament of the ankle joint, which as noted in Chapter 12 probably includes the precursor of the therian posterior talofibular ligament. It will soon become apparent that the posterior talocalcaneal ligament also has clear, frequently overlooked, therian homologues.

The distal calcaneal surface on the talus is continued in a semicircular arc (Fig. 13.2A) as articulations for the cuboid and navicular, and beyond that is a small independent facet for the tibiale. The non-articular area bounded by these facets carries the attachments of strong ligaments intruding into the acetabulum pedis from the cuboid and the navicular, and insinuating itself between these latter two bones and their ligamentous continuations, is another flattened ligament tethering the lateral cuneiform to the talus.

Function in monotremes   Barnett (1970) appreciated something of the significance of the rotatory movement between the juxtaposed talus and calcaneus in monotremes when he suggested that the movements produced alternate cupping and flattening of the sole. He did not relate this, however, to any particular functional need.

Now Rewcastle (1981) has shown that in reptiles with their sprawling attitude (Fig. 3.1) the gait involves femoral (and humeral) retraction in a near horizontal plane associated with some rotation of these bones. This demands some rotation between knee and the emplaced pes, effected by rotation of crus upon the foot: clearly the adaptive reason for the rotatory capacity of tibia and fibula both about their long axes and in the same direction, which is typical of reptiles (Haines 1942) and which was described in the previous chapter.

Jenkins (1970*a*), however, in a cineradiographic study of limb movements in the echidna refuted the idea that monotreme limbs are reptilian in posture and that they sprawl in a manner comparable to lizards. Indeed their attitude is well advanced towards the semi-sprawling attitude adopted by primitive therian mammals, and with the developmental limb rotations described in Chapter 3 partly effected, so that knee points forward, and elbow backwards. Although foot movements were not specifically referred to by Jenkins (1970*a*) in this cineradiographic study, they may be deduced with confidence from his informative illustrations of the phases of gait in the hindlimb. As would be expected femoral retraction and medial rotation are accompanied by a reptilian type of rotation of the long bones of the crus, clearly involving movement of these bones at the ankle. However, it is also apparent that as the body moves forward over the implanted pes there is an effective dorsiflexion at the ankle accompanied by medial angulation of crus in relation to foot. Whilst the ankle would seem capable of coping with a

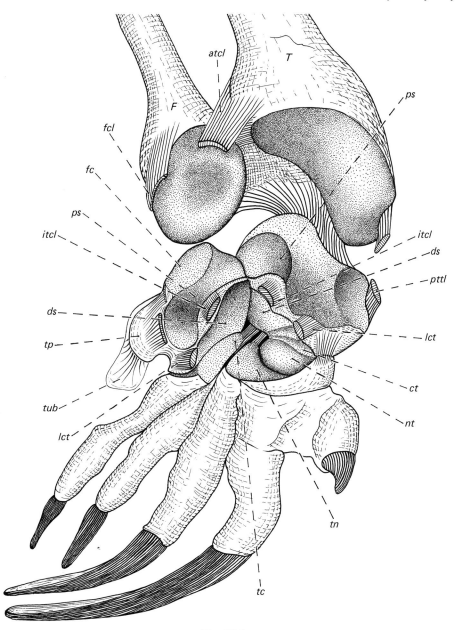

Fig. 13.1

The right ankle and foot of the echidna, *Tachyglossus aculeatus*, viewed from dorsally with tibia (*T*) and fibula (*F*) partly detached and folded back revealing the articular surfaces of the ankle joint including the fibular facet on the calcaneus (*fc*), the posterior ligament, the medial ligament (posterior tibiotalar ligament, *pttl*), and the anterior tibiocalcaneal ligament (*atcl*) and the lateral fibulocalcaneal ligament (*fcl*), both attaching distally to the trochlear process (*tp*) of the calcaneus which lies dorsal to the calcaneal tuber (*tub*). The talus and calcaneus are shown partly disarticulated and separated anteriorly revealing: *ps*, proximal articular surfaces on calcaneus and talus; *itcl*, interosseous talocalcaneal ligament; *ds*, distal articular surfaces on calcaneus and talus; *lct*, ligamentum cervicis tali; *nt*, navicular articular surface on talus; *ct*, cuboid facet on the talus. The talar articular surface on the navicular (*tn*) and the talar surface on the cuboid (*tc*) are joined posteriorly by ligaments to the talus, embracing between them a further ligament from the lateral cuneiform. (From Lewis 1983.)

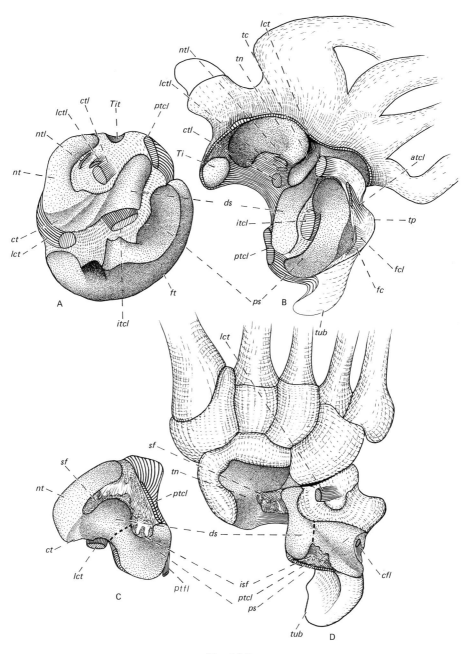

Fig. 13.2

Above, a dorsal view of the right lamina pedis (B) of *Tachyglossus aculeatus* with the talus (A) rolled away to reveal the under-surface. Below, comparable views of the right lamina pedis (D) and talus (C) of *Didelphys marsupialis*. *cfl*, calcaneofibular ligament; *ctl*, talocuboid ligament; *ft*, fibular articular surface on talus; *isf*, synovial folds representing the site of the interosseous talocalcaneal ligament (the broken line represents the site where in *Caluromys lanatus* the proximal (*ps*) and distal (*ds*) talocalcaneal surfaces are completely separated); *lctl*, talus–lateral cuneiform ligament; *ntl*, naviculotalar ligament; *ptcl*, posterior talocalcaneal ligament; *sf*, synovial fold representing the site of ligaments from cuboid, lateral cuneiform and navicular to talus; *Ti*, tibiale; *Tit*, tibiale facet on talus; *pcnl*, plantar calcaneonavicular ligament; *ptfl*, posterior talofibular ligament. (From Lewis 1983.)

proportion of the sideways movement it is obvious that if the sole is to remain applied to the ground a longitudinal torsion of the whole foot must occur. This seems to be effected by rotational movement between talus and calcaneus which in effect 'raises' the inner border of the foot at the end of propulsion. This is achieved because in monotremes (Fig. 13.1) it is primarily the talus which is functionally anchored to the forefoot, and made an integral part of it, by the ligaments uniting it to cuboid, lateral cuneiform, and navicular. The 'raising' of the inner border of the foot in the terminal phase of propulsion is analogous to supination or inversion in the therian foot.

Jenkins study was expanded by Pridmore (1985), who did a comparable cineradiographic study on the monotreme *Ornithorhynchus*, and reached similar conclusions to Jenkins' findings for the echidna. Pridmore did, however, specifically note the 'twisting' of the emplaced pes, although without venturing into discussion of the actual anatomical basis for it.

The marsupial joints

It is well known (Schaeffer 1941*a*) that therian mammals are distinguished by having the talus superimposed on the calcaneus, the intervening subtalar joints then providing for the movements of inversion and eversion. It is now apparent that these movements (and the joints involved) really only represent an elaboration of those already established in monotremes. At first sight the form of the talus and calcaneus of monotremes, and the articulations between them, seem far removed from the pattern characteristic of therian mammals. Two monotreme peculiarities contribute to this: the hemispherical form of the talus and the way in which the tuber calcanei is deflected distally into the sole. The foot has no real heel and the tuber calcanei, with its attached tendo calcaneus (Figs 13.1, 14.19) forms the lateral boundary of a deep channel for the entry of the massive flexor fibularis tendon into the sole.

If now the monotreme tuber calcanei could be imagined deflected back into a more advanced mammalian orientation, and comparison is made with certain transitional marsupial forms, the similarity in detailed aspects of the topography of articular surfaces and associated ligaments springs

into striking relief (Figs 13.2A, B). In fairly generalized marsupials the heel is incompletely bent backwards, thus providing a transitional stage heralding the characteristic eutherian foot plate.

The footplate of *Didelphys marsupialis* (Fig. 13.2D) contains comparable representatives of all the features found in the monotreme (Fig. 13.2B) obscured only by the fact that there is a partial confluence of proximal (posterior) and distal (sustentacular) talocalcaneal surfaces. The site of the canalis tarsi (and its interosseous talocalcaneal ligament) is, however, represented by an intrusion of fibrous and synovial tissue. In the opossum *Caluromys lanatus*, however, a complete canalis tarsi (containing an insubstantial interosseous talocalcaneal ligament) is retained and the resemblance to the echidna is even further heightened.

Posterior to the canalis tarsi, or its incomplete remnant, the articular surface for the talus is convex and cylindrical. Anterior to the canalis tarsi is a complex, concave articular cup providing the remainder of the articulation for the talus; this cup may appropriately be called the acetabulum pedis. This term was first applied to the same region in the human foot by MacConaill (1945) although Smith (1896) had earlier noted the resemblance of the whole region to the hip joint. Behind, this cup is formed by the distal talocaneal articular surface borne partly upon a sustentaculum tali, sculpted from the medial surface of the calcaneus by the course of the massive flexor fibularis tendon into the sole, and more anteriorly extending onto the body of the calcaneus. No sustentaculum, as such, is discernible in monotremes; its emergence is a consequence of deflection backwards of the heel, redirection of the flexor fibularis into sole, and superposition of talus upon calcaneus. In front of the calcaneal facet the cuboid intrudes into the cup forming part of its wall just as it did in echidna. The tarsus is therefore of the so-called 'alternating' type which is the primitive mammalian condition (Schaeffer 1947). The remainder of the bony cup is formed by the navicular, which of course incorporates the tibiale. Bridging the floor of this semicircular bony articular cup is the massive plantar calcaneonavicular (spring) ligament. This ligament was already strongly represented in monotremes despite the absence of a sustentaculum tali (Fig. 13.2B).

Springing from the surface of the ligament in the marsupial is a collection of synovial folds attaching to the talus, and clearly representative of the three ligaments found in this situation in echidna, which join the talus to navicular, cuboid, and lateral cuneiform.

The opossum talus (Fig. 13.2C), lacks a really clear-cut head and neck is transitional between the hemispherical echidna bone and the typical eutherian condition. To its under surface are attached the synovial remnants of the monotreme ligaments tethering it to navicular, cuboid, and lateral cuneiform and posterior to this are the synovial walls of the complete or incomplete canalis tarsi, with or without a recognizable interosseous ligament.

Whereas on the echidna talus there is a posterior cleft, the marsupial bone presents a broad groove, still for the flexor fibularis tendon, but in the marsupial bracketed by emergent posterior and medial talar tubercles. Groove and tubercles are strongly united to the calcaneus by the posterior talocalcaneal ligament. This often massive attachment of the talar tubercles to the calcaneus including the posterior part of its sustentaculum persists as a characteristic but seldom appreciated feature of the eutherian tarsus and is even found in Primates. As will be seen the ligament tends to be subdivided in Primates into medial and posterior talocalcaneal ligaments attaching to the twin tubercles on the talus. Dorsally and laterally the talus is attached to the calcaneus by the ligamentum cervicis tali, which was already well differentiated in monotremes. The talus is attached above to tibia and fibula by the posterior tibiotalar and talofibular ligaments.

In Australian marsupials the situation is rather different. The trend towards confluence of the proximal and distal (sustentacular) talocalcaneal surfaces, already initiated in *Didelphys*, is complete and there is no tarsal canal and no interosseous talocalcaneal ligament. Thus, in *Trichosurus vulpecula* (Fig. 13.3) the lamina pedis bears a single compound articular surface for the talus. Behind, this surface is formed by a cylindrical articular convexity which reverses its curvature when traced anteriorly, becoming concave and continuous with an articular cup for the reception of the head of the talus. Otherwise, the ligamentous apparatus, and so on, is as in *Didelphys*. Other

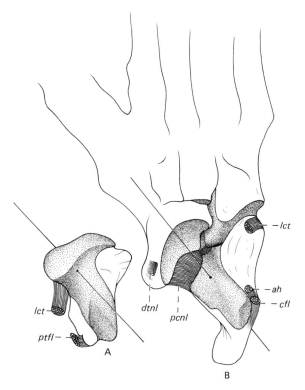

Fig. 13.3
A dorsal view of the right lamina pedis (B) of *Trichosurus vulpecula* with the talus (A) rolled away medially from its articulations. In each case the compromise subtalar axis, determined experimentally, is shown. *ah*, anterior fibrous horn of ankle meniscus; *dtnl*, dorsal talonavicular ligament. Other labels are as in Figs 13.1, 13.2. (From Lewis 1980*b*.)

Australian marsupials without extreme specializations of the feet (*Pseudochirus, Phascolarctos, Sarcophilus*) show essentially similar features.

There appears to be no reason why this confluence of articular surfaces should particularly affect the type of movement since the pattern of changing curvatures remains essentially similar.

Function in marsupials The compromise axis (Fig. 13.3) for the subtalar joint in marsupials, other than those with very specialized feet, lies very obliquely being almost in line with the widely divergent hallus and in a quite flat plane, rising relatively little when traced anteriorly. About this axis occur the movements of inversion–eversion, which have a self-evident application in adapting the attitude of the pes to a variety of grasping positions. The correspondence of the axis with the divergent hallux may be mechanically advantageous, for the grasping hallux thus remains relatively stationary as the remainder of the foot inverts into gripping posture. The strong ligamentum cervicis tali is advantageously disposed to limit inversion, and the very tough dorsal talonavicular ligament acts to retain the head of the talus within its articular cup. There is at least a suggestion of a screwing action accompanying the movements, with inversion being accompanied by slight forwards advance of the lamina pedis.

But there is much more to it than this. In the quadrupedal position the axis strikingly parallels the midcarpal axis in the forelimb (Fig. 5.3): both run backwards, laterally, and somewhat downwards. There can be little doubt that the subtalar axis functions to adjust the attitude of the emplaced pes during locomotion, in a similar manner to that described for the midcarpal axis in the forelimb (Chapter 5). Jenkins' (1971*b*) excellent cineradiographic study of the phases of gait in *Didelphys* makes it clear that important adjustments between crus and pes are obligatory, although these are not specifically commented on. As noted above such adjustments are also necessary in monotremes, where they are achieved by torsion of the footplate. The notable advance in marsupials over monotremes is the freeing of the talus from restrictive attachments to the forefoot within the acetabulum pedis, even though flimsy remnants of these attachments may persist.

As the marsupial femur is retracted (this is abduction, in contrast to adduction for the humerus) the bones of the crus become angulated on the emplaced pes, which effectively becomes everted, abducted, and dorsiflexed, the movement dictated by the oblique subtalar axis (Fig. 13.4). But movement about the oblique subtalar axis can transmit torque to the leg bones. In the attitude described the leg would medially rotate. Yet it is clear from Fig. 13.4 that abduction of the femur necessitates lateral rotation of leg upon foot. Something must decouple the leg bones from the action at the subtalar joint. Here then seems to be a key role for the ankle joint meniscus—while eversion occurs at the subtalar joint, lateral rotation occurs at the ankle. The same mechanical problem does not arise in the forelimb. Here

internal rotation of forearm upon hand is required, and this is just the direction of rotation which is imposed by the midcarpal axis (Fig. 5.3). In marsupials such as *Macropus*, with a different specialized type of gait, the subtalar joint is drastically remodelled (Fig. 12.4A) to form a 'lower ankle joint'.

The eutherian joints

As noted in Chapter 12 the treeshrews retain echoes of metatherian morphology at the ankle, for the talocalcaneonavicular joint still communicates medially with the ankle joint, and of course the posterior talocalcaneal joint is in continuity laterally with the ankle joint. Between the two subdivisions of the treeshrew subtalar joint complex there is a tarsal canal containing an interosseous talocalcaneal ligament. At least in *Tupaia* the talocalcaneonavicular joint partly retains the pattern of an alternating tarsus since, although the calcaneus reaches forwards to meet the navicular in the floor of the receptive articular cup for the talar head, more dorsally the cuboid participates in the joint and contacts the talus in eversion. A ligamentum cervicis tali and a dorsal talonavicular ligament are present. The treeshrew joints form an adequate model for the likely emergent eutherian structure and for the probable architecture which was antecedent to the primate pattern. Movements of inversion and eversion occur at the subtalar joint complex about an oblique axis very similarly disposed to that in marsupials. Jenkins (1974) has shown how advantageous this movement is in arboreal locomotion: the curvature of the branch requires that for maximum contact the foot must be inverted with the sole facing medioventrally, instead of ventrally as on flat surfaces; supination achieves the same effect for the manus. Inversion in treeshrews can be carried to such an extreme that complete foot reversal is allowed when descending vertical surfaces (Jenkins 1974; Jenkins and McClearn 1984), and this action is augmented by rotation at the transverse tarsal joint. The mechanical solution for foot reversal in this particular eutherian, as in others, is quite different from that favoured in marsupials (see Chapter 12).

In insectivores such as *Tenrec*, essentially the same morphology is seen with the same ligamen-

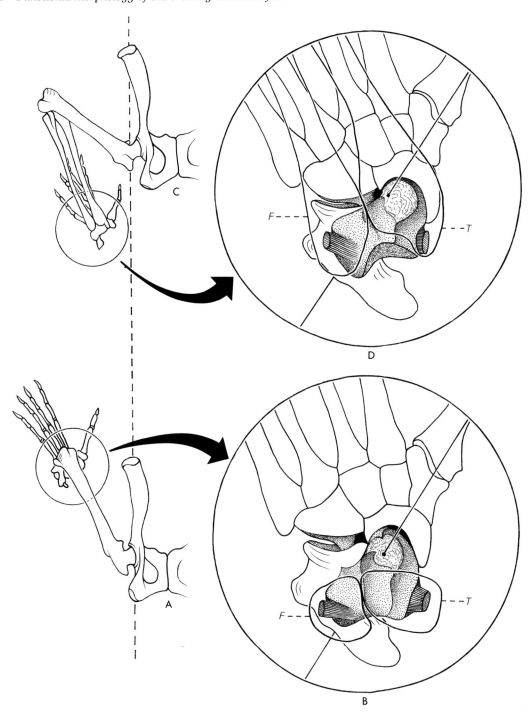

Fig. 13.4

The attitude of the left hindlimb bones of *Didelphys marsupialis* at the earlier (A) and later (C) parts of the stance or propulsion phase of locomotion are shown, as demonstrated by cineradiography (redrawn from Jenkins 1971). The interpretation described in the text of the way in which movement about the oblique subtalar axis maintains the contact of the pes with the substrate during these movements is indicated in B and D.

tous apparatus, and a partial intrusion of the cuboid into the acetabulum pedis. In other insectivores, such as *Suncus*, the tarsus is no longer of alternating type for the calcaneus is prolonged forward to articulate with the navicular, completely excluding the cuboid from contact with the talus. This parallels an advance characteristic of the Primates.

The subhuman primate joints The essentials of morphology and function are fairly constantly present among the primates, with the noteworthy

exception of *Homo sapiens*. There are variations in emphasis of certain features, and some grotesque distortions but the underlying pattern seems to be a basic primate heritage. This morphology is plainly shown in arboreal platyrrhine monkeys such as *Cebus* (Fig. 13.5) or *Pithecia* (Fig. 13.6).

An interosseous talocalcaneal ligament and a canalis tarsi are present subdividing the subtalar joint complex into the typical eutherian posterior and anterior components. Posteriorly the calcaneus bears a convex facet for the body of the talus and this facet is markedly prolonged backwards

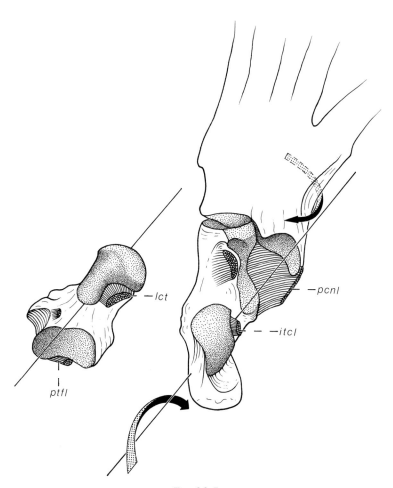

Fig. 13.5
A dorsal view of the left lamina pedis of *Cebus albifrons* with the talus rolled away laterally from its subtalar articulations. The conarticular surfaces of the subtalar joint complex are thus displayed and in each case the compromise subtalar axis is shown. The calcaneocuboid joint has been opened dorsally to expose the articular surfaces. The lower arrow indicates the screwing motion of the lamina pedis about the subtalar axis during inversion; the upper arrow indicates the accompanying movement of the forefoot during supination at the calcaneocuboid joint. Labels are as in Figs 13.1, 13.2. (From Lewis 1981.)

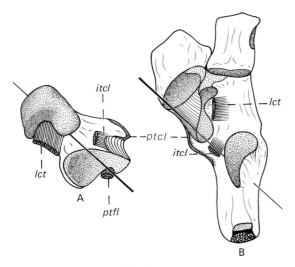

Fig. 13.6
A dorsal view of the right lamina pedis (B) of *Pithecia monachus* with the talus (A) rolled away. In each case the compromise subtalar axis is shown. Labels are as in Figs 13.1, 13.2. (From Lewis 1980*b*.)

and medially. Anteriorly is the receptive surface on the lamina pedis for the head and neck of the talus, which together are essentially cylindrical and capped at the extremity by a convexity for articulation with the navicular. On the surface of this cylinder is an L-shaped area for articulation with facets, which are variably confluent, on the body and sustentaculum tali of the calcaneus; the ligamentum cervicis tali inserts into the angle of the L-shaped area. The receptive cup for the head of the talus, the acetabulum pedis, is completed by the plantar calcaneonavicular (spring) ligament. The calcaneus extends forwards and achieves a considerable articulation with the navicular thus excluding the cuboid from the talocalcaneonavicular joint—the tarsus is no longer an alternating one.

The canalis tarsi is closed off medially by a ligamentous attachment between the talus and the back of the sustentaculum tali on the calcaneus. This fibrous sheet is clearly the homologue of the posterior talocalcaneal ligament of monotremes and marsupials (Fig. 13.2), where it lines the channel conducting the massive flexor fibularis tendon into the sole. In Primates, however, the talus has elaborated posterior and medial tubercles flanking this tendon. The great part of this

posterior talocalcaneal ligament attaches to the medial tubercle of the talus (Fig. 13.6), little if any spreading its attachment across to the posterior tubercle; this primate structure might perhaps be more accurately called the medial talocalcaneal ligament.

The compromise subtalar axis of movement, as determined experimentally, lies in a flat plane rising little when followed anteriorly but it is markedly deviated from the long axis of the foot, traversing the cylindrical talar neck whose angulation to the body of the bone is thus an expression of the obliquity of the axis (Figs 13.5, 13.6). Its relationship to the posterior part of the tarsus is not unlike that in *Trichosurus* (Fig. 13.3), but the overall relationship to the foot is rather different because of the relative lengthening of the anterior part of the pes in the monkey. This tarsal elongation is presumably a relic of the early terrestrial (or semiterrestrial) apprenticeship of the eutherians. The lamina pedis inverts and everts about the axis just as it does in marsupials, but whereas in the latter there is only a poorly defined screwing motion, in the Primates, this has become a major component of the movement. The posterior talocalcaneal articulation has a helical orientation in relationship to the subtalar axis—in a left foot it is a segment of a left-hand screw. Thus, as the lamina pedis moves into inversion the calcaneus screws forward beneath the talus, its exposed extremity then moving into substantial contact with the navicular (Fig. 13.5). As will appear, this characteristic of the movement is biomechanically advantageous in the overall function of the foot.

In catarrhine monkeys the morphology and function are in general comparable to the arrangements in platyrrhine monkeys (Figs 13.7, 13.8) but the screwing action at the joint complex is reduced, and in correlation with this, articular contact of the calcaneus with the navicular is relatively restricted and the posterior calcaneal articular surface is almost transversely disposed with respect to the axis. Although the screwing action is diminished, particularly in the more terrestrial cercopithecines, the contention that this particular type of movement has been lost in most cercopithecoids (Szalay 1975), seems to be in error.

In monkeys, just as in marsupials (Fig. 13.3) a

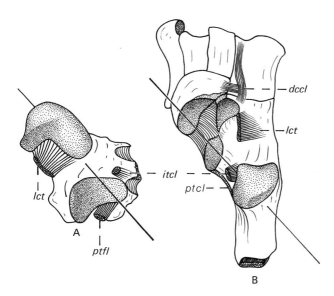

Fig. 13.7
A dorsal view of the right lamina pedis (B) of *Colobus polykomos* with the talus (A) rolled away from its articulations. In each case the compromise subtalar axis is shown. *dccl*, dorsal calcaneocuboid ligament. Other labels are as in Figs 13.1, 13.2. (From Lewis 1980*b*.)

dorsal talonavicular ligament is present (Fig. 13.8) and in some a band of this is prolonged forwards towards the metatarsal bases to form a dorsal talometatarsal ligament.

The essence of the morphology seen in New World monkeys, which doubtless is close to the ancestral primate structure, is also clearly apparent in fairly generalized lemuroids such as *Lemur* (Fig. 13.9). The calcaneus is prolonged further forward in *Lemur*, achieving a considerable articular contact with the navicular but this is probably an exaggerated specialization of extant lemuroids. There is the usual ligamentous apparatus—a ligamentum cervicis tali, and interosseous talocalcaneal, dorsal talonavicular and talometatarsal ligaments, and a posterior (or perhaps better, medial) talocalcaneal ligament. In function, the subtalar joint complex operates like that of New

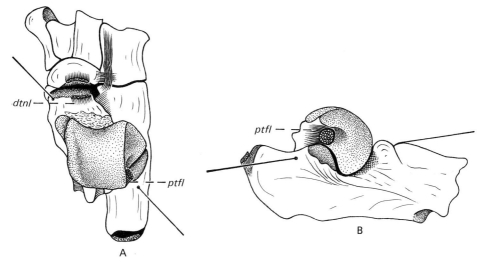

Fig. 13.8
The right foot of *Colobus polykomos* viewed from dorsally (A) and laterally (B) showing the compromise axis of the subtalar joint complex. Labels are as in Figs 13.1, 13.2, 13.3. (From Lewis 1980*b*.)

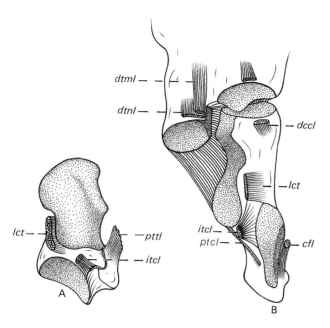

Fig. 13.9
A dorsal view of the right lamina pedis (B) of *Lemur catta* with the talus (A) rolled away. In each case the compromise subtalar axis is shown. *dtml*, dorsal talometatarsal ligament. Other labels are as in Figs 13.1, 13.2, 13.3. (From Lewis 1980*b*.)

World monkeys with a marked 'screwing' action.

A derivative prosimian pattern, which is a gross exaggeration of certain morphological features characterizing the lemurs, is seen in the galagines (Fig. 13.10). There is an enormously exaggerated and disproportionate anterior elongation of the cancaneus, which is matched by a comparable distortion of the navicular. However, if consideration is restricted to the subtalar joint complex, the morphology and function are not unlike *Lemur*. In *Galago*, however, subtalar movement is a minor adjunct to the more extensive movement occurring at the calcaneocuboid joint, as will be described later.

In the Asiatic apes (*Hylobates* and *Pongo*) the essence of the arboreal primate heritage, shown in New World monkeys is retained, although contact between the calcaneus and the navicular is limited to a narrow articular strip. The head and neck of the talus retain the conservative essentially cylindrical form, bearing an L-shaped calcaneal articular surface. The posterior articular surface has a helical orientation about the subtalar axis which is again very oblique and lies in a flat plane. A quite marked screwing action accompanies the movements of inversion and eversion. The ligamentous apparatus is as in monkeys. Interestingly, in at least some gibbon specimens a synovial fold connects the undersurface of the talar head to the floor of the acetabulum pedis, just as was noted in *Didelphys* (Figs 13.2C, D), a persistent relic of the ancient ligaments in this situation (Figs 13.2A, B).

The African great apes (Figs 13.11, 13.12A, B, 13.13) show certain derived features which are clearly grafted onto a morphology essentially similar to that described in the primates above. The subtalar axis is markedly oblique but its anterior end is now more elevated. The talus, perhaps in association with this feature, is more squat in form with a short, broad neck, this being particularly evident in *Gorilla*. The component limbs of its L-shaped articular surface are partly merged into a somewhat confluent single area, and in both *Pan* and *Gorilla* the head bears a clear-cut triangular facet for the ligamentous part of the floor of the acetabulum pedis—that is, for the plantar calcaneonavicular ligament. The posterior talocalcaneal joint retains an obvious helical disposition in relation to the subtalar axis and during the screwing motion accompanying inversion the calcaneus is close-packed onto a small articular contact with the navicular. In juvenile specimens of *Pan* the cuboid may intrude slightly into the talocalcaneonavicular joint, the tarsus thereby retaining a semblance of the ancestral alternating character.

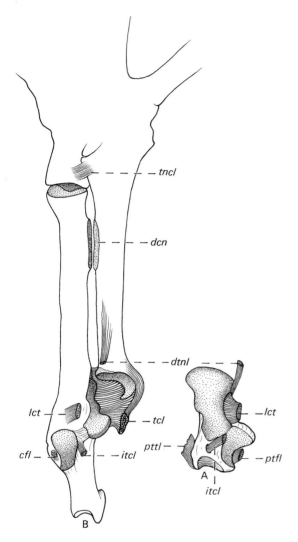

Fig. 13.10

A dorsal view of the left lamina pedis (A) of *Galago senegalensis moholi* with the talus (A) rolled away. *tnc*, transverse naviculocuboid ligament; *dcn*, diarthrosis between calcaneus and navicular. Other labelling is as in Figs 13.1, 13.2, 13.3. (From Lewis 1980*b*.)

The usual ligamentum cervicis tali and interosseous talocalcaneal ligaments unite the bones. Medially the derivative of the ancient posterior talocalcaneal ligament binds the two bones together. In *Pan* (Fig. 13.12B) it attaches above to the medial tubercle of the talus, but extends this origin into the floor of the groove for flexor fibularis—it is effectively a medial talocalcaneal ligament. In *Gorilla* there is a similar massive

union between the two bones, but additionally a strong posterior talocalcaneal ligament (of presumably similar derivation) runs down from the posterior tubercle of the talus (Fig. 13.13).

The human joints The subtalar joint complex of man (Figs 13.12C, D) retains the hallmarks of the basic arboreal primate heritage but this has been drastically remodelled, although the unique structural attributes represent little more than the accentuation of trends already incipient in *Pan* and *Gorilla*. The anterior part of the calcaneus has been retracted so that it no longer usually contacts the navicular, yet despite this the cuboid does not usually intrude into the joint, as in the primitive alternating tarsus. However, as will be seen in the discussion on the midtarsal joint, a small linear contact may be achieved between the calcaneus (a bevelled off margin of its anterior talar facet) and the navicular. An occasional contact between these bones has also been noted by Langelaan (1983), and illustrated radiographically by Leonard (1974) but it does not appear in textbook accounts; from the comparative point of view such articulation is of course not at all surprising.

The constitution of the acetabulum pedis is basically as in other primates. The calcaneus may exhibit two articular facets for the talus—a medial one on the sustentaculum tali and an anterior one on the body—but usually these are united into a single arcuate confluent one (Fig. 13.12C). Comparable separate or confluent facets are exhibited on the talus (Sewell 1904b, 1906). These facets are the lineal descendants of the dual facets having a rather L-shaped configuration in monkeys, and which are sometimes separate, and sometimes confluent in them also. The progression towards a more human appearance is apparent in *Pan* (Fig. 13.12B) and *Gorilla* (Fig. 13.13). The distribution of the confluent or dual facet pattern in different races has been documented by Bunning and Barnett (1965), but the differences presumably have little functional importance. Rarely (except apparently in the Veddah where it is quite common) the confluent anterior facets may also be united with the posterior talar facet obliterating the tarsal canal, and convergently resembling the morphology seen in *Trichosurus* (Fig. 13.4). The more medial part of the floor of the acetabulum pedis between calcaneus and navicular is formed

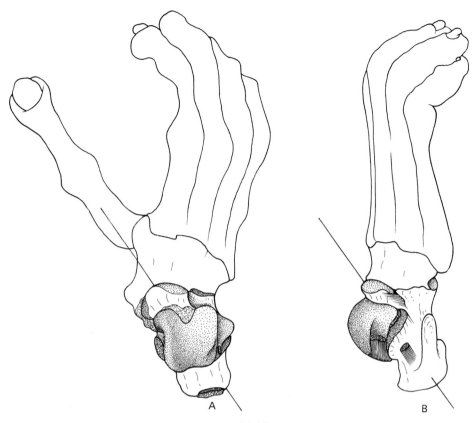

Fig. 13.11

The right foot of *Pan troglodytes* viewed from dorsally (A) and laterally (B), with the foot in inversion, showing the compromise subtalar axis. (From Lewis 1980*b*.)

by the fan-shaped plantar calcaneonavicular (spring) ligament arising from the anterior border of the sustentaculum tali. The ligament impresses a triangular facet on the head of the talus, already noted in *Gorilla* and *Pan*. This, the canonical plantar calcaneonavicular ligament was designated by Smith (1896) as only the 'superointernal part' of that ligament. Lateral to it, in the apex of the triangular area between the bones forming the acetabulum pedis, is a bunch of synovial folds below which lies another fasciculated ligament: Smith's 'inferior part' of the spring ligament. Occasionally, however, the cuboid may intrude here into the acetabulum pedis and stamp an additional cuboid facet on the talar head adjacent to that for the spring ligament (Sewell 1906); in these variants an alternating tarsus is re-established just as is sometimes found in *Pan*. Moreover,

in this situation a synovial or even ligamentous fold may attach to the underside of the talar head, as also noted by Smith (1896). Then, just as sometimes in *Hylobates*, representatives of the ancient ligaments occupying this site are re-established. Even sometimes, as figured in a dry bone by Schmidt (1981) a deep nonarticular enclave may cut into the head of the talus from the lateral side, as in *Didelphys* (Fig. 13.2C). In these cases the name 'acetabulum pedis' is particularly apt; as Smith noted, there is even an analogue of the ligamentum teres. Some of the fibres of the 'superointernal part' of the spring ligament sweep dorsally onto the talus and have been named the ligamentum neglectum (Volkmann 1973).

The ligamentous apparatus is directly derivative from the primates already described, but the human structures had received detailed attention

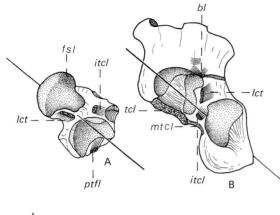

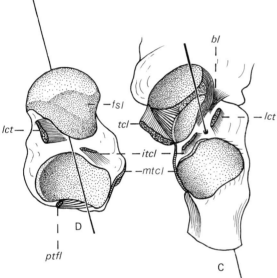

Fig. 13.12
Above, a dorsal view of the right lamina pedis (B) of *Pan troglodytes*, with the talus (A) rolled away. Below, a dorsal view of the right lamina pedis (C) of *Homo sapiens* with the talus (D) rolled away. In each case the compromise subtalar axis is shown. *tcl*, talocalcaneal part of deltoid ligament; *bl*, bifurcated ligament, *mtcl*, medial talocalcaneal ligament. Other labels are as in Figs 13.1, 13.2, 13.3. (From Lewis 1980*b*.)

long before any comparative insights were available. The interosseous talocalcaneal ligament and the ligamentum cervicis tali (Figs 13.12C, D) show no significant departures from the homologous structures found in other primates. These ligaments, together with the attachments of the fundiform ligament in the tarsal sinus and canal

(Fig. 15.5B) have been described in detail by Smith (1896) and more recently by Cahill (1965). A strong dorsal talonavicular ligament is present, attached to a rough arcuate tubercular ridge on the dorsum of the talar neck; to the ridge also attaches the anterior capsule of the ankle joint, as noted by Jones (1944), but it is clear that the former ligament is the significant structure responsible for producing the ridge. Ossification may extend into the ligament producing the radiological image of a 'talar beak' (Meschan 1975), also described as the 'processus talaris' by Sewell (1904*c*). Sewell noted that a vestigial 'ligamentum talometatarsale' may extend forward from this same ridge, that it was well-marked in the human fetus, and in apes; as remarked above, it is an ancient mammalian structure. The talus is attached to the calcaneus medially by a ligament arising from its medial tubercle and extending towards its posterior tubercle, thus flooring the groove for flexor hallucis longus. It is the homologue of the posterior talocalcaneal ligament of monotremes (Figs 13.2A, B). Although without this comparative insight it was so named in man by Sewell (1904*a*). It is the medial talocalcaneal ligament of official terminology (Figs 13.12A, B); some of its substance may reach the posterior talar tubercle—the occasional posterior talocalcaneal ligament.

The human subtalar axis (Figs 13.12C, D; 13.14, 13.15) runs upwards, forwards and medially through the calcaneus and talus, from the lateral side of the heel (Shephard 1951). In its exaggerated upwards inclination, and in its reorientation towards the long axis of the foot, it differs markedly from that of other Primates, and the morphological remodelling underlying this change must be of considerable functional and phylogenetic importance. The posterior calcaneal surface is inclined to this axis and in the right foot forms a segment of right-handed screw. This was elegantly demonstrated by Manter (1941) for the human foot and this was apparently the first suggestion of a screw-like action at this joint in any primate. It is thus particularly surprising that Szalay (1975) should have maintained that this type of action has been lost in the hominid joint. Another feature of obvious functional significance in the human foot is, however, highlighted by the comparative observations recorded above. The

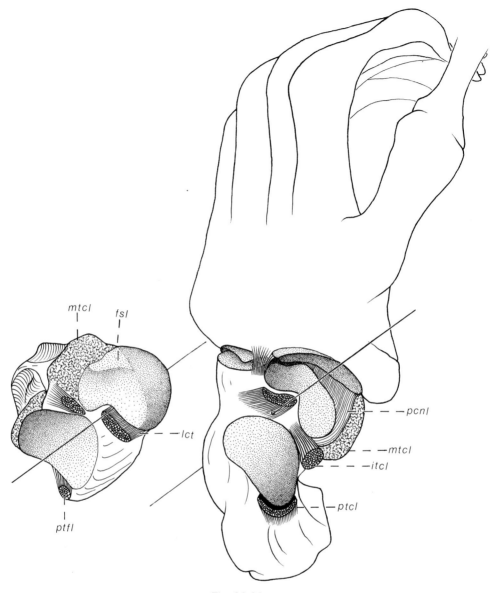

Fig. 13.13

A dorsal view of the left lamina pedis of *Gorilla gorilla* with the talus rolled away, revealing the conarticular surfaces of the subtalar joint complex; in each case the compromise subtalar axis is shown. *ptcl*, posterior talocalcaneal ligament. Other labels are as in Figs 13.1, 13.2, 13.3, 13.12. (From Lewis 1981.)

talocalcaneal surface (or surfaces) entering into the acetabulum pedis has been refashioned from an L-shaped arrangement, each segment having its own curvature related to the subtalar axis, into a smoothly confluent arc forming a segment of a screw with the thread left-handed in the right foot.

This is also clearly revealed by observing the corresponding articular surface on the underside of the talus.

The combination of these counter rotating screwing movements (Fig. 13.26) on the talus—forwards from behind, and backwards on the

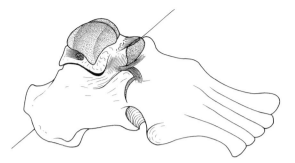

Fig. 13.15
The right foot of *Homo sapiens* (the same foot as shown in Fig. 13.14) viewed from laterally showing the compromise subtalar axis. (From Lewis 1980*b*.)

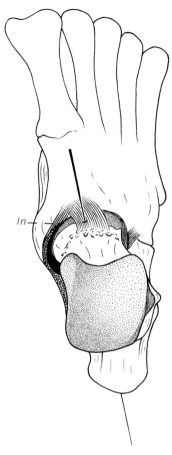

Fig. 13.14
The right foot of *Homo sapiens* viewed from dorsally showing the compromise subtalar axis. (From Lewis 1980*b*.)

head—produces a rotatory torque about a vertical axis. This component is reflected in the much more upright orientation of the compromise subtalar axis of man, compared to that of other Primates. Another feature elevating the axis has been that the anterior articular platform on the human calcaneus has been built up and effectively raised (Morton 1924). The movements of inversion and eversion are of course significant in adjusting the attitude of the human foot to inclined surfaces, a function not unlike that in other Primates. In bipedally walking man it is even more instructive to visualize the movement of the talus, and with it the leg, on the foot. Jones (1945) has noted that the extreme declination of the human subtalar axis

has added an element of rotational torque of leg (with talus) on the foot: as the foot is inverted the leg is externally rotated; as the foot everts the leg internally rotates. Inman (1976), also noting this called the subtalar joint 'a directional torque transmitter'. This function can be seen in one leg balancing, for the supporting leg laterally rotates as its foot inverts and the whole body then turns medially with respect to that foot (Jones 1945; Hicks 1953). The importance of this torque of leg (with talus) on the foot during walking will be considered later. In its new role as a pivotal link between leg and foot, providing for rotation about an increasingly vertical axis, the human talus has taken on a squat, dumpy character when compared with other Primates. As noted above, however, this trend is already initiated in *Pan* and *Gorilla*.

The subtalar axis described and illustrated here is a 'compromise axis' in the sense used by Fick (1904). It is apparent, however, that this axis could only be fixed and stationary if the component rotations of the inversion–eversion movement in the three body planes take place simultaneously and in constant ratio. For most joints it is improbable that this would be strictly true. In fact, as Langelaan (1983) has shown, a fairly closely aggregated bundle or sheaf of axes is necessary to depict the movement with accuracy throughout its different phases. Although a single axis may not define the movement with mechanical precision it nevertheless usefully characterizes the pattern of the movement as a whole.

The transverse tarsal joint complex

Again this joint complex is not a single anatomical entity but includes the talonavicular part of the talocalcaneonavicular joint together with the calcaneocuboid joint. Further, in subhuman primates (and as a variant in man) there is an articulation of variable extent between the calcaneus and navicular. The functionally effective part of this composite articulation, determining the direction and extent of movement, is the calcaneocuboid joint; the talonavicular part is a passive appendage, the navicular being carried across the head of the talus by its strong attachment to the cuboid. Attention, therefore, will mainly be directed to the form and function of the calcaneocuboid joint.

Rigorous definition of the movements involved is essential. Hicks (1953) recognized the role of twisting of the forefoot, and defined a pronation twist as being effected by flexion of the first ray of the foot, with progressively diminishing flexion laterally of the remaining rays; for such a foot to remain flat on the ground it is clear that a compensatory movement in the opposite direction (inversion) must occur at the subtalar joint complex. MacConaill and Basmajian (1969) likewise recognized twisting of the whole of the subtalar skeleton of the foot, which they called the 'lamina pedis', a term adopted in this book. By analogy with forelimb movements they called twisting 'pronation', and untwisting 'supination'. They reserved the terms 'inversion' and 'eversion' for movements at the subtalar joint complex, a similar convention to that adopted here.

MacConaill and Basmajian (1969) correctly directed attention to movement at the calcaneocuboid joint, rather than independent movement of the rays themselves, as the main focus of torsion (pronation) or untwisting (supination) of the lamina pedis. Their terminology will also be adopted here, but only for movements of the forefoot upon the hindfoot at the calcaneocuboid joint. Thus, movement increasing twist of the lamina pedis (and thus effectively depressing the first ray and elevating the lateral border of the foot) is called pronation; supination is the converse: untwisting of the lamina pedis by forefoot movement. Descriptive difficulties arise, however, when

movements of the calcaneus upon the stationary cuboid and forefoot are considered. Inward rotation of the heel and the calcaneal part of the sole increases twisting of the lamina pedis ('pronation' in the sense used by MacConaill) but the actual calcaneal movement is analogous to the supinatory component of inversion, it would be illogical to define this as pronation and it will here be termed endorotation of the calcaneus. The converse movement producing an untwisted lamina pedis will be called exorotation. The configuration of the lamina pedis will be described as 'twisted' and 'untwisted' rather than as pronated and supinated respectively: the former is achieved by pronation of the forefoot or by endorotation of the cancaneus; the latter is achieved by supination of the forefoot or by exorotation of the calcaneus.

The metatherian joints

In a marsupial such as *Caluromys lanatus* the proximal articular surface on the cuboid bears a convex peg-like prominence on its plantar aspect and this is surrounded dorsally and on the sides by a semicircular flat area. A complementary surface on the calcaneus (Fig. 17.5A) completes the joint which is related laterally to the trochlear process of the calcaneus. An essentially similar morphology of the joint surfaces is seen in *Didelphys marsupialis* (Fig. 17.1D) but here the trochlear process is retracted from the immediate vicinity of the calcaneocuboid joint. In Australian marsupials such as *Sarcophilus harrisi* or *Phascolarctos cinereus* the morphology is very different and in place of a peg-like protrusion on the cuboid there is only a low convex plantar prominence, blending into a flattened surface which encompasses it dorsally and on the sides. In *Trichosurus vulpecula* (Fig. 13.16A) or *Pseudochirus laniginosus* this type of morphology has been somewhat modified and the whole surface has become virtually flat, by suppression of the plantar prominence, and rather kidney shaped. This flat surface is rimmed by an articular cuff overlapped by a stubby projection of the calcaneus, which represents a consolidated trochlear process. What can be the morphocline polarity of these divergent joint types—is the didelphid morphology primitive or is the Australian type ancestral? It seems likely that the latter is

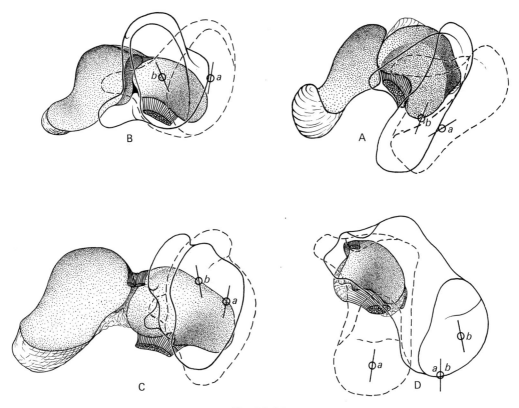

Fig. 13.16

Diagrams illustrating the movements at the calcaneocuboid joint, drawn from superimposed radiographs of right-sided ligamentous preparations of *Trichosurus vulpecula* (A), *Colobus polykomos* (B), *Pan troglodytes* (C), *Homo sapiens* (D). In each case the form of the distal articular surface—navicular and cuboid in A, B and C, and cuboid alone in D—is shown together with the associated ligaments; the outline of the calcaneus is superimposed in endorotation (broken line) and exorotation (solid line). A marker inserted in the tuberosity of the calcaneus allows estimation of rotation of the bone in its long axis between the positions of full endorotation (*a*) and full exorotation (*b*); *ab* is a position intermediate between these two attitudes. (From Lewis 1980*b*.)

true, and that in the emergent marsupials (and indeed in the ancestral Eutherians) the cuboid possessed only a low central elevation. In parallel this joint type seems to have been elaborated several times into a peg-and-socket pivot. In all these marsupial species a very strong plantar calcaneocuboid ligament radiates forwards from the anterior plantar tubercle on the calcaneus to attach to the ventral surface of the peg on the cuboid, or in the case of a flattened surface (*Trichosurus*) to the 'hilum' of the kidney shaped area.

Regardless of the degree of expression of the plantar prominence, all these joints function as pivots, although there may be differences in stability. A plantar prominence is not a prerequisite for pivotal action as Szalay (1977) supposed. At the joint the movement consists essentially of rotation about an axis approximating to the long axis of the foot, thus producing supination and pronation of forefoot upon hindfoot. When the movement is analysed a little more closely by considering movement of the calcaneus upon the stationary cuboid in a ligamentous preparation (Fig. 13.16A) it is clear that an element of sideways displacement or swing is allied to the rotatory movement or spin. As exorotation of the calcaneus occurs (equivalent to supination of the forefoot) a medial folding of the footplate occurs—the calcaneus adducts. The fulcrum of the movement is the tough plantar calcaneocuboid ligament.

The eutherian joints

In the insectivore *Tenrec ecaudatus* (Fig. 17.5B) the cuboid is again mildly convex and movement is pivotal; the calcaneus presents a complementary conarticular surface for the cuboid, and the trochlear process, although not overlapping the joint, is adjacent to the distal extremity of the calcaneus. Other insectivores possess an essentially similar morphology.

In *Tupaia* the peroneal process is retracted posteriorly as in Primates, but it should be noted that such a posterior displacement of the process has occurred convergently in a number of groups including certain rodents and marsupials (Stains 1959). In *Tupaia* also the joint surface of the calcaneus is also usually concave, and that on the cuboid gently convex, and the movement largely rotatory. In some specimens the convexity on the cuboid may be prolonged quite far forward at its plantar extremity providing perhaps for some dorsoventral hinging between forefoot and hindfoot. The source of the primate morphology can be envisaged as a compromise between the structure of these two extant quite generalized Eutherians—one in which the trochlear process was retracted as in *Tupaia*, but with a cuboid retaining a slightly convex pivotal conformation as in *Tenrec* (or better, as in the marsupials *Phascolarctos* and *Sarcophilus*). Early in primate phylogeny, however, the joint must have been modified by the enlargement forwards of the calcaneus to contact the navicular and thus exclude the cuboid from the subtalar joint. As with the subtalar joint, form and function show only minor variations throughout the order Primates, with the noteworthy exception of man, and to a lesser extent the great apes.

The subhuman primate joints

In lemurs, platyrrhine monkeys, catarrhine monkeys, and gibbons, the calcaneal surface on the cuboid is kidney-shaped with a slightly protuberant ventral convexity adjacent to the 'hilum', whilst the periphery of the surface is flattened or even slightly concave (Fig. 13.16B). The navicular and cuboid are bound together dorsally by a tough transverse ligament and a longitudinal ligamentous band from the calcaneus to cuboid again overlies this dorsally (Fig. 13.8A); this latter band presumably corresponds to the calcaneocuboid part of the human bifurcated ligament (there is no calcaneonavicular band). Ventrally the joint is strengthened by a very tough plantar calcaneocuboid ligament radiating symmetrically forwards from the anterior plantar tubercle on the calcaneus to the ventral concave margin of the cuboid. This ligament is clearly homologous with the short plantar ligament of human anatomy, but its more superficial fibres are prolonged forward across the groove for the peroneus longus tendon, towards the cuboid ridge and the metatarsal bases; this is the distal attachment of the long plantar ligament of human anatomy. This more superficial lamina has, however, not prolonged its attachment posteriorly, as in *Homo sapiens*, to the area of the calcaneus between its anterior tubercle and the calcaneal tuberosity. The joint is 'stepped' so that the calcaneus has a variable articulation with the lateral margin of the navicular.

The most obvious aspect of the movement is rotation about the long axis of the foot producing in the arboreal setting habitual movements of supination and pronation of forefoot upon hindfoot which complement the movements occurring at the subtalar complex and even the ankle joint, in a highly effective way. It is apparent however, that movement is really more complex for as the forefoot supinates it becomes angulated, or folded, medially (Fig. 13.24B).

If instead we consider motion of the calcaneus upon a stationary cuboid (Fig. 13.16B) it is apparent that the habitual movements consist essentially of rotation (spin) of calcaneus upon cuboid, with a slight element of sideways displacement (swing) on the calcaneus. As the calcaneus is exorotated (position 'b' in Fig. 13.16B) its distal surface engages firmly on to the more medial part of the cuboid facet and fits snugly in against the articular margin of the navicular. As the calcaneus is endorotated (position 'a' in Fig. 13.16B) it rotates onto the more lateral part of the cuboid facet. The fulcrum of the movements is the tough plantar calcaneocuboid ligament and alternately the medial or lateral parts of this radiating band are tightened. As the calcaneus moves into endorotation its posterior aspect is translated laterally.

In essence the movement is similar to that in marsupials (Fig. 13.16A) although the Metatheria, having an alternating tarsus, lack the specialization allowing the calcaneus in exorotation to close-pack against the navicular. In galagos (Fig. 13.10) the calcaneocuboid joint has been displaced distally by a greatly disproportionate lengthening of the anterior part of the calcaneus. The basic structure and ligamentous apparatus of the joint are recognizably primate but the joint surfaces show a derived specialization which is convergent to the pattern characterizing the hominoids. The plantar convexity on the cuboid is greatly accentuated into a peg-like projection which is eccentrically disposed lying towards the medial aspect of the joint. This structure is closely attuned to the pivoting function of the joint. Movements at the joint, allied with those at a proximal articulation between navicular and calcaneus, provide for movements of supination and pronation of the forefoot and are strikingly analogous to the movements of pronation–supination in the human forelimb. The small movements of inversion–eversion of the calcaneus below the talus at the subtalar joint complex supplement the pivotal movements and are analogous to the secondary movements of the ulna which occur in pronation–supination of the forearm (Kapandji 1966). The great apes all possess derived morphologies in the form of the conarticular surfaces of calcaneus and cuboid, and that shown by the chimpanzee appears to be central to the understanding of these differing shapes. Whereas lemurs, monkeys and gibbons show a relatively unobtrusive ventral convexity on the cuboid, *Pan* (Fig. 17.7A) exhibits a prominently exaggerated articular beak-like projection, fitting into a corresponding deep depression in the calcaneus (Fig. 17.8A). The articular beak is usually rather eccentrically situated, being displaced to a varying degree towards the medial side and it blends insensibly into the remainder of the cuboid surface which is flat or slightly concave. The surface on the calcaneus is reciprocally shaped with a deep excavation lodging the cuboid projection, and the calcaneus contacts the navicular along a narrow linear facet. Ventrally a massive plantar calcaneocuboid ligament radiates forward from the anterior calcaneal tubercle; there is still no separately differentiated long plantar ligament extending this

attachment backwards. Dorsally, there is a calcaneocuboid ligament but now associated with it is a calcaneonavicular band (apparently a dissociated part of the dorsal talonavicular ligament which has become secondarily attached to the calcaneus)—a bifurcated ligament has emerged. Despite this altered form, function appears to be essentially similar to that in monkeys (Fig. 13.16C) with the joint providing for the rotatory motion involved in supination of the forefoot, which again includes a component of angular displacement—slight adduction of the forefoot. The situation in *Gorilla* indicates derivation from a morphology similar to that seen in *Pan*. In most gorillas the joint surfaces are almost flat (Figs 17.7C, 17.8C) with only the slightest residual hint of an elevation on the cuboid, occupying a comparable site to the marked prominence found in chimpanzees. In occasional specimens, however, a very different morphology is found (Figs 17.7D, 17.8D). The calcaneus is excavated (to accommodate a marked prominence on the cuboid) to a greater extent even than is seen in chimpanzees and this deep depression is surrounded by a relatively flattened articular rim. It seems likely that this represents the ancestral *Gorilla* pattern and that the common flattened form is derived from it by suppression of the prominence on the cuboid. The calcaneus again contact the navicular along a linear strip and the ligamentous apparatus is similar to that of the chimpanzee with a very strong plantar ligament and with a bifurcated ligament dorsally. In both variant forms a pivotal action occurs which is essentially like that in *Pan*.

The orang-utan joint surfaces (Figs 17.7B, 17.8B) present a characteristically unique form of their own, but with obvious familial resemblances to the other great apes. The calcaneus bears a deep cup-like depression, which lodges a prominent eccentrically located hemispherical protuberance on the cuboid, and is sharply demarcated from a broad flat articular surface located to its lateral side. Plantar calcaneocuboid and bifurcated ligaments are present. A pivotal action occurs, as in the other great apes, but as will be shown later the calcaneocuboid joint has a unique relationship with the naviculocuneiform joints in this ape.

The human joints In man, despite persistent evidence of affinity with the great ape type of structure, the joint architecture has been subtly remodelled, producing a dramatic change in the character and range of movement.

One could hypothetically remodel a cuboid articular surface, such as that seen in *Pan*, by displacing the plantar protuberant beak medially so as to form a ·columnar inner margin to the surface. If, correlated with this, there is expansion of the lateral flattened or concave part of the surface into a widely flaring concave area, the resulting markedly asymmetrical surface has attained the sellar or saddle-shaped form characteristic of *Homo sapiens* (Figs 17.7F, 17.8F). The medially displaced convex component carries with it the plantar calcaneo–cuboid ligament to form the canonical short plantar ligament of human anatomy, which is thus very obliquely disposed in contrast to the fan-like arrangement found in apes (Fig. 13.16D). Dorsally lies the calcaneocuboid part of the bifurcated ligament. The calcaneonavicular part forms the lateral wall of the talonavicular joint.

If movement of the calcaneus on the cuboid is considered, as happens in the weight-bearing foot, it is seen (Fig. 13.16D) that rotatory movement still occurs, as in subhuman Primates, but now as a concomitant (conjunct rotation) of a greatly amplified angular or swing movement. Further, the lateral swing is now correlated with exorotation of the calcaneus: the reverse direction of that occurring in lower Primates. In the position shown at *a* (endorotation of the calcaneus) the joint surfaces are quite incongruent and closed-packed only dorsally; at *b* (exorotation of the calcaneus) the whole joint is maximally congruent and the plantar convexity on the cuboid is impacted into its slot under the sustentaculum tali. In this latter position as the calcaneus moves laterally with a rotatory motion, the lamina pedis is untwisted, tensioning the oblique short plantar ligament, the bifurcated ligament, and the plantar calcaneonavicular (spring) ligament. The excursion of the calcaneus as whole occurs along a path (Fig. 13.26) which is part of a right-handed helix in the left foot. At the same time the cuboid counterrotates against the calcaneus; again, its path is along a right-handed screw in the left foot, moving it into a supinated and adducted attitude. These movements bring the lamina pedis into its untwisted and close-packed condition, straighten-

ing out the arched lateral border of the foot skeleton. It is clear that no single axis could represent this movement. If a purely hypothetical axis to express the movement of calcaneus upon cuboid were to be postulated it would run upwards, backwards and somewhat medially; that for cuboid upon calcaneus would be upwards, forwards (and perhaps somewhat medially).

MacConaill (1945) and MacConaill and Basmajian (1969) were aware that untwisting of the lamina pedis, such as occurs when standing with the feet wide apart, involves movement of the calcaneus of the type referred to here as exorotation, and that it includes lateral displacement of the heel. They further appreciated that the movement must occur at the calcaneocuboid joints although details of the mechanics involved were not investigated by them.

Many other investigators have attempted to explain midtarsal movement in terms of single or multiple axes. Manter (1941) described the movement at the calcaneocuboid joint as occurring about dual axes, but these were not comparable to the purely notional ones mentioned above. Manter's concept was quite artificial and consisted, in effect, of resolving the habitual movement into two components: a rotation (spin) about an axis longitudinally disposed in the foot and an angulation (swing) about a transverse axis running upwards and medially through the joint. However, Manter also visualized the composite movement as having a screw-like action about the longitudinal axis, left handed in the right foot; quite a useful aid to appreciating the true nature of the movement. Elftman (1960) also described the joint as a saddle-shaped one with two axes, and deduced on theoretical grounds that the resultant transverse tarsal axis should run upwards and medially. This is clearly not in accord with the experimental findings shown in Fig. 13.16D.

Bojsen-Møller (1979), like Manter and Elftman, also described dual axes for the joint and followed the essentially artificial convention of trying to resolve the habitual movements shown in Fig. 13.16D into components about these two axes. Surprisingly he proposed that the joint is in close-packed position when the forefoot is pronated, or the calcaneus endorotated. In fact, in this attitude the protuberant beak on the cuboid (his calcaneal process of the cuboid) is retracted from its nidus in the calcaneus, and the joint surfaces are congruent only dorsolaterally. As I have described it, the joint is clearly in its position of maximum congruence when the forefoot is supinated, or the calcaneus exorotated, either movement or the two together causing the whole lamina pedis to be maximally untwisted. Herein lies the real functional significance of the characteristically human condition of an eccentric disposition of the beak-like process on the cuboid and the associated obliquity of the human short plantar ligament. Bojsen-Møller's observation that occasional human cuboids may be found to lack a 'calcaneal process' and apparently possess a flat surface for the calcaneus requires special comment, for in these cases it would seem that a uniquely human attribute was not developed. In fact, occasional human cuboids may be seen to have a separately ossified calcaneal process (the so-called cuboideus secondarius); it would seem likely that in Bojsen-Møller's macerated specimens such a separate ossicle had become detached.

Langelaan (1983) described a bundle of axes which are quite closely grouped together and proceed from anterosuperior to posteroinferior for the relative cuboid–calcaneal movements. As he noted these codify cuboid movement which has 'a distinct adduction as well as a distinct inversion component' (Fig. 13.26); these axes, however, completely fail to express the characteristic movement of the calcaneus (Fig. 13.16D) which plays such an important functional role in the human foot.

The tarsometatarsal and distal intertarsal joints

The first tarsometatarsal joint

When mammals possess a grasping abducted hallux the tarsometatarsal joint at its base plays a critical functional role. Thus, in the marsupial *Trichosurus vulpecula* concavoconvex conarticular surfaces are found, and a similar type of joint is characteristic of prosimian primates, such as *Lemur catta* or *Perodicticus potto*. This, of course, might be attributed to convergence. At the very least there is an a priori case for considering that such a joint was already established in the last

common ancestor of Metatherians and Eutherians, then retained in arboreal marsupials, and also among the Eutheria especially in the primate line, but lost in the more terrestrially committed placentals. This theme was previewed in Chapter 3.

The subhuman primate joints   In the higher Primates this joint has been the locus for a complex suite of evolutionary modifications (Lewis 1972c), leading to incorporation of the prehallux as a joint component, and its subsequent dissolution. As will be described in Chapter 14 the prehallux is primitively found in mammals as a nodal ossification in the tendon of the flexor tibialis at the site where it sends divergent attachments into the sole of the foot. These divergent attachments are varyingly disrupted in different mammals. Typically in primates, however, a remnant of this complex persist as a prominent naviculometatarsal ligament.

In the Ceboidea this ligament (Fig. 14.4C) is a prominent, although variable feature of the foot, and contains the prehallux which is displaced forwards to contact the first metatarsal and participate in the joint at the base of that bone. Moreover, the naviculometatarsal ligament bridges over the primitive insertion of the tibialis anterior tendon to the medial cuneiform and here the prehallux has captured part of the tendon converting the insertion into a dual one, with resultant splitting of the tendon. In *Cebus nigrivittatus*, for instance, the prehallux is large, participates in the joint, and receives the attachments of the naviculometatarsal ligament, and of the secondarily derived tendon of insertion of tibialis anterior which is supernumerary to the primitive cuneiform one (Fig. 13.17A). The ossicle forms about one quarter of a sphere with the flattened surfaces being articular, one applied to a bevelled off facet at the ventromedial corner of the base of the first metatarsal and the other contributing part of the composite distal articular surface of the first tarsometatarsal joint. The whole distal articular surface thus constituted is cylindrically concave, with an approximately vertical axis.

The articular surface on the medial cuneiform for the first metatarsal faces somewhat medially and is indented laterally by a roughened area for attachment of the first interosseous cuneometatar-

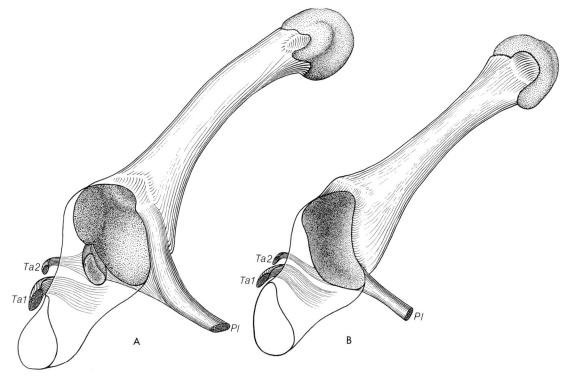

Fig. 13.17

The right first metatarsal of (A) *Cebus nigrivittatus* and (B) *Procolobus verus*, showing its articulation with the medial cuneiform, which is indicated in outline. *Pl*, tendon of peroneus longus; *Ta1*, main tendon of tibialis anterior inserting into the medial cuneiform in each case; *Ta2*, second tendon of tibialis anterior inserting into the independent prehallux in (A), and into the base of the first metatarsal in (B). (From Lewis 1966.)

sal ligament (homologue of the human ligament of Lisfranc) binding the medial cuneiform to the tibial side of the base of the second metatarsal, ventrally to an articulation between these bones (Figs 13.19B, 13.20A). The hallux is thus left free to move independently of the rest of the foot at its tarsometatarsal joint. On the medial cuneiform surface a deep concave furrow spirals down the articular surface from above and medially. This groove is smoothly confluent above with a convex area, but is demarcated below and medially by a rounded ridge which separates the large upper convexoconcave area of the surface from a gently convex facet occupying the lower, medial corner of the articular surface (Fig. 13.18).

The efficient grasping capability of the platyr-rhine hallux is largely dependent on the form of these joint surfaces. When the hallux is adducted the concave distal articular surface (including the prehallux) articulates with the medial part of the

entocuneiform articular surface, leaving the lower medial facet exposed; the articular surfaces match well. When the hallux diverges into the attitude of opposition, a movement which may be decomposed into the diadochal sequence of abduction from the axis of the foot together with some flexion, the two components of the distal articular surface splay apart. The prehallux then rides down to fit closely onto the ventromedial facet of the cuneiform and the exposed bevelled margin of the metatarsal engages into the groove above this. The joint is then close-packed. As is usual with such concavoconvex joint surfaces, this movement includes a considerable element of conjunct rotation, which is effective in screwing the bones into a locked position. The hallux is thus stabilized at its base for its grasping action, and its plantar aspect is advantageously rotated towards the other digits, an effect which is enhanced by the fact that the metatarsal shaft itself shows a considerable

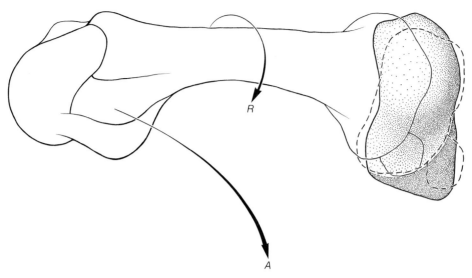

Fig. 13.18
The mechanism of the right hallucial tarsometatarsal joint in *Cebus nigrivittatus*. The distal articular surface of the medial cuneiform is stippled. The direction of the divergent movement of the hallux into the attitude of opposition is indicated by the arrow *A*; the component of conjunct rotation occurring during this movement is indicated by arrow *R*. The final close-packed position achieved by the base of the metatarsal, and its articulating prehallux, is indicated by the broken line. (From Lewis 1972*c*.)

degree of torsion in this direction. The peroneus longus tendon maintains it in this rotated position and holds the hallux against the grasped support; the tendon of insertion of tibialis anterior into the prehallux, acts as the antagonist of peroneus longus.

Among the New World monkeys as a whole a bony prehallux forming a significant joint component is constantly found in the family Callithricidae (except for *Leontideus* where the prehallux is cartilaginous) and in the subfamilies Aotinae and Cebinae of the suborder Ceboidea. In the Pithecinae a prehallux is similarly present but fails to ossify, remaining cartilaginous. A cartilaginous prehallux is also found in *Ateles* (and presumably in *Brachyteles*) but in the other genus of the Atelinae (*Lagothrix*) it is reduced to, at best, a mere remnant and the tibialis anterior thus effectively acquires a direct metatarsal insertion; a parallel state of affairs has been achieved in the Alouattinae.

There is little doubt that a prehallux-containing hallucial tarso–metatarsal joint is ancestral for the Ceboidea, but what of the Cercopithecoidea? The retrogressive trends affecting the prehallux of certain New World monkeys, by which it is degraded into a cartilaginous nodule or even further into the virtually unmodified metatarsal tendon of insertion of tibialis anterior, provide the clues to the derivation of the Old World monkey pattern. The Cercopithecoidea lack a prehallux yet they possess dual tibialis anterior tendons, the anterior one inserting into the metatarsal base in common with the naviculometatarsal ligament. At its insertion the metatarsal tendon is exposed to the interior of the joint, actually forming part of its lateral wall; such an arrangement is the logical consequence of regression of a prehallux. The metatarsal articular surface is concave about a vertical axis and of course lacks any prehallux facet (Fig. 13.17B). The medial cuneiform surface is convex above or furrowed medially and below by a spiral groove which in turn is limited by a prominent ridge forming the articular margin. Below and behind this ridge is found the cuneiform insertion of the tibialis anterior but, of course, no prehallux facet. The bone is like a New World monkey bone deprived of its prehallux facet.

In Old World monkeys abduction of the hallux is more nearly a simple hinge movement with less accompanying conjunct rotation than in New World monkeys, and apposition of the hallux to the other digits is therefore less effective. As the hallux diverges the spiral groove on the cuneiform comes to lodge the margin of the metatarsal base and the joint is then in close-packed position. As the hallux is adducted towards the axis of the foot, the groove on the cuneiform is occupied by the exposed metatarsal tendon of tibialis anterior. The only noteworthy modification of this morphology is found in the basically terrestrial cercopithecines, *Papio*, *Mandrillus*, and *Erythrocebus*, where the concavoconvex moulding of the articular surface of the medial cuneiform is minimal, with only a suggestion of a spiral groove and with the limiting ridge of the articular surface blunt and ill-defined. In these species the joint is remodelled into what is essentially a simple hinge joint.

There is therefore persuasive evidence that the Ceboidea and Cercopithecoidea shared a common ancestry with a primate possessing a prehallux-containing joint at the base of its hallux. Further, since the Hominoidea were doubtless derived from primitive catarrhines, it would be reasonable to expect evidence of the pre-existence of this same character complex in the apes and man.

In gibbons the presence of a prehallux, dual tendons for tibialis anterior, and an overall morphology similar to that of *Cebus*, seems to be a constant feature. The prehallux is, however, relatively smaller than in *Cebus* and the medial cuneiform has its various undulations less clearly defined than in New World monkeys, but the joint seems to function in an essentially similar fashion. The remaining apes, and man, show a parallel development to that found in Old World monkeys, with the clear signs of dissolution of a previously existing prehallux.

In the orang-utan no prehallux is present but what is perhaps a vestigial remnant may be found adjacent to the metatarsal attachment of the more anterior of the dual tibialis anterior tendons. The articular surface of the medial cuneiform lacks any prehallux facet but it is concavo-convex and bears an oblique groove, more pronounced than that in monkeys, and limited below and medially by a ridge.

Chimpanzees also show a joint morphology paralleling the derivative pattern of Old World monkeys. The metatarsal articular surface, in the absence of a prehallux, is cylindrically concave about a vertical axis and has the small anterior tendon of tibialis anterior inserted directly to its ventromedial corner in association with the naviculometatarsal ligament. Close to this insertion the tendon is exposed to the joint interior, here actually replacing the fibrous capsule. The articular surface on the medial cuneiform is cylindrical above, becoming concave below and medially. This concave furrow spirals obliquely down the articular surface and is again demarcated from the non-articular part of the medial cuneiform by a ridge (Fig. 13.19A). The form of the medial cuneiform articular surface shows a considerable range of variation and sometimes the inferomedial furrow is almost imperceptible and the whole surface then is cylindrically convex. Biomechanically the chimpanzee joint functions like that of the Cercopithecoidea. With the hallux adducted the oblique groove on the medial cuneiform lodges the anterior tendon of tibialis anterior; movement of the hallux into opposition is accompanied by a minor degree of conjunct rotation and the lower medial margin of the metatarsal base then lodges in the oblique groove on the cuneiform, close-packing the joint surfaces. The stability of the joint in this position is assisted by a cord-like accessory ligament which is obliquely disposed on the lateral aspect of the joint, and then becomes taut. Such a ligament has not been noted in any of the other Primates already described.

In gorillas the medial cuneiform also possesses the basic catarrhine pattern showing a general convexity which blends below and medially into an oblique groove. The ventral rim of this groove bounds the hollowed lower part of the articular surface and the upper convex portion is prolonged onto the medial aspect of the bone. As in Pan, abduction of the gorilla hallux is accompanied by conjunct rotation, locking the margin of the metatarsal into the inferior concavity of the medial cuneiform articular surface, bringing the joint into the close-packed position. In the single wet specimen which I have dissected, however, I could detect no oblique ligament comparable to that of the chimpanzee.

The human joint In *Homo sapiens* the tibialis anterior tendon again has a dual insertion into the medial cuneiform and onto the first metatarsal, but

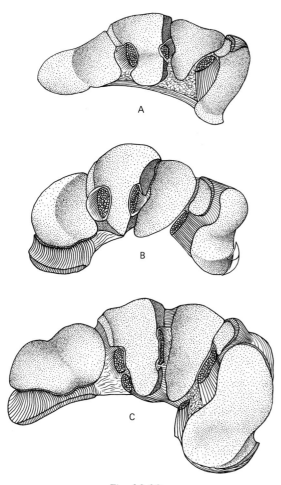

Fig. 13.19
The distal aspects of the right tarsus of wet specimens of
Pan troglodytes (A), *Saimiri sciureus* (B), and *Homo sapiens*
(C). In each case the metatarsals have been disarticu-
lated to reveal the articular surfaces and ligaments of the
tarsometatarsal joints. The position of the prehallux is
shown in line, articulating with the medial cuneiform of
Saimiri sciureus (B). (From Lewis 1980*b*.)

the tendon itself rarely shows a clear-cut subdivision and there is, of course, no prehallux. The
metatarsal tendon actually forms the lower medial
portion of the wall of the joint, just as it does in
other catarrhine Primates, and in this location a
synovial pocket balloons proximally between the
medial surface of the cuneiform and the tendon,
which here lies in a groove on the bone. This
oblique groove may be lined with articular cartilage (Fig. 13.19C) or merely clothed with synovial

membrane. The synovial diverticulum may extend
some distance into the substance of the tibialis
anterior tendon splitting off the metatarsal part
from the deeper cuneiform insertion, some of
whose fibres may extend down into the floor of the
groove. The cuneiform insertion itself is often
separated from the bone by a bursa, and this is also
a common finding in many other Primates. The
synovial-lined cul-de-sac is clearly homologous
with the upper portion of the oblique groove which
in other catarrhine Primates forms the lower part
of the cuneiform articular surface. In man, restriction of the articular surface proper to the anterior
aspect of the cuneiform (and its progressive
flattening) has relegated the upper part of the
groove to the medial aspect of the cuneiform where
it forms a mere appendage to the joint. The groove
may be prolonged onto the articular anterior
aspect of the cuneiform, producing a very shallow
cupping of the lower part, and in such cases the
upper part of the cuneiform articular surface is
typically rather convex. In these cases the morphology is only slightly modified from that which is
characteristic of *Gorilla* or *Pan*. More commonly
the whole anterior surface of the cuneiform is
almost flat. As in other Primates it is indented
laterally adjacent to the site of attachment of the
ligament of Lisfranc. From this region a non-
articular groove (or ridge) may traverse the surface
subdividing it partially, or completely, into upper
and lower parts; this arrangement may also be
found in *Gorilla*. Laterally there is a strikingly
obvious accessory oblique lateral ligament crossing the joint line just above the insertion of
peroneus longus (Fig. 13.19C), and this is clearly
the homologue of the comparable structure found
in *Pan*.

The variable form of the articular surfaces in
man, which has long been realized (Fick 1857),
dictates that movement in the human hallucial
tarsometatarsal joint, although always restricted,
varies somewhat in extent. When the joint surfaces retain some semblance of the primitive
concavoconvex form a very limited degree of the
fundamental movement of opposition involving
conjunct rotation can presumably occur, locking
the joint (assisted by the taut oblique lateral
ligament) into the close-packed position. The
resulting stability of the joint is of obvious benefit
when the medial longitudinal arch of the foot is

thus raised. The flatter the joint surfaces are, however, the more limited will be this movement of opposition. The affinity of the human joint in form and function to those of *Pan* and *Gorilla* seems obvious. Yet the topography of the joint at the base of the human hallux has provided fuel in the often acrimonious debates on human origins. The flattened form of the human joint surfaces has been sited as evidence for the view that the hominids had a very early derivation from a primitive primate stock and it has been held (Jones 1929, 1948) that this pattern, and its very early ontogenetic appearance, are incompatible with any close phylogenetic affinity between man and apes. This argument was effectively quashed by Schultz (1930, 1950) who demonstrated that the curvature of the human joint surfaces varies considerably among different individuals, being sometimes almost plane at one extreme, and occasionally nearly as convex as in some gorillas. He stressed that the exact conformation of the joint differs between man and ape only in degree but not in principle.

The distribution of the prehallux in the Anthropoidea   There are good grounds for believing that the earliest members of the Anthropoidea, the presumably arboreal ancestors of both the platyrrhine and catarrhine radiations, possessed as an apomorphic character, a complex prehallux-containing joint at the base of the hallux. The retention of the prehallux in most of the exclusively arboreal platyrrhine monkeys is not surprising, in view of the contribution which it makes to the stability of the basal joint of the grasping hallux. Nor should its loss in the evolving cercopithecines occasion surprise in view of indications (Napier 1970) that ground-living adaptations were a feature of their early evolution. It is less clear however why the arboreal colobines should have dispensed with this useful character.

Primitive catarrhines, still endowed with a prehallux-containing joint, seem also to have provided the source of the diverging hominoid lineage, with loss of this feature after the divergence of the gibbons; loss of the prehallux must then have been realized independently in the cercopithecoid and hominoid lineages. There is no evidence that a prehallux ever occurs even as an anomaly within the hallucial tarsometatarsal joint

of the great apes or of man but a presumptive prehallux does sometimes occur at another site in man (Chapter 14).

It was perhaps inevitable that this scenario should be challenged. Wikander *et al.* (1986) have proposed that the ossicles in gibbons and in New World monkeys are not homologous and are just functionally derived sesamoids in the tendon of tibialis anterior. This overlooks the fact that sesamoids in joints, such as the metacarpophalangeal ones described in Chapter 6, are of a quite different nature, and there are no glenoid plates at the tarsometatarsal joints which could provide a favourable basis for the development of sesamoids. In a sense, whether it was known to Wikander *et al.* or not, this recalls the now discredited idea, prompted by the appearance and situation of the primate ossicles, that the first metatarsal of the hallux represented in reality the 'missing' third phalanx of that digit (Testut 1904). The reactionary view of Wikander *et al.* ignores a wealth of detailed morphological evidence about the prehallux during the long history of mammals (Chapter 14) and obscures the realization that it is the anchor point for understanding an array of ligamentous and myological features at the medial side of the foot.

The remaining tarsometatarsal joints

A basic pattern for the remaining tarsometatarsal joints pervades the whole of the Primate order, and is doubtless primitive for the order; moreover it appears also to represent the ancestral therian pattern. It is subject to significant variation only in certain primate species. I have personally observed this stem morphology in wet specimens of the prosimians (*Lemur catta, Perodicticus potto, Galago moholi*), in New World monkeys (*Saimiri sciureus, Pithecia monachus, Cebus nigrivattatus*) and some Old World monkeys (*Procolobus verus, Cercopithecus nictitans*). The essence of the arrangement is as follows (Figs 13.19B, 13.20A).

The base of the third metatarsal articulates with the large lateral cuneiform, which protrudes distally beyond the two bones flanking it, and the joint is braced either side by interosseous tarsometatarsal ligaments joining the two bones. Anterior and posterior to these ligaments the bases of the second and fourth metatarsal bones articulate

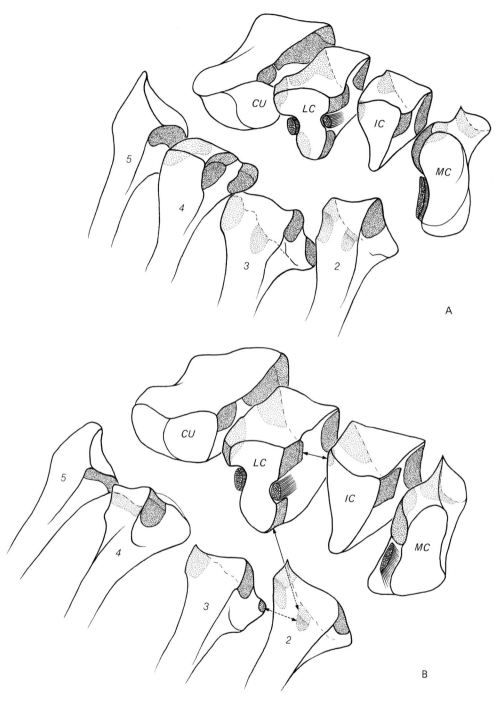

Fig. 13.20
'Exploded' diagrams of the distal tarsals and metatarsal bases of right feet of *Cebus nigrivittatus* (A) and *Pan troglodytes* (B) showing the intervening articulations. *MC*, medial cuneiform; *IC*, intermediate cuneiform; *LC*, lateral cuneiform; *CU*, cuboid; *2–5*, bases of metatarsals 2–5. In both diagrams the tarsometatarsal ligaments are shown. In *B* inconstant facets are indicated by arrows. (From Lewis 1980*b*.)

with the third, the adjoining sides of these metatarsal bases thereby possessing dual facets. The second metatarsal is deeply indented into the tarsus and articulates laterally with the adjacent lateral cuneiform by proximal extensions of the dual intermetatarsal articulations. The second metatarsal also bears a facet on the dorsomedial aspect of its base for the medial cuneiform. A massive ligament radiates from the medial cuneiform to the second metatarsal on the plantar aspect of this articulation; this corresponds to the so-called ligament of Lisfranc in man, so well described and illustrated by Jones (1944). Another substantial ligament walls this joint dorsally. Laterally the distal prolongation of the lateral cuneiform beyond the cuboid, results in the dual facets on the fourth metatarsal (as on the second) articulating not only with the third metatarsal, but also with the sides of the lateral cuneiform.

The fourth and fifth metatarsals articulate with the cuboid, the former typically being received into a concave depression on that bone. The articular surface on the cuboid for the fifth metatarsal is usually slightly convex from side to side but, as a variant, may be found to be dorsoventrally concave in some monkeys and apes. Between the fourth and fifth metatarsals there is a single intermetatarsal articulation. Intermetatarsal ligaments unite the bases of the second to fourth metatarsals just distal to their intervening articulations, as occurs at the comparable site between the metacarpals in the hand.

A similar alternating stepped character of the articulations, with identical ligaments and facets, is found in *Didelphys marsupialis* and in *Tupaia*; in *Tenrec* the arrangement differs only by the lack (or more probably loss) of the interosseous ligament on the medial side of the lateral cuneiform. There seem to be good grounds for believing that this fundamental morphological pattern was already established in the last common ancestors of marsupials and placentals—the Therians of Metatherian–Eutherian grade. The existence of this pattern as the basic architecture for Primates is then a symplesiomorphic character for that order. While showing great resemblance to the basic pattern for the carpus described in Chapter 6, it differs in that in the primitive carpus it is the hamate (homotype of the cuboid) which protrudes distally beyond the capitate (homotype of the lateral cuneiform).

As noted above, *Lemur* possesses the basic stepped character for these articulations, together with the complete ligamentous apparatus. Inexplicably Gebo (1985) has maintained that in lemurids and in indriids, the situation is quite different from that in other prosimians, in that the lateral cuneiform fails to protrude distally and consequently the third to fifth metatarsal bases are received into a smoothly contoured, continuous, dished out socket. On this basis he suggested that these prosimians possessed a specialized grasp ('the I–II grasp'), and on this shaky foundation he built a theory of anthropoid origins (Gebo 1986). There is no morphological basis in fact for this; in my experience *Lemur* retains the primitive tarsometatarsal pattern. In one point, however, Gebo (1985) is correct: the plantar facet on the second metatarsal for that on the third, faces largely distally in *Lemur*. In fact, it is crowded into this orientation by the large peroneal process on the first metatarsal. This is a minor variant on the basic theme.

Among other Primates the most common variant from this ground plan involves suppression of the ventral articulation between the medial side of the base of the fourth metatarsal on the one hand and the base of the third metatarsal and lateral cuneiform on the other. This joint is lacking in some monkeys (*Colobus polykomos*) and is characteristically reduced or absent in hominoids: in *Hylobates lar* it is rudimentary; in *Pan troglodytes* (Figs 13.19A, 13.20B), and *Gorilla gorilla* (Fig. 13.21) it is absent. In *Homo sapiens* (Fig. 13.23) it is similarly absent and a strong slip of the tibialis posterior tendon is insinuated between the metatarsal bones, here reinforcing the interosseous tarsometatarsal ligament. In *Homo sapiens* the residual dorsal articulation in this situation is commonly purely intermetatarsal and the lateral cuneiform then fails to contact the fourth metatarsal (Jones 1944). On the other side of the third metatarsal the ventral articulation is less commonly absent. The intermetatarsal part, even including the ventral articulation between the lateral cuneiform and the second metatarsal may sometimes be lacking in *Pan troglodytes* (Fig. 13.20B). However, the whole articulation here is absent in *Gorilla gorilla* (Fig. 13.21).

The distal intertarsal joints between cuneiforms and cuboid similarly present a basic primate pattern (Fig. 13.20A). Typically the adjoining surfaces of the cuneiforms and the cuboid bear

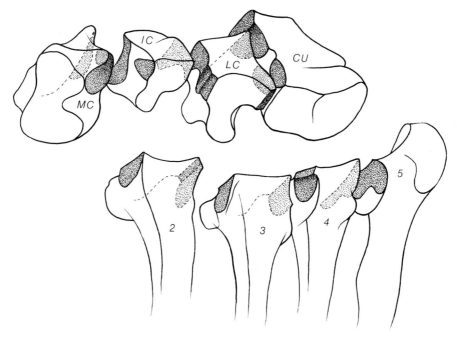

Fig. 13.21
An 'exploded' diagram of the distal tarsals and metatarsal bases of the left foot of *Gorilla gorilla*. Facets are indicated by stippling but ligaments are now shown. Labels are as in Fig. 13.20.

dual articulations the proximal one being usually a dorsoplantar articular strip whilst the distal one is dorsally located and here communicates with the articulations of the protruding medial and lateral cuneiforms with the bases of the second and fourth metatarsals. Between proximal and distal articulations are strong interosseous ligaments. This pattern is apparently a stable primate heritage and only occasional variations are observed in prosimians, monkeys and apes. In *Pan* the distal articulation between intermediate and lateral cuneiforms may be lacking, although a large articulation between the lateral cuneiform and the second metacarpal is retained; in effect, in these cases the protrusion of the lateral cuneiform is exaggerated. It is likely that the basic primate pattern is also the primitive mammalian arrangement, since in *Didelphys*, although proximal and distal joints between intermediate and lateral cuneiforms and cuboid are confluent, they show residual indications of subdivision; those between medial and intermediate cuneiforms are, however, merged in a single dorsally located strip.

Pongo pygmaeus (Fig. 13.22) departs strikingly from the basic primate morphology both in the architecture of the intertarsal joints between cuneiforms and cuboid, and in the structure of the tarsometatarsal articulations. The lateral cuneiform is, in effect, impacted back into the navicular, so that it no longer projects distally beyond the intermediate one; the tarsometatarsal joint line is straightened out from its primitive alternating stepped conformation. The lateral cuneiform is accepted into an excavated hollow in the navicular and this articulation becomes the effective transverse tarsal continuation of the calcaneocuboid joint. The bevelled off proximal articular surface of the lateral cuneiform thus incorporates the proximal lateral intercuneiform facet, thus leaving only a single distal true intercuneiform articulation. Laterally, however, the lateral cuneiform has the typical dual facets for the cuboid. The dual articulations between the intermediate and medial cuneiforms are, at least in some specimens, confluent.

The straightened tarsometatarsal joint line ensures that the second metatarsal is no longer deeply indented into the tarsus and it is thus devoid

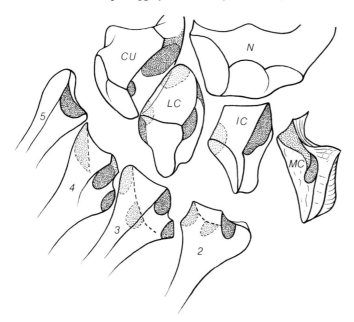

Fig. 13.22
An 'exploded' diagram of the distal tarsals and metatarsals of the right foot of *Pongo pygmaeus*. Facets are indicated by stippling but ligaments are not shown. N, navicular. Other labels are as in Fig. 13.20.

of articulation with the lateral cuneiform, although it retains the usual dorsal and ventral articulations with the third metatarsal. The primitive pattern of dorsal and ventral articulations between the third and fourth metatarsal persists, and there is only a sliver of participation of the lateral cuneiform in the dorsal one only. The cuneiform here shows a slight outcrop beyond the cuboid.

The functional meaning of this odd set of apomorphies in *Pongo* is readily seen. The orang-utan foot is habitually held in a position of marked inversion at the subtalar joint, such that the lamina pedis is virtually in a parasagittal plane, with the diminished hallux projecting from it almost at right angles, Pivotal capacity at the calcaneocuboid joint is enhanced, and in this movement the cuboid carries with it the lateral cuneiform and the third to fifth metatarsals, into a grasping attitude against the relatively fixed second metatarsal.

Among the Primates man possesses notably restructured tarsometatarsal joints (Fig. 13.23). Typically the distal articulations between the human intermediate and lateral cuneiforms, and between the lateral cuneiform and the cuboid are totally absent. This is due to considerable elaboration of the interosseous ligaments uniting the bones with fibres running predominantly upward,

distally and laterally. Another characteristically human derived condition is that the proximal and distal facets between the medial and intermediate cuneiforms are dorsally confluent encircling the interosseous ligament between the bones, and also continuous distally with the articulation between the medial cuneiform and the second metatarsal; in this feature man resembles *Didelphys*. However, as noted by Jones (1944) there may be interruption of the articulation into proximal and distal facets—a reversion to the primitive morphology; conversely, in *Pan troglodytes* the two facets may be very nearly confluent, closely approaching the human condition.

Function of the foot in arboreal Primates

Movements at the subtalar and transverse tarsal joint complexes are neatly integrated in the grasping feet of arboreal primates, and are particularly strikingly displayed in New World monkeys (Figs 13.5, 13.24), where the joint surfaces of the posterior talocalcaneal joint are markedly helical and a screw-like action accompanies movement about the compromise axis.

During terrestrial locomotion the foot is everted (Fig. 13.24A), and the helical motion retracts the calcaneus beneath the talus; conversely, one

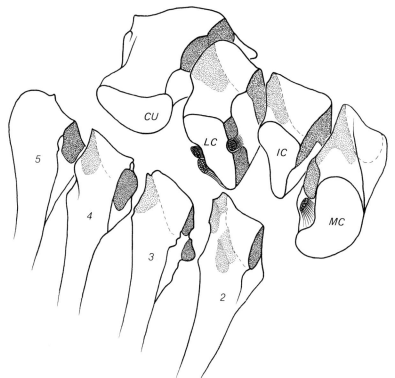

Fig. 13.23
An 'exploded' diagram of the distal tarsals and metatarsal bases of a right foot of *Homo sapiens* showing the intervening articulations and the tarsometatarsal ligaments. Labels are as in Fig. 13.20.

might consider that the talar head is firmly impacted into the acetabulum pedis, and so against the navicular. The retraction of the calcaneus thus tensions the plantar calcaneonavicular (spring) ligament enabling it to fulfil its role in supporting the talar head.

During arboreal activities the foot is characteristically dorsiflexed at the ankle, particularly when grasping near vertical supports. Because of the characteristic movement at the ankle joint the foot is abducted but the sole is also turned inwards. Congruent movement of inversion accompanies this at the subtalar joint complex.

The screw-like action at the posterior talocalcaneal joint, moving the lamina pedis forwards and into inversion, advances the calcaneus and increases its marginal contact with the navicular. The head of the talus rises up laterally from the acetabulum pedis, exposing a considerable part of its distal articular surface (Fig. 13.24B). A congruent movement of supination occurs at the midtarsal joint complex and as noted above this movement involves medial folding, or adduction of

the forefoot. This forefoot adduction is accommodated only because the helical movement at the subtalar complex has provided scope for it. As supination of the forefoot occurs the navicular rides up onto the exposed part of the talar head, the calcaneus is impacted against its articulation with the navicular, and the joints achieve a stable close-packed condition.

Szalay and Decker (1974) and Szalay *et al.* (1975) similarly had some insight into the correlation between helical action at the subtalar joint and posterior movement of the navicular during supination at the transverse tarsal joint complex. Szalay (1975), however, maintained that such helical movement was a derived apomorphic character unique to the Primates; as noted above there are persuasive indications that even the mammalian precursors of the Primates possessed at least a rudimentary facility of this sort although of course, having an alternating tarsus, they lacked the primate refinement whereby the calcaneus is protruded from under cover of the talus to contact the navicular. There is no foundation for

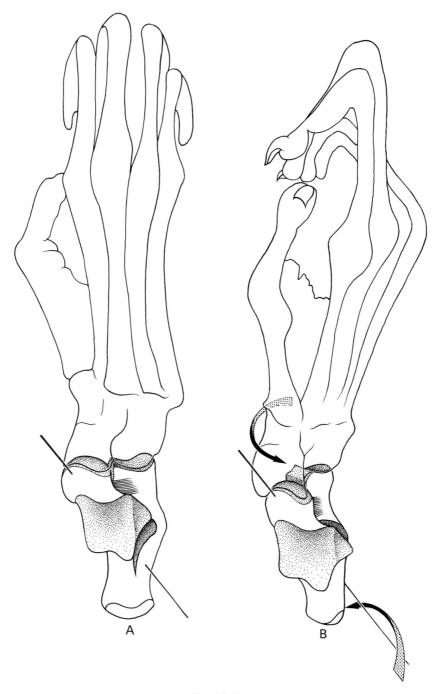

Fig. 13.24
Dorsal views of the right foot of *Pithecia monachus* in the everted position (A) and in the inverted grasping attitude (B), with the subtalar axis indicated in each case. The lower arrow in B indicates the screwing motion of the lamina pedis about the subtalar axis during inversion; the upper arrow indicates the accompanying movement of the forefoot during supination at the calcaneocuboid joint. (From Lewis 1980*b*.)

the view also expressed by Szalay (1975) that helical subtalar movement has been independently lost in most cercopithecoids and in hominids. Certainly it is considerably reduced in the more terrestrial cercopithecines, but it is quite well marked (and not of different nature) in colobines. The statement that it has been lost in hominids is particularly mystifying, since it was first most elegantly demonstrated by Manter (1941) in man.

The remodelling of the ape foot into a human form

It is a traditional view among physical anthropologists that the human foot was evolved from a grasping type not dissimilar to that of the extant African apes. The usual story depicting the evolutionary conversion of such an arboreally adapted foot to the human condition is said to involve rearrangements such as the following. The hallucial metatarsal has been adducted towards the functional axis of the foot (Morton 1935), which axis is essentially the line of progression shown in relation to the skeleton of the stance phase foot, and is not synonymous with the axis defining the interossei attachments; some change along these lines has already occurred in the massive terrestrial highland gorilla where the hallux lies closer to the functional axis than does the second metatarsal. Moreover, the hallucial metatarsal has undergone torsion so that its head no longer faces towards those of the other metatarsals but is directed downwards. The foot as a whole is said to have been everted so that the sole faces downwards rather than inwards (Gregory 1916). With such a model, however, the key first tarsometatarsal joint would be in an unstable, loosely-packed position. This joint, in fact, is most stable and attains its position of maximum congruence when the hallux is in the arboreal grasping posture: abducted, somewhat flexed, and pronated. When the basal joint of the hallux is screwed into this position, the first ray of the foot is effectively flexed and thus would be created a stable foot which moreover possessed a well-developed medial longitudinal arch. It seems that the attainment of such a habitual close-packed position of the hallucial tarsometatarsal joint was a central requirement in the elaboration of the arched human foot. This

accords with the fact that the human subtalar axis, although no longer very obliquely disposed towards the long axis of the foot is nevertheless still more or less in line with the hallux, just as in subhuman Primates; it is, however, markedly more vertical.

There is no doubt that some physical realignment of the human hallux itself towards the other digits has occurred but this change in orientation seems to have been brought about largely by a changed disposition of the joint surface of the medial cuneiform (Schultz 1930). However, it seems that divergence of the hallux has largely been masked by a realignment of the forefoot (Figs 13.25A, B, C) towards the stabilized hallux (and so towards the subtalar axis) rather than vice versa. In the lateral part of the forefoot this remodelling has been achieved by refashioning of the cuboid, and of its calcaneal articulation. In the feet of *Pan* and *Gorilla* the cuboid is distorted in such a way that its distal portion is deflected dorsally and laterally. In man the distortion is quite the reverse so that the distal aspect of the bone is bent plantarwards and medially carrying the lateral two metatarsals towards the already reorientated hallux. This means that these two bones form an arched complex bowed in the medial and plantar direction; in *Pan* and *Gorilla* the complex is arched in the opposite direction. Elftman and Manter (1935) perceived something of this when they suggested that the human transverse tarsal joint has become relatively fixed in a position of plantarflexion.

The lateral cuneiform is remodelled in a somewhat comparable way. Its distal portion, projecting beyond the other tarsals as in the African apes, is deflected medially and somewhat plantarwards. This is accompanied by an altered pattern of articulations between the intermediate and lateral cuneiforms and the cuboid, as described earlier in this chapter. The second and third metatarsals are further approximated to the first by an obvious angulation of their bases. Significantly, this angulation develops progressively during embryonic life (Cihak 1972) and for a time it is even exaggerated. The shafts of the lateral four metatarsals have undergone torsion, in the opposite direction to that of the hallux so that the plantar surfaces of their heads contact the substrate.

The hinder part of the lamina pedis has been

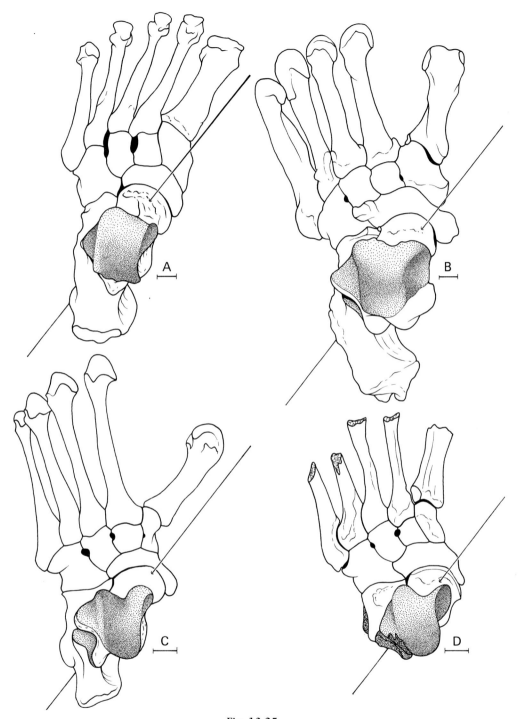

Fig. 13.25
Dioptograph tracings of the articulated tarsus and metatarsus of *Homo sapiens* (A); *Gorilla gorilla*, (B); *Pan troglodytes*, (C); and the casts of OH8 (D). In each case the subtalar axis is shown and the specimens are so orientated that these axes are parallel. The specimens are not all drawn to the same scale and in each case the bar represents 1.0 cm. (From Lewis 1980c.)

remodelled about the subtalar axis by what is effectively a bending laterally of the heel, altering the orientation of the subtalar articular surfaces, in relationship to the calcaneus as a whole (Fig. 13.12C) and thus the talus has been more completely superimposed upon the calcaneus. The remodelling of the functional anteroposterior axis of the foot to come closer to the line of the subtalar axis is completed by what is effectively a medial rotation of the trochlea of the talus in respect to the remainder of the bone, thus diminishing the talar neck angle; this characteristic of the human talus was previously noted by Elftman and Manter (1935).

These remodelling changes in the hindfoot affect the manner in which the flexor fibularis and flexor tibialis tendons enter the sole and they also change the constitution of the flexor retinaculum as will be described in Chapter 14 In the apes the posterior tubercle of the talus is located laterally in relation to the trochlea of the talus (Figs 13.25B, C) and the massive flexor fibularis tendon then runs below the talar body to reach a deep groove below the sustentaculum tali where it is flanked medially by another groove for the flexor tibialis tendon (Fig. 14.19A). In man the posterior tubercle of the talus is effectively medially relocated on the talus (Fig. 13.25A) as is the entry of the flexor fibularis tendon (the flexor hallucis longus of human anatomy). This tendon is not only redirected but it is relatively small because a large portion of the substance of flexor fibularis has descended into the widened tarsal canal to form the characteristically human medial head of flexor accessorius (Figs 14.19C, D).

The changed mode of entry of the human flexor tibialis and flexor fibularis tendons into the human foot was published some time ago (Lewis 1980c). Latimer et al. (1987) perceived something of the changed course of these tendons, at least in relation to the posterior and medial tubercles of the talus, but were unaware of the above explanation. Remarkably enough they attributed the oblique course in the apes to a varus attitude of the knee. It is mystifying how such a knee position, even if true (see Chapter 12) could affect the trajectory of a muscle arising below the knee.

So that the plantar surfaces of the metatarsal heads may contact the substrate, torsion has occurred in their shafts as noted by Elftman and

Manter (1935) and Morton (1935). In *Pan troglodytes* (Fig. 13.26, upper) there is some torsion of the first metatarsal in the sense that the head is rotated relatively to the base, so as to face towards the ground and in *Gorilla* this is even more obvious. Similar but exaggerated rotation is found in *Homo sapiens* (Fig. 13.26, middle). Since the direction of torsion in apes and man is similar this is not a particularly striking feature of contrast in the human foot. In the other metatarsals, however, it is. In *Pan* and *Gorilla* the second metatarsal has its head rotated to face the hallux whilst in *Homo sapiens* the comparable bone has its head rotated in the opposite direction. For the third metatarsal in *Pan* and *Gorilla* the torsion is in the same direction as that of the second, but of less marked degree. In *Homo sapiens*, the head of the third metatarsal is again twisted in the opposite direction to the apes so that its plantar surface is apposed to the ground. The heads of the fourth and fifth metatarsals of *Pan* and *Gorilla* show little torsion relative to the base but the arched form of the transverse tarsal arch means that nevertheless their plantar surfaces are directed somewhat towards the hallux. In the fourth and fifth metatarsals of *Homo sapiens* the direction of torsion is similar to that of the human second and third metatarsals but it is even more accentuated than in those bones. The external rotation of the heads of the human bones means that, despite the transverse tarsal arch, the weight-bearing plantar surfaces of the metatarsal heads are firmly applied to the ground.

Despite the remodelling of the human tarsus when the talus is removed (Fig. 13.12) the disposition of the navicular is quite similar in man to that in *Pan* and *Gorilla* and the form of the bone is also quite similar (Manners-Smith 1907) with a characteristically projecting tubercle which in apes may include the same supernumerary elements as it can in man (Chapter 14). In *Homo sapiens* the bone has a more squat form and its articular surface for the talus is less transversely elongated but its orientation is not markedly different from the apes. This would lead one to suspect that the attitude of the head of the talus—its torsion relative to the neck—would not differ very significantly in the three species. How then can this be reconciled with reports in the literature, indicating a considerably increased torsion in man when compared with the great apes? The solution to the

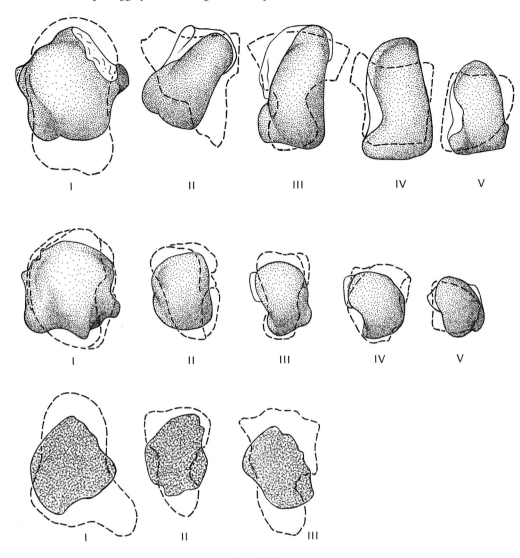

Fig. 13.26

Diagrams illustrating the direction and degree of torsion of the metatarsals of the left foot of *Pan troglodytes*, (upper); *Homo sapiens*, (middle); Olduvai Hominid 8 (lower); the numbers of the metatarsals are indicated in Roman numerals. For *Pan troglodytes* and *Homo sapiens* the heads of the metatarsals have been accurately superimposed by a photographic method on the outlines of the articular surfaces of the bases of the same bones (indicated by broken lines). For OH8 the broken surfaces of the shafts at the junction with the missing heads of the bones are similarly superimposed upon the bases, giving a reasonable indication of the torsion; this is only possible for metatarsals I, II and III. (From Lewis 1980c.)

discrepancy lies in the method by which the neck torsion angle has traditionally been measured: by taking the angle between the trochlear head plane and the median axis of the head. This traditional measurement is satisfactory for the anthropological purposes for which it was devised but is quite inappropriate for between-species comparisons. The trochlear head plane, entering into the ankle joint, varies with little direct relationship to changes in function of the intrinsic joints of the foot. Thus, in the great apes, particularly *Gorilla*, the medial margin of the trochlea is quite depressed whereas in man this region of the bone is elevated. A much more valid method of estimating the degree of talar neck torsion in the reality of its setting in the foot would be to relate it to the plane of the inferior surface of the talus, where the bone enters into the subtalar articulations. When the tali are thus orientated on their basal surfaces (Fig. 13.27) it can be seen that the neck torsion angles vary little between *Pan* and *Gorilla* and differ relatively slightly from *Homo sapiens*. The neck torsion angles have also been measured with the tali positioned with the subtalar axes parallel (Elftman and Manter 1935) but this introduces new sources of error.

Function in the human foot

Although direct visualization of joint movements in the human foot during walking has not so far been technologically possible, much data is available about the human gait from indirect methods, and a plausible reconstruction of the sequence of events affecting the joints can be deduced, based on a knowledge of the mechanics of the relevant joints.

It is well known that during the swing or recovery phase of walking the foot is carried in slight inversion so that at heel strike the postero-

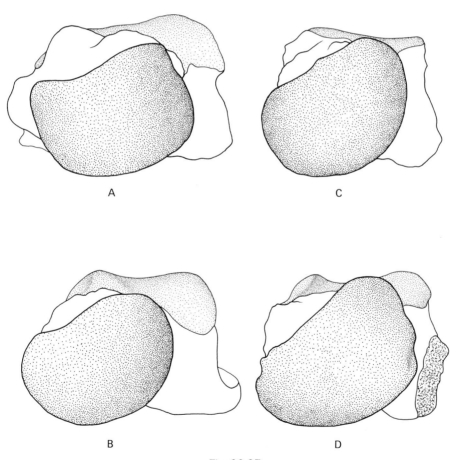

Fig. 13.27
Distal views of the tali of *Gorilla gorilla*, (A); *Pan troglodytes*, (B); *Homo sapiens*, (C); OH8 cast (D). In each case the bones are resting on the standard basal talar plane. (From Lewis 1980c.)

lateral part of the heel is first loaded and later also the medial part. Thereafter the outer margin of the sole, and then successively the metatarsal heads commencing with the most lateral and terminating with that of the great toe, bear the weight (Barnett 1956). During this stance or support phase the centre of gravity first moves laterally towards the support side and then deviates back towards the midline (Carlsöö 1972).

The motions of the leg bones, and of the foot as a whole during this weight-bearing support phase have been recorded (Wright *et al.* 1964). The first part of the stance phase extends from heel contact to foot flat and during this the ankle plantar flexes until the foot is flat on the ground. The foot now moves into a position of eversion ('pronation' in the jargon of some authors) and the tibia medially rotates. During the second interval of the stance phase the body moves forward over the weight-bearing leg, the ankle dorsiflexes and simultaneously the foot progressively inverts and the tibia externally rotates. The third interval starts as the heel begins to rise and terminates with toe-off; rapid plantar flexion of the ankle occurs while the foot continues to invert. Morris (1977) perceptively saw that a key part of this sequence of rotations was the way in which the oblique or 'mitred' hinge movement at the subtalar joint complex allows the leg to rotate on a fixed foot, internal rotation being accompanied by eversion, and external rotation by inversion. He saw this as a mechanism by which the foot in the early stance phase could 'absorb' the internal rotation of tibia and femur which then occurs, yet also 'absorb' the external rotation that occurs during the latter part of stance phase and push-off. It might have been more realistic if he had said that inversion and eversion really actuate the tibial external and internal rotation (rather than absorb it): Inman's (1976) concept of the subtalar joint as a 'directional torque transmitter'.

From what has been said above about the structure and function of the human joints it now seems possible to reconstruct the probable sequence of events in the foot during the stance phase. The two extreme attitudes of the foot are shown in Fig. 13.28. In Figure 13.28A the lamina pedis is maximally twisted by movement occurring at the calcaneocuboid joint; this consists of endorotation of the calcaneus, or pronation of the

forefoot, so that the medial arch is raised and the first ray is effectively depressed. This is compensated (Hicks 1953) by movement at the subtalar joint complex. The leg (with the talus) and the foot are effectively folded about the oblique hinge formed by the subtalar axis so that the lamina pedis is brought into an attitude of inversion; the movement also involves lateral rotation of the talus and so of the leg, upon the footplate. In Fig. 13.28B the lamina pedis is flattened, or untwisted, the major part of this movement occurring at the calcaneocuboid joint by exorotation of the calcaneus. The lamina pedis is now in close-packed position with the plantar calcaneocuboid (short plantar) ligament, plantar calcaneonavicular (spring) ligament, bifurcated ligament, and plantar aponeurosis all tensed.

At heel strike the foot is essentially in the attitude shown in Fig. 13.28A. As it enters the flat-foot phase it takes up the attitude shown in Fig. 13.28B. Weight is being transferred to this flattened footplate and femur and tibia are then medially rotating (Morris 1977). As the ankle is moved into dorsiflexion this is accompanied by some medial rotation at the ankle and what is effectively eversion of the foot (Fig. 12.8B). Congruent motion occurs at the subtalar joint complex bringing it into eversion, and the talus is screwed home into the acetabulum pedis. The lamina pedis is close-packed and the ligaments and plantar aponeurosis are tensioned.

In this close-packed position of the lamina pedis it has become increasingly apparent that the stretched ligaments store elastic strain energy (Wright and Rennels 1964; Ker *et al.* 1987). This energy is then returned as elastic recoil in the latter part of the stance phase. The foot again takes up the attitude shown in Fig. 13.28A. This spring of the arch of the foot has been estimated to act at an efficiency of 70 per cent—better than the best running shoes (McMahon 1987). The foot is then in an inverted position and the talus, and with it the leg bones, are externally rotated. This external rotation of the leg accommodates the movement of the centre of gravity over the implanted foot and moreover complements hip rotation in swinging the other foot forwards. In the latter half of the stance phase the twisted bunched up character of the lamina pedis, and its high arched form is even further accentuated by activity of the intrinsic

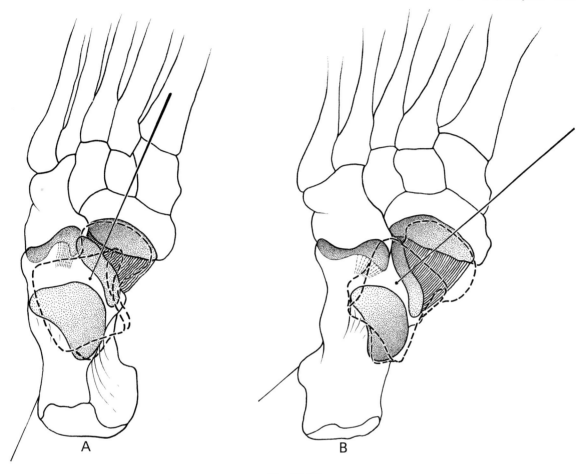

Fig. 13.28

Diagrams illustrating the joint movements in the human foot at the two extreme attitudes during the stance phase of gait. In A the foot is shown with the lamina pedis twisted at the midtarsal joint and inverted at the subtalar joint; in contrast at B the footplate is shown untwisted and everted at the subtalar joint. In each case the subtalar axis is shown and the talus is indicated in broken line superimposed on the lamina pedis. The plantar calcaneonavicular (spring) ligament is shown and the plantar calcaneocuboid (short plantar) ligament is indicated in broken line as though visualized through the foot; both ligaments are under tension in B. (From Lewis 1980*b*.)

muscles in the sole, which are then called into play (Mann and Inman 1964).

As the foot goes to 'toe-off' the twist of the lamina pedis and inversion are further increased. The triceps surae which is powering the plantar-flexion at this phase is also a powerful inverter of the foot (Mann and Inman 1964), but the propulsive tensing of the longitudinal arch of the foot is augmented by the windlass-like action of the plantar aponeurosis about the first metatarsopha-langeal joint which has been demonstrated by Hicks (1951, 1954). The living foot has been

photographed by Bojsen-Møller (1979) at this stage of the gait defined by him as 'high gear push off'—that occurring during the normal striding gait. As he has shown, the actual push off is about a transverse axis through the heads of the first and second metatarsal bones. There is a concentration of flexor muscle power on the glenoid plates of the first and second metatarsophalangeal joints which can be seen to be a complementary specialization. In the case of the first joint it is provided by the two-headed flexor hallucis brevis, augmented by its continuity behind with the powerful crural tibialis posterior (Chapter 14); in the case of the second joint it is a consequence of the remodelling of the interossei about an axis through the second digit (Fig. 16.3).

Bojsen-Møller has also described and photographed 'low gear push off' occurring from an oblique axis through the heads of the lateral four metatarsals. It is clear that such a push off, occurring during slow deliberate walking, occurs from an earlier part of the stance phase when the joints of the foot are orientated more as in Fig. 13.28A. It is clear then why the living foot as photographed during low gear push off shows the leg to be externally rotated and the foot inverted at the subtalar joint; the calcaneus is endorotated and the forefoot adducted and pronated at the transverse tarsal joint; the medial arch is high but the plantar aponeurosis shows no evidence of tension, since the windlass mechanism is much less effective about the smaller lateral metatarsal heads.

The significance of the remodelling of the joint surfaces in the human subtalar joint complex to produce a screwing action of the talus (Fig. 13.29), and so of leg and body about a comparatively upright axis, is now apparent. This increased obliquity of the subtalar axis of man when compared with the apes, compensates for diminished obliquity, and so of rotatory motion at the ankle joint (Fig. 12.8). The rotatory movement of the talus, carrying with it the supported body, is clearly a significant factor in the transfer of weight to the support side. The remodelling of the calcaneocuboid joint provides a mechanism for tensioning important ligaments so bringing the

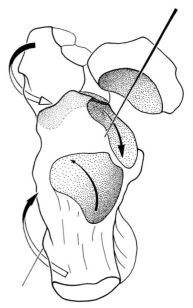

Fig. 13.29
A dioptograph tracing of the hinder part of the articulated foot of *Homo sapiens*, after removal of the talus to expose the articular surfaces of the subtalar joint complex whose compromise axis is indicated; the direction of the helical movements at the anterior and posterior joint surfaces are indicated by small arrows. The large upper and lower arrows indicate the helical directions of movement of cuboid and calcaneus during untwisting of the lamina pedis. (From Lewis 1981.)

lamina pedis into close-packed position, providing a stable platform for the pivotal movements of the talus, and providing important and energy-conserving spring to the step.

14

The extrinsic muscles of the foot—the long flexors

Theories of homology in mammals

The basic tetrapod substrate for the cruropedal muscles is very similarly arranged to that already described in Chapters 3 and 8 for the forelimb. In the leg McMurrich (1904) has identified the same basic three layers (with the middle one, plantaris profundus subdivided into a further three) and quite similar layers in the sole of the foot (McMurrich 1907) as those described for the hand. As in the forelimb the leg musculature primitively forms a pronator–flexor mass fanning down into the limb from the postaxial side; the extensor musculature in both limbs is similarly obliquely disposed and acts in an opposing supinator–extensor action. As Humphry (1872) put it, this in accordance with the 'feathering' action accompanying movements in the primitive limbs: during the propulsive phase of walking the foot is pressed into the ground in an internally rotated (adducted) position (Schaeffer 1941) by the pronator–flexor group.

With the emergence of the mammals this basic layered arrangement has been subdivided and reassembled rather differently from the forelimb. Attempts to find strict serial homologues are thus doomed to failure and have greatly confused the issue; for example, Taylor and Bonney (1905) attempted to identify a 'condylo-tibialis' in the hindlimb (homotype of the forelimb condyloradialis) when in fact none exists.

The cruropedal muscle group presents a bewildering spectrum of patterns among the Eutheria. The impression to be gained from the literature is that these present many instances of major evolutionary novelties, apparently discontinuous phylogenetic jumps which on the face of it appear to demand some sort of saltationist explanation. However, as will be seen, this exercise in myology bears out the contention of Mayr (1960) that the emergence of 'new' structures is due normally to the acquisition of a new function by modification of an existing structure. Thus, the apparently widely different mammalian patterns in the long cruropedal flexors can be seen as the result of minor variations of emphasis in derivation from a basic morphotype which possessed great potentialities for divergent modification.

This spectrum of mammalian muscular morphotypes has spawned a variety of theories concerning their phylogenetic history. Thus, Parsons (1894a, 1898b) considered that the plantaris was originally prolonged into the foot as the plantar fascia (aponeurosis) and was hence a primitive flexor of the proximal pedal phalanges, and that flexor brevis digitorum was the distal portion of soleus, divorced therefrom by the growth of the heel. Plantaris was, by this reasoning serially homologous with palmaris longus, while the soleus and flexor brevis digitorum corresponded to the forelimb superficial flexor (flexor digitorum superficialis). Thane (1894) entertained a similar opinion, and this is still current (for example, Last 1954).

As noted by Parsons (1898a,b) and by Windle and Parsons (1898, 1899, 1903), in various mammalian orders (Insectivora, Ungulata, Edentata, Hyracoidea, Carnivora) the plantaris continues into the sole and therein divides into (a) a superficial layer (plantar aponeurosis proper) and (b) a deep layer, fibrous or fleshy, which gives rise to the perforated tendons passing to the intermediate phalanges. This deep layer is the flexor digitorum brevis. Though Parsons (1898a,b) had recognized that this arrangement had arisen from the establishment of a secondary connection between flexor digitorum brevis and plantaris,

Windle and Parsons (1899) paradoxically identified this as a generalized mammalian arrangement. A similar assumption that this is the primitive mammalian disposition seems to have been the basis for the hypothesis advanced by Humphry (1872), Chauveau (1891) and Jones (1944) that the human flexor digitorum brevis is the divorced distal part of a flexor perforatus, whose proximal portion persists in the leg as the plantaris. The continuity between plantaris and flexor digitorum brevis noted by Jones (1949b) in those marsupials (Phascogalinae) which he regarded as the most primitive seems at first sight to be a vindication of this view. Jones (1944) states 'there is however no animal in which a flexor perforatus co-exists with a plantaris' in the leg: were such a co-existence demonstrable (and, in fact, it is), Jones' hypothesis would be untenable.

Keith (1948) maintained a third opinion. He agreed with Parsons that plantaris was primitively continuous with the plantar aponeurosis and was the homotype of palmaris longus. Regarding flexor digitorum brevis he merely stated that in lower Primates this muscle is confined to the sole and arises from the long flexor tendons. McMurrich (1907), discussing flexor digitorum brevis, also stated 'there are no grounds for assuming a descent of the muscle from the crural region. It is from the beginning an intrinsic muscle of the foot'—a statement which does not, however, accord with other of his observations (McMurrich 1904).

Glaesmer (1910) considered that the flexor digitorum brevis was primitively a composite structure composed of two heads, a superficial and a deep, an arrangement which is the typical primate form (Sawalischin 1911). His view combined features of both Jones's and Keith's hypotheses: he considered that the superficial head was derived by separation of a fleshy continuation of plantaris into the sole (cf. Jones), while the deep head was primitvely an intrinsic muscle of the foot arising from the surface of the long flexor tendons (cf. Keith). He suggested that soleus arose by a longitudinal splitting from the lateral head of gastrocnemius.

Testut (1884) and Wiedersheim (1895) considered that plantaris, together with the plantar aponeurosis, was the serial homologue of palmaris longus but, without providing any supporting evidence, they suggested that flexor digitorum brevis was the homotype of the flexor digitorum superficialis of the forelimb, and primitively a crural muscle which had later descended to the foot. This, in fact, was a perceptive hypothesis.

None of these views, except perhaps the last, stands up to critical analysis when the whole range of muscular morphotypes is considered. It is clear that no living eutherian mammal preserves the arrangements that must have existed in the stem form of the Eutheria. However, the primitive marsupial morphotype is readily determinable, and the whole gamut of specialized derivatives found among the metatherian radiation are paralleled in a remarkable way among the Eutheria where they must clearly be surviving end products derived from a comparable primitive pattern. The primitive marsupial pattern then is the key to understanding the hitherto obscure stem form for the Eutheria.

The primitive marsupial arrangement

The considerable literature on the cruropedal flexors of the marsupials has previously been reviewed in detail (Lewis 1962a) and this need not be repeated here. Most of these studies were monographic, done on a single species, or on a group of species. The overriding impression when they are looked at anew, with the benefit of hindsight, is how often they are marred by superficiality or gross inaccuracy—muscles are often incorrectly identified, whole muscles are overlooked, insertions are incorrectly given. This is despite the fact that some of the great names of comparative anatomy—Meckel, Owen, Cunningham, Parsons, Dobson—are represented here. This highlights the fact that purely descriptive studies are rarely accurate when they are performed without the framework of a hypothesis which would focus attention on particular details, and often restricted to inadequate material. Even trivial anatomical details, easily overlooked as in these monographs, may help to round out the phylogenetic picture. Much of this literature was reviewed by Leche (1900) and the confused state of knowledge is there clearly apparent: the impression to be gained is of scant interspecific affinity, of the absence of any common ground plan of hindlimb organization, and of little relationship between the marsupial anatomy and that of eutherian mammals.

The likely primitive stem pattern of this muscle group in marsupials (Fig. 14.1, 14.2) was deduced (Lewis 1964c) from the diverse arrangements in extant members of the Australian radiation. I have since noted that this arrangement is closely approximated in the living opossum, *Caluromys lanatus*. The picture is rounded out by the arrangements in the Phalangeridae of Australia which seems to have retained a primitive pattern, despite their pedal specializations consisting of reduction and syndactylism of the second and third digits.

In general the muscles radiate downward into the limb from the postaxial side. Lying most deeply is a muscular sheet, passing obliquely from fibula down to tibia: the rotator fibulae. The next plane consists of three muscles, arising primarily from the fibula and reaching the sole: flexor fibularis, tibialis posterior, and flexor tibialis. Most superficially lie gastrocnemius and plantaris. These three planes are clearly derived from the three strata identified by McMurrich (1904) in the primitive tetrapod limb: interosseous, plantaris profundus, and plantaris superficialis. Soleus is a progressive mammalian apomorphy whose derivation will be considered later. Also, sandwiched between the superficial and middle layers is a flexor perforatus, not unlike that (flexor digitorum superficialis) in the marsupial forelimb; its derivation will also be considered later.

Gastrocnemius (and soleus)

Gastrocnemius has two heads, a medial one arising directly from the femur above its medial condyle and a lateral one arising from the upper part of the bony parafibula (fabella) articulating with the head of the fibula, and having a ligamentous connection to the femur above its lateral condyle. This sizeable bone is an integral part of the marsupial skeleton, and was clearly present at the dawn of mammalian history. It has been considered as a complex of cyamella (the sesamoid in the popliteus tendon) and the lateral fabella (the sesamoid in the lateral head of gastrocnemius); this misleading notion will be considered later.

The presence of the medial head obscures the

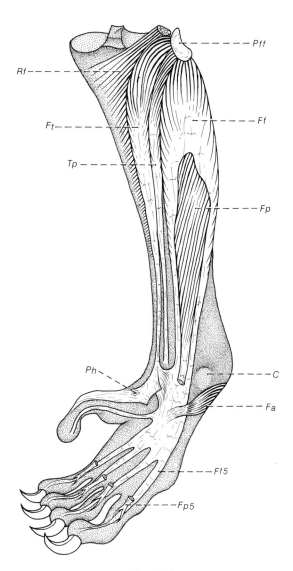

Fig. 14.1

The deeper layers of the musculature of the primitive marsupial right leg and foot. Only those features of the intrinsic foot musculature relevant to the present discussion are shown in this and subsequent figures. Except for their terminations the flexor perforatus tendons in the foot have been removed, and the connection between the flexor tibialis tendon and the plantar aponeurosis is cut. *Pff*, parafibular facet on fibula; *Ff*, flexor fibularis; *Fp*, flexor perforatus; *C*, calcaneus; *Fa*, flexor accessorius; *Ff5*, deep flexor tendon to digit 5; *Fp5*, flexor perforatus tendon to digit 5; *Ph*, prehallux; *Tp*, tibialis posterior; *Ft*, flexor tibialis; *Rf*, rotator fibulae.

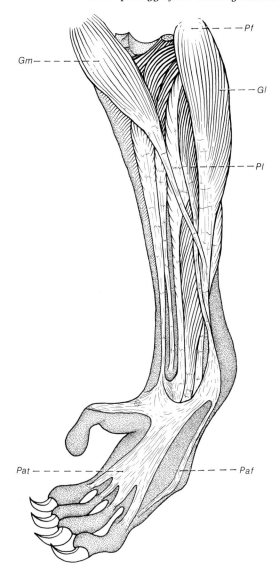

Fig. 14.2
The complete musculature of the primitive marsupial leg and foot. The deeper layers, shown in Fig. 14.1, are not labelled. *Pf*, parafibula; *Gl*, lateral head of gastrocnemius; *Pl*, plantaris; *Paf*, fibular part of plantar aponeurosis; *Pat*, tibial part of plantar aponeurosis; *Gm*, medial head of gastrocnemius.

general obliquity of the pronator–flexor mass, which fans into the limb from the postaxial side, and contrasts with the situation in the forelimb. In general in marsupials this head is smaller than the lateral one and makes a less significant contribution to the tendo Achillis, reflecting the fact that it

is probably a secondary accession to the superficial layer (Humphry 1872): no muscle is present in this situation in Amphibia and its precursor first appears in reptiles as a thin muscular sheet.

These marsupials already show the characteristic twist of the tendo Achillis, as described by Parsons (1894a), which was to become a hallmark of the mammals in general: the medial head crosses the lateral head superficially to insert more laterally on the calcaneus. In the adult some continuity is shown between the tendon and the long plantar ligament but in fetal specimens there is complete continuity prolonging the insertion forward to the cuboid and fibular-sided metatarsals. Even in the human embryo the tendo Achillis continues into the sole (Jager and Moll 1951) and this is likely to have been the primitive insertion. It has been suggested (Barnett and Lewis 1958) that it acquired its new insertion to the calcaneus by first developing a sesamoid thereabouts, which subsequently fused to form a part of the calcaneal epiphysis; the presence of the calcaneal epiphysis is thus explained, but the idea remains mere supposition.

In these primitive marsupials a third component of the muscle is distinguishable lying deep to the lateral head, with which it is somewhat united. Typically (Fig. 14.3A) it arises from the parafibula by a tendinous origin, which may be separated from the rest of the more superficial fleshy fibres by a branch of the posterior tibial nerve. Below it merges with the remainder of the lateral head. Its subsequent history proclaims it as an emergent soleus.

Plantaris

The muscle belly arises from the parafibula under cover of the lateral head of gastrocnemius. The course of its tendon, emerging low down in the leg from the medial side of the tendo Achillis, reflects the twist of the latter. The tendon enters the sole over the posterior aspect of the calcaneus and there expands to form the plantar aponeurosis. This aponeurotic sheet consists of a large tibial portion passing to the region of the second to fifth metatarsophalangeal joints, acting as a flexor of the proximal phalanges, and a slender fibular part, presenting a fibrous cap sliding over the tuberosity of the fifth metatarsal, and continuing thence to

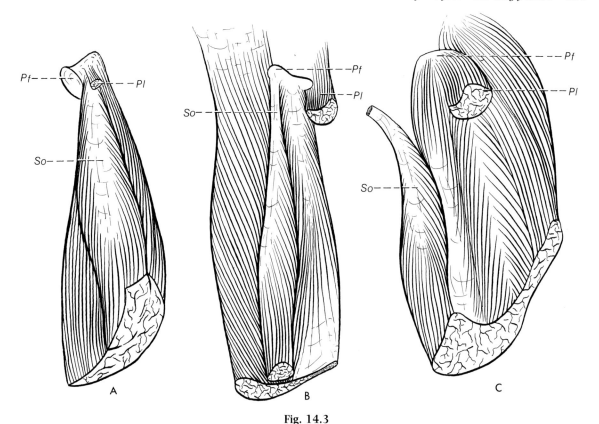

Fig. 14.3
The emergence of soleus from the deep surface of the lateral head of gastrocnemius in *Pseudochirus laniginosus* (A), *Macropus major* (B), and *Perameles gunnii* (C). *So*, soleus; other labelling as in Fig. 14.1.

the region of the fifth metatarsophalangeal joint. The aponeurosis is laterally continuous with an expansion from the flexor tibialis tendon into the hallux.

Flexor tibialis

This muscle is derived from the intermediate stratum (Plantaris profundus) in the primitive tetrapod limb. It arises from the uppermost part of the tibial border of the fibula, just below the parafibular facet, with additional fibres secondarily taking origin from the aponeurotic posterior surface of rotator fibulae. The tendon enters the foot behind the medial malleolus and has a complex insertion which undoubtedly represents the primitive therian morphology, for it provides the ideal pattern from which the various arrangements found in other Metatheria, and in the Eutheria for that matter, could easily have been

derived. The insertion assembly is tripartite. The direct continuation of the tendon passes as a fascial sheet into the hallux, crossing the insertion of tibialis anterior to the medial cuneiform, and investing the tendon of extensor hallucis longus; this tendinous sheet contains an ossicle, the prehallux, which articulates with the medial cuneiform (Figs 14.1, 14.4A). The tendon also sends a deeper prolongation which joins the flexor fibularis tendon, and another more superficially to merge with the plantar aponeurosis.

Flexor fibularis and flexor accessorius

Flexor fibularis is also a derivative of the reptilian Plantaris profundus. It arises from a considerable part of the shaft of the fibula and enters the sole as a common tendon plate which divides into deep digital tendons inserting on the distal phalanges of all five digits. The tendons are similarly derived to

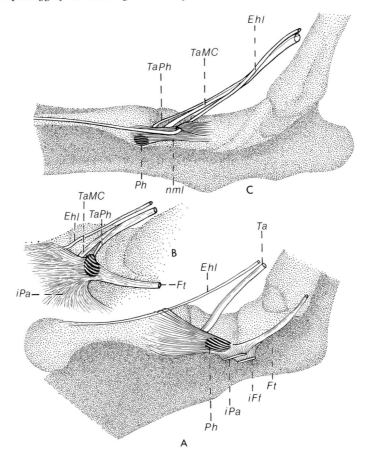

Fig. 14.4

A, the sole of the right foot in *Didelphys marsupialis* showing the triple insertion of the tendon of flexor tibialis (*Ft*): deeply to the flexor fibularis tendon (*iFf*); to the plantar aponeurosis (*iPa*); a hallucial continuation containing the prehallux (*Ph*) and covering the single cuneiform insertion of tibial anterior (*Ta*) and the tendon of extensor hallucis longus (*Ehl*). B, the modified arrangement in *Trichosurus vulpecula*, with loss of the deep insertion of flexor tibialis on flexor fibularis, and splitting of the tendon of tibialis anterior, one slip attaching to the prehallux (*TaPh*) and one (*TaMC*) retaining an insertion to the medial cuneiform; other labelling as in Fig. 14.4A. C, the right foot of *Pithecia monachus* showing the naviculo-metatarsal ligament (*nml*) bridging over the two tendons of insertion of tibialis anterior and which is pierced by the tendon of extensor hallucis longus; labelling as in Figs 14.4A, B. (A and C from Lewis 1972c.)

those in the forelimb, from the deep layer of the reptilian plantar aponeurosis (McMurrich 1907), and show the same double helical arrangement of their fibres.

The flexor accessorius is a single-headed muscle arising from the under surface of the trochlear process of the calcaneus, and it inserts by a flat tendon onto the surface of the flexor fibularis tendon at about the level where the hallucial tendon diverges. The muscle was probable derived,

as suggested by Humphry (1872), from the lowest fibres of the pronator–flexor mass, trapped within the foot, and separated from the remainder by the projecting heel.

Flexor perforatus

This muscle is the counterpart of the flexor digitorum superficialis in the forelimb. It is a crural muscle (Fig. 14.1) arising from the aponeurotic surface of flexor fibularis and terminating as perforated flexor tendons, quite similar to those of its forelimb homotype, on the middle phalanges of the four postaxial digits. There is little doubt that these tendons were derived, like those in the palm, from the slips of the flexor brevis superficialis in the reptilian foot. (McMurrich 1907).

The crural belly is presumably derived from the superficial muscle stratum, but unlike the situation in the forelimb embryological confirmation of this is lacking. However, the crural belly does not undergo the progressive evolutionary elaboration shown by the forelimb flexor digitorum superficialis, and so the embryological history is not so significant.

In their passage down the digits the perforated and perforating tendons lie within fibrous flexor sheaths lined by synovial membrane and which include specialized retaining loops or pulleys. These are simpler in arrangement than those in the forelimb, and consist of a very strong loop opposite the more distal part of the proximal phalanx and a less strong one on the middle phalanx; there is some thickening of the fibrous sheath opposite the glenoid plate (with its sesamoids) and the base of the proximal phalanx.

Tibialis posterior

This muscle arises from the upper part of the fibular shaft where its belly lies between those of flexor fibularis and flexor tibialis. Indeed it is derived from the same reptilian stratum as those muscles—plantaris profundus (McMurrich 1904). Its tendon enters the sole behind the tibial malleolus, deep to that of flexor tibialis, and inserts into the navicular tuberosity. This insertion represents the tibiale of the primitive tetrapod foot, and it is noteworthy that the amphibian tibiale already receives the insertion of the amphibian precursor of the muscle, the plantaris profundus I (McMurrich 1904).

Rotator fibulae (and popliteus)

This muscle is derived from the deepest interosseous stratum in amphibia and reptiles. Its fibres pass obliquely inferomedially from fibula to tibia, occupying virtually the whole length of the leg. Its highest fibres, however, arise from an intra-articular disc situated between the articulating fibula and femur.

This disc is a standard part of the locomotor equipment in marsupials. The disc initially appears in reptiles as an appendage of the lateral meniscus of the knee joint and was described in a variety of reptiles and marsupials by Furst (1903). In some of the details of attachments, Furst's lengthy descriptions are somewhat suspect, but these descriptions have been refined for a number of reptiles, and for the marsupial *Didelphys* by Haines (1942). It was Haines, and later Barnett and Napier (1953), who pointed out that the disc facilitates rotatory motion between femur and fibula, producing movements analogous in result to pronation–supination in the forelimb, although the mechanical solution in the two limbs differs in the way in which particular joints are pressed into service. But it was Furst who first had the inspired notion that the femorofibular disc (inappropriately called by him a 'meniscus') had provided the basis for the conversion of the upper part of the rotator fibulae into a typically eutherian popliteus, complete with its oddly located intracapsular tendon of origin from the femur.

The generalized marsupial arrangements are well shown in the opossum *Caluromys* or the Australian 'possums' *Trichosurus* or *Pseudochirus* (Fig. 14.5A). The whole fibrocartilaginous structure intervenes between the articular surfaces on the fibula and the lateral femoral condyle. Anteriorly it is firmly attached to the lateral meniscus, not surprisingly in view of its reptilian derivation; from this situation a little additional meniscus projects down into the superior tibiofibular joint. Laterally the disc is firmly tethered to the head of the fibula, and anteriorly it is prolonged as a stout ligament attaching to the femur just below and behind the lateral collateral ligament of the knee joint. Behind it gives rise to a somewhat discrete upper lamina of the rotator fibulae, which in view of its subsequent history to be detailed later, merits separate distinction as a popliteus muscle.

Metatherian and eutherian modifications

This basic marsupial morphology shows a whole gamut of specializations among the remarkable

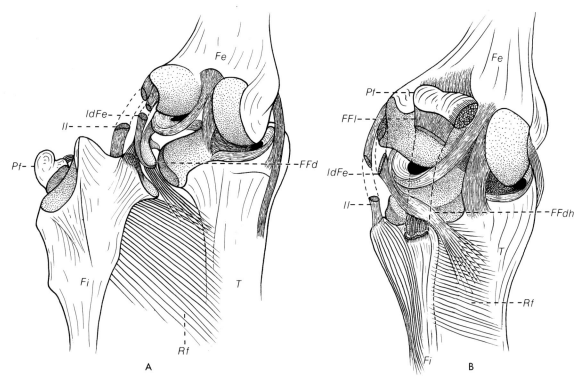

Fig. 14.5

Right knee joints of *Pseudochirus* (A) and *Wallabia* (B) with the bones sprung apart laterally to reveal the internal structures. *Fe*, femur; *Fi*, fibula; *T*, tibia; *Pf*, parafibula; *ll*, lateral collateral ligament of knee joint; *FFd*, femorofibular disc; *ldFe*, ligament from femorofibular disc to femur; *FFl*, fabellofibular ligament; *FFdh*, homologue of femorofibular disc in *Wallabia*; *Rf*, rotator fibulae.

Australian marsupial radiation producing end products which parallel in a remarkable way the various patterns of the cruropedal flexor musculature which are found in extant Eutheria.

Gastrocnemius and soleus

In marsupials the medial head of gastrocnemius is quite conservative and shows little significant variation. This is not so with the lateral head. As a variant in some specimens of *Vombatus* and *Sarcophilus* the parafibula may be fused with the fibula, thus creating a fibular crest, which then gives rise to the lateral head. More significantly, in the kangaroos and wallabies, the parafibula has been translocated proximally across the joint line of the knee and become firmly tethered by a short syndesmosis to the femur above its lateral condyle. It has become a fabella (or more accurately a

lateral fabella) and here gives rise to the lateral head of gastrocnemius (Fig. 14.5B). In drifting proximally it has drawn out behind it a tough fabello–fibular ligament, connecting it to its old site of articulation, the apex of the fibula. This strikingly parallels the change which has also characterized the Primates. Moreover, in the Primates a similarly located medial fabella, giving rise to the medial head of gastrocnemius has appeared, apparently *de novo*, with no historical skeletal antecedent. These two fabellae are found as a rule in most prosimians and monkeys (Vallois 1914), fairly common in *Hylobates*, but rarely in the great apes. In man the medial fabella is very rare, but the lateral one occurs in 13–21 per cent of cases (Weinberg 1929). Even in the absence of the lateral fabella, the fabello–fibular ligament is represented by the short lateral ligament of the knee joint (Kaplan 1961). With regression, or suppression, of the fabella the two heads of gastrocnemius take their origin directly from the femur; in man at least the medial head raises a prominent supracondyloid tubercle on the femur (Stopford 1914).

The derivation of soleus as a delamination from the deep surface of the lateral head of gastrocnemius is clearly seen in marsupials (Fig. 14.3). In *Pseudochirus* this separation is incomplete, but in, for example, *Macropus* it has become virtually a separate muscle, though still retaining a fabella origin. In *Perameles*, as also described by Glaesmer (1910), the origin has shifted by secondary attachment to the fibular head, and it there presents the typical appearance of a mammalian soleus; Coues (1872) has recorded an apparently similar fasciculus in *Didelphys virginiana*. Such a soleus is present in many of the Rodentia (Parsons 1894b), most Carnivora (Windle and Parsons 1898), *Gymnura rafflesi* (Parsons 1898a), many of the Ungulata (Windle and Parsons 1903) and most Primates (Hepburn 1892; Osman Hill 1953). There can be little doubt, therefore, that the deep fasciculus of the lateral head of gastrocnemius, seen in many Metatheria, really represents soleus in its most primitive emergent form. Doubtless the transfer of origin to the fibula has been independently acquired in the Peramelidae on the one hand and in the majority of the Eutheria on the other hand; indeed, some of the Ungulata (Windle and Parsons 1903) retain the more primitive arrangement, the soleus being closely associated with the lateral head of gastrocnemius and arising therewith from the femur. Also in the orang (Hepburn 1892) and in *Hylobates*, although arising from the fibula, the soleus is intimately associated with the lateral head of gastrocnemius. The muscle has expanded its origin from the fibula towards the tibia in the Hominoidea. This is heralded by a trivial tibial origin in the chimpanzee (Hepburn 1892) and the gorilla (Raven 1950) and is maximally achieved in man, where it has acquired a fibrous arch bridging the interval between fibula and tibia, giving it a spurious resemblance to the flexor digitorum superficialis of the forelimb. It is this highly derived appearance which was the source of the clearly erroneous, but still current, scheme of homologies which considers that soleus is the serial homologue of flexor digitorum superficialis of the forelimb and that the

flexor digitorum brevis of human anatomy is its distal part, isolated by the growth of the heel.

The development of the muscle in man confirms this phylogenetic interpretation. Bardeen (1906) and Jager and Moll (1951) have shown that the human soleus emerges from an anlage common to it and the lateral head of gastrocnemius, and that later it gradually extends its origin from fibula onto tibia.

Plantaris

In Metatheria this muscle arises from the parafibula and, thus, indirectly from the femur which, with reduction of the fabella, becomes its descriptive origin in other mammals. In its primitive form the tendon passes over the projecting calcaneus (sometimes presenting a sesamoid here) to become continuous with the plantar aponeurosis; this occurs in the Metatheria and in many Eutheria, e.g. *Gymnura rafflesi* (Parsons 1898a), most Carnivora (Windle and Parsons 1898), Edentata (Windle and Parsons 1899), Ungulata (Windle and Parsons 1903), and *Ptilocercus* (Clark 1926). Even in some Primates (Loth 1908) this continuity is retained. In other Primates, however, and in *Hyaena crocuta* (Windle and Parsons 1898), the continuity is interrupted at the heel, perhaps by fusion of the sesamoid (Barnett and Lewis 1958) which contributes to the calcaneal epiphysis; in such cases the muscle belly in the leg may disappear (Hepburn 1892; Raven 1950; Osman Hill 1955). In some marsupials (*Trichosurus, Vombatus, Phascolarctos*) the aponeurosis in the sole contains a large plantar sesamoid (Fig. 14.11) which articulates with the navicular tubercle and lies superficial to the long flexor tendons as they enter the sole. The presence of such a sesamoid in the plantar aponeurosis appears to have no particular morphological significance and a similar ossicle is found in some rodents (Parsons 1894b). *Sarcophilus harrisi* shows what is apparently a primitive mammalian arrangement of the plantar aponeurosis (Fig. 14.12)—a large tibial portion passing to the second, third, and fourth metatarso-phalangeal joints and a slender fibular part gliding over the projecting basal tuberosity of the fifth metatarsal and then continuing down the sole. Such a fundamentally primitive arrangement persists as the basic Primate type, as described by

Keith (1894*a*) and by Loth (1908), although in various families of this order the arrangement is often modified: the fibular part, for instance, frequently gains a secondary attachment to the fifth metatarsal tuberosity. However, contrary to the usual descriptions, even in man, as Loth showed, the primitive arrangement is frequently retained. The fibular part (the 'lateral cord' of the plantar aponeurosis) may co-exist with an abductor ossis metatarsi digiti quinti muscle in man and many other Primates (Clark 1926; Straus 1930; Raven 1950), just as it does in *Sarcophilus* (Figs 14.12, 14.13). The lateral cord, then, does not represent a degenerate form of this muscle, as is sometimes maintained, but is truly a part of the plantar aponeurosis.

The association of flexor tibialis and flexor perforatus with the plantar aponeurosis, found in many mammals, is considered in the sections dealing with these muscles.

Flexor tibialis

The primary origin of this muscle is from the fibula but it has a tendency to expand its origin onto the aponeurotic surface of rotator fibulae. With regression of the latter muscle the flexor tibialis tends to take root from the exposed tibia obscuring the primitive arrangement, whereby the muscles fan into the limb from the postaxial side. This is maximally achieved in man where the tibial origin has become the predominant one, and often the only one described in textbooks. There is, however, often a strong indication of the ancient origin in the form of a tough aponeurotic attachment to the upper part of the fibula, which has been described by Poirier and Charpy (1912), Jones (1944) and Last (1954). Confirmation of this phylogenetic interpretation is afforded by human ontogeny which shows the tibial attachment to be a late development. Thus, in man tibialis posterior has come to be buried beneath the flexor tibialis (flexor digitorum longus) and the flexor fibularis (flexor hallucis longus). These three human muscles in turn lie under cover of the greatly expanded soleus with its secondarily acquired tibial origin. The human calf, then, has acquired a uniquely specialized arrangement which at first sight obscures its affinity with that of other mammals.

There is little doubt that the primitive termina-

tion of the flexor tibialis tendon was a tripartite one. The main prolongation of the tendon continues down the medial margin of the foot, superficial to the tendon of tibialis anterior and enters the hallux as a fascial expansion which wraps over the tendon of extensor hallucis longus (Figs 14.2, 14.4A). This tendinous sheet contains within its substance the bony prehallux articulating with the medial aspect of the medial cuneiform. Offshoots leave the tendon more proximally, one passing deeply to join the flexor fibularis tendon and one fanning out superficially to join the plantar aponeurosis.

This basic pattern is modified in other marsupials. In *Thylacinus* and *Didelphys crassicaudata* (Glaesmer 1910) the flexor tibialis tendon is joined only to that of flexor fibularis in the sole. In most Australian marsupials (Lewis 1962*a*) and in *Didelphys cancrivora* (Glaesmer 1908) the connection to the flexor fibularis tendon is lost, leaving only the attachment to the plantar aponeurosis and the hallucial continuation, the latter containing the prehallux. Both these derivatives of the stem form are paralleled in Eutheria. In some rodents (Parsons 1894*b*) flexor tibialis does not join flexor fibularis but presents a small prehallux, as in Australian marsupials, with similar attachments; a comparable arrangement is found in some Insectivora and Edentata (Dobson 1883). In most Eutheria, however (Windle and Parsons 1898; Dobson 1883), flexor tibialis is entirely joined to flexor fibularis in the sole, (as in *Thylacinus*), and from the common tendinous mass arise the deep flexor tendons for the five digits. The remainder of the more superficial pedal portion of the flexor tibialis, divorced from the tendon, tends to become secondarily adherent to the navicular; this portion contains the prehallux and sends a prolongation down the hallux and one to join the plantar aponeurosis; such an arrangement is seen in *Tupaia* (Fig. 14.6). From such an arrangement would also seem to be derived the morphology seen in certain prosimians e.g. *Perodicticus* or *Lemur* (Fig. 14.7). Here the prehallux (confusingly identified as a supernumerary tarsal element, the 'os ventrale' by Osman Hill 1953) is displaced rather ventrally yet it still retains attachments to the naviculometatarsal remnant of the flexor tibialis insertion and to the plantar aponeurosis. In *Lemur* the bone articulates with the medial cuneiform but

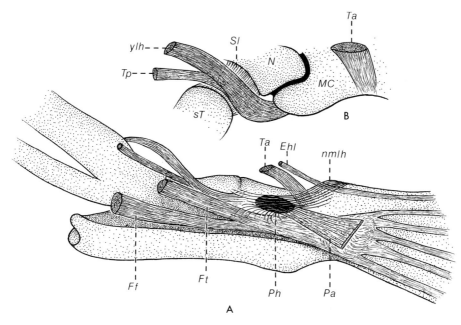

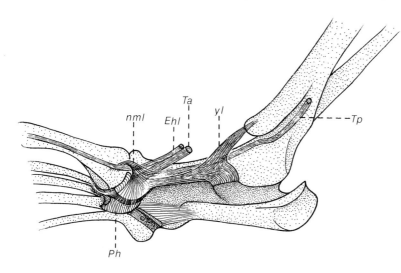

Fig. 14.6

A, the sole of the left foot of *Tupaia sp.* showing a prehallux embedded in the homologue of the primate naviculo-metatarsal ligament (*nmlh*), but which retains a connection to the plantar aponeurosis. B, a detail showing the conjoined insertion of tibialis posterior and the homologue (*ylh*) of the primate Y-shaped ligament (in this case a tibial–cuneiform band attaching to a projection of the medial cuneiform). *MC*, medial cuneiform; *N*, navicular; *sT*, sustentaculum tali; *Sl*, spring ligament; other labelling as in Figs 14.1, 14.4.

does not participate in the first tarsometatarsal joint; here, however, it is grooved to form a pulley for the long flexor tendon of the hallux.

In the higher Primates the residual superficial part of the flexor tibialis insertion assembly persists as a well-marked naviculometatarsal ligament,

which doubtless primitively contained the prehallux. In fact, in the rat a very similar prehallux-containing naviculometatarsal ligament is also found, severed from the remainder of the flexor tibialis tendon, which is left with its sole attachment to the plantar aponeurosis. The prehallux in

Fig. 14.7

The sole of the right foot of *Lemur catta* showing the Y-shaped ligament (*yl*) and the naviculo-metatarsal ligament (*nml*) attaching to the prehallux (*Ph*). Other labelling is as in Figs 14.1, 14.4.

the ancestral higher Primates then seemingly gained an attachment to the underlying tibialis anterior tendon. That this is not merely fanciful, is shown by the condition in *Trichosurus vulpecula* (Fig. 14.4B) where just such a thing has happened, and a subdivided tendon of tibialis anterior is peeled off from the main primitive insertion to the medial cuneiform. Two unique specializations must have then occurred in the emergent Anthropoidea: the contained prehallux was displaced forward to contact the first metatarsal and participate in the joint at the base of that bone; secondly, the extensor hallucis longus tendon, already traversing the most superficial part of the ligament in *Didelphys marsupialis*, is drawn back into very close association with that of tibialis anterior, and thence is redirected on its course down the length of the divergent hallux (Fig. 14.4C). This pulley-like disposition of the ligament, associating the extensor hallucis longus with tibialis anterior, is found throughout the order, with the exception of man.

As noted in Chapter 13, the Anthropoidea invariably possess a tibialis anterior with two tendons of insertion (and are apparently unique in this regard, with the exception of *Trichosurus*). One attaches, conservatively, to the medial cuneiform; the other inserts into the prehallux when present, and in its absence merges with the naviculometatarsal ligament to reach an attachment to the base of the first metatarsal. It has been suggested (Chapter 13) that the latter arrangement is secondarily derived from the former, the prehallux having either disappeared or become incorporated with the epiphysis of the first metatarsal.

Pfitzner (1896) has described what is clearly the prehallux in a number of mammals under the name of 'Paracuneiforme' or 'Präcuneiforme' but he makes no mention of it in Primates including man. It has, however, been described in man by Brailsford (1953) and by Morrison (1953) and named by them either 'prehallux' or 'os paracuneiforme'. Significantly these atavistic human bones occupy the primitive mammalian position, adjacent to the medial cuneiform, and are not incorporated in the hallucial tarsometatarsal joint, which represents the derived higher primate state.

Everything seems to point to the fact that the above interpretation of morphocline polarity is correct, with a tripartite flexor tibialis insertion as

primitive, and leading to the various derivative patterns, which could easily result by mechanisms which are well established. The alternative idea, that the usual eutherian arrangement with fused flexor tibialis and flexor fibularis tendons was primitive (Dobson 1883), bristles with difficulties.

Flexor fibularis and flexor accessorius

The term flexor tibialis and flexor fibularis are adopted here for those muscles whose crural portions correspond to the flexor digitorum longus and flexor hallucis longus of human anatomy. The former terms, as observed by Dobson (1883) and by Windle and Parsons (1897a), are preferable in view of the varied terminations of these muscles in any comparative series. In different higher primate species the digital flexor tendons, with which flexor accessorius is associated, are derived in varying fashion from these two muscles. Following Keith (1894b), it is usually stated that in the primitive prosimian condition each muscle supplied a tendon to each digit, with varying definitive muscular patterns arising thence by loss of one or other member of these pairs of tendons, in an apparently haphazard manner. When presented in tabular or diagrammatic form, as it usually is, this notion may seem convincing, but in fact it conveys an entirely false idea of the true course of evolutionary change.

A much more realistic view of events, originally suggested by Glaesmer (1910), is that in the primitive primate condition both flexor tibialis and flexor fibularis entered into a twisted common tendon plate and the different anthropoid arrangements result from variable longitudinal splitting, occurring both ontogenetically and phylogenetically. Indeed a fused condition of the tendons is the rule in prosimians (Murie and Mivart 1872; Osman Hill 1955) and the contributions of flexor tibialis or flexor fibularis can only be ascertained by shredding this common tendon. In the treeshrews the common mass is even ossified into a bony sesamoid plaque (Clark 1924, 1926).

The architecture of the common tendon plate largely determines how it segregates into individual tendons. The common tendon shows a twisted fibre arrangement, reminiscent of the pattern seen in individual deep digital flexor tendons (Martin 1958) of the hand or foot: the most superficial fibres diverge towards the marginal tendons arising from the distal end of the plate, leaving the deeper fibres to come to the surface and largely form the more central tendons. Some of the most proximal of the superficial fibres spiral onto the deep surface of the plate from whence they then contribute to the central tendons. Primitively associated with such a tendon plate is the flexor accessorius muscle (also known as 'quadratus plantae' or 'caro quadratus'). This muscle is commonly present in mammals and has been described in at least some members of the orders Monotremata, Marsupialia, Edentata, Rodentia, Carnivora, Tubulidentata and Primates. Among the Primates it is lacking in all Prosimii and some specimens of *Tarsius spectrum*, but it is present in the Cercopithecoidea and most Ceboidea, although it is frequently absent in the Pongidae.

In all these mammals it is a simple, single-headed muscle slip with its origin laterally on the calcaneus from the under surface of the trochlear process. With a single exception, mentioned below, a complex two-headed muscle is unique to man and this contrasts markedly with the single-bellied muscle of other mammals, so much so that some authors have failed to identify the latter as the homologue of even part of the human muscle.

It is generally assumed that Humphry (1872) was correct in stating that the muscle represents the distal part of the obliquely disposed crural pronato/flexor muscle mass of the pre-mammalian vertebrate limb, severed from the remainder by the projecting heel. Keith (1948) and Winckler and Giacomo (1955), however, considered the muscle to be a part of flexor hallucis longus which had migrated to the sole, thus implying that it had emerged as a separate entity after the initial establishment of typical mammalian segmentation in the limb musculature. The phylogenetic progress of the muscle from a simple single-headed slip to its human complexity has been far from clear. Winckler and Giacomo (1955), however, considered that the anomalous flexor accessorius longus of human anatomy, the medial head, and the lateral head represented successive stages in the descent of part of flexor hallucis longus to the sole. Acceptance of this view implies that man alone retains the more primitive medial head. Jones (1944) held such a view and used this as 'evidence' showing man to be more primitive than the

monkeys or the anthropoid apes. On the other hand, Weidenreich (1922) considered that each of the heads of the human muscle represented a different morphological element, the lateral head being the homologue of the quadratus plantae of other mammals, and the medial head a new development, derived from that deep head of flexor digitorum brevis present in many mammals and arising from the surface of the long flexor tendons.

It is probable that in the primitive marsupial (and therian) condition the muscle was disposed as in Fig. 14.1: a single muscle belly arising from the under surface of the trochlear process with a tendon merging onto the surface of the common flexor tendon (here essentially that of flexor fibularis), holding the common tendon laterally and thus preventing bowstringing, and approximately in line with the divergent hallucial tendon derived from it, and so directing the pull of this tendon. Such an arrangement indeed has been described in specimens of *Didelphys* by Glaesmer (1910), McMurrich (1904), and Coues (1872), although the last named author failed to appreciate the true identity of the muscle. In some specimens of a *Didelphys* a more specialized arrangement has been achieved (Fig. 14.8A) for the flexor accessorius has attained direct continuity with the hallucial tendon by largely appropriating its substance. it may even in some marsupial species (*Phascogale tapoatafa*) constitute the whole of the hallucial tendon, and even in species with a diminished hallux (Fig. 14.13) such a tendon may persist, fading out by merging into the fascia on the medial side of the foot. This specialization, as will be seen, has commonly been paralleled in other mammals.

A basic sort of arrangement, characteristic of eutherian mammals where a fully-fledged common tendon plate (common to both flexor tibialis and flexor fibularis tendons) has been elaborated, is seen in a carnivore such as *Felis domestica* (Fig. 14.8B): both the plate and its derivative tendons (that for the hallux is, of course, lacking) show the characteristic twisted arrangement and flexor accessorius merges onto the surface.

In the higher Primates the twisted tendon plate shows a variable degree of splitting. A fairly generalized primate arrangement, which is variously modified in other species, is shown by *Cebus nigrivattatus* (Fig. 14.9A). The flexor tibialis now

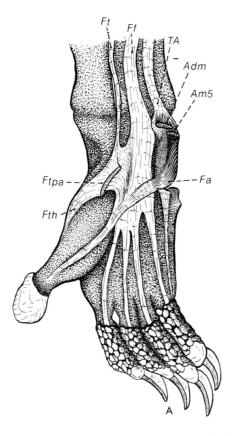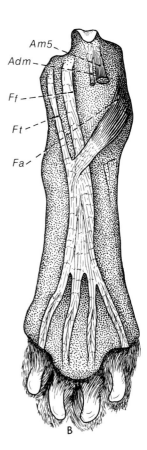

Fig. 14.8

A, the sole of the right foot of *Didelphys marsupialis*. B, the sole of the right foot of *Felis domestica*. *TA*, tendo Achillis; *Adm*, abductor digiti minimi; *Am5*, abductor ossis metatarsi quinti; *Ftpa*, insertion of flexor tibialis to plantar aponeurosis; *Fth*, continuation of flexor tibialis down hallux. Other labelling is as in Fig. 14.1. (From Lewis 1962*b*.)

largely annexes the superficial fibres of the common plate which diverge towards the marginal tendons. Thus, the flexor tibialis tendon, lying here superficial to the other tendons, is mainly continued laterally to form the tendon of the fifth digit, but it nevertheless gives a substantial contribution to the tendon of the hallux and smaller slips to those of digits three and four; flexor tibialis is therefore distributed mainly to the marginal tendons. Flexor fibularis forms the larger part of the hallucial tendon and almost all the tendons of digits two, three, and four. Flexor accessorius arises from the under surface of the trochlear process of the calcaneus and, entering the sole, gives rise to a tendon which lies freely between the tendons of flexor fibularis and flexor tibialis and which terminates more medially by merging with

the surface of the flexor fibularis tendon, thus contributing to the digital tendons. It must be emphasized that the flexor accessorius tendon lies sandwiched but quite freely between the crossing tendons of the flexor tibialis and the flexor fibularis and has no attachment to the lateral border of the flexor tibialis tendon. The three tendons of this triple-layered arrangement are therefore wholly independent on the fibular side.

Much the same arrangement is seen in *Lemur*, but the crossing tendons are particularly well fused; no flexor accessorius is present to separate them.

In other New World and Old World monkeys also the tendon plate is frayed out into individual tendons but in a variable fashion (Fig. 14.10) so that the contribution of flexor tibialis or flexor

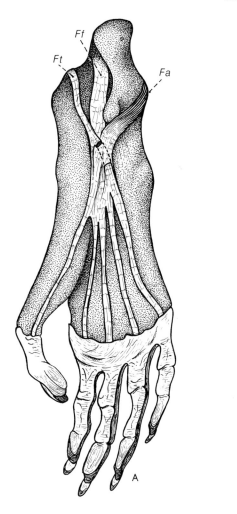 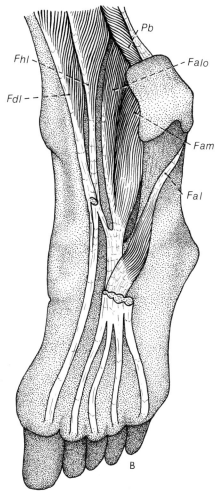

Fig. 14.9
A, sole of the right foot of *Cebus nigrivittatus*; part of flexor tibialis tendon has been removed and is indicated by a broken line. B, sole of the right foot of *Homo*, in which a flexor accessorius longus (*Falo*) is present and the tendon of flexor digitorum longus (*Fdl*) gives a contribution to the tendon of flexor hallucis longus (*Fhl*); part of flexor digitorum longus tendon has been removed to show the trilaminar arrangement of the tendons present in this foot. *Pb*, peroneus brevis; *Fam*, medial head of flexor accessorius; *Fal*, lateral head of flexor accessorius. Other labelling is as in Fig. 14.1. (From Lewis 1962*b*.)

fibularis to the individual digits varies; these various patterns are, however, all determined by the basic architecture of the common plate. In some species (Fig. 14.10F) the whole of the hallucial tendon may be appropriated by flexor accessorius.

It has been stated (Osman Hill 1955) that flexor accessorius arises from the medial side of the calcaneus in Colobidae. This was not so in the

specimens illustrated here (Fig. 14.10A, E) nor in one dissected by Straus (1930). However, the plantar disposition of abductor digiti minimi and abductor ossis metatarsi quinti, together with the fact that the muscular belly proper of flexor accessorius arises from the flattened tendon of origin medial to these muscles, may give an easily mistaken impression of a medial origin. Indeed, after removal of the abductor digiti minimi and

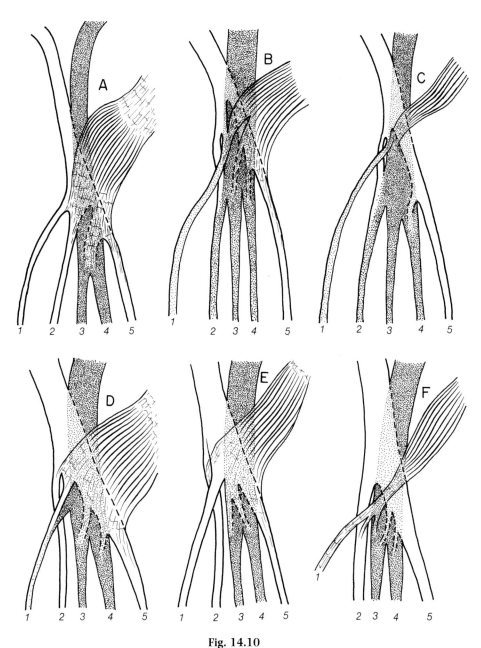

Fig. 14.10

The tendon arrangements in the right foot of *Colobus polykomos (A)*, *Pithecia monachus (B)*, *Saimiri sciurea (C)*, *Cercopithecus nictitans (D)*, *Procolobus verus (E)*, *Mystax midas (F)*. The flexor fibularis tendon is stippled; the overlying flexor accessorius is represented as though semi-transparent, as is the most superficial flexor tibialis tendon; where the latter crosses the other tendons it is represented by a broken line. (From Lewis 1962b.)

abductor ossis metatarsi quinti, the thin aponeurosis lying on the surface of the long plantar ligament may very easily escape detection. In some species this flat aponeurosis may even perhaps become incorporated into the long plantar ligament bringing the muscle origin of accessorius secondarily more medially (apparently, as discussed below, the lateral head of *Homo* may suffer a similar fate). There is little doubt, then, that in *Colobus* the same morphological element is concerned as that in other monkeys and mammals.

Among the apes flexor accessorius is always lacking in gibbons, usually in orang-utans and gorillas, and frequently in chimpanzees (Hartman and Straus 1933). Nevertheless the breaking up of the common tendon plate, while showing some variation, presents the same sort of patterns dictated by the architecture of the plate, as it does in monkeys. Thus, in gibbons the flexor fibularis typically forms the whole of the tendon of the third digit and the major parts of the tendons of digits one, two, and four. The flexor tibialis lies superficially, dividing into medially and laterally divergent parts which embrace the flexor fibularis tendon. The lateral portion continues as the whole of the fifth digital tendon; a small slip from its lateral margin winds dorsally to tendon five and joins tendon four. The medial portion continues as a sizeable contribution to the hallucial tendon; a small slip from its medial margin winds around dorsally to the hallucial tendon to join digital tendon two. In *Pan* at least one arrangement is that the flexor tibialis lies superficially dividing into digital tendons two and five and the flexor fibularis tendon, lying deeply, divides into digital tendons one, three, and four; there is no connection between the tendons.

The human arrangement   Despite statements to the contrary, it is clear that a medial head for flexor accessorius is a uniquely human apomorphy, and that the double-headed condition is unique to man. The relationship of flexor accessorius to the long flexor tendons in man, according to most formal accounts, would seem to be quite unlike the generalized primate arrangements described above. Standard text-book accounts attributing the attachment of flexor accessorius to the lateral border of the flexor digitorum longus tendon are invalid, for only in some cases does this arrangement obtain. Though Turner (1864) long ago indicated the truth of this matter, such misleading accounts still persist. Any accurate description of flexor accessorius must separately consider the two heads of the muscle which represent morphologically different components: the lateral head is the homologue of the muscle of other mammals, including other primates, but the real key to understanding is the realization that the medial head is no more than part of flexor fibularis which has 'descended' to the sole.

Winckler and Giacomo (1955) analysed in detail, and figures, the tendon arrangements in fifteen feet; their findings, but not their morphological interpretation, are comparable to those described below. Barlow (1949, 1953) has also described several examples of flexor accessorius which, together with the present series, are readily interpretable in the light of the phylogenetic hypothesis advanced herein.

The flexor accessorius and the long flexor tendons exhibit a wide range of variation in man, but the variational pattern in reality often shows clear and unmistakable affinity to that described above in monkeys. Again flexor tibialis forms the superficial layer and passes laterally to form digital tendon five and at least part of the tendons of digits two, three, and four on the way. Little diverges to the medial side of the sole, though a contribution may pass to the hallucial tendon, as in *Cebus*. The deep layer is constituted differently from that in other Primates, flexor fibularis forming the whole (or most) of the hallucial tendon with a contribution (the connecting slip between flexor hallucis longus and flexor digitorum longus) being give to certain of the other tendons. A new element, the medial head of flexor accessorius, constitutes the remainder of the deep layer, and this also gives contributions to the digital tendons. These three components together represent the deep layer seen in monkeys. The medial head of flexor accessorius arises far back from the calcaneus, often marking the bone and flanking the flexor hallucis longus as it enters the sole. There can be little doubt that it is a part of the flexor hallucis longus which has 'descended' to the sole. Indeed, sometimes in man it continues to have a partial origin (flexor accessorius longus) in the leg, usually from the fibula in close association with the flexor hallucis longus (Testut 1884; Le Double 1897). Wood

(1868) recorded its presence in 8 of 204 legs. The fundamental similarity between the human and generalized primate arrangements is most emphatically shown when flexor accessorius longus is present (Fig. 14.9B). The basic tri-laminar arrangement of the tendons in the sole is also often seen, however, when no flexor accessorius longus is present, that is when this part of flexor hallucis longus has completed its 'descent' to the foot to become the medial head of descriptive anatomy. In man the lateral head of flexor accessorius often has the same relationship to the deep lamina as in other Primates. It crosses the deep lamina and merges with its surface so contributing to the digital flexor tendons. It frequently has no attachment to the lateral border of the flexor digitorum longus tendon, which may therefore be lifted from its surface, though a casual inspection may give the impression of constant termination here. This head generally has a flat, narrow tendon of origin which passes between the long plantar ligament and abductor digiti minimi before expanding into a fleshy belly, a condition similar to the generalized primate arrangement. This head is not infrequently absent, though in some such cases remnants of its tendon of origin are seen fused to the surface of the long plantar ligament; it has thus shifted its origin medially to become incorporated with the medial head.

These human plantar tendons present a range of variations, all attributable to a varying pattern of break-up of a common tendon plate. Indeed it is known that flexor tibialis and flexor fibularis in the human embryo terminate in a common tendon plate which later breaks up. The 'descent' of the medial head is clearly brought about by a portion of the anlage of flexor hallucis longus in the human sole developing into muscle instead of into tendon as happens in most other mammals. Apart from man a bicipital muscle has apparently been described in but one orang-utan (Hafferl 1929), where again the medial head was clearly derived from flexor hallucis longus. I agree with Weidenreich (1922) that the human muscle is a composite structure whose lateral head is the homologue of the flexor accessorius of other mammals. It cannot be agreed, however, that the medial head represents the deep head of the flexor digitorum brevis, for the history of this muscle is quite different, as will be seen.

The primitive single-headed muscle has several functional potentialities. In Primates the frequent use of the foot in the inverted position necessitates a loop, the fundiform ligament, to hold the extensor tendons laterally and prevent bowstringing. Flexor accessorius may perform a similar function for the flexor tendons. The absence of the muscle in mammals such as *Trichosurus vulpecula* or *Lemur fulvus*, which have a markedly inverted grasping foot, might be raised as an objection to this view, but in such species the flexor fibularis tendon is firmly held in a deep groove between the talus and calcaneus rendering unnecessary any stay-like action by flexor accessorius. Doubtless flexor accessorius also correctly aligns the pull of the flexor tendon for the widely divergent hallux and as has been seen may, in some species, become directly continuous with that tendon.

Kaplan (1959), elaborating on the classical study of Winckler (1929), considered that flexor accessorius must be an active pronator of the forefoot, and correlated the ability to perform this movement with the presence of the muscle. Its usual attachments, however, seem totally inappropriate for this task. Jones (1944) suggested that the functional role of the human muscle was to maintain flexion of the toes during walking, so that the leg extensors might enjoy an unopposed action in pulling the body forward over the foot. The uniquely human specialization which has converted part of flexor hallucis longus into a medial head of flexor accessorius, intrinsic to the sole, accords well with this view.

Flexor perforatus

This term is adopted for the muscle which is represented in man by the flexor digitorum brevis. It is primarily a crural muscle (Fig. 14.1), homologous with the flexor digitorum superficialis of the forelimb and of similar derivation, which secondarily descends to the foot. The alternative view (Glaesmer 1908) that it is primitively pedal and may secondarily show ascent to the leg is improbable.

Marsupials exhibit all stages in the descent of the muscle to the foot: the muscle belly may be crural (*Trichosurus*, *Pseudochirus*, *Caluromys*), partially descended (*Phascolarctos*, Fig. 14.11), or intrinsic to the foot (*Dasyurus*, *Sarcophilus*, Fig. 14.13). In in the water-opossum (Sidebotham 1885) the deep attachment is lost and the whole muscle then arises from the calcaneus. In the Macropodidae and Peramelidae the deep attachment to the flexor fibularis tendon is lost and the remainder of the muscle has become fibrous and incorporated with the deep aspect of the plantar aponeurosis, forming a deep fibrous layer from which the perforated digital tendons arise (Fig. 14.14). In the Eutheria, only the later phases in an apparently parallel series of changes are found. In all the muscle is intrinsic to the foot.

In the more primitive Primates the muscle is two-headed, rather like that of *Sarcopilus harrisi* (Fig. 14.13) (Sawalischin 1911; Straus 1930) but within the order there is a progressive transition, with the muscle transferring itself more and more

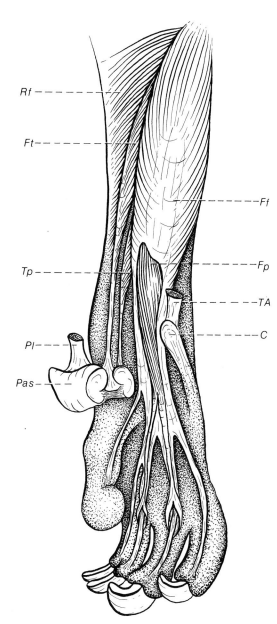

Fig. 14.11
The right leg of *Phascolarctos cinereus* with the muscle bellies of gastrocnemius and plantaris removed. The sesamoid in the plantar aponeurosis (*Pas*) is shown folded away from its articulation with the navicular. The intrinsic foot musculature is now shown. Labelling is as in Figs 14.1, 14.2. (From Lewis 1962*a*.)

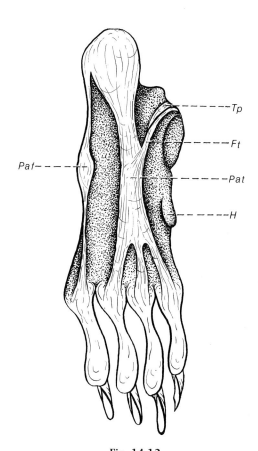

Fig. 14.12
A superficial dissection of the left foot of *Sarcophilus harrisi*. *H*, reduced hallux; other labelling as in Figs 14.1, 14.2. (From Lewis 1962*a*.)

the latter species it has gained a secondary attachment to the overlying plantar aponeurosis and therefore presents deep and superficial heads;

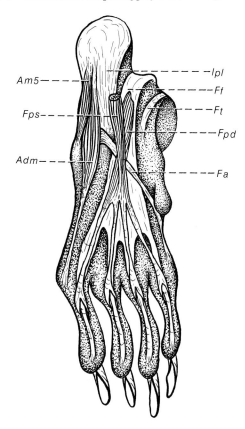

Fig. 14.13
The same foot as in Fig. 14.12 after removal of the plantar aponeurosis. *Fps*, superficial head of flexor perforatus; *Fpd*, deep head of flexor perforatus; *lpl*, long plantar ligament; other labelling as in Figs 14.1, 14.2, 14.7. (From Lewis 1962a.)

to the superficial attachment, which also tends to gain its origin directly from the calcaneus, rather than indirectly through the plantar aponeurosis (Hepburn 1892; Clark 1924, 1926; Woolard 1925; Raven 1950; Osman Hill 1953, 1955). This transference reaches its highest expression in man. That this primate condition is derived similarly to that of *Sarcophilus* gains some support from the finding of Straus (1930) that the deep head in *Gorilla beringei* apparently has its origin within the leg. In man the transfer to the superficial origin is not always complete and a deep head arising from the surface of the common flexor tendon and giving rise to perforated tendons to the fifth (and even the fourth and third) digits is quite often

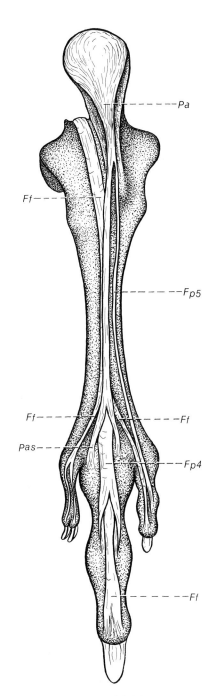

Fig. 14.14
The right foot of a juvenile specimen of *Macropus major*. *Pas*, slip of plantar aponeurosis; *Fp4,5*, flexor perforatus tendons to digits 4 and 5; *Ff*, flexor fibularis and its tendons to digits 2, 3, 4 and 5. (From Lewis 1962a.)

found, apparently more commonly in negroes (Turner 1864; Flower 1867; Hallisy 1930; Straus 1930; Wells 1952). Indeed, in the human embryo (Bardeen 1906) the muscle is initially attached to the surface of the long flexor tendons, and only later does it transfer to its more superficial calcaneal attachment.

In the remaining eutherian orders the whole muscle has transferred its origin to the overlying structures (calcaneus or plantar aponeurosis) just as in the higher Primates. In some Edentata and in *Ursus americanus* (Windle and Parsons 1899) the whole origin is from the calcaneus but in most the muscle is incorporated into the plantar aponeurosis. In *Gymnura rafflesi* (Parsons 1898*a*), most Carnivora (Windle and Parsons 1898) *Hyrax, Hippopotamus* (Windle and Parsons 1903), and *Castor* (Parsons 1894*b*) it persists as a deeper fleshy stratum of the plantar aponeurosis which gives rise to the perforated tendons of the digits; it is, thus, brought into secondary continuity with the plantaris. This deep stratum may become fibrous, in which case the perforated tendons arise from the deeper part of the fibrous plantar aponeurosis, as occurs in the Hyaenidae and Canidae (Windle and Parsons 1898), in some Edentata (Windle and Parsons 1899), in most Ungulata (Windle and Parsons 1903) and in many of the Rodentia (Parsons 1894*b*). There seems little doubt that the muscle has 'descended' to the foot and then progressively become attached to the overlying plantar aponeurosis and calcaneus, the two-headed condition representing an intermediate stage in this transfer. Further support for the notion of descent is afforded by similar changes in the flexor digitorum superficialis of the forelimb in some mammals, which have been noted in Chapter 7.

It is clear that the completely pedal flexor digitorum brevis of Primates and the corresponding crural muscle of certain marsupials must develop from different parts of the embryonic limb anlage and must also represent different components of the amphibian and reptilian limb. Nevertheless, in marsupials at least, clear phylogenetic continuity is shown in the 'descent' of the muscle, or rather in its progressively more distal differentiation in the limb. Clearly, the muscles at varying stages in this descent should be considered as homologues. Perversely, there is some truth in the contention of McMurrich (1907) that the flexor digitorum brevis of man corresponds to the flexor brevis superficialis of the reptilian foot.

Tibialis posterior

It is held that the feet of man and of other Primates differ strikingly in the manner of termination of the tibialis posterior tendon. The commonly quoted phylogenetic interpretation for this follows Keith (1894, 1929), who maintained that this tendon, like its supposed serial homologue in the forelimb (flexor carpi radialis), inserted primitively into the bases of the second, third, and fourth metatarsals and that it reached this insertion in Primates by passing deep to a ligament spanning the interval between the calcaneal sustentaculum tali and the navicular tuberosity. The human condition could then plausibly be derived by fusion of this sustentaculo-navicular ligament (part of the primate internal Y-shaped ligament, as interpreted by Keith) to the underlying tendon, conferring on tibialis posterior a secondary navicular insertion, which Keith (1929) held to be important in the evolution of the longitudinal arch of the human foot. It is apparent that a calcaneal attachment must also result from such a mechanism of transfer.

This theory is quite untenable on two major counts. Firstly the tibialis posterior and flexor carpi radialis can in no way be considered as serial homologues; they are even derived from different strata in the primitive tetrapod limb! Secondly, Keith was in error when he described the distal attachment of the Y-shaped ligament to the navicular: it is to the medial cuneiform and the ligament could not then be implicated in transfer of the insertion to the navicular. In fact, Keith got the morphocline polarity wrong.

Rather, it seems clear that the primitive insertion of the tendon is to the mammalian homologue of the tibiale. In most mammals this is the tubercle of the navicular, but in rodents it is the 'tibial navicular' bone, and in insectivores and tree-shrews it is apparently the posterior process of the medial cuneiform. It is among the Primates, in particular, where this insertion is drastically modified. Alternatively, in those mammals which have undergone digital reduction (and presumably with a consequent correlated disappearance of the marginal tarsal tibiale), the distal insertion is aborted: in the dog, for example, the slender tibialis posterior tendon terminates by merging with the medial ligaments of the tarsus; in ungulates a muscle belly is present but its tendon, failing to reach the sole, merges in the leg with the flexor fibularis tendon; in Macropodidae the whole muscle is lacking and it is noteworthy that here, unlike other marsupials, no cartilaginous anlage of the tibiale appears during development.

In the primitive mammalian condition, as can be seen in rodents (Fig. 14.15A), the tibialis posterior tendon inserts entirely into the 'tibial navicular' (tibiale). Prior to its insertion the synovial-covered tendon lies deep to an arcuate ligament passing from the tibial malleolus to the medial cuneiform bone. This ligament is no mere indefinite fascial thickening but is a very tough, discrete structure (even containing a hemisesamoid) which, after entering the sole from its tibial attachment, skirts the lateral part of the tibiale to attach to the medial cuneiform. Its lateral convexity in the sole is in contact with the sustentaculum tali, but with only a flimsy attachment to it. In generalized marsupial feet (*Trichosurus, Didelphys*) a similar ligament spans the tendon which here inserts to the tibial tuberosity (tibiale). In insectivores and treeshrews, the same ligament is present but the tendon inserts to the posterior prong on the medial cuneiform (tibiale). In *Tupaia* (Fig. 14.6) ligament and tendon are fused into a fibrocartilaginous mass as they

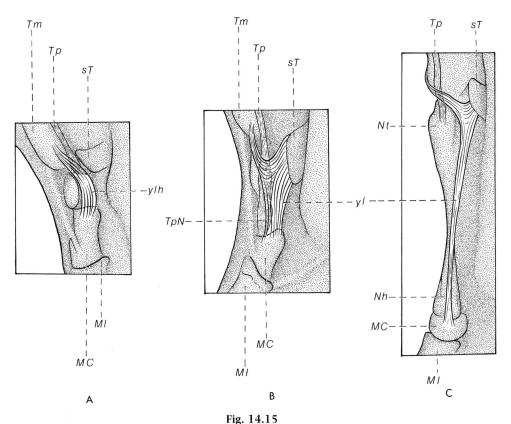

Fig. 14.15

The arrangements in the sole of the right foot of A, *Mus norvegicus albinus*, B, *Lemur catta*, C, *Galago moholi*. In each case the region of the foot illustrated corresponds to that shown in Fig. 14.14B. *MI*, first metatarsal; *Nh*, head of navicular; *MC*, medial cuneiform *sT*, sustentaculum tali of calcaneus; *ylh*, ligament between tibial malleolus and medial cuneiform, precursor of the Y-shaped ligament; *Tm*, tibial malleolus; *Nt*, tubercle of navicular; *Tp*, tibialis posterior tendon, *TpN*, navicular insertion of tibialis posterior; *yl*, Y-shaped ligament. (From Lewis 1964*b*.)

approach their common insertion. In *Tenrec* the ligament has a more substantial attachment to the sustentaculum but it is retracted from it, taking on a Y-shaped form which foreshadows the primate condition.

A tough internal Y-shaped ligament bridging over the tibialis posterior tendon is a characteristic feature of the primate foot. It has substantial attachments to tibial malleolus, sustentaculum tali, and medial cuneiform. It has had an unfortunate history. It is very obvious that the Y-shaped ligament of Primates has not appeared as an evolutionary novelty within this order, but is a simple modification of the generalized mammalian ligament, described above, running from the tibial malleolus to the medial cuneiform. Since its distal attachment is to the medial cuneiform, and not to the navicular, as described by Keith, it cannot be implicated in any mechanism accounting for a shift of the tibialis posterior insertion to the navicular. Keith (1929) incorrectly illustrated the Y-shaped ligament: the structure he depicted includes two quite separate and distinct primate pedal entities, namely, the true Y-shaped ligament and the naviculo-metatarsal ligament.

In prosimians a quite simple primitive arrangement is preserved (Fig. 14.15B): the tibialis posterior tendon passes in a synovial-lined tunnel deep to the tough Y-shaped ligament, to insert entirely into the navicular tuberosity. In *Galago* (Fig. 14.15C) the arrangement is grotesquely distorted by the lengthening of the navicular but this serves to highlight Keith's error and emphasize the independent attachments of tendon and ligament to navicular and medial cuneiform respectively.

In New World monkeys an essentially similar arrangement is usually retained (Fig. 14.16A) but in some species an incipient continuity may be established between the tendon and the ligaments of the tarsus, producing slight prolongation of the tendon into the central region of the sole (Fig. 14.16B) by the mechanism shown in Fig. 3.7A.

This trend is more advanced in Old World monkeys: the tibialis posterior tendon, where it lies deep to the Y-shaped ligament, attains a dual insertion, attaching to the navicular tuberosity and through a lateral prolongation into the sole to the lateral and intermediate cuneiform bones, the cuboid, the sheath of the peroneus longus and the bases of the second, third and fourth metatarsals (Figs 14.16C, 14.17). This trend culminates in the condition in some species where the navicular insertion is absent (Fig. 14.16D). The prolongation of the tendon into the sole may perhaps enhance the grasping action of the foot by pulling the digital portion against the hallucial portion.

The ape arrangement The progressive specialization characterizing the hominoids involves adherence between the tibialis posterior tendon and the overlying Y-shaped ligament. In *Hylobates* the arrangement is not unlike that shown in Fig. 14.16C but there is some commencing adherence between the terminal part of the tendon and the overlying ligament.

This trend has progressed even further in *Pongo pygmaeus*: there is still an obvious Y-shaped ligament but its upper band, from the tibia, is partly subsumed into the joint capsules of the ankle and talocalcaneonavicular joints: it is the precursor of the anterior (tibio-navicular) band of the human deltoid ligament.

The process of loss of autonomy of the Y-shaped ligament is even more advanced in *Pan*. The tibialis posterior tendon has a considerable primary navicular attachment which is somewhat prolonged forward to the medial cuneiform presumably by the tendon here merging with the ligamentous connexions of the two bones and a lateral part of the tendon is prolonged into the central region of the sole as far as the bases of the second, third, and fourth metatarsal bones. The Y-shaped ligament no longer exists as a separate entity. Its upper limb, from the tibial malleolus, is fully incorporated in the deltoid ligament of the ankle joint as its anterior fasciculus; but the remains of its lower part, bridging the interval between the sustentaculum tali and the medial cuneiform, can still be clearly recognized, although merged with the surface of the tibialis posterior tendon, which gains thereby a secondary insertion onto the sustentaculum tali and the medial cuneiform.

The human arrangement With one additional modification the arrangements are not unlike those in *Pan*. The tibialis posterior tendon has a considerable attachment to the navicular tuberosity (prolonged forward to the medial cuneiform) and also a lateral continuation into the sole.

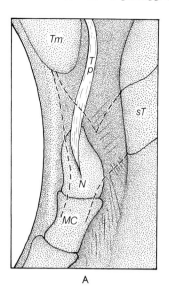

A

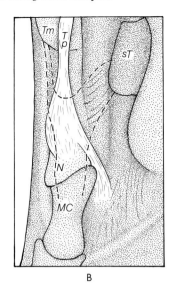

B

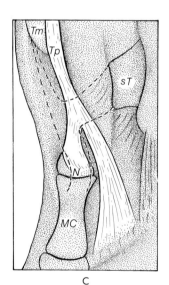

C

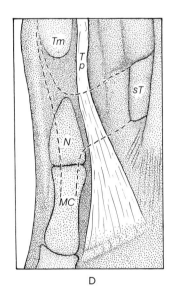

D

Fig. 14.16
The arrangements in the sole of the right foot of A, *Pithecia monachus*; B, *Cebus nigrivittatus*; C, *Cercopithecus nictitans*; D, *Colobus polykomos*. In each case the region of the foot illustrated corresponds to that shown in Fig. 14.17B. In each case the position of the overlying Y-shaped ligament, which has been removed, is indicated by the broken line. N, navicular; other lettering as in Fig. 14.15. (From Lewis 1964*b*.)

Superficial to this prolongation, and thereby merged with the tendon, is a thick tendinous bundle, clearly derivative from the lower sustentaculo-cuneiform part of the Y-shaped ligament, which confers upon the tendon its strange, tough, and recurrent secondary insertion to the sustentaculum tali, and a large part of its insertion to the medial cuneiform bone. As in *Pan* the upper band of the Y-shaped ligament is no longer separately identifiable, since it has become the anterior fasciculus (canonical pars tibionavicularis) of the

medial ligament of the ankle joint; despite the official terminology this fasciculus to a large extent spreads down over the surface of the tibialis posterior tendon, as would be expected from its derivation.

The prolongation of the tendon entering the sole laterally has acquired attachments, differing from those of *Pan*, and which are rarely recognized in textbook descriptions. A considerable part of it has been pressed into service as a tendon of origin for the flexor hallucis brevis muscle (Fig. 14.18). This

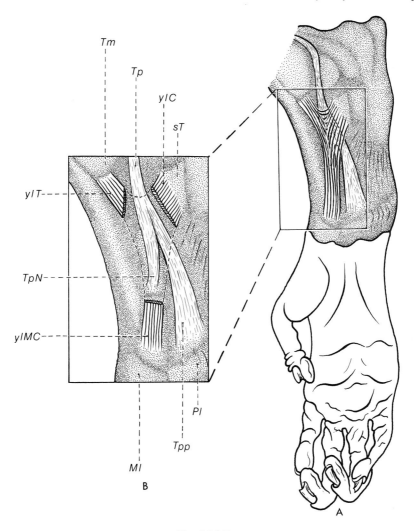

Fig. 14.17

A, a diagrammatic representation of the sole of the right foot of *Procolobus verus* illustrating the tibialis posterior tendon and the overlying Y-shaped ligament. B, an enlarged area of the foot, with part of the Y-shaped ligament removed (broken line) to expose the underlying tibialis posterior tendon. *MI*, first metatarsal; *Pl*, position of peroneus longus tendon; *Tpp*, plantar prolongation of tibialis posterior tendon; *sT*, sustentaculum tali; *Tm*, tibial malleolus; *Tp*, tibialis posterior tendon; *TpN*, navicular insertion of tibialis posterior; *ylMC*, attachment of Y-shaped ligament to medial cuneiform; *ylC*, attachment of Y-shaped ligament to the sustentaculum tali; *ylT*, attachment of the Y-shaped ligament to the tibial malleolus. (From Lewis, 1964*b*.)

tendon is drawn laterally by another attachment to the cuboid and lateral cuneiform bones and their associated ligaments, thus giving the flexor hallucis brevis a Y-shaped tendinous origin. Further tendinous bundles derived from the lateral part of the tibialis posterior tendon usually run in the depths of the interval between, on the one hand, the slip in continuity with flexor hallucis brevis

and, on the other, the navicular and medial cuneiform bones; these bundles retain the insertion, noticed above in the more advanced Primates, into the intermediate cuneiform and the bases of the second, third, and fourth metatarsals. With only minor variations this arrangement is almost constant.

The strong, tendinous band in continuity with

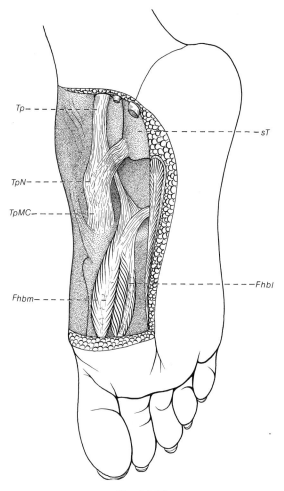

Fig. 14.18

A diagrammatic representation of the arrangements in the human foot. Note that the attachment of the tibialis posterior tendon (*Tp*) to the sustentaculum tali (*sT*) and the origin of flexor hallucis brevis by the Y-shaped tendon whose medial limb is in continuity with the tibialis posterior tendon. *TpN*, navicular insertion of tibialis posterior; *TpMC*, medial cuneiform insertion of tibialis posterior; *Fhbm*, medial head of flexor hallucis brevis; *Fhbl*, lateral head of flexor hallucis brevis. (From Lewis 1964*b*.)

flexor hallucis brevis is free of attachment to the related bones and ligaments, other than that attachment anchoring it laterally. Free movement of the whole musculotendinous apparatus is facilitated by a large bursa which is usually interposed between the commencement of the

muscular part of flexor hallucis brevis and the underlying structures—the medial cuneiform, the first tarsometatarsal joint, and the terminal part of the sheath of the peroneus longus tendon. Another bursa may be found more proximally.

In most recent British textbooks flexor hallucis brevis is incorrectly described and illustrated, as merely running obliquely across the sole from a fibular-sided origin from the cuboid, lateral, and (perhaps) intermediate cuneiform bones and their ligamentous investments, including here the tendinous attachments of tibialis posterior to these bones. French anatomists, however (see Poirier and Charpy 1901; Le Double 1897), have stressed the constant continuity between a part of the tibialis posterior tendon and flexor hallucis brevis. Wood (1868) and more recently Kaplan (1955) noted this continuity, but believed that it was an anomalous condition; indeed Kaplan thought it was present only in subjects with hallux valgus and that it was a variation predisposing to this condition.

The direct continuity which is established in man between the tibialis posterior tendon and flexor hallucis brevis is a unique human apomorphy and is perhaps associated with the evolution of an arched, weight-bearing foot. It clearly provides a mechanism for enhancing the contraction of the short hallucial flexor and for projecting the pull of tibialis posterior forwards to the metatarso-phalangeal joint of the great toe. This is precisely where the pull of an arch-raising muscle is most required; as Hicks (1953) has shown, arch-raising is brought about by flexion of the first ray of the foot (pronation of the forefoot) accompanied by inversion (at the subtalar joint). The continuity of the two muscles would be effective in contributing to that elevation of the medial propulsive part of the longitudinal arch of the foot which occurs during extension of the great toe (as in walking) and which adds to the spring of the step. Such continuity then provides an active mechanism for this effect, additional to the passive windlass-like action of the plantar aponeurosis about the first metatarso-phalangeal joint; which has been demonstrated by Hicks (1951, 1954).

Considerable confusion (see Jones 1944) surrounds the nomenclature and explanation of sesamoid ossicles located in the tendon of insertion of the human tibialis posterior; these have been variously named as 'naviculare secondarium' and 'os tibiale externum' (Pfitzner 1896). The true nature of the situation seems to have been clarified by Manners-Smith (1907), who pointed out that two quite different entities are involved. The true tuberosity, the primary and primitive insertion of the tibialis posterior, may be separate; it is said that this part of the navicular develops as a separate cartilage in the second month of fetal life and it is also said that on occasion the tuberosity of the navicular can ossify from a separate centre (Flecker 1933). It seems obvious that this ossicle represents an atavistic reappearance of the tibiale by a mechanism of 'putting the clock back' (Fig. 3.2). The second entity is a true sesamoid in the tendon of tibialis posterior located at its point of division, just proximal to the navicular tuberosity (Fig. 14.18). I have seen ossicles in this situation, and their presence is not surprising, for sesamoids not uncommonly occur in tendons at the point of division into radiating slips—they appear to function rather in the way in which a knot prevents fraying of a rope. As Manners-Smith noted, these sesamoids may fuse with the navicular prolonging it backwards as a posteriorly directed prong. One or other of these ossicles have been not infrequently confused with the prehallux (Kidner 1929), a quite different skeletal element, as noted earlier.

Rotator fibulae and popliteus

In eutherian mammals the upper part of rotator fibulae is remodelled to become the popliteus, with its characteristic intracapsular tendon of origin from the femur. Part of the primitive rotator sheet, however, commonly persists as a peroneotibialis muscle, forming a bundle arising from the head of the fibula and inserting on the tibia deep to popliteus itself. It has been recorded in a wide range of eutherian mammals, including prosimian primates and monkeys; it is apparently less common in the apes, but does occur in about 8 per cent of human limbs (Le Double 1897). It is said that it is constant in the human fetus (Weinberg 1929).

A telling point in Furst's (1903) theory of the origin of the femoral tendon of popliteus from an intra-articular femorofibular disc (which he inappropriately termed a 'meniscus') was the situation in kangaroos and their relatives. He pointed out that they showed a beautiful transitional stage in the conversion of disc to tendon, apparently consequent upon retreat of the fibula from direct articulation with the femur.

In the wallaby shown in Fig. 14.5B the disc has become represented by what is really just a firm nodular thickening in the course of the emergent femoral tendon of origin; as befits its derivation this thickening is continued up as the flattened femoral tendon of origin, but has an additional strong attachment down to the head of the fibula, where it attaches in common with the fabellofibular ligament, and a further one joining onto the more anterior part of the lateral meniscus. Kaplan (1961) gave a description for *Macropus giganteus* which was similar in essentials to this, but it is perhaps unfortunate that he referred to the relic of the disc as the 'fibular meniscus'. These arrangements provide a very reasonable model for the likely pattern in emergent Eutherians; it is, of course, an incomplete model: no ancestor–descendant relationship is implied.

Eutherian mammals commonly present a prominent sesamoid bone (the cyamella) in politeus at the point of its transition from muscle belly to intracapsular tendon. It appears to be almost invariably present in prosimians and New World monkeys, rarely if ever present in Old World monkeys, and only found among the apes in *Pongo*; in man there is no satisfactorily documented account of its existence (Vallois 1914). It was in dealing with the cyamella that Furst's theorizing went sadly awry. He considered that it was the detached upper epiphysis of the fibula, and that this disruption had brought about continuity of the upper fibres of rotator fibulae with the femorofibular disc, and so the future tendon of popliteus. This idea was elaborated by Pearson and Davin (1921*a,b*) who reasoned that a fibular crest (present in fact only as a variant in certain marsupials) separated to become the free parafibula, characteristic of marsupials, which in turn subdivided to form the lateral fabella and cyamella. This was an influential theory and was followed, for instance, by Weinberg (1929). It was, in fact, a classical case of misinterpretation of morphocline polarity long before the fashion for cladistic terminology.

But if Furst's theory is untenable, what is the

explanation for the cyamella? Haines (1942) roundly rejected Furst's viewpoint and pointed out that some lizards, but not marsupials, have a bony lunula in the femorofibular disc. He was loath to go as far as to suggest a lineal relationship between the two structures but felt that it was 'preferable to speak of a derivation of a lacertilian femorofibular lunula and of a mammalian cyamella from the femorofibular disc which formed part of the ancestral heritage of tetrapods'. Although not entirely explicit this seems to be along the right lines; it would not be unreasonable to see the development of a sesamoid ossification in the nodular enlargement of an ancestral eutherian stage similar to that shown in Fig. 14.5B.

With this sort of history it would be reasonable to expect that part of the popliteal tendon (or the cyamella) derived from the disc would retain a substantial ligamentous attachment to the head of the fibula. Such, in fact, is the case and it appears to be universal in Primates (Vallois 1914). There is even a substantial attachment of popliteus here in man, although often unnoticed in textbooks, and from time to time 'rediscovered' (Lovejoy and Harden 1971).

A connection to the lateral meniscus should also be reasonably expected. Indeed Furst (1903) maintained that it was universal except for certain rodents. This has not been confirmed by subsequent work—Vallois (1914) mentioned it only in some Primates. Weinberg (1929) carried this further noting such a membranous connection (his 'membrane popliteo-méniscale') in *Cebus* and *Macaca* and even in fetal and newborn *Homo* but not the adult. This is quite different from the attachment of popliteus to the posterior part of the lateral meniscus, which is a human apomorphy. Weinberg (1929) noted such a well developed posterior 'capsular fasciculus' only in man. It has, of course, long been known that by this fasciculus the popliteus muscle is intimately connected to the lateral meniscus (Higgins 1895*a,b*); again it is from time to time re-emphasized (Last 1951).

One of the more confused and confusing papers on the morphology of popliteus was by Taylor and Bonney (1905). They noted fibres of flexor tibialis arising from the surface of rotator fibulae (this expansion of the origin has been noted above). They, however, jumped to the quite unwarranted conclusion that it must be a serial homologue of

condyloradialis in the forelimb naming it 'condylo-tibialis'. By an extraordinary chain of reasoning they decided that popliteus belonged in the same stratum as flexor tibialis and flexor fibularis and corresponded to the deep head of pronator teres in the forelimb (this has been noted in Chapter 7).

The flexor retinaculum

The deflection laterally of the human heel dramatically alters the architecture of the portal of entry for the long flexor muscles and the neurovascular bundle into the sole. This passageway in the human has been called the tibiocalcaneal tunnel, Richet's tunnel, calcaneal tunnel or tarsal tunnel; by analogy with the forelimb carpal tunnel the last term is the one of choice.

Subhuman Primates have retained a simple arrangement for this tarsal tunnel and for its overlying flexor retinaculum which seems to be no more than an inheritance from the earliest Therians. A similar simple arrangement is found in the marsupials *Pseudochirus* and *Trichosurus*.

In *Pan*, for example, the general similarity to the carpal tunnel is quite apparent (Figs 14.19A, B). Because of the inwardly inflected heel there is no separation in the calf between the triceps surae muscle bellies (or the tendo calcaneus) and the underlying deep muscles, and no real separation of the investing fascia of the leg into superficial and deep layers as in man, although continuations of this single fascia pass as intermuscular septa between the different muscle layers. The flexor retinaculum is a localized thickening of this single deep fascial sleeve, extending from tibia to the tuber calcaneus. Within the tarsal tunnel so formed with the underlying bones (Fig. 14.19A) the tendons of tibialis posterior, flexor tibialis and flexor fibularis are isolated by tough fascial septa which form individual compartments, and the neurovascular bundle lies in the loose tissue between the latter two tendons. The massive bipennate flexor fibularis enters the sole far laterally, adjoining the tuber calcaneus. Initially its canal passes between the medial and posterior tubercles of the talus, the latter of these being offset laterally in relation to the trochlea of the talus (Fig. 14.19A). It then courses down an obvious groove in the undersurface of the sustentaculum

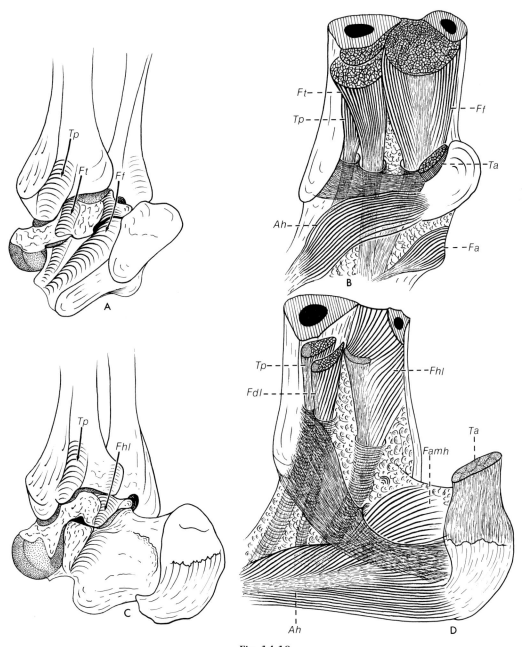

Fig. 14.19

The bony basis of the tarsal tunnel in *Pan troglodytes* (A) and the soft tissue components and contents of *Pan* (B). The similar bony arrangement in *Homo sapiens* (C) and the soft tissue relationships in *Homo sapiens* (D). *Ah*, abductor hallucis; *Ta*, tendo achillis; *Fdl*, flexor digitorum longus; *Famh*, medial head of flexor accessorius; *Fhl*, flexor hallucis longus. Other labelling is as in Figs 14.1, 14.2.

tali, bound here by a continuation of its fibrous tunnel. The flexor tibialis lies in an adjoining groove on the sustentaculum tali.

The deflexion laterally of the human heel splits the lower crural deep fascia into superficial and deep layers, opens up the tunnel, and relocates the long flexor tendons forward to the medial side of the ankle (Figs 14.19C, D). Moreover, provision is thus made for the descent of a large part of flexor fibularis to the foot to form the uniquely human medial head of flexor accessorius. Both superficial and deep layers of the crural fascia contribute to the flexor retinaculum. The deep layer forms the individual tunnels for the three tendons. That for the flexor fibularis (flexor hallucis longus) passes between medial and posterior tubercles on the talus, which are relocated in relation to the trochlea, and then in a groove underlying the sustentaculum tali. The flexor tibialis skirts the free margin of the sustentaculum but there is no clear-cut groove as in apes. The delaminated superficial layer of the retinaculum bounds the subdivision of the tunnel for the neurovascular bundle. Below, a septum (the interfascicular ligament) attaching to the calcaneus at the upper border of the origin of the medial head of flexor accessorius, divides this channel into upper and lower chambers for the medial and lateral plantar neurovascular bundles. Essentially the same description of the human tarsal tunnel has been given by Sarrafian (1983).

The monotreme arrangement

Several particular features of the monotreme hindlimb, specialized and primitive, are the keys to understanding the cruropedal musculature in these mammals. These were not appreciated at the time when the classical nineteenth century studies were done, with disastrous results.

In monotremes the fibula is prolonged proximally as a fibular crest. In fact, it represents a fused parafibula and not surprisingly ossifies by a separate centre (Barnett and Lewis 1958). In fact, the arrangement is comparable to the situation in these variant specimens of the marsupials *Phascolomys* and *Sarcophilus*, where the parafibula is similarly fused. There is little doubt that this is a monotreme apomorphic character: in the very primitive mammal *Eozostrodon* it is almost certain

that a free parafibula existed (Fig. 17.4A). Despite this specialization the monotreme fibula retained an articulation with the femur. It has been said that there is no trace of a femorofibular disc (Haines 1942). However, in the echidna at least a remnant of this disc persists as an appendage of the lateral meniscus, wedged in between femur and fibula and attached to the former bone by the ligament to be expected in this situation. As described in Chapter 11 monotremes also possess a free tibiale in the tarsus.

The musculature in *Tachyglossus* is shown in Figs 14.20 A, B; *Ornithorhynchus* differs from this in certain respects, which will be specified. The medial head of gastrocnemius arises as usual from the femur, but the lateral head has its origin from the fibular crest, and the two insert by a twisted tendo Achillis on the calcaneus; no soleus is separable from the lateral head. Plantaris is lacking.

Flexor fibularis arises from the upper two-thirds of the fibula and provides the deep flexor tendons to only the first, second, third, and fourth digits in *Tachyglossus*, but to all five in *Ornithorhynchus*. In the latter species the common tendon shows the typical helical fibre arrangement and includes a sesamoid within its substance (a similar parallel ossification has occurred in treeshrews). In both species a flexor accessorius is present (Figs 14.20A; 14.21A). No flexor perforatus is present in *Tachyglossus* but in *Ornithorhynchus* it is present, descended to the foot, and arising from the common tendon of flexor fibularis in the region of its sesamoid, and the adjacent calcaneus, and giving tendons to the second to fifth digits. Tibialis posterior also arises from the upper two-thirds of the fibula and its tendon adheres somewhat to the talus on entering the sole, but terminates on the tibiale. Flexor tibialis arises from the upper margin of the fibular crest and its small tendon enters the sole on the surface of that of tibialis posterior. It then expands into the plantar aponeurosis and also sends a fascial sheet down the hallux, which contains a fibrocartilaginous thickening in *Tachyglossus*, but a bony prehallux, articulating with the medial cuneiform, in *Ornithorhynchus*; this is exactly the derived type of insertion of flexor tibialis seen in so many Australian marsupials.

Rotator fibulae in *Tachyglossus* occupies the upper three-quarters of the interosseous tibio-

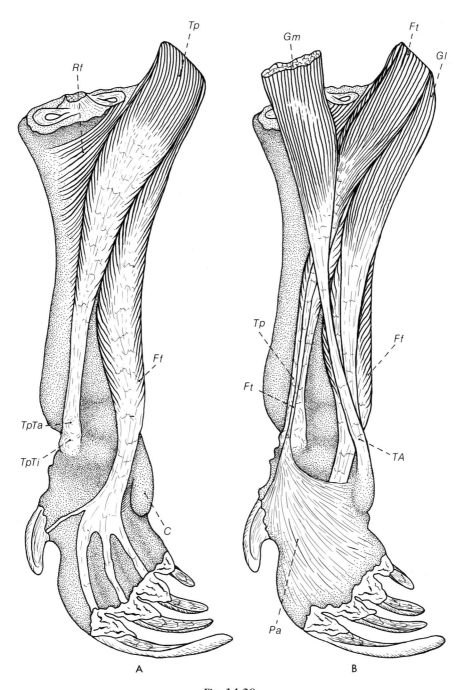

Fig. 14.20

A, the deepest muscle layer in the flexor aspect of the leg of *Tachyglossus aculeatus*. B, the complete musculature. *TpTa*, attachment of tibialis posterior to talus; *TpTi*, terminal attachment of tibialis posterior to tibiale; *TA*, tendon Achillis; other labelling as in Figs 14.1, 14.2. (From Lewis 1963.)

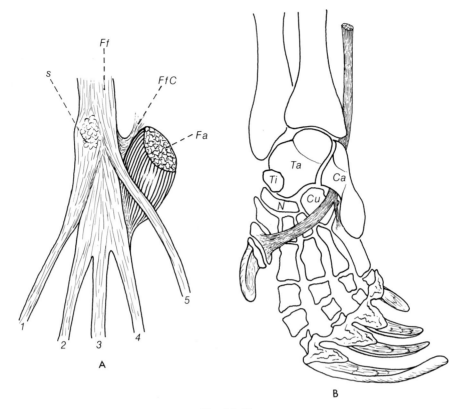

Fig. 14.21
A, the relationship of flexor accessorius to the flexor fibularis tendons in the right foot of *Ornithorhynchus anatinus*. B, the relationship of the right peroneus longus tendon to the tarsus in *Tachyglossus aculeatus*. *FfC*, attachment of flexor fibularis to calcaneus; *s*, sesamoid in the flexor fibularis tendon (*Ff*); *Fa*, flexor accessorius; *Ta*, talus; *Ti*, tibiale; *Ca*, calcaneus; *Cu*, cuboid; *N*, navicular. (From Lewis 1963.)

fibular space and its uppermost fibres attach to the lateral meniscus of the knee joint (which has been seen above to include the residual femoro-fibular disc). The muscle is restricted to the upper part of the space in *Ornithorhynchus*.

The muscle group under consideration has been described in *Ornithorhynchus* by Meckel (1826), Coues (1871), and Manners-Smith (1894), and in *Tachyglossus* by Mivart (1866) and Westling (1889). The identification of the various muscles given by these authors is shown in Table 14.1, and clearly many muscles have been incorrectly identified. These authors were unaware that the fibular crest included the parafibula. Also confusion over the complex history of the insertion assembly of the flexor tibialis led to errors in the identification of the two tendons passing behind the medial malleo-

lus. Furthermore, none of these authors recognized the tibiale as the ultimate and most distal attachment of the tibialis posterior tendon. Indeed, this bone has had an unfortunate history in the literature. Meckel (1826), in his osteological description, had clearly noted it, but by the latter part of his monograph had apparently come to confuse it with the spur-bearing os calcaris overlying the tibialis posterior and flexor tibialis tendons in male monotremes. This unhappy confusion between the os tibiale and the os calcaris persisted throughout the literature and the true os tibiale virtually disappeared from consideration.

When the correct identification of the muscles is appreciated, it is apparent that the monotremes present a specialized pattern of the basic mammalian arrangement shown in Figs 14.1, 14.2.

Table 14.1

Present study	Meckel	Mivart	Coues	Westling	Manners-Smith
Gastrocnemius, lateral head	Gastrocnemius, caput laterale	Soleus	Gastrocnemius, caput laterale	Soleus	Gastrocnemius, caput laterale
Gastrocnemius, medial head	Gastrocnemius, caput mediale	Gastrocnemius	Gastrocnemius, caput mediale	Gastrocnemius	Gastrocnemius, caput mediale
Rotator fibulae	—	Popliteus	Popliteus	Popliteus	—
Flexor fibularis	Flexor digitorum longus	Flexor digitorum longus	Flexor digitorum longus	Flexor digitorum longus	Flexor digitorum longus
Flexor perforatus	Flexor perforatus and lumbricals	—	Flexor digitorum brevis	—	Flexor digitorum brevis
Tibialis posterior	Soleus	Tibialis posterior	Tibialis posticus	Tibialis posticus	Soleus
Flexor tibialis	Tibialis posterior	Plantaris	Plantaris	Plantaris	Tibialis posticus
Flexor accessorius	—	Flexor accessorius	Dismemberment of flexor fibularis	Flexor accessorius	Flexor accessorius
Peroneus longus	Peroneus longus	Peroneus longus	Peroneus longus	Peroneus longus	Peroneus longus

Indeed Gregory (1947, 1951) used the monotreme hind-limb musculature as a significant point in formulating his so-called 'palimpsest theory' which postulated that the monotremes were derived from the arboreal Australian phalangeroid stem; this theory at first sight seems to provide a plausible explanation for similarities both in geographical distribution and in many structural features between the monotremes and Australian marsupials. In fact, Gregory took his data from Mivart. He adopted Mivart's figure but oddly enough gave the name peroneus longus to the muscle identified by Mivart as plantaris (the true flexor tibialis) and even added a broken line to give the impression of insertion of the tendon in that site typical of peroneus longus. The result was the remarkable and anomalous appearance of a leg with 'peroneus longus' on the tibial side and the flexor aspect! Clearly such a description of the limb does little to substantiate Gregory's thesis of affinity between marsupials and monotremes.

In fact, the explanation of affinity, which appears real enough, seems not to be that there was a dramatic reversal in evolution as envisaged in the palimpsest theory, but rather that a clear and recognizably therian pattern had already been established before the divergence of the ancestors of the highly aberrant and specialized extant monotremes.

There are even indications that the predecessors of the Monotremata already possessed a somewhat divergent grasping hallux. In the primitive marsupial foot peroneus longus clearly has the important function of approximating the divergent hallux to the rest of the foot. A large part of this tendon takes a similar course in the sole of the monotreme foot (Fig. 14.21B) and the obvious inference is that in the ancestors of both the Monotremata and the Marsupialia it had a similar action.

15

The extrinsic muscles of the foot—the extensors

The reptilian–mammalian transition

The primitive tetrapod arrangement of this muscle group, as determined by Brooks (1889) for *Sphenodon punctatum* (Fig. 15.1), is startlingly similar in basic plan to the comparable musculature in the forelimb (Fig. 8.1). It has a similar bilaminar arrangement with the superficial stratum divided longitudinally into three sectors and the deep stratum running obliquely down to the foot. However, only the intermediate sector of the superficial stratum in Sphenodon retains a femoral origin (from the external condyle). The tibial and fibular sectors have an origin below the knee, but this is doubtless a secondary specialization for a femoral origin for these marginal sectors is retained in certain primitive amphibia. The intermediate sector inserts into the metatarsal bases, but these tendons in *Sphenodon* are reduced to two; the likely primitive insertion, as in the forelimb, was to metatarsals two to five. The tibial sector arises from the tibia and inserts into the first metatarsal base, with a slip running onwards to the base of the proximal phalanx of the hallux. Likewise, the fibular sector arises from the fibula, inserts to the base of the fifth metatarsal, and sends a slip on down the digit.

The highest part of the obliquely disposed deep stratum arises from the fibula and the lowest part below from the tarsus. The fibres with the highest origin terminate as a tendon on the tibial side of the shaft of the first metatarsal; the remainder of the muscle provides five tendons passing to the phalanges of all the digits. These extensor brevis tendons are joined by a set of metatarsal heads—each of the three central tendons are joined by a pair, that for the hallux by a single one on the tibial side, but no metatarsal heads join the tendon to the fifth digit.

Notwithstanding the striking similarity of this basic pattern to that in the forelimb, the way in which it has been restructured in the mammals is quite different. Comprehensive accounts by Ruge (1878a), Frets (1908), Ribbing (1909), and Hunter (1925), incorporating the findings of many other, more restricted studies, have adequately documented the descriptive anatomy of the cruro-pedal extensors in a wide variety of mammals. Yet, despite this wealth of data, the phylogenetic picture has been confused, for wholly contradictory hypotheses have been proposed by Brooks (1889), Ruge (1878a), Frets (1908), and Ribbing (1909).

Most of those authors who have ventured into phylogenetic speculation, seem to have taken as their starting point the recognition that the human extensor digitorum brevis provides no tendon to the fifth digit, but instead not uncommonly a slip leaves the peroneus brevis tendon and mimics such a tendon—the so-called peroneus digiti quinti tendon.

Brooks' interpretation of the human homologues of the various primitive muscular components is indicated in Table 15.1. It is noteworthy that no contribution from the deep layer to the human fifth digit is recognized and that the common peroneus digiti qunti tendon is simply interpreted as a supernumerary attachment of peroneus brevis; after all, he had recorded an apparently similar slip in *Sphenodon* (Fig. 15.1).

Ruge (1878a) proposed an entirely different derivation for the mammalian extensor digitorum brevis denying its affiliation with the deep stratum of reptiles. He was much influenced by his recognition that a number of mammalian orders exhibit additional muscle bellies in the peroneal region, furnishing pedal tendons having topographical relationships similar to those of the extensor digitorum

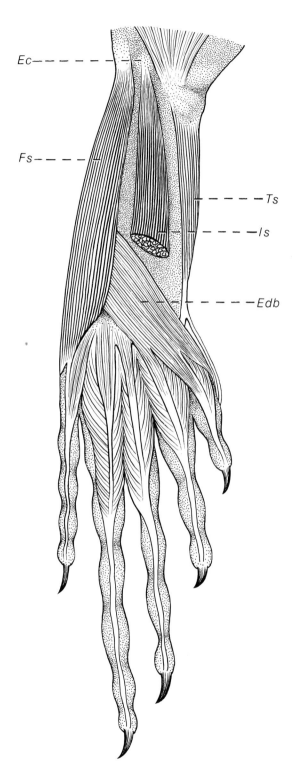

brevis of *Homo*. These bellies were not recognized by Brooks. Ruge regarded them as initially derived from the peronei, as homologous with extensor digitorum brevis, and as prone to undergo a progressive phylogenetic descent to the foot.

Frets (1908), however, denied a peroneal derivation of extensor digitorum brevis and concluded that the supernumerary muscle bellies often associated with the peronei, although inserting in the manner typical of extensor digitorum brevis, were of a totally different nature. Denying their phylogenetic descent to the foot, he suggested that they regressed during evolution and were subsequently replaced by a lateral enlargement of the true extensor brevis, invading the domain of the more lateral toes. The basis of Fret's hypothesis was the concept of nerve–muscle specificity. He reasoned that, since the superficial peroneal nerve supplies the peronei, together with their associated supernumerary bellies (which he called the peronei digitorum), and since the deep peroneal nerve supplies the true short extensor, then an extensor brevis traditionally described as innervated entirely by the latter nerve (e.g. in *Homo*) could not include descended peronei digitorum. He did admit, however, that the pedal extensor brevis of a few carnivores and edentates included a lateral component of peroneal derivation, for in these forms he demonstrated that the superficial peroneal nerve descended behind the lateral malleolus and innervated the lateral part of the extensor brevis. Such unswerving faith in the idea of nerve–muscle, specificity is unwarranted, as discussed in Chapter 3.

Ribbing (1909), like Frets, also maintained that the additional muscle bellies associated with the peronei, which he called collectively extensor digitorum lateralis, were not homologous with the

---

Fig. 15.1

The extensor aspect of the right hindlimb of *Sphenodon* (*Hatteria*), after Brooks (1889). *Ec*, external condyle of femur; *Ts*, tibial sector of superficial stratum; *Is*, intermediate sector of superficial stratum; *Fs*, fibular sector of superficial stratum. *Edb*, deep stratum (extensor digitorum brevis); the part labelled arises from the fibula, but the more postaxial heads arise lower from the tarsus; each of the central three tendons is joined by a pair of metatarsal heads and that for the hallux by a single metatarsal head on the tibial side; no metatarsal heads join the tendon to the minimus. (From Lewis 1966.)

Table 15.1 (Muscle homologies according to Brooks 1889; inconstant muscles are indicated in brackets)

	Hindlimb			Forelimb	
	Primitive tetrapod (Sphenodon)	Man	Man	Primitive tetrapod (Sphenodon)	
	Tibial sector	Tibialis anterior	Brachioradialis Supinator Extensor carpi radialis longus Extensor carpi radialis brevis	Radial sector	
Superficial layer	Intermediate sector	Extensor digitorum longus Peroneus tertius	Extensor digitorum longus	Intermediate sector	Superficial layer
	Fibular sector	Peroneus longus Peroneus brevis (Peroneus quinti digiti)	Extensor carpi ulnaris Anconeus	Ulnar sector	
Deep layer	Extensor digitorum brevis	Extensor hallucis longus Extensor digitorum brevis (Extensor hallucis brevis)	Abductor pollicis longus Extensor pollicis brevis Extensor pollicis longus Extensor indicis (Extensor medii digiti) (Extensor annularis) Extensor minimi digiti	Extensor digitorum brevis	Deep layer

lateral bellies of extensor digitorum brevis. Unlike Frets, however, he maintained that the extensor digitorum lateralis was a specialized development rather than a primitive mammalian feature, and had presumably replaced the lateral part of a primitively complete extensor digitorum brevis in certain mammals.

Both Ruge and Brooks based their respective hypotheses on inadequate data: Ruge, restricting his studies to mammals, took no cognizance of the primitive tetrapod deep layer, while Brooks, working only on an amphibian, a reptile and a monotreme, omitted consideration of the common supernumerary muscles associated with the true peronei. It would seem that the truth must include something of both views and I proposed (Lewis 1966) the following hypothesis to accommodate

all the data: (1) the fate of the primitive three subdivisions of the superficial layer, as recounted by Brooks (Table 1) may be accepted; (2) part of the deep layer extended proximally on the fibula as the extensor hallucis longus; (3) early in mammalian evolution the more lateral part of the deep lamina migrated proximally into the territory of the peronei, the digital tendons of these bellies becoming isolated along with those of the peronei behind an emerging lateral malleolus on the fibula; (4) subsequently these bellies underwent a progressive phylogenetic return to the dorsum of the foot, that for the fifth digit, however, achieving this in certain Prosimii only; (5) the intermediate sector of the superficial layer acquired new terminal tendons, fashioned from the deep fascia of the dorsum of the foot, extending the insertion from the metatarsal bases to the phalanges of the four postaxial digits.

The primitive marsupial arrangement

Clear echoes of the reptilian derivation are here retained and the pattern provides in most respects a sound prototype for the arrangements found in other mammals, both Metatheria and Eutheria. Only a rudimentary elaboration of a lateral malleolus is found on the fibula, and those tendons destined to be trapped behind it in the Eutheria here merely occupy a lateral groove in the expanded lower end of the bone. *Didelphys marsupialis* (Fig. 15.2) can be used as an example, acting as a prototype for discussion, although in certain respects it is an inadequate model and reference must be made to other mammals to fill out the primitive picture.

Superficial layer, tibial sector

This sector is unsegmented and is represented only by the tibialis anterior which arises from the upper quarter of the lateral surface of the tibia. Although a femoral origin of this sector doubtless occurred in the early tetrapods it seems likely that descent to the tibia had already occurred in the emergent mammals; the femoral origin found in certain mammals (tapir and hippopotamus), as described by Le Double (1897) is likely to be a secondary specialization. The muscle belly gives rise to a single undivided thick tendon passing anterior to the medial malleolus and inserting into the medial cuneiform bone. As noted in Chapter 14 the tendon reaches its insertion by passing deep to the marginal part of the terminal assembly of the flexor tibialis which here contains the bony prehallux (Fig. 14.4A). Interestingly, in the

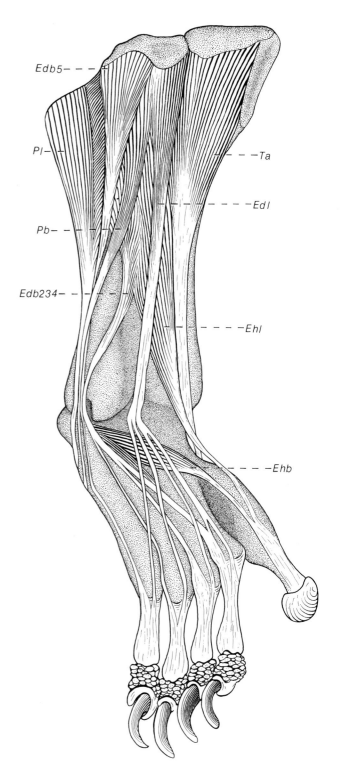

Fig. 15.2

The extensor aspect of the right leg and foot of *Didelphys marsupialis* showing the complete musculature. The peroneus longus and brevis have been parted to display the extensor digitorum brevis of digit five. *Edb 2–4*, extensor digitorum brevis of digits two, three, and four; *Edb5*, extensor digitorum brevis of fifth digit; *Edl*, extensor digitorum longus; *Ehb*, extensor hallucis brevis; *Ehl*, extensor hallucis longus; *Pb*, peroneus brevis; *Pl*, peroneus longus; *Ta*, tibialis anterior. (From Lewis 1966.)

marsupial *Trichosurus vulpecula* (Fig. 14.4B) the tibialis anterior tendon has become adherent to the overlying prehallux, thus acquiring an additional insertion: the muscle then terminates in dual tendons attaching to the medial cuneiform and prehallux respectively. Except for the higher Primates, such a dual insertion of tibialis anterior appears to be unique among mammals and probably parallels the apomorphic change which must have occurred in ancestral members of the Anthropoidea—the higher Primates.

Superficial layer, fibular sector

The peroneus longus and brevis muscles comprise the fibular component of the superficial stratum. Peroneus brevis arises high in the leg from the intermuscular septum between it and extensor digitorum longus, and by a few fibres from the anterior aspect of the fibula just below the passage of the deep peroneal nerve to the front of the leg. Its tendon is moulded on to the outer aspect of a bony prominence, proximal to the lateral malleolus, produced by lateral bowing of the fibular shaft in this species. Below this the tendon lies within the groove on the lateral aspect of the fibular lower extremity, there being no true lateral malleolus, and then passes to an insertion into the fifth metatarsal base. Peroneus longus arises from the fibular head and upper half of the lateral aspect of the shaft. This would suggest that the primitive femoral origin had already been relinquished in early therian mammals. However, in certain highly specialized mammals (beaver, otter, seal, mole) a femoral origin, which replaces the long lateral ligament of the knee joint, is found; moreover in a number of Primates there is continuity between the muscle and the ligament. It seems plausible to suppose that the ligament was derived by transformation from the muscle, probably at a premammalian grade, and has atavistically reverted to muscle in a few mammals (Vallois 1914). The peroneus longus tendon runs together with that of peroneus brevis to the foot, where it first gives a slip to the fifth metatarsal base and then traverses a groove in the cuboid to continue towards its ultimate insertion into the base of the first metatarsal. There is little doubt that the peroneus longus has shifted its major insertion across the foot from the fifth metatarsal (the primitive attachment of the fibular sector as a

whole) to the first metatarsal by a mechanism such as that shown in Fig. 3.7A; such a phylogenetic transfer is well established (Ruge 1878*a*). This is recapitulated in human ontogeny.

Superficial layer, intermediate sector

Any indications that the marginal sectors of the superficial layer retained a femoral origin in emergent mammals are dubious in the extreme; the few mammals showing such an arrangement are highly specialized and probably exhibit an atavistic reversion to the primitive tetrapod origin. The same cannot be said for the intermediate sector, where the sole mammalian derivative is the extensor digitorum longus. In *Didelphys* (Fig. 15.2) this muscle arises from the head of the fibula and from an intermuscular septum shared with the adjoining peroneus brevis, but in other marsupials (*Macropus*) a tendinous expansion of the origin is prolonged up to the lateral condyle of the femur, and in many eutherian mammals the origin of the muscle is by a cord-like tendon from the femur (Kaplan 1958). It seems that a femoral origin probably persisted in the emergent theria, although descent to the fibula has commonly occurred. In *Didelphys* the fibular belly of the muscle overlies the deep peroneal nerve as it winds around the fibula onto the front of the leg, and the muscle belly finally furnishes tendons which terminate in the extensor expansions on all the digits save the hallux. In the Australian marsupial *Trichosurus vulpecula* the origin of the muscle has extended half-way down the fibular shaft, bridging over the passage of the deep peroneal nerve, and thus paralleling the condition to be found in most Primates. Moreover, in this species the reduced and syndactylous second and third digits receive only a slender distally divided tendon and those for the fourth and fifth digits are united by a junctura intertendinea, comparable to those found in the forelimb.

There is little doubt that this mammalian transfer of insertion from the metatarsal bases to the phalanges has been effected by tendinous transformation within the deep fascia of the dorsum of the foot, just as has been described for the forelimb (Chapter 8). Indeed in *Homo*, and in other Primates, these tendons are still incorporated in this fascial sheet, and the embryological evidence further supports this view. Thus, Bardeen

(1906) showed that during ontogeny extensor digitorum longus is initially attached to the metatarsals, later extending towards the digits as an undivided tendon plate. Variation in the subdivision of this aponeurotic sheet into discrete tendons is therefore to be expected: in *Lemur catta* the whole sheet persists, and in *Trichosurus*, and occasionally in man (Testut 1884; Le Double 1897), juncturae intertendineae, comparable to those of the hand, are found; on the other hand a metatarsal insertion may be retained in mammals (sloth). The peroneus tertius, found only in man and occasionally in *Gorilla* (Straus 1930), seems to represent a partial retention of this ontogenetically and phylogenetically early attachment. There are no grounds for Jones' (1944) speculation that peroneus tertius represents the apparently missing short extensor of the fifth digit. Extension of this same mechanism may give rise to a hallucial tendon arising from extensor digitorum longus, occurring normally in the seal, and anomalously in man.

Deep layer

Just as the forelimb (Chapter 8) this stratum of short or deep extensors has migrated proximally into the leg, but this migration in the hindlimb is incomplete, for part remains in the foot. Furthermore, that highest part in the primitive tetrapods (Fig. 15.1), which inserts on the tibial side of the first metatarsal has regressed (the comparable part in the forelimb forms abductor pollicis longus). The highest surviving portion has ascended up to the middle third of the fibula as the extensor hallucis longus, and its derivation from the deep stratum is emphasized by the close association which it retains with the modified remainder of the deep layer. Its unipennate belly terminates in a tendon closely related to that of tibialis anterior, in front of the medial malleolus, and which inserts by means of an expansion into the terminal phalanx of the hallux. Part of the deep stratum associated with the hallux has conservatively retained an origin from the calcaneus, immediately below the lateral malleolus, forming the extensor hallucis brevis, whose tendon joins that of the long extensor to form a common extensor expansion for the hallux. As noted above the extensor digitorum longus anomalously in man, and normally in the seal, provides a hallucial tendon; this is clearly a

progressive specialization. Yet it has been interpreted as support for a quite unwarranted view, espoused by Ribbing (1909), Jones (1944), and Howell and Straus (1961), that the extensor hallucis longus is a derivative of the superficial extensor.

The remainder of the deep stratum gives tendons to the second to fifth digits, and most of the muscle substance of this part has ascended into the leg. The imperfectly separated bellies terminating as tendons to digits two, three and four arise from the middle third of the fibula, where they are in close association with extensor hallucis longus, although distally they are separated from the latter by the rudimentary lateral malleolus. A small part of the muscle substance retains a calcaneal origin, superficial to that of extensor hallucis brevis, and this belly is closely associated with the tendon of extensor digitorum brevis to the second digit, which overlies it and which is reinforced by its tendon.

The belly for the fifth digit, a more proximal extension of the same layer as the other bellies, arises from the proximal third of the fibular and is buried deeply between the peroneus longus and brevis. Its tendon, accompanying those of the two peronei, is separated in the lower leg from the tendons of the extensores breves of digits two, three, and four by the bony prominence produced by the angulation of the fibular shaft. Finally all the extensor brevis tendons come to lie deeply to those of the peronei in a lateral groove on the distal fibular extremity. Entering the foot, they radiate across its dorsum, deep to those of extensor digitorum longus, to reach insertions into the lateral wings of the extensor expansions on the lateral four digits.

The arrangements in this species leave little doubt that the extensor hallucis longus and the extensores breves two to five are ascended segments of the same muscle sheet—the oblique deep lamina of Amphibia and Reptilia. The elaboration of the lateral malleolus, and the entrapment behind it of part of this sheet, have not progressed to the point where this affinity is greatly obscured.

In *Trichosurus vulpecula* the extensores breves are modified in a way that is noteworthy, for it shows a trend paralleled in the Primates. The extensores breves two and three have returned to the dorsum of the foot where they arise, below and behind the lateral mammeolus, by a slender tendon of origin from the surface of the calcaneo-fibular ligament. The extensores breves four and five retain a fibular origin between the two peronei. Extensor hallucis brevis is lacking in this marsupial.

The whole sequence and mode of the ascent of the bellies in the deep stratum is in many respects very similar to that which occurred in the forelimb (Fig. 8.2) and presumably was achieved by a similar mechanism. The extensor hallucis longus has clearly ascended into the territory of the extensor digitorum longus and the extensores breves two to five into the territory of the peronei. Doubtless this has occurred by recruitment of myoblasts from the appropriate part of the superficial layer, which would otherwise have been incorporated into extensor digitorum longus or the peronei. Indeed, Bardeen (1906) described the development of the belly of extensor hallucis longus from the deep surface of the anlage of extensor digitorum longus.

The muscular pattern then has really only changed in the relative proportions of the component muscles. There is no logical basis for homologizing the muscle bellies of the extensores breves two to five, in their new crural situation, with the peronei, although in a sense they are probably so derived ontogenetically. Comparative embryology abounds with examples of structures changing their ontogenetic derivation, but nevertheless accepted as homologues: as has been repeatedly stressed, the concept of homology has implicit within it the notion of phylogenetic continuity. The mechanism which alters the domains of individual muscle bellies has already been noted in a number of other situations and is clearly a commonly operating evolutionary strategy. Recognition of this process provides the logical basis for weighting muscle insertions (which can also change of course) more than origins in determining homologies.

By setting such a process in reverse, the muscle bellies of the extensores breves two to five can descend from the primitive therian peroneal situation (Fig. 15.2) to the dorsum of the foot. This trend of descent seems to have occurred convergently several times. Marsupials must primitively have had crural bellies for the tendons to digits two to five, but *Trichosurus* retains crural bellies for the lateral two digits only. Such an arrangement has also been acquired in rodents, whilst in carnivores

only the belly for the fifth digit has retained a fibular origin. As will be seen, among the Primates generally the extensor digitorum brevis of digit five alone (but occasionally also the belly for digit four) arises in the leg, the others having descended. In the Hominoidea, however, the muscle belly of extensor brevis five (usually crural but sometimes pedal) is only rarely present, but its digital tendon, separating from the termination of the peroneus brevis tendon, is commonly found. That the true nature of this tendon has been poorly understood is indicated by the variety of names given to it—'peroneus quinti digiti', 'extensor proprius quinti digiti', 'peroneus medius', 'peroneus accessorius', 'peroneus parvus', and even most inappropriately 'peroneus tertius'.

The monotreme arrangement

This muscle group has been described in *Ornithorhynchus* by Brooks (1889). I have confirmed his observations and also observed a comparable pattern, modified by the loss of certain digital tendons, in *Tachyglossus aculeatus*. As noted above, a key aspect of the transformation of this muscle group into a typically mammalian pattern has been the proximal migration of the deep stratum. Comparably, in Monotremata the deep lamina has migrated into the leg, but in this case leaving no residual portion, no extensor hallucis brevis, arising in the foot. Moreover, this whole sheet retains its deep position and unity, for there is not even an emergent lateral malleolus to separate the part destined for the hallux (extensor hallucis longus) from the remainder of the muscle. In monotremes even the peroneus longus has achieved its characteristically mammalian arrangement, with most of the insertion transferred across the sole to the first metatarsal (Fig. 14.21B).

Yet certain ancestral reptilian characteristics are retained. Tibialis anterior sends a slip down the hallux to join the extensor expansion there, just as the tibial sector did in *Sphenodon* (Fig. 15.1); indeed the part of the muscle concerned is delaminated as a supernumerary belly. Similarly peroneus brevis sends a slip down the fifth digit, like the peroneal sector in *Sphenodon* (Fig. 15.1). These features prompted Ruge (1878a) mistakenly to identify the supernumerary tibial telly as extensor hallucis longus, and the whole peroneus brevis as extensor brevis digiti quinti; unlike Brooks, he was, of course, unfamiliar with the arrangements in *Sphenodon*.

Despite these reptilian characters, the overall appearance is compellingly mammalian. As with the comparable musculature in the forelimb, much of the characteristic mammalian pattern must have been evolved already, by the time of divergence of the monotremes.

The Primate arrangement

Bearing in mind the principles discussed, the diverse primate patterns can readily be derived from an ancestral arrangement similar to that shown in Fig. 15.2. Only those significant features in which a particular muscle group departs from this basic pattern will be specially described.

The prosimian pattern

Lemur catta As in all other Primates the lateral malleolus is well developed. Only minor changes have occurred in the superficial muscle layer, with the origins of extensor digitorum longus and peroneus longus reaching across from the fibula onto the lateral tibial condyle. The extensor digitorum longus expands into an aponeurotic sheet in the foot, from which the tendons to the lateral four digits are derived, again an indication of the secondary derivation of these tendons from deep fascia. The tibialis anterior has a single insertion into the medial cuneiform. Peroneus brevis inserts into the fifth metatarsal and peroneus longus into the first, with a mere fascial attachment to the fifth. Extensor hallucis longus also has extended its origin across to the lateral tibial condyle, and its tendon, passing to the usual insertion, is closely bound to that of tibialis anterior by the naviculo-metatarsal ligament (Fig. 14.7).

The most noteworthy changes concern the extensor digitorum brevis. Only the belly for the fifth digit remains a fibular origin, between the peronei, and its tendon as usual enters the foot behind the lateral malleolus. The bellies serving the remaining four digits arise from the calcaneus: significantly those for the fourth and third digit being most superficial and arising most posteriorly, below the lateral malleolus, where they are overlaid by the peroneal tendons. All the extensor brevis tendons take the usual deep course to their insertions, that for the hallux being very tenuous.

Galago mohili Despite the striking specializations of the foot of this prosimian—great elongation of the navicular and of the distal part of the calcaneus—no fundamental modifications have been imposed upon the basic prosimian muscular pattern represented by *Lemur*. The insertions of tibialis anterior and of peroneus brevis, which are unchanged, are carried far forward in the foot, as is the passage of the peroneus longus tendon to the sole. Again the extensor brevis of the fifth digit has a high fibular origin between the peronei, and its tiny belly terminates in a thread-like tendon which passes with those of the peronei behind the lateral malleolus. The extensores breves one to four arise from the calcaneus.

The monkey pattern

Only tibialis anterior and the extensor digitorum brevis group merit special description, because of the way in which they depart from the ancestral mammalian pattern.

Tibialis anterior It would seem that in the ancestral higher Primates tibialis anterior must have become adherent to the prehallux in a manner analogous to that found in *Trichosurus* (Fig. 14.4B), thus providing dual insertions for the muscle, and initiating subdivision of its tendon. The ossicle-containing part of the flexor tibialis tendinous expansion was divorced in higher Primates to form the naviculo-metatarsal ligament but the prehallux, whilst retaining its association with this ligament, became closely related to the base of the first metatarsal and entered into the composition of the first tarso-metatarsal joint. This situation persists in extant platyrrhine monkeys and in gibbons, as described in Chapter 13. In Old World monkeys, the great apes and man the prehallux has disappeared, either by suppression or by its incorporation into the first metatarsal base, so that the supernumerary tibialis anterior tendon has come to insert into the first metatarsal base, whilst still retaining an intimate relationship with the hallucial tarso-metatarsal joint.

In New World monkeys (e.g. *Cebus, Pithecia*) not only the tendon of the tibialis anterior is subdivided but even the muscle belly may be: one tendon, the larger, inserts on the medial cuneiform, and the other more slender one onto the prehallux (Fig. 13.17A). These two tendons, together with that of extensor hallucis longus, are bound in typical Primate fashion to the medial border of the foot by the naviculo-metatarsal ligament (Fig. 14.4C).

In Old World monkeys again the tibialis anterior has dual tendons with variably separate muscle bellies, but in this case the smaller of the tendons inserts onto the base of the first metatarsal (Fig. 13.17B) in common with the naviculo-metatarsal ligament.

Extensor digitorum brevis   In both New World and Old World monkeys the bellies of extensor digitorum brevis serving the hallux and digits two, three, and four arise from the calcaneus under cover of the peroneal tendons. Typically the extensor digitorum brevis of the fifth digit retains a peroneal location often arising quite high up from the fibula between the two peroneal muscles (Fig. 15.3B) but sometimes low down with its belly partly descended into the foot (Fig. 15.3A); in either case the tendon accompanies that of peroneus brevis, parting company from it only at the base of the fifth metatarsal, to reach the usual digital insertion. In at least some Old World monkeys (e.g. *Colobus polykomos*) there may be no persistent muscle belly, neither crural nor pedal, serving the fifth digit; a tendon with the typical insertion is then, however, present separating from that of peroneus brevis just below the lateral malleolus: the muscle belly itself must have either been suppressed or amalgamated with that of peroneus brevis.

The ape pattern

Again only tibialis anterior and extensor digitorum brevis merit special description, the rest of the muscle group conservatively conforming to the well-established basic mammalian pattern.

Tibialis anterior   In *Hylobates* the muscle belly is split in its lower portion and of its two derivative tendons the larger one inserts as usual into the medial cuneiform but the smaller terminates on a separate prehallux lying in a position comparable to those described above in the platyrrhine mon-

keys. These tendons, together with that of extensor hallucis longus, are bound down by a typical primate naviculo-metatarsal ligament, which is well developed in gibbons. Alone among the catarrhine primates, the gibbons have retained a prehallux. In the other apes (and man) it has regressed to leave a supernumerary metatarsal insertion for the muscle, just as in the Old World monkeys. This metatarsal tendon again inserts in association with the naviculo-metatarsal ligament (Fig. 15.5A) to the base of the first metatarsal in *Pan*; in *Pongo* the ligament is rather ill-defined, yet still ties the tibialis anterior tendons into close association with that of extensor hallucis longus; in *Gorilla* (Raven 1950) the ligament is well defined, and again overlies the tendons of extensor hallucis longus and tibialis anterior.

Extensor digitorum brevis   Extensor digitorum brevis of digits one to four is restricted to the dorsum of the foot. The extensor digitorum brevis of the fifth digit may be present in gibbons. The only example which I have seen arose as usual from the fibula between the peronei, but its tendon on entering the foot in the usual way, failed to attain the usual insertion, and finally merged with that of peroneus brevis; in other gibbon specimens neither muscle belly nor tendon has been found. In *Pan* the only remnant of extensor brevis five is a tendon, having the usual digital insertion, separating from that of peroneus brevis near its insertion, as noted above for *Colobus*. The belly of extensor digitorum brevis providing the hallucial tendon is largely separate as an extensor hallucis brevis, and its tendon terminates on the base of the proximal phalanx of the hallux, failing to reach the extensor expansion.

The human pattern

The arrangement described for *Pan*, where the phylogenetic history seems to be quite clear, differs from the corresponding disposition in *Homo* only in minor details. The splitting of the tibialis anterior tendon in man is generally restricted to its terminal part, although occasionally the muscle belly itself may show duplication. The relationship of the tendons to the first tarsometatarsal joint has been described in Chapter 13. With realignment of the human hallux to the other digits the extensor hallucis longus tendon is no longer held in

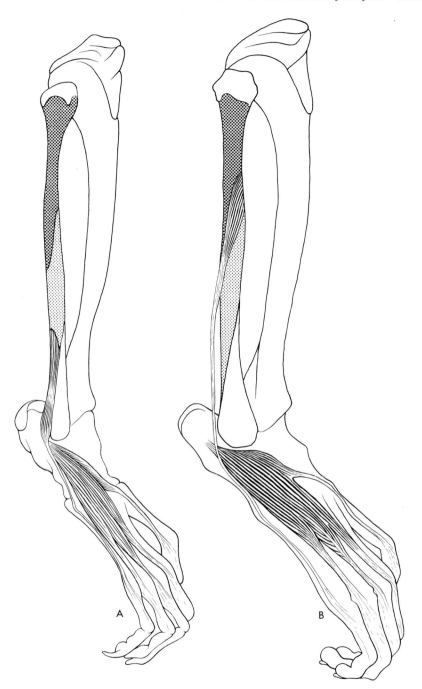

Fig. 15.3
The bones of the right leg and foot of (A) *Cebus nigrivittatus*, (B) *Procolobus verus*. The extensor digitorum brevis is shown in each case; the extensor digitorum brevis of digit five arises from high on the fibula in (B) and from the lower part of the fibula in (A). In each case the origin of peroneus longus is shown in dark stipple and that of peroneus brevis in light stipple. (From Lewis 1966.)

intimate association with the tibialis anterior tendons by the naviculo-metatarsal ligament, as in other Primates (Fig. 14.4C), and although no naviculo-metatarsal ligament as such is usually recognized, a probable homologue of it is sometimes found (Fig. 15.5B). The extensor hallucis longis in *Gorilla*, having passed deep to the naviculo-metatarsal ligament, peels off some fibres to attach to the base of the proximal phalanx (Raven 1950); similarly, it is a common finding in man to have a similar supernumerary insertion.

As in *Pan* an extensor hallucis brevis is largely separated from the remainder of the pedal part of extensor brevis associated with digits two, three, and four. Again as in *Pan* usually the only representative of the extensor brevis of the fifth digit is a tendon separating from that of peroneus brevis, but this is inconstant. Interestingly, in man the lateral part of the extensor brevis muscle is known to be often innervated (22 per cent) by a branch of the superficial peroneal nerve (the so-called accessory deep peroneal nerve) descending behind the lateral malleolus (Lambert 1969). This is quite reasonable in view of the derivation of this part of the muscle by descent from the leg.

Despite the essentially similar arrangement of the extensor muscle group in the forearm and in the leg of primitive tetropods, the restructuring in mammals has occurred along radically different lines in the two limbs. Brook's (1889) view (Table 15.1) detailing the derivation of the various mammalian muscles can be largely accepted, except for his confusion over the nature of peroneus digiti quinti. It is clear that any strict description of muscle homotypes between the two limbs is unreasonable. However, anatomists familiar only with the human cadaver have attempted this. For instance, there may be a plausible resemblance in origin and insertion between the abductor pollicis longus and the tibialis anterior. Yet from the phylogenetic perspective it is quite clear that these two muscles are derived from different myological strata in the two limbs, their evolution to the human condition has occurred along quite different lines, and they are in no sense homotypes.

The extensor retinaculum

The extensor retinaculum in the forelimb is an unremarkable structure, a mere thickening of the antebrachial deep fascia which overlies the extensor tendons as they enter the hand. Attachments of the sheet to bony excrescences between individual tendons, or groups of tendons, subdivide the region under it into compartments. In contrast, the extensor retinacular apparatus in the hindlimb is complex, with an interesting evolutionary history (Figs 15.4, 15.5).

Marsupials with fairly generalized feet (*Didelphys, Trichosurus*) show it in what is apparently its basal simplicity. In essence it is a quite complex pulley system, fashioned from deep fascia, and beautifully adapted for restraining the extensor tendons from bowstringing medially when the foot is in its characteristically inverted position. It consists of two independent parts (Fig. 15.4A). The first arises from the lower end of the fibula, crosses over the extensor tendons as a strong almost transverse band, which then loops back deep to the tendons in the region of the tibial malleolus (with no significant attachment thereto) as a deep reflected band which attaches independently to the upper aspect of the calcaneus. In this situation a second very tough loop arises. The first loop includes all the extensor tendons, the latter only the extensor digitorum longus. The upper loop is not unlike the forelimb flexor retinaculum in marsupials, and indeed the evolutionary modifications which it undergoes are rather similar.

In monkeys such as *Cebus*, this arrangement is largely retained but with some minor progressive modifications (Fig. 15.4B). The upper loop, or at least its upper part, has acquired a substantial tibial attachment, but the deep reflected band persists unchanged below. This band has acquired a small additional attachment to the talus as it passes on towards its main insertion to the calcaneus. The tough lower loop investing the extensor digitorum longus tendons is unchanged. Essentially the same arrangements have been described in Old World Monkeys by Stamm (1931).

Even in the chimpanzee the situation still shows relatively little further elaboration (Fig. 15.5A). The upper loop is substantially attached to the tibial malleolus, at least in its upper part: the transverse limb has become recognizable as a superior extensor retinaculum. Some of its constituent fibres peel away early to join the reflected band, thus creating a separate compartment for

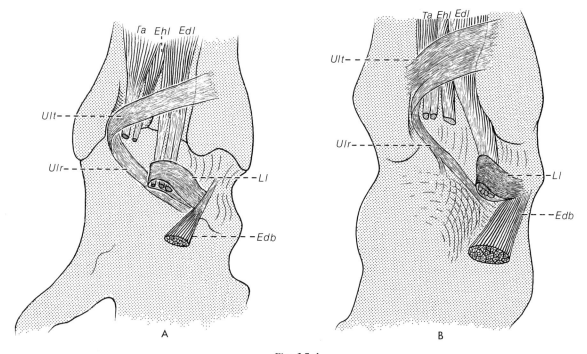

Fig. 15.4

The extensor retinaculum in the left foot of A, *Trichosurus vulpecula* and B, *Cebus nigrivittatus*. *Edb*, extensor digitorum brevis; *Ult*, transverse band of upper retinacular loop; *Ulr*, reflected band of upper retinacular loop; *Ll*, lower retinacular loop. Other labels are as in Fig. 15.2.

the tibialis anterior tendon. The reflected band is again present, again largely formed by recurrent fibres from the lower part of the transverse band, but now with some direct origin from the tibial malleolus. As usual this reflected part attaches independently to the calcaneus, but has a minor additional attachment to the talus as in monkeys. The tough lower loop for the extensor digitorum longus tendons is unchanged.

In man, the whole retinacular system is substantially modified, yet it still retains clear reminders of these past evolutionary stages. Its description has had a chequered history, and the relative simplicity of the arrangements have become greatly confused.

Descriptively there are two extensor retinacula: superior and inferior. The superior one is the clear derivative of the transverse band of the upper loop, and this was already foreshadowed in *Pan*. Spanning the interval between the fibular and the tibial malleolus, it blends above with the deep fascia of the leg. In about 25 per cent of cases there is a separate tunnel beneath it for the tibialis anterior

tendon, just as there was in the chimpanzee specimen shown in Fig. 15.5A.

The inferior extensor retinaculum is traditionally described as Y-shaped, having a lateral stem and oblique superomedial and inferomedial bands. It may sometimes, however, additionally have an oblique superolateral band which converts it into an X-shaped form. Weitbrecht (1742) so described and illustrated it, naming it the ligamentum cruciatum cruris; this name has persisted in the literature despite the fact that in most cases it is descriptively inappropriate.

The stem of the retinaculum retains the looped character it has had throughout mammalian evolution. Because of its sling-like morphology the French anatomists (following Retzius) called it the 'ligament en fronde'; in English works this has been translated as the frondiform ligament, although it is usually called, with similar implications, the fundiform ligament. The deep reflected band of the primitive upper loop has become fused to its deep surface. Thus, to the original calcaneal insertion of the deep limb are added new medial

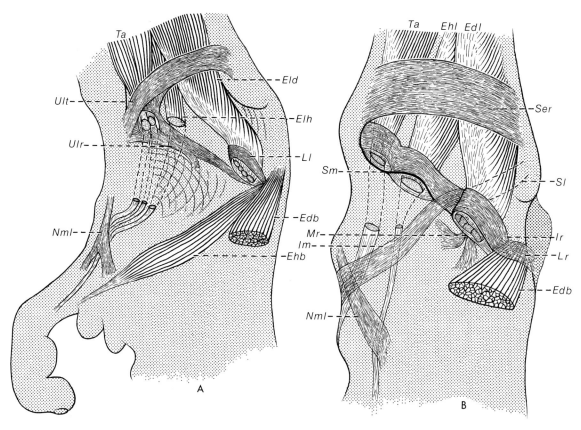

Fig. 15.5

The system of extensor retinacula in A, *Pan troglodytes* and B, *Homo sapiens. Nml*, naviculometatarsal ligament; *Ser*, superior extensor retinaculum; *Lr*, lateral root of fundiform ligament; *Ir*, intermediate root of fundiform ligament; *Mr*, medial root of fundiform ligament (three parts); *Sm*, superomedial band of inferior extensor retinaculum; *Im*, inferomedial band; *Sl*, the occasional superolateral band. Other lettering is as in Figs 15.2 and 15.4.

calcaneal and talar attachments penetrating deeply into the tarsal canal; this is the phylogenetic derivation of the three attachments of the deep limb, comprising the 'medial root', which have figured in purely descriptive human anatomical studies (Smith 1896; Cahill 1965; Sarrafian 1983). The superficial limb of the sling remains as the 'intermediate root' and this is prolonged down over the extensor digitorum brevis, by the incorporation of deep fascia (the 'lateral root') to merge with the inferior peroneal retinaculum.

Merging of the reflected band of the upper sling with the deep surface of the frondiform ligament has created the oblique superomedial band of the inferior retinaculum as an apparent prolongation of the stem, passing up to attach to the medial tibial malleolus. Man has here shown a unique modifica-

tion of this band: an additional layer, derived from the deep fascia, has become applied to the superficial surface of the band (Fig. 15.5B). Commonly, this layer is very thick over the extensor hallucis tendon and very thin over the tibialis anterior; conversely, the deep layer (the original reflected band) is commonly thick beneath tibialis anterior, and thin (or even absent) deep to extensor hallucis longus. The main tract of fibres in the superomedial band then weaves from deep to the tibialis anterior tendon onto the surface of the extensor hallucis longus tendon.

The inferomedial band is also an apomorphic human acquisition. Again it is an elaboration of the deep fascia covering the tendons. The inconstant oblique superolateral band has a similar derivation.

The many variations of the inferior extensor retinaculum (Sarrafian 1983), which to the purely descriptive human anatomist may seem to present a meaningless random array, are satisfyingly encompassed within the theme of an understanding of the phylogeny of this retinacular apparatus.

The aligning of the hallux with the other digits in the human foot has apparently brought about other changes in the fascial retaining systems. Throughout primate evolution the tendons of tibialis anterior and extensor hallucis longus have been held in intimate association right down to the base of the hallucial metatarsal by the naviculo-metatarsal ligament (Fig. 15.5A). In man these tendons part company as they pass down the dorsum of the foot. Yet there sometimes seems to be a remnant of the naviculo-metatarsal ligament, lying deep to the inferomedial band of the inferior extensor retinaculum (Fig. 15.5B). The structure figured by some authors (e.g. Sarrafian 1983) as the 'medial transverse retinacular band of the dorsum of the foot' is presumably the same thing.

16

Intrinsic muscles of the foot

The reptilian–mammalian transition

The early tetrapod foundation for the intrinsic musculature of the foot was essentially similar to that in the hand. McMurrich (1907) described the same layered arrangement as for the hand in the sole of the foot of amphibia and reptiles: a flexor brevis superficialis sandwiched between layers of the plantar aponeurosis; a flexor brevis medius, subdivided into a superficial and a deep stratum; a flexor brevis profundus consisting of a series of slips on the surface of the metatarsals; most deeply a series of intermetatarsal muscles. The main mammalian derivatives of these layers are as for the hand. The flexor brevis superficialis slips are converted into the flexor perforatus tendons and the layer of plantar aponeurosis deep to it is subdivided to form the long perforating flexor tendons. The flexor brevis medius, stratum superficiale, provides the lumbricals and the stratum profundum forms a layer of muscles adducting the digits. The flexor brevis profundus forms a series of short bicipital flexors for the digits; as they are the only short flexors remaining as such in the mammalian foot, they will now be referred to merely as flexores breves. The intermetatarsal muscles form a deep layer abducting the digits. This basic trilaminar arrangement of the musculature in the sole of the mammalian foot, identical in overall plan to that of the hand, was firmly established by Cunningham (1882), whose studies on the foot were much more extensive than his similar work on the hand.

From the records of Cunningham, supplemented by my own material, it is easy to establish the basic marsupial groundplan (Fig. 16.1). The superficial layer consists of four adducting muscles, or contrahentes, radiating into the palm and insert-ing onto the proximal phalanges and the sesamoid-containing glenoid plates on the fibular side of the first and second digits, and the tibial side of the fourth and fifth. There are flexores breves associated with the plantar aspect of all five metatarsals. Cunningham described these as five bicipital muscles; it is preferable, however, (as in the palm) to consider them as ten paired muscles, the members of each pair being confluent at their origins. These flexores breves insert largely on the sesamoid-containing glenoid plates but have some continuity dorsally with the extensor expansions; perhaps the latter, as in the hand, was the primitive insertion but it has certainly commonly been largely superseded by arrest of the attachment at the sesamoids. In the intermetatarsal spaces are four dorsal abducting muscles. Their metatarsal origins are more restricted than in the palm, and this a bipenniform character is less obvious. The muscles typically insert by tendons into the bases of the proximal phalanges, being so arranged as to abduct about an axis formed by the middle digit (Fig. 16.1B). The third and fourth muscles differ, however, in a manner which seems to be primitive, for their tendons bifurcate so as to pass to adjacent proximal phalanges either side of the interdigital space. Moreover, these muscle bellies tend to be delaminated, in order to service this dual insertion, and so a fourth muscle layer is effectively formed here. The muscle abductor hallucis has the core of its origin from the prehallux (cf. the corresponding muscle in the hand) but may spread onto adjoining bones, and it inserts onto the base of the proximal phalanx of the hallux. The abductor digiti minimi arises from the calcaneal tuberosity and inserts on the base of the proximal phalanx of the fifth digit. A deep bundle has its insertion arrested on the base of the fifth

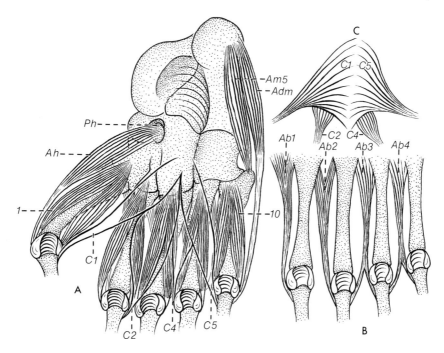

Fig. 16.1

A, the arrangement of the musculature in the sole of the primitive marsupial foot. *Ah*, abductor hallucis arising from the prehallux (*Ph*); *Adm*, abductor digiti minimi; *Am5*, abductor ossis metatarsi five; the ten flexores breves are shown, (numbers *1*, *3*, and *10* are specifically indicated; the dorsal abductors are largely obscured by the flexores breves; the four overlying contrahentes, for digits one (*C1*), two (*C2*), four (*C4*) and five (*C5*) are shown in dark outline only. B, the arrangement of the four dorsal abductors (*Ab1*, *Ab2*, *Ab3*, *Ab4*) after removal of the contrahentes and flexores breves. C, a common derived arrangement of the four contrahentes found in marsupials.

metatarsal as an abductor ossis metatarsi quinti digiti. The abductors of the marginal digits quite obviously belong to the same muscle layer as the intermetatarsal abductors.

The whole aspect is one of basic symmetrical simplicity, which is virtually identical in pattern to that in the palm. In contrast to more advanced mammals, this resemblance is largely a consequence of the fact that the calcaneal tuber of the generalized marsupials shows a relatively modest prolongation backwards as a heel (Fig. 16.1A). In the emergent mammals (Fig. 17.2) no doubt this similarity was even more accentuated.

Various marsupial species show specializations, peculiar to their own lineages, and which have little relevance to subsequent mammalian evolution. Thus in *Cuscus* (Cunningham 1882) and in *Trichosurus* (Barbour 1963) there is a so-called opponens minimi digiti arising from the sesamoid

in the plantar aponeurosis and inserting on the shaft of the fifth metatarsal. On dissection, however, it is clear that it is part of the substance of abductor digiti minimi which has acquired these aberrant attachments, and has nothing to do with the true eutherian opponens to be described later.

Other clear-cut trends, seen in various specialized marsupials, but particularly accentuated in Eutheria, mask this pattern of primal simplicity and of serial similarity between hand and foot. The marginal members of the adductor series (the contrahentes) show a conspicuous tendency to extend their origins distally along a median fibrous raphe, becoming broad triangular muscles (Fig. 16.1C) which largely submerge the contrahentes of the second and fourth digits. This same trend was noted in the palm. The marginal members of the abductor series—abductor hallucis and abductor minimi digiti—show an invariable tendency to extend their origins backwards to the tuberosity of the calcaneus, and when this bone is prolonged posteriorly as a prominent heel, the affinity of these muscles with the intermetatarsal representatives of the dorsal series is obscured. Moreover, when the hallux is absent the first dorsal interosseous muscle, now cast in the role of a marginal muscle, shows a similar migratory tendency and then mimics the abductor hallucis; this further emphasizes the true nature of the abductor hallucis as a derivative of the dorsal abductor layer. There is also a marked trend towards fusion of the intermediate muscle layer (the flexores breves) with the underlying dorsal layer (the dorsal abductors).

The basic marsupial pattern has been converted with stark simplicity, except for the trends noted above, into the Primate, and even the human arrangement. The derived patterns are uncluttered by the complex, and contentious, migrations which have masked the derivation of the higher primate thenar and hypothenar muscles. In broad plan, as Cunningham appreciated, the marginal members of the dorsal abductor series, with origins projected back to the tuber of the calcaneus, have become the abductor hallucis and abductor minimi digiti (the latter with an offshoot as abductor ossis metatarsi minimi digiti). Flexores breves one and two have become the two-headed flexor hallucis brevis, and flexor brevis ten has become the flexor brevis minimi digiti. The remaining flexores breves, together with the interosseous abductors are varyingly transformed into the palmar and dorsal interosseous muscles. The four adductors (contrahentes) are reduced until only one, the adductor hallucis, remains.

This simple plan should have provided the basis for an enlightened understanding of the intrinsic muscles of the sole—not so! Subsequent ill-conceived work, which will be described, did little more than to introduce needless confusion and a contrived complexity.

The mid-sole musculature

The contrahentes

Prosimian Primates and New and Old World monkeys invariably retain the full complement of four contrahentes, although in prosimians they may show aberrant modifications associated with shift of the foot axis to the fourth digit (Jouffroy 1971), so-called ectaxony. Equally invariably the contrahens of the hallux shows distal migration of its origin, coming to cover the contrahens of the second digit, but unlike the fairly common situation in marsupials (Fig. 16.1C), the contrahens of the fifth digit is not similarly affected. Thus the contrahens of the hallux comes to arise from the midline of the palm, from the fascia over the contrahens of the fourth digit, and most distally from the heads of the second and third metatarsals and the glenoid plates of their metacarpophalangeal joints. The muscle becomes the renamed adductor hallucis. The migratory portion is commonly separated by a small interval from the part conservatively retaining a proximal origin; these are the transverse and oblique heads of the muscle (Hartman and Straus 1933).

Reduction of the three lateral contrahentes occurs in hominoids, beginning with gibbons and being most advanced in *Gorilla* and *Homo*. Their phylogenetic order of disappearance is as follows: contrahens of fourth digit, then second digit, and finally that of the fifth digit (Straus 1930). The contrahens of the fifth digit at least is commonly present in *Hylobates* and in *Pan* but *Gorilla* shows only the adductor hallucis as does *Pongo* although apparently the orang-utan often shows some vestiges of the absent muscles. The muscle for the second digit has been reported as an anomaly in man (Cunningham 1882).

The human foot is distinctive in possessing a particularly wide separation between the transverse and oblique heads of the adductor hallucis. Ruge (1878) showed that the adductor transversus at an early stage in human development is in apposition to the adductor obliquus, but it migrates distally to take up its transverse position and then significantly regresses. Cihak (1972) showed that the full complement of contrahentes is represented in early development in the human foot, but the muscles for digits four and five are quite rudimentary and transient. That for digit two is even more insubstantial and its substance is soon assimilated into the predominant muscle, for the hallux. Cihak also showed that the transverse head of adductor hallucis is late in development, and apparently on this basis he denied that it could be homologized with any of the four contrahentes; this is quite at odds with the facts of comparative anatomy.

In the foot as in the palm, a deep transverse metatarsal ligament links the glenoid plates of the four postaxial digits in non-human Primates. There can be little doubt that, as in the palm, this ligament is derived by fascial condensation. Man, however, is unique in having the hallux similarly bound to the other digits. This has been interpreted (Jones 1948) as a potent argument showing that the human foot is of basal mammalian primitiveness and could never have been derived from that of an arboreal ape. The argument is quite specious, since chimpanzees can show (Raven 1936) condensed interdigital fascia overlying the distal border of the adductor transversus, in the same relative position as the hallucial portion of the ligament of man, which needs only to be shortened and thickened to give rise to the human condition.

The interossei

Theories about the phylogenetic history of the foot interossei (Fig. 16.2) have closely paralleled those which have been proposed for the comparable muscles in the hand (Fig. 9.1), but with the additional problem that they were required to explain a shift in axis from the third digit (usual in higher primates) to the second in man.

Cunningham (1882) proposed that in the foot the third, fourth, sixth, and eighth flexores breves disappeared leaving the fifth, seventh, and ninth as the three plantar interossei. The dorsal abductors were represented as the four dorsal interossei. (Fig. 16.2B).

Ruge's (1878) ideas were in fact the inspiration for comparable theories on the hand interossei. He suggested that no representatives of Cunningham's dorsal abductor layer were to be found in the human foot, but that flexores breves numbers three, four, six, and eight migrated into the intermetatarsal spaces to become dorsal interossei (Fig. 16.2C). The plantar interossei were seen to be derived *in situ* from muscle numbers, five, seven, and nine. Campbell (1939) supported this idea and further suggested that it provided a plausible mechanism also for shift of the axis to the fourth digit, as is found in prosimians; the migrating muscles would then be numbers three, five, seven,

and eight. Cunningham himself saw the attractions of Ruge's theory but it is difficult to see how he could reconcile this with his demonstration of a primitive trilaminar muscle arrangement.

McMurrich (1927) suggested that certain of the flexores breves in the foot fused with dorsal abductors (Fig. 16.2D) to form composite dorsal interossei. This is the counterpart to the evolutionary process which has occurred in the hand and has been explored in some depth in Chapter 9.

Cihak (1972), again from a purely embryological viewpoint, proposed a solution which was an odd amalgam of the views of Ruge and McMurrich. He accepted the latter's proposal for the derivation of third and fourth dorsal interossei. The first, however, he believed to be derived by migration and fusion of two flexores breves: numbers three and four (Fig. 16.2E). The second dorsal interosseous then could only consist of a dorsal abductor component.

All these theories merely represent plausible juggling of muscle components as seen in two dimensions, in order to account for changes in the foot axis. They take little account of the realities of myological evolution and leave open the tantalizing questions of how the muscle insertions could possibly accommodate the various types of reshuffling. Moreover, the purely embryological studies (Ruge, Campbell, Cihak) are deeply rooted in ideas of stereotyped recapitulation and take no account of the distorting influence of heterochrony. On those grounds, they can be discredited just as in the hand studies.

Monkey interossei The interossei of monkeys are orientated about the middle digit as an axis and are built up out of components comprising the four dorsal abductors and the flexores breves number three to nine; in the foot flexor brevis two is a constituent of the hallucial musculature. In the foot the arrangement is quite like that seen for the hand (Fig. 9.1D): flexores breves three, five, six, and eight are closely associated with the dorsal abductors and the remainder, numbers four, seven, and nine, persist as independent palmar interossei. The dual muscle partners of a dorsal abductor and an associated flexor brevis show fewer progressive trends involving their insertions than were noted for the comparable pairs in the monkey hand.

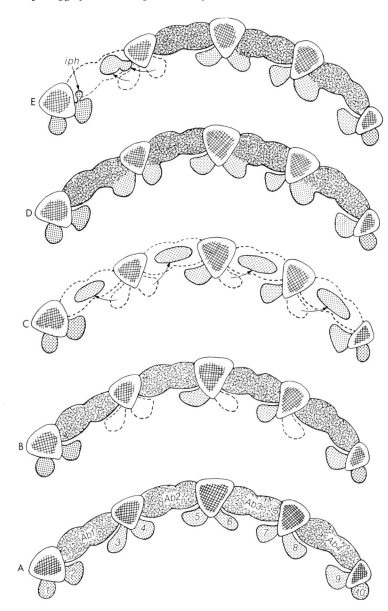

Fig. 16.2

A diagrammatic representation of the various suggested theories on the evolution of the pedal interossei. A, the arrangement of the dorsal and intermediate muscle layers in the primitive mammalian foot (the contrahentes layer is not shown); the four bipennate dorsal abductor muscles are labelled *Ab1*, *Ab2*, *Ab3* and *Ab4*; the ten flexores breves are stippled and indicated by Arabic numerals (numbers 1 and 2 by general consent give rise to the two heads of flexor hallucis brevis and are indicated by darker stippling as is number 10 which gives rise to the flexor brevis minimi digiti. The shading conventions, which are the same as those used for the hand in Fig. 9.1, are retained in all other parts of the figure and a muscular component which is believed to disappear is represented by a broken line, as is the primitive site of a muscle which has undergone a supposed migration; the arrows indicate the direction of such postulated movement. B, the view of Cunningham (1882). C, the view of Ruge (1878). D, the view of McMurrich (1927). E, the view of Cihak (1972); in this figure *iph* represents a supposed interosseous plantaris hallucis, derived from flexor brevis 2.

A New World monkey such as *Pithecia monachus* provides the key to the interpretation of this muscle group. The following is a description of a specimen of that species. Transverse laminae comparable to those in the palm extended as hoods from the extensor tendons around the metatarsophalangeal joints. In each case the dorsal abductors ended as phalangeal tendons passing deep to these hoods. The associated flexores breves (numbers three, five, six, and eight) were variably merged with the surface of the underlying abductor, with numbers five and six particularly affected, but nevertheless all these flexores breves retained some degree of independence and ended as wing tendons for the related digits. The remaining flexores breves (numbers four, seven, and nine) were the descriptive plantar interossei and similarly ended as wing tendons, although that to the tibial side of the fourth digit (number seven) was largely arrested at the glenoid plate. Insinuated between the second and third plantar interossei (flexores breves seven and nine) and the subjacent abductors were additional muscle bellies, apparently delaminated from the underlying abductors, and ending in phalangeal tendons to the tibial side of the fourth and fifth digits. The additional phalangeal tendons to the tibial sides of the fourth and fifth digits recall the similarly located tendons in marsupials (Fig. 16.1B) and the incipient delamination associated with them. It will be remembered that this was a primitive feature of the hand and that comparable additional muscle bellies and phalangeal tendons for the fifth digit were noted in *Pan* and occasionally in man in Chapter 9. Overall the picture is of conservative primitive simplicity, even less obscured by specialized overlays than was seen in the comparable hand muscles of any monkey.

A specimen of *Cebus nigrivattatus* showed a very similar simple arrangement but the insertion of all the flexores breves was largely restricted to the glenoid plate. This is a trend increasingly in evidence in other Primates. Moreover, in this particular New World monkey a supernumerary phalangeal tendon was only found for the tibial side of the fourth digit and its belly was imperfectly separate from the related muscles; as in the hand these delaminated bellies and supernumerary tendons are variable.

In Old World monkeys such as a specimen of *Colobus polyokomos*, the essentials of this arrange-ment are retained but flexores breves three, five, and six show an even greater tendency to merge with the subjacent abductors. In the specimen examined the last two of these flexor brevis muscles retained only trivial residual wing tendons. Flexor brevis number three interestingly showed the same specialization as the comparable muscle did in the hand of this species (Chapter 9): the substance of the flexor brevis had almost entirely been subsumed by the abductor and its phalangeal tendon. This is the specialization which modifies the insertion of the comparable muscle in the hands of *Pan* and *Homo*. In this species supernumerary bellies and phalangeal tendons for the tibial sides of the fourth and fifth digits were also found; clearly these are of common occurrence in monkeys.

Ape interossei In a specimen of *Hylobates lar* the characteristics of the monkey pattern were retained in a virtually unmodified form. However, the flexores breves, whether they were the three descriptive plantar interossei or those associated with a dorsal abductor, inserted virtually entirely on the sesamoid-containing glenoid plates. Again a phalangeal tendon, derived from a supernumer-ary belly located between abductor three and flexor brevis seven, passed to the tibial side of the fourth digit. A phalangeal tendon also passed to the tibial side of the fifth digit, but the belly of this one had amalgamated with flexor brevis nine and the tendon represented their joint insertion.

A specimen of *Pan troglodytes* threw particular light on the potential for evolutionary modification inherent in the arrangement of the hominoid interossei. Flexor brevis three had largely coa-lesced with the first dorsal abductor and scarcely retained any attachment to the glenoid plate. The other flexores breves associated with abductors number five, six, and eight were fairly discrete and inserted almost entirely on the glenoid plates, with only a tiny residual wing tendon insertion. It was flexor brevis number four which was most illumi-nating. It had enlarged its domain greatly at the expense of the subjacent abductor two, not quite sufficient to achieve exposure on the dorsum of the foot, but almost. Thus when the bipennate abduc-tors were viewed from dorsally they indicated a third digit axis; exaggeration of this trend, of course, could have dictated a descriptive axis through the second digit. In this particular speci-men an additional phalangeal tendon derived from the plantar surface of abductor number three passed to the tibial side of the fourth digit, but there was no such tendon to the fifth.

It is on record that the axis about which the interossei are disposed passes through the third digit in most gibbons, chimpanzees, and orang-utans but that in occasional forms it has shifted medially to the second toe, which is the norm for both gorillas and man (Hartman and Straus 1933). In *Gorilla* the second dorsal interosseous is usually bipennate (Straus 1930), and when the foot is viewed from dorsally indicates an axis through the second toe; but this is not always the case. Indeed Deniker (1886) recorded a case in *Gorilla* where the second dorsal interosseous had its origin restricted to the second metatarsal. This after all, only represents an exaggeration of the trend noted above in a specimen of *Pan*.

It is clear from what has been said that the shift of the axis is not an all-or-none phenomenon, and this doubtless accounts for some of the variation in interpretation of ape specimens. The mechanism of expansion of the domain of one muscle at the expense of a neighbouring one is well established (Chapter 3). Knowing the detailed insertions of the various muscular components described above, all the ingredients are provided for an instructive analysis of the muscles in man.

The human interossei A central issue in the consideration of the human interossei must be consideration of the mechanism of change of the abduction–adduction axis from the middle digit, as is usual in most primates, to the second. This regrouping of muscles has commonly been attri-buted to an exchange of insertions: the first plantar interosseous of the prehuman (flexor brevis four) shifting its attachment laterally from the second toe to the third, whilst the second dorsal interos-seous moved medially to a new insertion on the second toe. But such a near miraculous switching is not feasible by any of the known mechanisms for transfer of insertions (Chapter 3).

The clue to the truth has long been available in the literature in a study by Manter (1945) of variation of the origins of the human pedal interossei. His most significant observation was that the relationship of the dorsal and plantar muscles in the second interosseous space was particularly liable to variation, with the second dorsal and first plantar muscles varying consider-ably in their extent of origin in both the dorsal and plantar directions. The most common variation was that the second dorsal interosseous muscle was unipennate, arising only from the second metatarsal; in fact, just like the *Gorilla* specimen described by Deniker (1886). None of the other interossei showed such a degree of variability.

Manter was a supporter of the Campbell–Ruge theory that the interossei, both plantar and dorsal, were derived from a single layer of plantar flexors by differential migration (Fig. 16.2C). He therefore reasoned that the change in axis of this human foot would be achieved if the first plantar interosseous of an ape, inserting on the second toe, had become equal to the human second dorsal interosseous, by shifting its origin dorsally; similarly, if the second dorsal interosseous of the ape's foot, while retain-ing its insertion on the third toe, were to take its origin from a plantar position, it would in effect become the equivalent of the human first plantar interosseous. There is a germ of truth in this, but Manter had taken little account of the insertions of the muscles. Had he realized that the insertions of a plantar interosseous and of a dorsal interosseous (and indeed of the dorsal abductor and flexor brevis components of the latter) differed he would have seen his theory to be insupportable.

In fact, the situation in the chimpanzee des-cribed above provides the key to the true phylo-genetic progression. If flexor brevis four (the usual primate first plantar interosseous), by merging with dorsal abductor two, had annexed even more of its substance so that it reached the dorsal aspect of the foot, a situation would be attained like the variant human pattern described by Manter (1945) or the *Gorilla* described by Deniker (1886). If this expansionist tendency exhibited by flexor brevis four were to include most of the remainder of abductor two arising from the third metacarpal then the axis as seen from dorsally would be changed and the usual human type of second dorsal interosseous would be realized. But, signifi-cantly, unlike other dorsal interossei it would have no phalangeal tendon. This would leave flexor brevis five, incorporating only a residual part of the second abductor, but that part including its phalangeal tendon. This composite muscle is the new human first plantar interosseous, but unlike other plantar interossei it would have an additional phalangeal tendon. The fundamental process involved is a common phylogenetic strategy, by which one muscle annexes substance from an adjoining or overlying one and so extends its origin (Chapter 3); many examples have already been noted. Clearly this hypothesis can easily be verified by checking the insertions of these two muscles. This has been done, and the predictable insertions are, in fact, present (Fig. 16.3).

The flexor brevis components of the human interossei, whether they be independent plantar interossei or the associated parts of dorsal ones, invariably terminate at the glenoid plates with no significant continuation into the wing tendons. This suppression of the primitive distal insertion is already established in *Hylobates* and *Pan*. The first dorsal interosseous typically inserts entirely to the glenoid plate on the medial side of the second digit with no residual phalangeal tendon; this is a specialization comparable to that affecting the

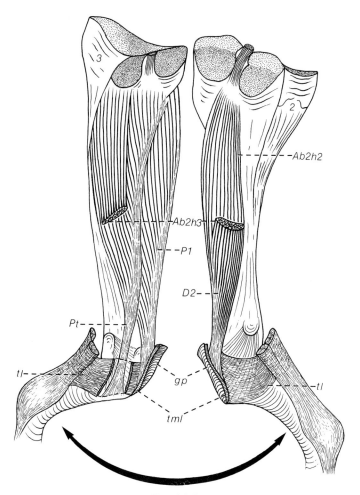

Fig. 16.3

A dorsal view of the human right second (*2*) and third (*3*) metatarsal bones, splayed apart to display the musculature in the interosseous space. *P1*, the first plantar interosseous (flexor brevis 5); *D2*, the second dorsal interosseous—a composite of flexor brevis 4, the second metatarsal head of the dorsal abductor 2 (*Ab2h2*) and part of the third metatarsal head (*Ab2h3*) of this muscle; *Pt*, phalangeal tendon derived from the residual part of the third metatarsal head of the second dorsal abductor; *gp*, glenoid plates of the metatarsophalangeal joints; *tml*, the cut deep transverse metatarsal ligament; *tl*, transverse lamina (hood).

fourth dorsal interosseous of the hand in man and some monkeys (Chapter 9). The second dorsal interosseous (Fig. 16.3), arising typically in bipennate fashion from the second and third metatarsals also inserts solely to the glenoid plate, for despite the composite nature of the muscle belly, its insertion is purely that of a flexor brevis (number four). The first plantar interosseous has a dual insertion: by much of its mass into the glenoid plate on the medial side of the third digit but also by a phalangeal tendon (derived from the second dorsal abductor) which passes to its insertion deep to the hood or transverse lamina. The third dorsal interosseous has dual insertions (representative of its two component parts—flexor brevis six and abductor three) into the glenoid plate and by a phalangeal tendon passing deep to the hood on the fibular side of the third digit. The second palmar interosseous is a composite of flexor brevis seven (the true palmar interosseous) and the superficial part of the abductor (number three) lying deep to it and furnishing a phalangeal tendon to the medial side of the fourth digit; this part of the abductor, often delaminated as a separate belly, was present also in all the monkeys and apes described. In the human example, however, the phalangeal insertion had appropriated the whole of the overlying flexor brevis and the two entities did not retain independent insertions. The fourth dorsal interosseous is unremarkable with dual phalangeal tendon and glenoid plate insertions for its constituent parts. The third plantar interosseous inserts entirely to the glenoid plate on the medial side of the fifth digit. Deep to it lies an imperfectly separated muscle belly (as in the apes and monkeys described, except for *Cebus*) giving a supernumerary phalangeal tendon to the same digit. These human muscles, of course, like those in the hand, show a spectrum of variation.

What can be the functional significance of these human specializations? The question of a shift in axis is, of course, really an irrelevance: abduction and adduction of the human toes, whether it be in relation to a second or third digit axis, is not a significant part of the human locomotor repertoire. The real question is the significance of the underlying muscular reshuffling, of which the arrangement of the dorsal interossei, as seen from the dorsal aspect, is a mere indicator. The really significant feature is the elaboration of massive

insertions onto the glenoid plate of the second digit. In this it has come functionally to resemble the hallux which similarly has massive flexor brevis insertions (of the flexor hallucis brevis) into the glenoid plate with its sesamoids. Now these two digits are the ones which bear the brunt of the push-off at the end of the stance phase of the bipedal gait, when the metatarsophalangeal joints are hyper-extended. Moreover, Sarrafian and Topouzian (1969) have shown that the long extensor tendons of man cause hyperextension of the metatarsophalangeal joints by the sling-like action of the extensor hood. Massive insertions to the glenoid plates, not only in the first and second digits but also in the others, oppose this action.

It has been noted in Chapters 8 and 9 how Cihak (1972) described 'interossei dorsales accessorii', amalgamating with the dorsal aspect of the dorsal interossei in the human hand. It was also noted that these embryonic muscle slips were the likely mammalian representatives of the dorsometacarpales of reptiles and the source of the anomalous short extensor brevis manus of man. Cihak described similar accessory primordia in the foot dorsal to the interosseous of the second interspace, and occasionally the third, contributing myoblasts to the dorsal interosseous. He conceded that in the foot this layer delaminated from the deep surface of the deep extensor blastema. This does not tally with his denial of a similar origin in the hand. As in the hand there is little doubt that these slips represent the reptilian dorsometatarsales (Fig. 15.1). In fact, Ruge (1878) had already clearly illustrated the slips in a human fetus, Lucien (1909) in the human adult, and Straus (1930) in *Gorilla*.

The marginal musculature

The muscles of the hallux

The muscles of the hallux retain a relatively primitive form even in man, uncomplicated by the migrations and partial subdivision which has affected the thenar muscles.

The abductor hallucis   McMurrich (1907) derived this muscle from the reptilian flexor brevis superficialis, and Cihak (1972) in an embryological study of the human foot also derived it from the superficial blastema. These authors similarly viewed the abductor pollicis brevis of the hand; the same counterarguments apply here as were used in the discussion on the hand.

There is, in fact, little doubt that Cunningham (1882) was right and that the muscle is a marginal counterpart of the interosseous dorsal abductor layer. Its primitive origin is from the prehallux (Fig. 16.1A) and in this it conforms with the origin of its serial homologue, the abductor pollicis brevis, from the prepollex. Its insertion is to the medial side of the proximal phalanx of the great toe. The primitive origin from the prehallux is retained in the treeshrew *Ptilocercus* (Clark 1926). Even in prosimians it retains this origin from the prehallux (the so-called 'os ventrale' discussed in Chapter 14) perhaps with spread onto the adjacent plantar aponeurosis (Jouffroy 1971). In the higher primates, however, its origin is prolonged back to the tuber of the calcaneus. This can, of course, only have occurred by appropriating some of the substance of the adjacent flexor digitorum brevis. In a sense, therefore, Cihak and McMurrich were right; however, it would be totally inappropriate to homologize the muscle in a treeshrew with the intermetatarsals of reptiles (the dorsal abductor layer), and that of man with the flexor brevis superficialis, implying some sort of saltatory evolution whereas a clear-cut thread of phylogenetic continuity exists.

The flexor brevis hallucis This muscle is clearly a two-headed muscle, representing two flexores breves (numbers one and two) inserting either side on the sesamoid containing glenoid plate of the hallux. Its counterpart in the hand was of course derived only from the first flexor brevis. It has been remarkably stable throughout evolution. Yet even this simple story has become needlessly obscured.

In monkeys the typical bicipital muscle has the fibular part supplied by the lateral plantar nerve and the tibial head by the medial plantar nerve. The lateral head may, however, be smaller and deeply pressed into the sole (Hartman and Straus, 1933). This trend may even be more emphasized in the apes and the lateral head may be very weakly developed and deeply located in *Gorilla* (Straus 1930) and *Hylobates* and even merged with the adductor obliquus in *Pongo* (Cunningham

1882). The human muscle is traditionally stated to receive its nerve supply from the medial plantar nerve and to have a single origin far across the sole from cuboid and calcaneus. The inevitable outcome of this (Jones 1944) was the belief that the muscle of man represented only the tibial head. Where then was the missing fibular head? Occasionally in the human foot a detached slip of the adductor obliquus (Jones 1944; Manter 1945) mimics in appearance a plantar interosseous of the hallux. Indeed it has been considered as the hindlimb counterpart of the first palmar interosseous (of Henle), thus apparently rounding out an appearance of symmetry of plan between the two limbs. This anomalous slip was supposed to be the missing head of the human flexor hallucis brevis and the counterpart of the fibular head of that muscle in monkeys. This tortuous tale rests on a fallacy—the idea of immutable nerve–muscle specificity. In fact, it is not even true that the flexor hallucis brevis is exclusively supplied by the medial plantar nerve for it commonly receives a contribution from the lateral plantar nerve. The anomalous muscle slip (figured by Manter and by Jones) and described by Le Double (1897) is apparently no more than an occasional separated slip of the adductor obliquus. Yet it figured in Cihak's (1972) general scheme for the derivation of the hallucial muscles as the 'interosseous plantaris hallucis' as a partial derivative of flexor brevis two. In fact, the human flexor brevis retains a quite clear-cut primitive simplicity, modified only by the way in which its origin has established continuity with the tibialis posterior tendon (Fig. 14.18).

In the apes, particularly *Pongo*, and sometimes in man the medial head of the muscle may have a partial attachment to the shaft of the first metatarsal thus constituting an opponens hallucis.

The fifth toe muscles

These muscles, even in man, show none of the controversial complexities which surround the hypothenar musculature.

Abductor digiti minimi McMurrich (1907) and Cihak (1972) derived this muscle from the superficial stratum, but there is little doubt that Cunningham (1882) was right in considering it as a marginal derivative of the dorsal abductor layer.

Its primitive origin is from the tuber of the calcaneus and it is only when this is displaced by the formation of an elongated heel that its resemblance to the other representatives of its layer is obscured. Its forearm counterpart arises from the pisiform. In fact, in the human embryo the tuber calcanei is strikingly similar in disposition to the pisiform (Olivier 1962). This is no argument, however, for serial homology between the two bony structures, but it does emphasize the correspondence between the two muscles. In the absence of any postaxial ray rudiment in the hindlimb the marginal muscle merely takes origin from the next adjacent bone, the calcaneus. Insertion is to the base of the proximal phalanx of the fifth digit.

Part of the muscle is typically arrested at the base of the fifth metatarsal as an abductor ossis metatarsi quinti. It is usually present from marsupials (Fig. 14.13) to apes and is commonly present in man (Le Double 1897).

Flexor brevis minimi digiti This muscle is the virtually unchanged derivative of flexor brevis ten. Usually in primates, and at least commonly in man (Le Double 1897) some of its substance peels away to insert on the shaft of the fifth metatarsal as an opponens digiti minimi, and the muscle develops in just this way in the human embryo (Cihak 1972). It is to be noted that it is quite different in nature to the so-called 'opponens digiti minimi' found in some marsupials, which was described above, and is a specialized derivative of the abductor digiti minimi.

17

Fossil foot bones

A fairly comprehensive phylogenetic scenario covering all aspects of the evolution of foot anatomy has been presented in the preceding chapters. The final arbiter for such speculations, however, must be the palaeontological record. Inevitably this is fragmentary, but the tantalizing glimpses given of the course of evolution can confirm or deny the more detailed insights derived from comparative anatomy, which, in turn, provides the essential underpinning for any reasoned analysis of the fossil record.

No comprehensive account of the burgeoning fossil record will be given; space alone would prohibit this, but availability also dictates the selection. Where possible the descriptions are based on original specimens or casts. In some cases only descriptions, often with imperfect illustrations, have been available. The often superficial formal descriptions of new finds, concerned mainly with overall appearance and with perceived resemblances to extant species, not infrequently gloss over or even fail to note key apomorphic features of particular interest; where possible an attempt will be made to highlight these. The overall picture, disjointed though it may be, bears witness to the anatomical discussions in preceding chapters.

The foot of therapsid reptiles and prototherian mammals

The cynodont foot

As noted in Chapter 2 there is little doubt that the mammals were derived from therapsid reptiles in the Triassic, and in particular from some cynodont stock. The consensus of palaeontological opinion would have us believe that cynodonts were already well on the way to achieving the essential characteristics of mammalian foot structure, and published illustrations and museum exhibits often portray these reptiles with hindfeet apparently not so very different from those of plantigrade therian mammals. This view is, in fact, based on a quite rudimentary level of osteological analysis. Reference is made to such characteristics as the elaboration of a sustentaculum tali leading to 'astragalar superposition', and to the acquisition of a backwardly directed tuber calcaneus (Jenkins 1979*b*, 1971*a*).

The cynodont talus and calcaneus at first sight appear totally dissimilar to those of extant therian mammals. However, familiarity with monotreme foot structure, and not merely the osteology, surveyed in previous chapters, provides the groundwork which clarifies the cynodont morphology. The roughly hemispherical cynodont talus (Figs 17.2, 17.3A) is reminiscent of that of the echidna (Fig. 13.2A); the strange, flattened, and semidiscoidal calcaneus, however, seems to belie any such affinity. The core of this odd looking tarsal (Fig. 17.1A) is a thickened bony column forming its medial margin and bearing the articular surfaces. A wing-like lateral flange is attached to this bony rod. In some of the larger cynodonts, however, this appendage was greatly reduced, and as will be seen, it was largely but not entirely eliminated in subsequent mammalian stages.

When the cynodont calcaneus is so oriented as to highlight the functionally important articular features and compared with an echidna calcaneus in comparable position (Figs 17.1A, C) it is apparent that the same basic articular surfaces are present. The most striking difference, which at first sight obscures the resemblance between the bones,

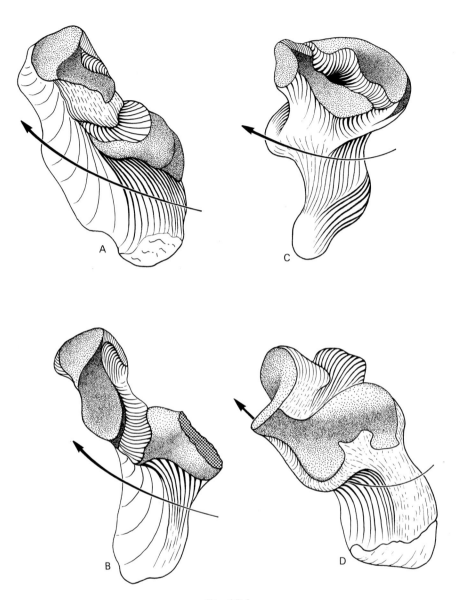

Fig. 17.1

The right calcanei of a cynodont, (TR8) (A), the Triassic morganucodontid *Eozostrodon* (B), *Tachyglossus aculeatus* (C), *Didelphis marsupialis* (D), all viewed from medially, and somewhat towards the plantar aspect; in all the diagrams the cuboid facet faces upwards and to the left and like other articular surfaces it is stippled (that for the fibula in B is partly broken off). The arrow in all the diagrams represents the known, or presumed, course of entry of the flexor fibularis tendon into the sole. The various articular areas and ligamentous attachments in C and D may be identified from Figs 13.1 and 13.2; corresponding features may be identified with some confidence in the fossils despite their rather different relative positions, as described in the text.

is that in the cynodont the distal and proximal talar facets are in a stepped configuration down the length of the core of the calcaneus. If the distal prolongation could be imagined as compacted into the main mass of the bone, so that the distal talar facet is in a more mediolateral relationship to the posterior talar facet, then something very similar to the monotreme arrangement would be realized. As already described in Chapter 13, the monotreme morphology can readily be conceived as being pre-adaptive to the typical therian condition.

There are also distinct resemblances between the orientation of the tuber of the calcaneus in cynodonts and monotremes. This tuberosity is the insertion of the tendo calcaneus but the bone bearing it is moulded around the large flexor fibularis tendon, the primary flexor of the toes, thus forming an obvious channel for the entry of this tendon into the sole. In the weight-bearing foot of the echidna this heel-like prominence is directed not only downwards towards the ground but also somewhat distally towards the toes. When the pes of a small cynodont (TR.8) is accurately articulated (Fig. 17.2), using the experience derived from knowledge of the articular and muscular arrangements in echidna, it is clear that the cynodont tuber is also directed downwards in the plantigrade position, although lacking the marked distal deflection seen in echidna (Fig. 13.1). There can be little doubt that the published reconstruction of this fossil foot, showing a backwardly projecting tuber (Jenkins 1971*a*), represents an unnatural position, apparently the result of deformation during fossilization, which moved the bones out of true articular contact, flattening the natural arched conformation of the foot.

It has already been described in Chapter 13 how the transition from a monotreme to a therian foot architecture has, as its essence, the bending outward and backward of the primitively downward projecting heel. A consequence of such realignment is that the massive flexor fibularis tendon now enters the sole by undercutting the distal talar facet (Fig. 13.2D). Only when this structural grade is reached is it truly accurate to speak of a sustentaculum tali and sustentacular facet; application of these terms to the cynodont condition is not, therefore, strictly appropriate.

Where the echidna flexor fibularis tendon enters the sole through its calcaneal furrow (Fig. 17.1C)

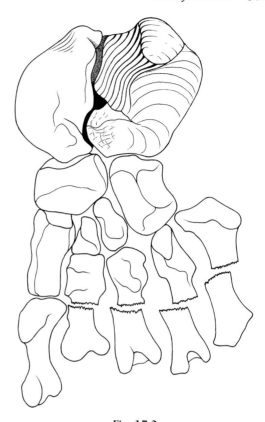

Fig. 17.2
The right pes (tarsus and metatarsus) of a small cynodont (TR8), viewed from the plantar aspect with the bones in what seems to be the correct articular relationships.

it is flanked laterally by a bony prominence on that bone to which the tendon is held by the substantial flexor accessorius muscle. The same bony tuberosity is crossed above by the peroneal tendons, with peroneus longus entering the sole in the groove between the tuberosity and the cuboid articulation. These topographical relationships identify this tuberosity as the homologue of the trochlear process of therian mammals. Although this stubby monotreme protuberance contrasts in form with the shelf-like therian trochlear process (to be considered later) there can be little doubt that both structures are remnants of the projecting semicircular flange on the cynodont calcaneus. Probably therefore, the cupped lower surface of this bony therapsid structure gave rise also to a flexor accessorius muscle, and its disposition in relationship to the flexor fibularis groove is precisely as would be expected (Fig. 17.2). Similarly its upper surface presumably supported the bundle of peroneal tendons entering the foot; indeed, the calcaneus of a large cynodont (DMSW R.191) shows a clear-cut groove here for these tendons, and especially for the entry of peroneus longus.

The cynodont talus (TR.8) shown in Figs 17.3A and 17.2 has been figured and described also by Jenkins (1971a) but in only the most general terms, such as recognition of the duality of facets for the calcaneus. With the benefit of hindsight provided by knowledge of the anatomy of the soft parts in monotremes it seems possible to go further and to propose that a considerable approximation to the basic structure exhibited by echidna had already been achieved in cynodonts. Recognition of the articular areas and ligamentous attachments is necessarily somewhat speculative but the interpretations given in Fig. 17.3A seem to be reasonable.

The triconodont tarsus

As described in Chapter 2 it is believed that the Triconodonta was the basal order of the emergent Triassic mammals and that the family Morganucodontidae provided the source of the more advanced Therians. The talus and calcaneus of a morganucodontid, *Eozostrodon* (*Morganucodon*), have been described and figured by Jenkins and Parrington (1976) but only in general terms. In the light of the descriptions given already for echidna, and for the cynodonts, it seems that deeper structural insights are possible. The hemispherical talus is again reminiscent of both monotremes and of cynodonts, and a plausible interpretation of its various features is shown in Fig. 17.3B. The main distinction shown by the morganucodontid bone is that the medial extremity of the groove between the two calcaneal facets is walled over to create an 'astragalar canal'.

The calcaneus (Fig. 17.1B) is also similar to that of cynodonts, with comparable articular surfaces and a similar thin semicircular lateral flange. The same basic arrangement of the articular surfaces and the same presumptive ligamentous areas are identifiable and there are also clear indications that the tuber was directed towards the ground, as it appears to have been in cynodonts.

A fascinating glimpse of the level of refinement of the leg and foot musculature of *Eozostrodon* is provided by a proximal fragment of fibula (Fig. 17.4A). It is strikingly, almost identically like the corresponding bone in *Pseudochirus*

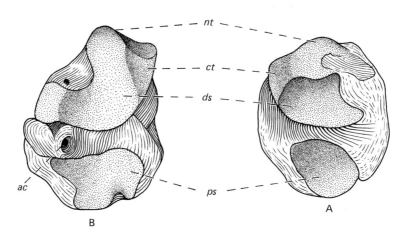

Fig. 17.3
The right talus of a cynodont (TR8) (A) and left talus of *Eozostrodon* (B) shown in positions similar to that of the echidna talus in Fig. 13.2A. The features labelled may be identified with confidence; related areas were probably comparable in function to those shown in Fig. 13.2A. *ac*, astragalar canal; *nt*, navicular articulating surface; *ct*, cuboid facet; *ds*, distal articular surface for calcaneus; *ps*, proximal articular surface for calcaneus.

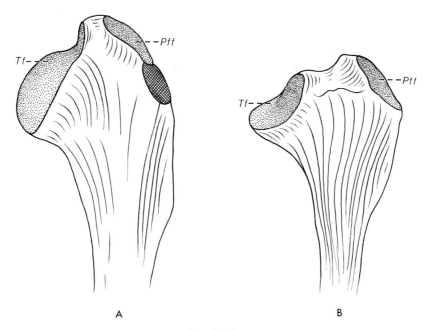

Fig. 17.4

The upper extremities of the right fibulae of *Eozostrodon* (A) and *Pseudochirus laniginosus* viewed from the medial aspect. *Pff*, parafibular facet; *Tf*, tibial facet.

(Figs 17.4B, 14.5A). It is apparent that a free parafibula must have been present and that the bone articulated with both tibia and a femorofibular disc, just as it does in *Pseudochirus*. Clearly, although the foot of *Eozostrodon* was not as advanced on the mammalian line as in echidna, already in this morganucodontid the mammalian pattern of flexor musculature must have been established. The bone was described by Jenkins and Parrington (1976) who completely missed the point, oddly enough they likened it to the fibula of monotremes; as described in Chapter 14 the monotreme fibula evolved from such a bone by fusion of the parafibula to create the specialized parafibular crest. Jenkins and Parrington further confused the issue by describing the tibial and discal articular facets on the fossil as a 'bulbous facet for articulation with the femur—its shape is evidence of fibular mobility'.

Joint function in therapsid and protherian feet

Palaeontologists in general seem to have over-emphasized the extent of the trend towards the establishment of relatively advanced mammal-like form and function in Triassic fossils. This applies particularly to analyses of the hindlimb, and interpretations of mechanisms of gait in cynodonts and early mammals (Kemp 1982) where a major innovative advance is attributed to these creatures, transforming the sprawling reptilian gait of their predecessors into a more erect and 'fully mammalian' posture, with parasagittal femoral movements and feet placed close to the midline. This scenario fails adequately to take account of Jenkins (1971*b*) important cineradiographic study, which showed that vertical limb orientation with parasagittal excursion, the traditional hallmark of the 'fully mammalian' type, is, in fact, characteristic only of the specialized cursorial mammals; more generalized therians retain much of the primitive sprawling attitude. The advanced cynodonts and the early mammals could scarcely have forsaken this mode of locomotion entirely, even though they might have had some facultive capacity for more erect running. This overriding notion of a dramatic conversion to a more erect gait has had a profound effect on interpretations of therapsid foot structure. It is generally acknowledged that therapsids must have retained that rotatory capacity of tibia and fibula, both about their long axes and in the same direction, which is typical of reptiles (Haines 1942) and which is a necessary concomitant of the femoral retraction and rotation which characterizes the sprawling gait (Rewcastle 1981). The reptilian ankle permits such rotatory motion, but it is argued that this is the only movement accommodated here. Foot extension providing thrust at the end of the stride is said to be effected by a new rotatory capacity at the talocalcaneal joints about a more or less transverse axis; such function has been attributed to gorgonopsids (Kemp 1982) and therocephalians (Kemp 1978) but is suggested to be of quite general application, the mechanism in therapsids being made particularly effective by the supposed backwardly projecting lever arm of the tuber calcaneus. Interestingly, the whole forefoot is envisaged as moving with the calcaneus (in a manner analogous to the lamina pedis of therian mammals) whilst the talus effectively acts as part of the crus. Yet there seem to be no sound morphological grounds for this assumption. There can be little doubt that in these therapsids, as in monotremes (Figs 13.2A, B) it is primarily the talus which is functionally anchored to the forefoot by the ligaments uniting it to cuboid, lateral cuneiform, and navicular; indeed, sites for the attachment of such ligaments are clearly indicated on the cynodont (and morganucodontid) talus (Figs 17.3A, B). In fact, it seems highly likely that cynodont and morganucodontid locomotion had much in common with that of the echidna and presumably required rather similar mechanical solutions, catering for adjustments between the plane of the sole of the foot and the crus, as described in Chapter 13. However, whereas the axis of talocalcaneal movement in echidna is approximately transverse, it is apparent that in cynodonts and morganucodontids it must have been more longitudinally orientated. Moreover, the distal talar facet on the calcaneus, like the proximal one, was convex and had not elaborated the cup-like concavity characteristic of Therians and even apparent in echidna. It seems that the movement must have been a rocking one between calcaneus and talus with the latter bone nevertheless transmitting its movement to its attached part of the forefoot. This torsional movement then would adjust the plane of the sole of the foot essentially as happens in echidna.

It is clear that in foot structure and function the monotremes were further along the path to a therian pattern than either cynodonts or Triassic prototherians; this endorses the view proposed in Chapter 2, that the monotremes were probably derived from eupantotheres. The monotremes, of course, possess their own quota of derived characters in the foot, such as the bulbous enlargement of the talocalcaneal articular regions, the probably accentuated distal deflection of the tuber calcaneus, and the lateral deviation of the postaxial digits.

The multituberculate tarsus

In an imaginatively illustrated account Jenkins and Krause (1983) have described form and function in the tarsus of a multituberculate, *Ptilodus*. From Chapter 2 it will be remembered that this hardy order of herbivorous 'Prototherians' survived longer than all others, although they were clearly aberrant and divorced from the mainstream of mammalian evolution. It is clear that they had capitalized on the type of tarsus described above in a remarkable way. Considerable rotatory capability was developed at the ankle joint, in a manner analogous to that described in Chapter 13 for *Didelphys*, although presumably without the interposition of a meniscus. This was coupled with rotatory movement between talus and calcaneus, similar to that described above for morganucodontids, cynodonts, and monotremes, producing a twisting of the foot in the sense of inversion. Together these movements were clearly capable of producing foot reversal analogous to that in *Didelphys*. Moreover, this Prototherian had a widely divergent hallux, articulating at a saddle-shaped joint with the medial cuneiform. The clear implication was that these mammals were arboreal, with a prehensile pes and the specializations providing for foot reversal and controlled headfirst descent.

This has obvious implications for the theories of arboreal origin of mammals discussed in Chapter 2. There is no indication that the multituberculates were anything other than an aberrant dead-end of mammalian evolution. Yet it does show that the triconodonts at large had an inherent capacity to evolve such specializations. This at least establishes the possibility that also the more advanced Theria of Metatherian–Eutherian grade could also have evolved comparable specializations.

The derivation of the therian tarsus

Calcaneus

General form   It has been noted in Chapter 13 that a significant apomorphy of the therian foot is deflection of the calcaneal tuber (heel) backward and outwards. In many mammals significant hints of its earlier disposition remain—in marsupials (Fig. 13.2D) and even in the African apes (Figs 13.13, 17.5C). In the human embryo at 54 days (McKee and Bagnall 1987) this feature, and the relationship of talus and calcaneus, are strikingly like those in the generalized marsupial shown in Fig. 16.1.

But more than this, the calcaneus has shown striking other changes in form during therian evolution. Many mammals show a striking lateral projection on the bone which is usually described as the 'trochlear process' or 'peroneal tubercle'. The presence or absence of such a process, and whether it is distally located adjacent to the cuboid articulation or whether it is posteriorly displaced on the lateral surface, has proved to be a useful taxonomic feature (Stains 1959). Inadequately understood use of this character, has unfortunately figured in studies on primate (Szalay *et al.* 1975; Szalay 1977) and tupaiid relationships (Szalay and Drawhorn 1980). The derivation of the process, however, has been obscure, although Jenkins (1971*a*) suggested that the lateral flange of cynodonts was lost to produce calcaneal proportions similar to those of mammals 'although some mammals (e.g. *Didelphis*) retain a slight lateral shelf on the calcaneum'.

Laidlaw (1904) gave an instructive, but seldom quoted, description of the surface features on the lateral aspect of the human calcaneus. He noted the presence of a processus trochlearis (peroneal tubercle) in 40 per cent and further noted that behind it lay an eminentia trochlearis connected by a faint ridge to the lateral process of the tuber.

Furthermore Laidlaw (1905) supplemented this study with some significant comparative observations, suggesting that the processus lateralis of the tuber calcanei is peculiar to man, and that it was derived by downwards and backwards migration of the posterior part of the retrotrochlear eminence; such posterior displacement, he pointed out, is manifested to varying degrees in different human bones. Essentially this same ground was covered by Weidenreich (1940), reaching the same conclusions, but strangely without any reference to Laidlaw.

Some examples will demonstrate the general principles involved in the evolution of this part of the calcaneus. A marsupial such as *Caluromys lanatus* (Fig. 17.5A), which seems to possess a foot morphology close in many respects to the basal pattern for the Metatheria (and for the 'Theria' as a whole), shows what may reasonably be considered the primitive therian form of trochlear process. A broad flattened flange, hollowed into a concave form on the plantar aspect, projects laterally from the calcaneus. Distally the margin of this flange extends to the lateral edge of the calcaneocuboid articular surface. Posteriorly, the thickened rim blends onto the lateral surface of the calcaneus below the talocalcaneal joint surface and fades away indeterminately towards the lateral surface of the inflected calcaneal tuber. In generalized Australian marsupials (*Trichosurus vulpecula* or *Pseudochirus laniginosus*) and trochlear process is similarly distally located but condensed into a thick stubby projection overlapping the lateral aspect of the calcaneocuboid joint. *Didelphys marsupialis* retains a trochlear process with a shelf-like character (Figs 13.2D, 17.1D) but here it is displaced backwards behind the lateral margin of the calcaneocuboid joint. As will be seen, a similar posterior location of the trochlear process has evolved in parallel in a number of mammalian orders.

In insectivores such as *Tenrec ecaudatus* (Fig. 17.5B), or *Suncus caeruleus*, the trochlear process is a prominent laterally projecting shelf so distally located that its anterior border adjoins the margin of the calcaneocuboid joint. In elephant shrews, however, (*Elephantulus sp.*), it is rather reduced and more posteriorly located. In *Tupaia sp.* it is again a prominent shelf but is markedly retracted along the lateral aspect of the calcaneus.

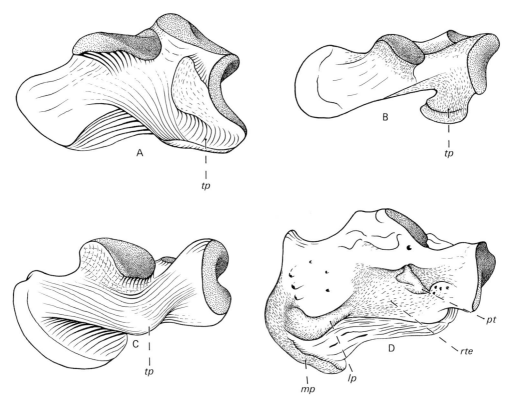

Fig. 17.5
The lateral aspects of the right calcanei of *Caluromys lanatus* (A), *Tenrec ecaudatus* (B), *Pan troglodytes* (C) and *Homo sapiens* (D). The human calcaneus is tilted somewhat medially to reveal the plantar aspect, and this particular specimen shows the lateral process of the tuber still associated with the other components of the trochlear process— the peroneal trochlea and the retrotrochlear eminence. *tp*, trochlear process; *pt*, peroneal trochlea; *rte*, retrotrochlear eminence; *lp*, lateral process of tuber calcanei; *mp*, medial process of tuber calcanei.

Rodents invariably show a well-marked trochlear process, (Stains 1959) which may be either distally located, or posteriorly displaced, as it is in *Sciurus carolinensis*.

The functional role subserved by the trochlear process, when it is well developed as in the above species, only becomes apparent when soft tissue anatomy is understood. The plantar surface provides a substantial platform for the origin of the muscle flexor accessorius, as described in Chapter 14. The upper surface of the process forms a supportive shelf underlying the bundle of peroneal tendons, described in Chapter 15, which are here often bound down by a peroneal retinaculum. The peroneus longus lies usually along the margin of the shelf and runs around this pulley-like formation into the sole to reach its insertion into

base of the hallucial metatarsal. The other tendons are variable in number, for besides the peroneus brevis they include tendons of bellies of the deep extensor stratum, extensor brevis digitorum, when those are located in the peroneal region of the leg.

These myological relationships confirm the identification of the trochlear process in echidna (Fig. 13.1) for flexor accessorius arises from its plantar surface and the peroneus longus and brevis course over its dorsal aspect. There can be little doubt that the therian trochlear process is a remnant of the wing-like lateral flange on the calcaneus of cynodonts and Mesozoic mammals, and it follows that probably this bony plate had similar myological relationships. Its under-surface is appropriately located to provide origin for a flexor accessorius muscle, lying as it does alongside the entry of the flexor fibularis tendon into the sole. Its upper-surface may well, however, be more than just a bony shelf supporting the peroneal tendons. The deep extensor layer of reptiles, the homologue in part of the accessary peroneal bellies in many mammals, is at least partly pedal in origin (Chapter 15): the upper-surface of this broad flange may have provided for such an origin.

It is probable that a prominent trochlear process, withdrawn well back along the lateral aspect of the calcaneus, was an ancestral character in the Primates. This is despite the fact that in one part of the extant primate radiation (prosimians) it is almost totally suppressed and in another (*Homo sapiens*) the process is fragmented into several parts. In prosimians (for example, *Lemur, Galago, Perodicticus*), the trochlear process is merely represented by a slight residual roughened elevation situated far posteriorly below the posterior talocalcaneal joint. New World monkeys (for example, *Cebus, Pithecia*), however, retain the trochlear process in well-developed form and the morphological arrangements here may be taken as a model for the Anthropidea as a whole. In these platyrrhine monkeys the process is a prominent shelf projecting laterally from the calcaneus with its summit lying below the anterior part of the posterior talocalcaneal joint. Its lower aspect gives origin to the m. flexor accessorius. The grooved upper aspect of the shelf supports the bundle of peroneal tendons: peroneus longus, peroneus brevis, and the extensor digitorum brevis of the fifth digit. The remaining extensor brevis bellies, having descended to the foot, arise under cover of these tendons from the medial part of the shelf (Fig. 15.3A). A very similar trochlear process is commonly found also in Old World monkeys (for example, *Procolobus, Cercopithecus*) but in the more terrestrial forms it may be reduced to a rudimentary tubercle. It is clear that living monkeys in general have retained a feature lost in extant prosimians. The fossil Adapidae (in particular notharctines) do present a well developed posteriorly located retracted trochlear process (Decker and Szalay 1974) as befits their likely relationship to the ancestry of the Anthropoidea.

The apes (*Hylobates, Pan, Gorilla, Pongo*), all possess well developed trochlear processes resembling those of New World monkeys. In the chimpanzee there is a particularly prominent trochlear process (Fig. 17.4C) and it has the usual morphological relationships. In *Pan* the posterior surface of the lateral malleolus of the fibula is broad, giving the bone a triangular cross section and with the subcutaneous surface facing antero-laterally, because the peroneal tendons here lie rather side by side during their course towards the trochlear process. The peroneus brevis crosses the upper surface of the process while peroneus longus curves around its lateral margin to continue its transverse across the sole. Extensor digitorum brevis extends its origin onto the upper surface of the process. The flexor accessorius muscle (not always present) arises from the plantar aspect of the trochlear process. In *Gorilla* the trochlear process is similar but less prominent. It is apparent that during terrestrial locomotion, with the foot held in its habitually inverted posture, the trochlear process of the African apes is one of the main weight-bearing points of the foot, sharing this function with the tuber calcanei. This seems to be the key to the rather strange derivative condition of the trochlear process in man.

The lateral aspect of the human calcaneus is highly variable and can present an appearance ranging from a morphology with a trochlear process of almost chimpanzee type down to the classically described type of human anatomy. Laidlaw (1905), despite the limited comparative material at his disposal, and the absence then of any relevant fossils, came very close to a clear understanding of these morphological features. It now seems clear that all three of the main features recognized by Laidlaw (1904, 1905) on the lateral aspect of the human calcaneus (his trochlear process, retrotrochlear eminence, and lateral process of the tuber calcanei) are, in fact, together homologous with the single unified trochlear process of *Pan* and other mammals. The commonly used term 'peroneal tubercle' (PNA trochlea peronealis) is perhaps most appropriate for the anterior of these elements, lying as it does between the two peroneal tendons. Before reaching it these tendons overlie one another behind the lateral malleolus, which therefore has a narrower posterior surface than in apes, giving a more quadrilateral cross section, with the subcutaneous surface facing laterally. The retrotrochlear eminence behind the peroneal tubercle is another derivative of the ape process and should not be confused, as is often done in textbooks, with the attachment of the calcaneo-fibular ligament, which lies higher, as Laidlaw (1904) emphasized. The uniquely human lateral tubercle of the human calcaneal tuber is obviously derived by what is effectively a migration of the posterior part of the ape trochlear process to the heel (Fig. 17.5D), a natural progression from the weight-bearing function of the whole process foreshadowed in *Pan*. As Weidenreich (1940) noted, this variably situated 'external tubercle' in man may sometimes have its own epiphysis, a not surprising finding for such a bony apophysis. The origin of the lateral head of flexor accessorius in man, by a flattened tendon from the lateral tubercle of the tuber, testifies to the derivation of the tubercle from the trochlear process. It is as though an ape-like trochlear process had melted or merged into the lateral aspect of the calcaneus, leaving only this trio of variable excrescences. Such deformations of bones are quite in accord with general biological theory. They are just the sort of stuff studied by Thompson (1952) using his system of transformed Cartesian coordinates. For example, if we study an evolutionary series of humeri or femora, such as those illustrated by Gregory (1951), even more dramatic translations and resorptions of ridges and tuberosities are strikingly apparent.

Calcaneofibular articulation   As noted in Chapter 12, in mammal-like reptiles from pelycosaurs to cynodonts the fibula participates in the ankle joint as a significant weight-bearing component, articulating with the calcaneus and often with the talus as well. The monotremes retain the same morphology (Fig. 13.1) as did the fossil protptherian *Eozostrodon*. It has therefore become an automatic assumption that any calcaneofibular contact in extant or fossil mammals represents a persistence of this primitive condition. Szalay (1984) assumed this for the living insectivore *Erinaceus*, although, in fact, the morphology is quite like that shown for *Oryctolagus* in Fig. 12.4B; in both, as described in Chapter 12, it is clearly a derived condition. There is ample evidence that the emergent therian mammals had no such articulation, but a secondary calcaneofibular contact has secondarily appeared in a number of mammalian orders, perhaps most notably the ungulate orders.

This misunderstood pedal character has figured in a number of phylogenetic analyses with unfortunate results. Under the misapprehension that calcaneofibular contact is necessarily primitive, Szalay and Decker (1974) postulated that the condylarth tarsus was an appropriate model for that of the precursors of the Primates. The ungulate orders arose from the condylarths, and as described in Chapter 12 calcaneofibular contact in therians is apomorphic, and has been evolved independently a number of times—in kangaroos, rabbits, elephant shrews, and artiodactyls. Luckett (1980) and Novacek (1980) also misunderstood the true nature of the morphocline polarity of calcaneofibular contact, and of its correlation with tibiofibular fusion, in cladistic based studies aimed at assessing the relationships of treeshrews; Novacek (1980) moreover, incorrectly denied calcaneofibular contact in Macroscelidae but asserted its presence in Tenrecidae. These studies were also biased by inadequate understanding of the nature of the calcaneal trochlear process.

The tarsus and the marsupial–placental dichotomy

The evolution of the foot skeleton has particular relevance to the discussion in Chapter 2 on the question of the arboreal origin of mammals. The general theme developed in Chapter 12 was that the last common ancestors of the marsupials and placentals, the advanced therians of the early Cretaceous, probably possessed an ankle joint similar to that of the more generalized living marsupials, complete with a meniscus and with those functional attributes which equipped them for scansorial or arboreal life. This does not necessarily mean exclusive canopy-dwelling; it could equally well have included the spatially complex interface between arboreal and terrestrial habitats, represented by the forest floor and margins, with its tangle of roots, vines, and bush-like growth. It seems clear, however, that the emergent marsupials became committed to arboreal life, and capitalized on this particular morphology. The secondarily terrestrial members of the Australian radiation refashioned the ankle joint into a hinge joint which was a striking analogue of that characterizing eutherians; some even established a secondary calcaneofibular articulation. The first placentals, however, are likely to have colonized the more terrestrial niches on the forest floor and at the bushy forest margins. There they must have dispensed with the talocrural supinatory capability, facilitated by the marsupial meniscus, in favour of re-fashioning the lower extremities of the tibia and fibula into a stable mortise for articulation with the remodelled talar body. From early members of this group the secondarily arboreal Primates were presumably derived. Other Eutherians, invading the wider reaches of the terrestrial environment, became cursorial or saltatory and in some cases, as with their marsupial counterparts, evolved a secondary calcaneofibular contact.

Central to this scheme is the notion that the more generalized arboreal Australian marsupials form a good model, not only for the structure of the ancestral marsupials, but also for the ancestors of the Eutherians. In one feature, however, they are admittedly (Lewis 1980) specialized: there is confluence of the primitive dual articulations of the subtalar joint complex. Virtually on the basis of this one derived character, Szalay (1982a,b) reappraised the classification of the marsupials and proposed a monophyletic cohort, Australidelphia, to include all the Australian marsupials, together with the Chilean marsupial *Dromiciops*. Confluence of the subtalar joints is a shaky criterion, however, since it can also occur in American didelphids (Figs 13.2C, D). Szalay seems also to have been wrong in considering that quite deeply etched malleolar and medial tibial facets, and fibular facet, on the didelphid talus (Fig. 12.2), were primitive for marsupials. The smoothly rounded contours, characteristic of say *Trichosurus* (Fig. 12.1) can, however, also be found in American didelphids and the deeply sculpted form is correlated with an obviously derived type of lower tibial articular surface. In particular he was totally wrong (Szalay 1982a) in asserting that a calcaneofibular articulation was a primitive characteristic of didelphids. In one point, however, he was probably right: a calcaneocuboid joint with the calcaneal surface of a centrally depressed ovoid or kidney shape, as described in Chapter 13 for certain Australian marsupials, was primitive for the Metatheria as a whole. Yet there was a germ of truth in Szalay's tale; it is not unlikely that *Dromiciops* represents a surviving descendant of stragglers along the route of marsupial migration down through the Americas.

Szalay (1984) elaborated on these ideas, to mount an attack on what he believed to be proponents of the view that arboreality was 'homologous' in metatherian and eutherian mammals. His reasoning was deeply rooted in the idea that a calcaneofibular articulation was primitive for the Eutheria, and that the condylarth *Protungulatum*, the possessor of such an articulation, was a satisfactory ancestral eutherian morphotype; significantly, in this later publication he had abandoned any reference to such an articulation in didelphids.

Astonishingly he turned the arguments given above on their head, but the position is not sustainable. In brief, he believed that the unknown ancestral therian morphotype was little advanced beyond the monotreme condition, and that from this base true astragalar superposition had evolved independently in the Eutheria and in the Metatheria, and that loss of calcaneofibular articulation had also occurred independently in these two groups. Most importantly he postulated that the tenon and mortise eutherian ankle was derived independently from this monotreme-like precursor and not from one similar to that in generalized marsupials. How much more likely it is that the striking resemblance between the hinge-like ankles of eutherians and those of kangaroos and bandicoots, were derived in both groups from a basically similar precursor morphology. In a lengthy discussion Szalay particularly attacked papers (Lewis 1980a,b) giving essentially the view presented above and that given in Chapters 2, 12, and 13; the understanding of the views expressed in those papers, and of the morphology described were, however, seriously flawed. Paradoxically, and for the wrong reasons, Szalay reached (at least in part) the view expressed above: that the earliest Eutherians were basically terrestrial; this, of course, says nothing about their more distant ancestry.

Fossil primate foot bones

The emergent Primates

As described in Chapter 2, the first supposed Primate, *Purgatorius ceratops* appears in the late Cretaceous of Montana and is said to be a paromyid (a member of the Paromomyiformes) with supposed resemblances to condylarths and leptictid and erinaceoid insectivores (Van Valen and Sloan 1965). The leptictid and erinaceoid insectivores (Van Valen and Sloan 1965). The leptictids are generally believed to have had a central role in insectivore evolution (Van Valen 1967) and many believe that tupaiids and Primates were independently derived from them (McKenna 1966; Van Valen 1965; Bown and Gingerich 1973).

The primitive presumptive Primates, the Paromomyiformes, produced a radiation of four families in the Palaeocene–Paromomyidae, Plesiadapidae, Carpolestidae and Picrodontidae (Savage 1975). It is believed that all were probably arboreal, the central family being the Paromomyidae. The Plesiadapidae showed cranial specializations excluding them from direct primate ancestry but they are the best known postcranially. There are strong arguments, however, against the primate status of the suborder as a whole and even whether they were arboreal (Kay and Cartmill 1977).

The primate status of *Plesiadapis* has been particularly advocated by Szalay *et al.* (1975), Sazlay and Decker (1974) and Szalay (1975) and the argument rests in the final analysis on certain key tarsal characters, said to represent emergent primate synapomorphies. These are a screwlike or helical subtalar joint, a calcaneocuboid joint of pivotal nature, and a posteriorly retracted trochlear process (peroneal tubercle). As noted in Chapter 13, however, the first of these is merely an ancient therian character, which has been secondarily refined and elaborated in the Primates. The second is similarly an ancient therian character (Szalay and Decker 1974, themselves note it in fossil Metatheria); Szalay (1977) appeared to consider that a pivot joint requires the elaboration of a peg-like plantar prominence on the cuboid— this is not so, for many monkeys (and *Hylobates*) have no more of a protuberance than such diverse forms as *Sarcophilus*, *Phascolarctos*, and *Tenrec*. The last character, a posteriorly located trochlear process is certainly a common feature in Primates, but is also seen in *Tupaia* and occurs in some rodents (Stains 1959) and even in Didelphis (Fig. 13.2D); its value in cladistic analysis is diminished, therefore, as it has occurred convergently on a number of occasions.

The pulley-like trochlea of the talus in *Plesiadapis* is merely primitive for Eutheria—an early terrestrial modification—and this morphology is paralleled by a number of the more terrestrial marsupials. Simpson (1935) early argued for the inclusion of *Plesiadapis* among the Primates on the grounds of various resemblances to *Tupaia* and *Lemur*; this argument is undermined now that *Tupaia* is by fairly common consent banished from the primate order. The Tupaiidae, however, doubtless arose from some group of leptictid insectivores and the Paromomyiformes may be fossils in a similarly related category. However, from such stock undoubted Primates certainly arose, in North America and/or Asia (Szalay 1973). With little doubt these early unequivocal Primates were the fossil lemuriforms, the Adapidae.

The adapid tarsus has been described in some detail by Decker and Szalay (1974) and it appears to show the transition from a form not unlike that of *Plesiadapis* towards undoubted lemurid morphology. A trend can be shown in the two adapid subfamilies, adapines and notharctines, for lengthening of the anterior part of the calcaneus, with the attainment of a quite considerable articulation of the calcaneus with the navicular: in effect the primitive type of alternating tarsus (talus articulating with cuboid) has been reversed to produce a new type of alternating tarsus. The functional importance of this in the arboreal primate foot has been shown in Chapter 13. Correlated with this the helical action of the posterior talocalcaneal joint is accentuated. The slightly concave and convex conarticular surfaces of the calcaneocuboid joint also achieve a typical primate form by the elaboration of a low conical plantar protuberance on the cuboid. The absence of lengthening of the calcaneus to achieve contact with the navicular in adapines, reflected in a high calcaneal index, means that this group lacks a characteristic primate attribute. As with *Plesadapis* their Primate status is therefore arguable, as already noted by Martin (1979).

There is little doubt, however, that the living Strepsirhini and the Haplorhini were derived from some early adapid stock, with the tarsal modifications characteristic of Primates already evolved, and the notharctines meet this requirement.

The emergent Anthropoidea

The earliest known Anthropoidea are from Oligocene deposits in the Fayum of Egypt. As noted in

Chapter 2 there is an emerging consensus that collectively they are primitive catarrhines, achieving neither true ape nor monkey status, but close to the source from which those diverging radiations were derived. The Fayum Primates show many resemblances to platyrrhine monkeys, not least in foot structure, which is in accord with the strengthening view that these South American Primates were derived from a similar but somewhat earlier late Eocene African stock. Despite the generally primitive anthropoid morphology of the Fayum fossils there is little doubt that this minor radiation included specialized offshoots which were evolutionary dead ends.

Five primate tali and five calcanei have been known for some time from the Fayum and these have been described by Conroy (1976a). Access to this material has not been possible and the published illustrations are of unsatisfactory quality. However, some tentative comments are worth making on the published data. The tali have a well angulated neck, indicative of an oblique subtalar axis, and a well-marked cup for the medial malleolus. The calcanei also fit the structure to be expected of the ancestors of the apes and of the monkeys. The posterior talocalcaneal joint appears to have had a marked helical action and associated with this one would expect a reasonable articular contact of the calcaneus with the navicular; no data on this latter feature are available but the published illustrations make it seem probable. According to Conroy (1976a) 'the articular surface for the cuboid is a crescentic-shaped, gently concave facet'; it seems thus to have possessed the likely ancestral anthropoid form as preserved today in extant monkeys, and not to have elaborated the marked plantar excavation which apparently characterized early phases of ape evolution.

A particularly interesting finding is that a hallucial metatarsal attributed to *Aegyptopithecus* seems to show clear indications of an articular surface for a prehallux (Conroy 1976b), an expected finding for a primitive catarrhine with platyrrhine affinities. These same fossils, together with several recovered more recently, have also been described by Gebo and Simons (1987). They allocated the *Apidium* and *Parapithecus* specimens to the Cercopithecoidea and *Propliopithecus* and *Aegyptopithecus* specimens to the Hominoidea. It must be said, however, that the criteria which they

used are very dubious and carry little conviction in the light of the descriptions given in Chapters 12 and 13. Yet it is true that the tali attributed to *Apidium* had a laterally compressed trochlear which would have fitted into an almost vertical-sided mortise formed by the malleoli. As noted in Chapter 12 this is not unlike the specialized arrangement in the cercopithecine ankle joint. But it is known that about the lower two fifths of the tibia and fibula in *Apidium* were virtually fused (Fleagle and Simons 1983) suggesting that this species was a specialized leaper; despite resemblances to the cercopithecine talus there seem to be no real grounds for referring the fossil *Apidium* tali to the Cercopithecoidea. The tarsals of *Aegyptopithecus* and *Propliopithecus* retain a much more generalized anthropoid morphology, but this is no real reason for referring them to the Hominoidea.

If it is accepted that the Fayum Primates represent an early catarrhine stock then residual members seem to have lingered on well into the Miocene, alongside the burgeoning ape and emerging monkey radiations. Many believe now that European *Pliopithecus* and African *Dendropithecus* (formerly *Limnopithecus*) *macinessi* were in this category. Both, however, have been widely touted as gibbon ancestors. Some time ago I suggested (Lewis 1972) on the basis of published illustrations, that the *Pliopithecus* medial cuneiform showed no evidence of an articulation for the prehallux, thus effectively eliminating it from any plausible place in gibbon ancestry. Instead the bone showed the morphology expected when a prehallux had been lost, as has been realized independently at least twice before in the cercopithecoid and hominoid lineages. Fleagle (1983), however, maintained that 'the hallucial metatarsal shows evidence of a prehallux bone (contra Lewis 1972), a primitive sesamoid in the tendon of peroneus longus . . .'. Wikander *et al.* (1986) seized on this disagreement to doubt the validity of attempting to assess this character from dry bones. It is not surprising that there was disagreement: Fleagle had cited the wrong tendon, which is even on the wrong side of the joint!

Early Miocene African apes

Fossil tali and calcanei have been known for some time from two early Miocene inter-rift localities in East Africa: Songhor (labelled KNM-SO) and

Rusinga Island (KNM-RU). Cranial and dental remains suggest that the common species present at Songhor were *Proconsul major*, *Rangwapithecus gordoni*, and *Limnopithecus legetet*. At Rusinga the common species were *Proconsul africanus*, *Proconsul nyanzae* and *Dendropithecus macinnesi* (Andrews and Van Couvering 1975). At neither site were there apparently any monkey fossils. Casts of a number of these foot bones, graded in size from the smallest *Proconsul africanus* to *Proconsul nyanzae* and *Proconsul major* are readily available.

The tali all show the trochlear form characteristic of the arboreal primate ankle joint, with a deep articular cup at the anterior end of the tibial malloelar facet. The functional significance of this was described in Chapter 12. In the fossils this feature is well accentuated, perhaps even more so than in the extant great apes with the possible exception of *Pongo*.

Taken together the tali and calcanei give clear insights into the functional anatomy of the subtalar joint complex. In the basic arboreal primate condition the anterior articulating surfaces on the talus for the calcaneus, corresponding to the anterior and medial facets of human anatomy (Sewell 1904*b*) are arranged as an L-shaped area on the surface of a cylinder formed by the head and neck of the talus. The axis of this cylinder corresponds to the axis of the subtalar joint complex and the ligamentum cervicis tali attaches in the angle of the L-shaped area. The posterior articulating surfaces are aligned obliquely to this axis so that they impart a helical action to the joint movement. This subtalar axis is set very obliquely to the functional anteroposterior axis of the foot and lies in a quite flat plane. As described in Chapter 13 the essentials of this morphology are conserved in *Hylobates* and *Pongo*. *Pan* and *Gorilla*, however, share certain derived features grafted onto this basic arboreal morphology. In both of these largely terrestrial great apes the talus is much more squat in form with a shortened neck and broadened head and with the component limbs of its L-shaped articular surface for the calcaneus partially blended to form an almost continuous semilunar facet. All the fossils (Figs 17.6A, B, C) possess the basic Primate arrangement, devoid of the specializations distinguishing *Pan* and *Gorilla*.

No fossil cuboids are known from these sites but all the calcanei quite clearly show the form of the articular surface for that bone. As has been shown in Chapter 13 evolutionary change in the calcaneocuboid joint must have played a highly significant role in primate evolution. As noted in Chapter 13 in monkeys and prosimians the articular surface on the cuboid is kidney-shaped with the variable expression of a protuberant ventral convexity; the calcaneus is reciprocally shaped and movement is largely rotatory although with some angular movement. A similar morphology is retained in gibbons. In the great apes, however, the morphology is modified with chimpanzees apparently occupying a central and relatively conservative status in this spectrum of evolutionary change. In *Pan* the ventral beak on the cuboid (and the corresponding calcaneal concavity) is considerably enlarged and generally is displaced somewhat medially (Figs 17.7A, 17.8A). In the majority of gorillas the form of the articulating surfaces is in striking contrast to that shown by the chimpanzee, since the joint surfaces are almost flat, with only the slightest suggestion of an elevation on the cuboid which is accommodated by a correspondingly shallow depression on the calcaneus (Figs 17.7C, 17.8C). Quite uncommonly, however, gorillas possess a very different morphology in which the calcaneus may be excavated to accommodate the cuboid even perhaps to a greater extent than is seen in chimpanzees, and this deep depression is surrounded by a relatively flattened articular rim (Figs 17.7D, 17.8D). It is probable that this variant represents the ancestral *Gorilla* pattern, being easily derived from a morphology like that of *Pan*, and that the common flattened form is a progressive derivation. In both variants a pivotal action occurs which appears to be essentially like that in *Pan*. The orang-utan joint surfaces present a characteristically unique pattern of their own. The calcaneus bears a deep hemispherical pit which lodges a conical protuberance on the cuboid, and is sharply demarcated from a broad flat articular surface located to its dorsal and lateral sides (Figs 17.7B, 17.8B). It forms the classical design for a pivotal mechanical action, with the flat area mimicking the thrust plate on a drive shaft.

The calcaneal fossils from Rusinga and Songhor, with one notable exception, show a morphology close to that of *Pan* in the form of the articular

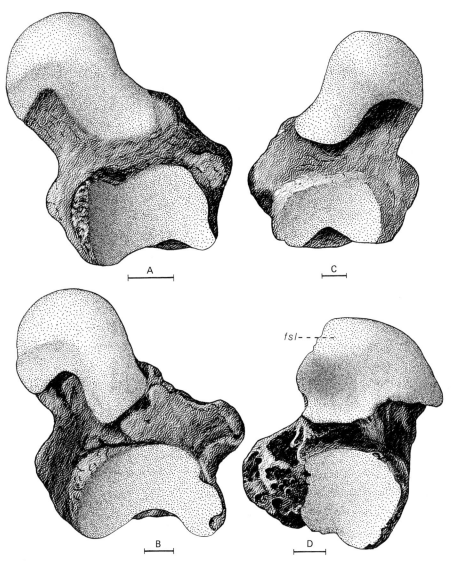

Fig. 17.6

The undersurfaces of casts of a right talus attributed to *Proconsul africanus*, KNM-RU-1745 (A); a right talus attributed to *Proconsul major* KNM-SO-389 (B); a left talus attributed to *Proconsul nyanzae*, KNM-RU-1743 (C); and OH8 (D). In each case the bar represents 0.5 cm. *fsl*, facet for the plantar calcaneonavicular (spring) ligament.

surface for the cuboid. Indeed the excavation may be rather more accentuated than that usually seen in *Pan*. This morphology (Fig. 17.9B) is the likely precursor to that seen in *Pan* and seems also to represent a plausible ancestral morphology for those unusual variants among *Gorilla* calcanei in which there is a deep excavation for the cuboid. Progressive flattening of this surface then appears

to have been the evolutionary trend which is most characteristic of the *Gorilla*.

The notable exception among the fossils is the largest specimen (Fig. 17.9A) which has reasonably been attributed to *Proconsul major*. This calcaneus must have articulated with the cuboid in a joint strikingly resembling that found in *Pongo*. Could it, therefore, be that here is represented the

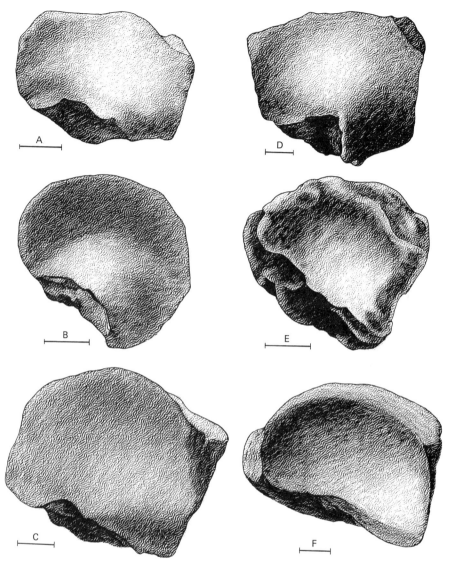

Fig. 17.7
Proximal views of the left cuboids of *Pan troglodytes*, (A); *Pongo pygmaeus*, (B); *Gorilla gorilla*, (C); another specimen of *Gorilla gorilla*, (D); OH8 cast (E); *Homo sapiens* (F). In each case the bar represents 0.5 cm.

ape stock for the Asiatic orang-utan whose emigration from Africa could then have been easily effected, since a land bridge to Eurasia is known to have formed at the beginning of the middle Miocene. Some support for this notion comes from the description of a palate, also from Songhor, which is said to show suggestive orang-utan affinities (Andrews 1970). An alternative view, although perhaps less likely, must be consid-

ered. The form of the cuboid surface on the calcaneus of *Proconsul major*, although very like *Pongo*, is also not markedly dissimilar to that found in the variant form of *Gorilla* (Fig. 17.8D), to which it may have been ancestral. Whatever the truth of this, these fossil apes have the requisite morphology among them to make them an acceptable pool from which the extant great apes could have evolved.

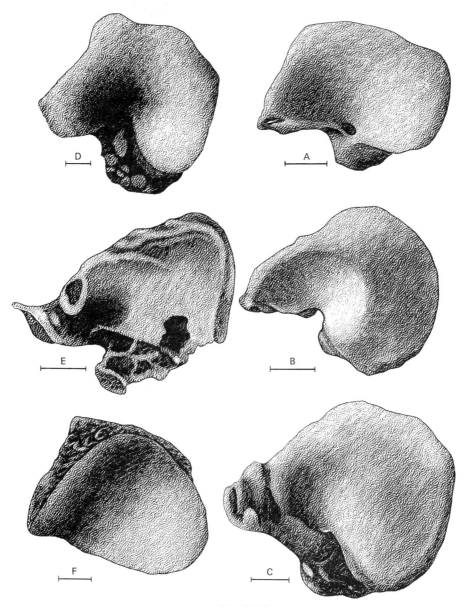

Fig. 17.8

Distal views of the left calcanei of the same specimens as shown in Fig. 17.7: *Pan troglodytes*, (A); *Pongo pygmaeus*, (B); *Gorilla gorilla*, (C); *Gorilla gorilla*, (D); OH8 cast (E); *Homo sapiens* (F). In each case the bar represents 0.5 cm.

This has not always been the interpretation. Clark and Leakey (1951) and Clark (1952) previously described some of these specimens and surprisingly they came to the conclusion that these tarsal bones indicated a habitually everted posture of the foot associated with largely terrestrial, quadrupedal locomotion, and that the nearest modern counterparts were to be found among the cercopithecoid monkeys. Among the features which they included in support of their thesis was the cup-shaped hollow at the anterior end of the medial malleolar facet and the form of the articular

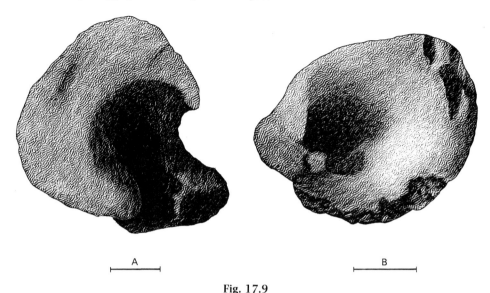

Fig. 17.9
The distal aspect of the cast of the right calcaneus of *Proconsul major*, KNM-SO-390 (A) and of the left calcaneus from Songhor, KNM-SO-427 (B). In each case the bar represents 0.5 cm.

surface on the under aspect of the head and neck of the talus; in reality the pattern of these features is clearly correlated with arboreal activities, and among the apes these features are especially evident in the arboreal *Hylobates* and *Pongo*. Similarly Clark and Leakey interpreted the considerable lateral extent of the articular surface on the talar head as reflecting provision for a greater degree of eversion at the forefoot whilst in reality this feature is concerned with just the opposite, namely providing for movement of the navicular during supination of the forefoot in an arboreal setting. The dorsal tubercle of the talar neck was interpreted as a further cercopithecoid feature supposedly forming a locking mechanism for the dorsiflexed ankle joint, but it is merely a universal primate feature—the site of attachment of the strong dorsal talonavicular ligament. However, they did note the remarkable form of the calcaneo-cuboid joint surface of *Proconsul major* and did perceive its resemblance to that of the orang-utan, but then oddly enough interpreted it as a specialization ensuring stability in quadrupedal terrestrial activities. They also noted the posteromedial extension of the posterior talocalcaneal articulation and interpreted this as a provision for occasional assumption of an erect posture whereas it merely indicates a well developed helical action

at the subtalar joint complex, to be expected in an arboreal foot.

Several of these tali have had a chequered history in multivariate statistical studies, which have offered conflicting results: they are similar to modern African apes (Day and Wood 1969); unique and well separated from the African apes (Oxnard 1972); close to cercopithecoid quadrupeds (Wood 1973); rather close to the orang-utan (Lisowski *et al.* 1974, 1976). This diversity of opinions is, however, not surprising when it is appreciated that the eight or so measurements utilized reflect little more than overall shape and are correlated only at the crudest level with the biomechanical apomorphies which characterize the different groups.

The position has been, or should have been, transformed by the discovery of almost complete foot skeletons of *Proconsul africanus* and *Proconsul nyanzae*. These specimens have been briefly described by Walker and Pickford (1983). It is clear from the illustrations, and such is the conclusion of the authors, that there are close overall resemblances to the living apes, particularly *Pan*. The description is disappointingly deficient in that anatomical detail, highlighted in previous chapters, that would have been so useful in establishing the affinities of these important fossils.

Yet fascinating snippets about the anterior foot skeleton are revealed, almost incidentally. The tarsometatarsal joint line is stepped with a deeply inset second metatarsal and a strongly projecting lateral cuneiform. This is of course like *Pan*, but at this level of analysis it is also merely a primitive primate feature (Fig. 13.20A). However, in *Pan* (Fig. 13.20B) there is a derivative condition, since the strongly projecting lateral cuneiform has at best a tiny distal articulation with the intermediate cuneiform and this is often lacking; by Walker and Pickford's testimony this was the state in *Proconsul nyanzae*! Another interesting finding by Walker and Pickford was the presence of a single large sesamoid bone, apparently adjacent to the first metatarsal in the matrix containing the foot skeleton of *Proconsul africanus*. They suggested that it could have been a 'peroneal sesamoid' but this, of course, should be at the lateral margin of the foot; could it, in fact, be a prehallux? In fact, the previously known medial cuneiform and first metatarsal of this fossil were reported (Lewis 1972) to show suggestive indications of the presence of a prehallux. This would not be surprising at this stage of ape evolution.

These new *Proconsul* feet, together with the other fossil tali and calcanei were included in a comprehensive study by Langdon (1986) but again there is little depth in the anatomical treatment although comprehensive and useful measurements are given.

Middle–Late Miocene Eurasian apes

Several important isolated foot bones have recently been described from Pakistan from this period and are presumably referrable to the genus *Sivapithecus*. Their significance centres on the postulated ancestral relationship of *Sivapithecus* to the orang-utan, an idea based hitherto entirely on cranial specimens. Only published descriptions have been available and these, and the accompanying illustrations, are a tenuous basis for any conclusions, but some points are of interest.

Left lateral cuneiform (GSP 17118) This specimen was described by Rose (1984), with the traditional comparison to the bones of other apes and the not surprising finding of a general similarity to chimpanzee lateral cuneiforms 'in many characters'.

The particular apomorphies of the orang-utan tarsus noted in Chapter 13 were, however not addressed. As has been seen these seem to have strikingly changed function in the orang-utan foot skeleton, but the changes in a single bone (for example, the lateral cuneiform) are quite subtle and could easily be overlooked. The lateral aspect of the lateral cuneiform of *Pongo* (Fig. 13.22) is quite like that of *Pan* (Fig. 13.20B). But when the medial aspect is viewed the situation is quite different: bevelling off the posteromedial corner (Fig. 13.22) has brought the naviculocuneiform and proximal intercuneiform facets into continuity in *Pongo* and the retraction of the bone as a whole into the tarsus leaves only a distal intercuneiform facet without any articulation for the second metacarpal which is not here indented, although in wet specimens virtual contact between the cartilage clad bones is retained: it is not unlikely that in some osteological specimens a tiny facet would be found. This is in total contrast to the situation in *Pan* (Fig. 13.20B) where there is a large facet for the second metatarsal, with or without a tiny contact with the intermediate cuneiform.

In these functionally significant features (contra Rose) the fossil seems to be *Pongo*-like and moreover, there is at least incipient bevelling-off of the posteromedial corner of the bone. The implications are that retraction of the fossil lateral cuneiform into the tarsus was well under way giving a *Pongo*-like functional complex. It is interesting to note that in an earlier publication (Conroy and Rose 1983) this specimen was said to be 'in all respects like a chimpanzee lateral cuneiform'; this it most certainly is not.

Calcanei (GSP 17152 and GSP 17606) and cuboid (GSP 19905) Buried within the formal description of these specimens by Rose (1986) are critical points about the nature of the calcaneocuboid joint, which, although adequately described and illustrated, have had their significance blunted by inadequate consideration of published comparative findings, which were in fact available. It has been shown in Chapter 13 how the orang-utan has acquired an essentially unique calcaneocuboid joint in which crescentic planar surfaces surround a sharply defined peg and socket part of the joint. Just such an arrangement was clearly present in these fossils assigned to *Sivapithecus*.

The evolutionary relationships of Sivapithecus   It is generally agreed that the *Ramapithecus/Sivapithecus* group must have been more ground dwelling than earlier Miocene hominoids, owing to the change in the middle Miocene from a closed forest environment to one of open forest woodland. This was a persuasive factor in assigning a hominid ancestral status to this group. Laporte and Zihlman (1983) reject the newer idea of orang-utan ancestry on these same palaeoecologic grounds. Consideration of the fragmentary postcranial fissils can contribute something to this debate. The strangely offset and diminutive hallux of the orang-utan may well be heritage of a largely terrestrial ancestry, its exclusive arboreality at present being secondary. It is fairly clear that the hallux of *Sivapithecus* was not reduced, and one would not expect fully blown orang-utan morphology in this fossil genus. Yet there are tantalizing indications that the apomorphic features of this morphology were well on the way to being established by the late Miocene.

Fossil hominids

The australopithecine foot   Apart from one fragmentary talus, no significant foot fossils are known which could have belonged to the two species of South African ape-men, *Australopithecus africanus* and *Australopithecus robustus*. Speculation about the locomotor capabilities of these species is based almost entirely upon fossil innominate bones and femora. Robinson (1972) believed that the robust form was incompletely adapted to the erect posture, had an ape-like lifestyle, but was ancestral to the gracile form which he believed had further perfected the specializations required for habitual erectness. Such locomotor differences between the two species are not generally accepted, but most would agree with Zihlman and Brunker (1979) that the australopithecines were well on the way to being suitably adapted for bipedalism. But whether this capability differed significantly from modern humans, and whether substantial anatomical specializations for arboreality were retained, can only be resolved by the analysis of the foot skeleton; this should be clear from the preceding chapters.

Some significant foot fossils have been described for *Australopithecus afarensis*, and these should provide useful clues to the locomotion of this particular hominid species, which is widely considered as ancestral to man (Chapter 2).

Several specimens of the lower extremities of the tibia and fibula, together with a talus, have been found and have been analysed by Stern and Susman (1983) and Susman (1983). They discussed the tibia and fibula in terms of those osteological criteria which were considered at the end of Chapter 12, but did not discuss the talus, although it was figured articulated with the relevant tibia and fibula (AL 288–1, 'Lucy') by Susman (1983). Yet, as noted in Chapter 12, these features of the tibia and fibula are really rather peripheral issues; what one really wants to know is whether the ankle functioned as in subhuman Primates or as in man.

These authors noted that the inferolateral part of the distal articular surface of the tibia in the fossils faced inferiorly, as in man, rather than inferolaterally, as in apes. Despite the disclaimer by the joint authors, Susman (1983) considered this as a general characteristic of the Afar hominids. As noted in Chapter 12 this feature is merely a conarticular compensation for an elevated lateral rim on the talar trochlea. Moreover, despite the suppression of this talar feature in man, some such tibial hollowing may persist. Taken in isolation as a tibial characteristic, not much can be read into this feature, nor did the authors dwell on it.

They noted that the distal surface of the tibia, at least in the specimen called 'Lucy', showed a posterior tilt as in apes, and that the proximal border of the talar facet on the fibula was directed anterodistally as in apes, correlated with such a tilt of the tibial surface. Susman (1983) rather oddly described this as a talocrural joint which 'opens posteriorly'. They interpreted it as indicating a joint with a 'plantarflexion set', a trait which would 'be useful in reaching for branches with the feet and in hindlimb suspension'.

They also noted that the fibular malleolus was short with its articular surface shelving away laterally, so that it had a more downward inclination than in man. Strangely they suggested that this primitive primate feature indicated significant transmission of vertical force by the fibula. They noted that the subcutaneous surface of the australopithecine fibula faced more anteriorly than laterally, as in apes, and this was suggestive to them of large peronei, playing an important role in arboreal use of the foot. The real significance of this ape characteristic has been described above: it reflects the way in which the peroneal tendons enter the foot on their way to straddling a large trochlear process.

Altogether, on this tenuous evidence Stern and Susman (1983) and Susman (1983) concluded that the behaviour of *A. afarensis* included a significant component of arboreality. Latimer *et al.* (1987) considered the ankle bones of AL288-1 ('Lucy') in their study of the African ape talocrural joint, which was discussed in Chapter 12. Rightly enough they summarily rejected the idea of a 'plantar flexion set' to the *A. afarensis* foot, but they went further, asserting that the Hadar hominids had 'forfeited any significant arboreality' and that the fossil ankle joint displayed 'a functional pattern equivalent to *H. sapiens*'. As noted in Chapter 12 they failed to take account of any difference of the ankle axis in the transverse plane and thus failed to consider the one critical human distinction.

A talus of 'Lucy' has been recovered and illustrated (Johanson *et al.* 1982; Latimer *et al.* 1987) and this bone puts the described features of the tibia and fibula into proper perspective. The trochlear surface entering the ankle joint is strikingly like that of *Pan*: the lateral border of the trochlea is elevated especially anteriorly, the lateral malleolar surface is flared laterally, and the medial malleolar surface shows an obvious cup-like depression anteriorly. Taken together with the various features of the tibia and fibula there can be little doubt that the ankle joint operated in a

manner like that of apes (Fig. 12.8A) rather than in the human manner where eversion accompanies dorsiflexion (Fig. 12.8B), a significant determinant in the human gait, described in Chapter 13. Moreover, it seems clear that the mode of entry of the flexor fibularis and flexor tibialis tendons into the sole must have been very much as it is in *Pan* (Fig. 14.19A). Latimer *et al.* (1987) asserted exactly the opposite—that it was as in *H. sapiens*—but their illustrations carry little conviction in the light of the talar morphology, and their odd interpretation of this feature has been discussed in Chapter 13.

Significant other foot fossils of *A. afarensis* are known, and perhaps most important amongst these are three damaged calcanei (Latimer *et al.* 1982). The best of these (A.L.333.8) is lacking the anterior portion bearing the cuboid facet and most of the sustentaculum tali. It is clear however (Fig. 17.10A) that it retains virtually the same morphology as that of the chimpanzee bone (Fig. 17.10B) having an inflected tuber and a large trochlear process which was not consolidated into the side of the calcaneus, and thus inevitably there was no lateral tubercle as such. Deloison (1985) also reported the absence of the lateral tubercle; inexplicably Latimer *et al.* (1982) maintained that this essentially human feature was present in the Afar calcanei. In correlation with the existence of the ape-like peroneal trochlea it is not surprising that the fibular malleolar fragments of *A. afarensis* should show the characteristically ape-like feature of a subcutaneous surface facing anterolaterally.

Just as the *A. afarensis* calcaneus is remarkably

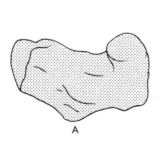 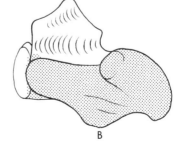

Fig. 17.10

A, the right Hadar calcaneus, A L 333-8, viewed from the plantar aspect, redrawn from a photograph by Latimer *et al.* (1982); weathering has abraded much of the surface detail and the sustentaculum tali and cuboid facet are broken off. B, the right calcaneus of *Pan troglodytes* with the part of the bone corresponding to the fossil stippled.

Pan-like in general form, so the articular surfaces of the talus which enter into the subtalar joint are quite like those of *Pan* (Fig. 13.12A), complete with a similar triangular impression for the spring ligament. The contention of Lamy (1986) that such an impression for the plantar calcaneonavicular ligament is absent in apes, but present in fossil hominids (including *A. afarensis*, and as will be seen later in OH8) and is indicative of a human-like medial longitudinal arch is quite untrue.

The Homo habilis *foot*   An almost complete foot (OH8), consisting of the tarsals (but with the calcaneal tuber broken off) and metatarsals (lacking the heads), was recovered in Bed I of Olduvai Gorge, dated at about 1.7 MY before the present. The foot was associated with hand bones (OH7) and the juvenile skull bones and mandible which form the Type specimens of the taxon *Homo habilis*, (Leakey 1971). This taxon is now widely accepted, and the foot bones have been assigned to this same taxon. Somewhat earlier a tibia and fibula, lacking the proximal extremities, had been found at a nearby site (Davis 1964). It has now been realized that all these fossils almost certainly belong to the same individual (Susman and Stern 1982).

Thus, an almost complete ankle joint of *Homo habilis* is available. This has been studied by Susman (1983) in the same sort of way as for the Afar fossils, using similar rather dubious criteria. Not surprisingly he concluded that in essentials these ankle bones resemble the human condition. This view cannot be sustained, although it is true that the distal articular surface of the OH8 tibia shows a slight tilt anteriorly. However, the trochlear surface of the talus (Fig. 13.25D) is strikingly like that of *Pan*, with an elevated lateral margin, and the malleolar facets on it are also almost identical to those of the chimpanzee. When the articulated casts of the ankle are manipulated, it seems almost certain that the motion is that typical of subhuman Primates, with dorsiflexion involving components of internal rotation of the crus and inversion of the foot. The anterior colliculus on the fossil tibia has unfortunately been broken off and thus there is no engagement into the cup-like receptive area on the talus, which rather detracts from the conclusiveness of this experiment.

In the subtalar joint complex of the OH8 foot the joint surfaces retain the ape-like form and the axis, as found experimentally, is very obliquely disposed (Fig. 13.25D) like that in *Pan*. The talus itself shows the squat foreshortened appearance seen in the extant African apes, and the calcaneal articular surfaces on its head and neck are similarly merged (Fig. 17.6D), no longer conserving the primitive L-shaped pattern; there is a well marked facet for the plantar calcaneonavicular ligament like that of *Pan* or *Gorilla*. Logically, from the obliquity of the axis, it would be expected that the fossil heel would not be realigned towards the axis by being twisted laterally as it is in *Homo sapiens*, where this leads to opening out the calcaneal canal. The heel is missing but the broken surface clearly indicates that it had an orientation more like that of *Gorilla* and *Pan* than of *Homo*. Similarly, the trochlea of the talus has not been medially rotated towards the subtalar axis; the talar neck is thus angulated in relation to the body rather as it is in *Pan*. Although the posterior tubercle of the talus is missing in OH8 its disposition can readily be inferred from the site of the broken surface. It is clear from this that the mode of entry of the tendons of flexor fibularis and flexor tibialis into the sole (Fig. 17.11B) was very like the arrangement in *Pan* (Fig. 17.11A) and *Gorilla* and quite unlike the derived condition characteristic of *Homo sapiens* (Fig. 14.19C). It may reasonably be assumed from this that OH8 lacked the uniquely human medial head of flexor accessorius, which plays a significant functional role in the human gait. The posterior part of the foot is thus strikingly conservative in OH8 and in functional morphology was presumably very similar to that of the great apes.

It has been noted that the calcaneocuboid joint of man has had striking modifications apparently grafted onto the type of morphology generally characteristic of apes. New dimensions have been introduced into the movement in man which provide for the untwisting of the lamina pedis, bringing it into the close-packed position, during the support phase of the gait. In OH8 rotatory movement is greatly limited and is far less than that found in *Pan*, *Gorilla* and *Pongo*; *Pan*, *Gorilla* and *Pongo* despite prominently projecting articular beaks on the cuboid have quite free rotatory movement. Movement in OH8 is also much more limited than that found in *Homo sapiens* although there is a superficial resemblance between the form

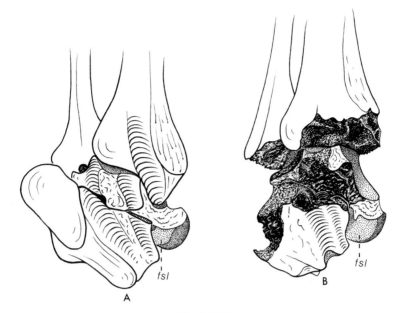

Fig. 17.11
The bony basis of the tarsal tunnel in *Pan troglodytes* (A) and in OH8 (B). The broken off posterior parts of tibia, fibula, talus, and calcaneus in the fossil are darkly shaded.

of the articular surfaces (Figs 17.7E, F; 17.8E, F). Thus, the lateral expansion of the cuboid is less than that in man and the rotatory motions between calcaneus and cuboid, in directions similar to those occurring in man (Fig. 13.29), quickly bring the joint surfaces into maximal congruence, arresting movement before the heel is laterally deviated, as happens in *Homo sapiens*. In this close-packed position the joint is very stable, for the projecting tongue on the cuboid is impacted under the sustentaculum tali, and the expanded flat lateral areas in contact. The fossil joint had thus acquired a new found stability, which is in striking contrast to the situation in *Pan*. The joint in *Pan* and *Pongo* does not lock in such a way, whilst in man it not only locks but is also modified to provide for lateral deviation of the heel. This stable locking joint could be envisaged as a reasonable transitional modification between the morphology of arboreal apes and bipedal man. Its structure could not have been predicted from the study of extant primates alone. Elftman and Manter (1935) postulated that during the evolution of man the transverse tarsal joint had become relatively fixed in a position of plantar flexion. This was a quite inspired prediction, since in one form at

least close to the lineage of man (OH8), it now appears that the joint had certainly become stabilized, but not in an attitude of the plantar flexion: this appearance in man is in fact largely the result of further remodelling of the cuboid.

The joint surfaces of the calcaneocuboid joint of OH8 could readily have evolved from a morphology like that shown by *Pan troglodytes* (Figs 17.7A, 17.8A). The distal aspect of the OH8 calcaneus with its deep excavation, is even more like that of certain of the early Miocene apes, for example that shown in Fig. 17.9B. This does not, however, presuppose an early Miocene derivation of an independent hominid line, for the derived features in *Pan troglodytes* are likely to be of relatively recent acquisition. Moreover, in at least some specimens of the pygmy chimpanzee, *Pan paniscus*, there is a calcaneal articulation for the cuboid strikingly like that shown in Fig. 17.9B.

The forefoot (Fig. 17.12), in contrast to the hinder part, presents a rather more human appearance, at least in its medial portion. This impression largely results from the comparative lack of divergence of the hallux, largely the result of less extreme medial bending of the medial cuneiform. There is a single confluent articulation

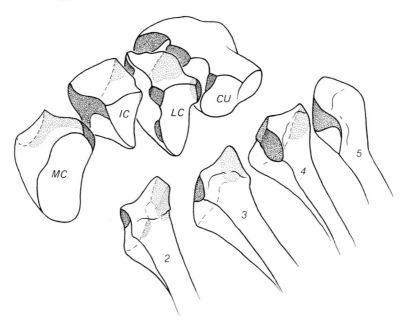

Fig. 17.12
An 'exploded' diagram of the distal tarsals and metatarsal bases of OH8 showing the intervening articulations. *MC*, medial cuneiform; *IC*, intermediate cuneiform; *LC*, lateral cuneiform; *CU*, cuboid; *2–5*, bases of metatarsals 2–5.

between the medial and intermediate cuneiforms as in man, rather than dual ones as in *Pan* (Fig. 13.20B) and *Gorilla* (Fig. 13.21). Not too much emphasis however, should be put on this as reflecting a human resemblance, for *Pongo* (Fig. 13.22) with its strikingly deviated hallux also shows such confluence. The degree of adduction of the fossil hallux has generally been over-emphasized, largely because there are indications of a neomorphic joint between the bones of the first and second metatarsals.

The apparent presence of this joint in OH8 between the first and second metatarsals has been cited as a feature demonstrating 'unequivocally' or 'indisputably', the lack of divergence of the hallux (Day and Napier 1964; Day 1978). Articulations between these two bones certainly are quite frequently encountered in human feet, and indeed one was present in the fresh state in the macerated foot shown in Fig. 13.25A. These anomalous joints vary considerably in the refinement of their structure, often being more of the nature of a pseudarthrosis, with a synovial cavity of bursal type and quite thick fibrous or fibrocartilaginous pads, fashioned from ligamentous attachments,

forming the emergent articular surfaces on the apposed bones. Synovial joints are known to be formed as a variant at a number of anatomical sites, for example, between the coracoid process and clavicle (Lewis 1959), and in most cases the articular surfaces are remodelled from the ligamentous attachments to the bones, and the cavity is derived from a bursa. In the case of the foot the articular surface on the second metatarsal is a progressive elaboration of the raised protuberance which is the site of attachment of the ligament of Lisfranc; that on the first metatarsal is remodelled from the neighbouring elevation which gives attachment to converging oblique ligaments on the medial side of the hallucial tarsometatarsal joint. These ligamentous markings are present even when there is no synovial joint, but if a direct articulation is present they show a varying degree of progressive specialization even sometimes achieving the appearance of smooth, elevated articular facets. Similar closely opposed ligamentous tuberosities are present in *Gorilla gorilla* (Fig. 13.25B). In the specimen shown no synovial articulation was present in the fresh state, although this might have been the conclusion reached from a study of the dry bones. It is quite apparent that the impressions on the adjacent first and second metatarsals of the OH8 foot similarly are basically ligamentous ones. However, by analogy with comparable human examples, there can be little doubt that a rather rudimentary form of synovial joint was indeed present. When, however, the foot is reconstructed with due allowance being made for the probably quite thick fibrous investments of the bones, it is clear that the hallux would be somewhat divergent (Fig. 13.25D). Moreover, when the first tarsometatarsal joint is brought into maximum congruence some degree of divergence of the hallux is apparent, and the form of the conarticular surfaces is quite in accord with some residual grasping function.

The remainder of the anterior tarsus and of its articulations with the metatarsal bases (Fig. 17.12) is essentially conservative and resembles particularly *Pan troglodytes* (Fig. 13.20B). The form of the cuboid lacks the human apomorphic features and is somewhat bowed dorsally as in *Pan* and *Gorilla*, and no lateral longitudinal arch is therefore present. The lateral cuneiform is similarly like that of the African apes, in that it retains the distal articulations with the adjoining cuboid and intermediate cuneiforms, which are lost in man because of the encroachment of interosseous ligaments. It does not, however, show the reduction in length shown in the extant apes.

The bases of the medial four metatarsals of OH8 also show little of the angulation characteristic of man and the joints between them are like those of the chimpanzee. The fifth metatarsal articulates with a dorsoventrally concave surface on the cuboid, unlike the situation in man where the cuboid surface is convex or sellar. The suggestion has been made that this condition in OH8 was pathological and the result of arthritic changes (Day and Napier 1964). Although in the African apes the cuboid usually articulates with the fifth metatarsal by a convex surface examples can be found where the cuboid bears a dorsoventral concavity strikingly like that of OH8. Taken overall the forefoot, like the hindfoot, lacks the remodelling which would align its functional axis towards the subtalar axis, and which is so characteristic of *Homo sapiens* (Fig. 13.25A).

Torsion of the fossil metatarsals (Fig. 13.26) is intermediate between that shown by apes and that of man. In OH8 the first metatarsal shows torsion of a direction and degree comparable to that of *Pan*, and possibly even less than that of *Gorilla*; this is to be expected when the incomplete adduction of the hallux is taken into account, with its implications for a probable retention of some grasping capability. Thus, in *Homo sapiens*, *Pan*, *Gorilla* and OH8 the direction of torsion is the same in this bone, but in *Homo sapiens*, it is more accentuated than in the other three. The remaining metatarsals, or at least the second and third, appear to have a direction and degree of torsion comparable to that of man, and this contrasts with the ape condition. The extent of damage to the fourth and fifth metatarsals makes it impossible to estimate the degree of torsion, just where it would have been especially informative.

Estimates of the talar neck torsion angle, measured according to the traditional method, have produced conflicting results: Lisowski (1967) found that it is within the range of subhuman Primates, Day and Wood (1968) suggested that it was similar to man. It is apparent, however, that orientation of the navicular within the lamina pedis (which must determine the true attitude of the talar head in the foot) differs little between OH8 apes, and man. When the OH8 talus is orientated on its basal surface (the standard basal talar plane) after reconstruction of the broken posterior talocalcaneal surface, it is apparent that the talar neck angle then differs little from either apes or man (Fig. 13.27).

The OH8 foot has been the subject of a large number of studies which in general are in considerable conflict with the interpretation just given, which is itself founded upon the comparative findings recounted in the previous chapters. The first description of OH8 was given by Day and Napier (1964) who concluded that the principal affinities of the fossil foot were with *Homo sapiens* and noted that 'this is particularly apparent in the anatomy of the metatarsal bones'. The features which particularly prompted this interpretation were the apparent absence of hallucial divergence and the robusticity formula of the metatarsal bones which, with one exception, shows a similarity to modern man rather than to the African apes. This difference they suggested might be an individual variation or an indication that the human pattern of weight distribution had not yet been fully evolved; indeed Archibald *et al.* (1972) did subsequently show that variants, including that found in the OH8 foot, occur in human populations, thus apparently endorsing this idea. Putting undue emphasis on the preponderant robusticity of metatarsals four and five as an indicator of advanced bipedal capability, however, is misplaced for the specimen of *Gorilla gorilla* illustrated in Fig. 13.25B showed a similar sturdiness of the postaxial metatarsals, with a particularly robust and buttressed fifth one. Overall, Day and Napier concluded that the foot was adapted to the upright stance and to a fully bipedal gait, and no concessions were made to the idea expressed in this chapter that it might retain conservative arobreal hallmarks.

A spate of studies involving measurements, angles, and indices (all conceived as being related in some vague way to functional morphology), often with multivariate statistical treatment, have yielded conflicting results. Day and Wood (1968) noted the ape-like talar neck angle, as had Lisowski (1967) previously and found this feature rather perplexing since in the fossil foot (of their reconstruction) the hallux was fully adducted, yet a high neck angle has been commonly considered, for some inexplicable reason, to be correlated with a divergent hallux. They sought a solution to this problem by suggesting that compensation was achieved in the fossil by an altered orientation of head upon neck of the talus. As has been noted above, there is really no conflict here, for the talar neck angle is an expression of the orientation of the subtalar axis and has no direct causal correlation with the degree of divergence of the hallux. In contrast these same authors found that the talar neck torsion angle was of similar extent to that found in man (contra Lisowski 1967). They used these and other talar parameters in a multivariate statistical study which purported to reveal that the fossil talus was intermediate in form between that of bipedal man and the African apes. Oxnard (1972) reinterpreted exactly the same data concluding that the OH8 talus was uniquely different from both the African apes on the one hand, and modern man on the other, but was closest to the *Proconsul* tali; Lisowski *et al.* (1974) claimed to show that the OH8 talus (together with the *Proconsul* tali) was closest in form to the orang-

utan. Wood (1974), in a reappraisal of the OH8 talus, cautiously inclined to the view that it should be attributed to the genus *Australopithecus*, thus further downgrading the assumption that it had specifically hominine affinities. The culmination of this reaction, which was based almost entirely on these dubious statistical studies, has been reviewed by Oxnard (1975) with the suggestion that the australopithecines (including here OH8) were at least partly arboreal Primates, which retained efficient climbing capabilities, although they had evolved some degree of a type of bipedal capability. This he maintained is reflected in some morphological resemblances to the orang-utan, which are apparently the result of functional parallels. These multivariate statistical studies challenging the initial view of the essentially human character of the OH8 foot have been greeted with increasing scepticism, largely because it is difficult to trace any causal link between the dimensions used, which are merely standard measures reflecting overall shape, and the functional morphology. In a departure from their usual studies, Oxnard and Lisowski (1980) performed a reconstruction of the OH8 foot which seemed to endorse their multivariate studies by indicating a foot that is not arched like that of man (as shown in previous reconstructions), but displays features allying it more closely with those of various apes.

Locomotion in Plio-Pleistocene hominids Despite the various inconclusive studies just cited the trend has been to view the *Homo habilis* foot (OH8) as one belonging to a polished bipedal performer differing little in its gait from modern humans (Susman 1983). For the australopithecines the current situation is rather different. Although some adhere to the earlier fashionable view that they too had a gait like that of *Homo sapiens* (Lovejoy 1978), the emerging view (Susman *et al.* 1985; McHenry 1986) accredits them with a considerable retention of traits indicative of arboreal activities, despite a newly evolved capacity for bipedality which was of a less perfected type than that found in modern humans.

Rather different data described in this chapter go some way to confirm that idea. But, unlike the currently fashionable view, it seems clear that at least quite a lot of the same ape-like morphology was retained in the OH8 foot. Grafted onto that, however, are some apomorphic characters presaging those in *Homo sapiens*. In particular the medial cuneiform is obviously remodelled with a consequent diminution in divergence of the hallux. Also a locking calcaneocuboid joint has been evolved preventing the midtarsal 'break' seen in bipedally walking chimpanzees, as the weight passes forward over the foot (Susman 1983). But the calcaneocuboid joint lacks the specialization which enables the footplate to be flattened during the stance phase, thus stretching the spring-like ligaments in the sole. Nor are the similarly functioning ligaments between the distal row of tarsal bones enlarged at the expense of the distal articulations between these bones. The heel seems to have retained much of the inflected ape character and the subtalar axis is not re-orientated as in man, where it provides for the important rotatory movements of crus (with talus) upon the footplate. Nor is the ankle joint remodelled to produce eversion in dorsiflexion, which together with the congruent movement at the subtalar joint, counteracts the tendency to tip the body to the swing side as the footplate is flattened.

In fact, the subtalar axis in OH8 is aligned essentially as in *Pan*. Important data derived by cineradiographic analysis (Jenkins 1972) is available for chimpanzee bipedalism and this may give some insight into the probable operation of the OH8 hindlimb. In the chimpanzee, weight is transferred to the support side by a considerable pelvic sway and tilt—the pelvis on the side of the swing phase is elevated, whereas in humans it sinks somewhat. It seems reasonable to assume that some similar mechanism operated in OH8, at least to some extent.

Taken overall, the OH8 foot represents a plausible transitional stage in the evolution of an ape foot towards one perfected for energy efficient bipedal locomotion of a refined type, although still perhaps retaining some arboreal capability.

Evolutionary mechanisms and the foot

The evolutionary history of the foot and of its extrinsic muscular mechanisms, which has been recounted in the preceding chapters, has been characterized by a steady accumulation of modifi-

cations. The sum total of this evolutionary tinkering with a basic plan has nevertheless culminated in a human appendage which is radically changed overall in structure and function. The misleading impression often conveyed by the literature is that various evolutionary novelties (such as a medial head to the flexor accessorius or a lateral tubercle at the heel) have somehow emerged, but this is an illusion which melts away with deeper study. The overwhelming impression is of a classically Darwinian progression and there seems to be no necessity to invoke episodes of dramatic change, as envisaged in the model of punctuated equilibria.

It seems inconceivable, however, that all these minor character changes could be independently determined genetically and individually subjected to selection. It was such complex, correlated changes in morphology which Thompson (1952) studied in his classical work on growth and form. He stated 'modifications of form will tend to manifest themselves, not so much in small and isolated phenomena . . . but rather in some slow, general, and more or less uniform graded modification, spread over a number of correlated parts'; and it was for such changes that he introduced the method of coordinates showing transformations and deformations. He noted that changes often represent 'the old figure under a more or less homogeneous strain'.

Could, in Thompson's words, 'some more or less simple and recognizable system of forces' have been in control in the evolution of the human foot? Although the picture can only be sketched in at the present state of the art, it seems that something like this has probably occurred and that the process is re-enacted during human development.

The deformations described in Chapter 13, which have apparently converted an ape foot into a human one, share a common denominator of change. In the left feet shown in Fig. 13.25 appropriate growth changes could presumably set up a clockwise stress in the tissues which might bend the heel outwards, compact and modify the trochlear process, internally rotate the talar trochlea, and bend the distal tarsals and metatarsals medially (Fig. 13.25A). A cascade of correlated changes might inevitably follow: opening up of the tarsal canal sets the stage for 'descent' of part of flexor fibularis to the foot; the changed direction of pull of the originally oblique crural muscles leads to major changes in origin; the same factors modify the insertions of these same muscles in the foot. Many, at least, of these changes could result from the inbuilt responses of the various tissues which have been described in Chapter 3. It is perhaps then not so surprising that the amount of genetic difference between man and *Pan* should be so apparently trivial.

List of references

Adams, M. B. (1980). Severtsov and Schmalhausen: Russian morphology and the evolutionary synthesis. In *The evolutionary synthesis* (ed. E. Mayr and W. B. Provine), pp. 193–225. Harvard University Press, Cambridge.

Aiello, L. C. and Day, M. H. (1982). The evolution of locomotion in the early Hominidae. In *Progress in anatomy*, Vol. 2 (ed. R. J. Harrison and V. Navaratnam), pp. 81–97. Cambridge University Press.

Albinus, B. S. (1734). *Historia musculorum hominis.* Haak and Mulhovium, Leyden.

Albrecht, P. (1884). Sur les homodynamies qui existent entre la main et le pied des mammiferes. *Presse medicale belge,* **36,** 329–31.

Alexander, R. McN. and Dimery, N. J. (1985). Elastic properties of the forefoot of the donkey, *Equus asinus. Journal of Zoology,* **205,** 511–24.

Andrews, P. (1970). Two new fossil primates from the lower Miocene of Kenya. *Nature,* **228,** 537–40.

Andrews, P. (1974). New species of *Dryopithecus* from Kenya. *Nature,* **249,** 188–90.

Andrews, P. (1977). The origin of the hominids. *Primate Eye,* **8,** 5–6.

Andrews, P. (1981*a*). A short history of Miocene field palaeontology in Western Kenya. *Journal of Human Evolution,* **10,** 3–9.

Andrews, P. (1981*b*). Species diversity and diet in monkeys and apes during the Miocene. In *Aspects of human evolution* (ed. C. B. Stringer). Taylor and Francis, London.

Andrews, P. (1982). Hominoid evolution. *Nature,* **295,** 185–6.

Andrews, P. (1985). Family group systematics and evolution among catarrhine primates. In *Ancestors: the hard evidence* (ed. E. Delson), pp. 14–22. Alan R. Liss, New York.

Andrews, P. (1986). Molecular evidence for catarrhine evolution. In *Major topics in primate and human evolution* (ed. B. Wood, L. Martin, and P. Andrews), pp. 107–29. Cambridge University Press.

Andrews, P. and Simons, E. (1977). A new African gibbon-like genus, *Dendropithecus* (Hominoidea, Pri-mates) with distinctive postcranial adaptations: its significance to origin of Hylobatidae. *Folia Primatologica,* **28,** 161–9.

Andrews, P. and Tobien, H. (1977). New Miocene locality in Turkey with evidence of the origin of *Ramapithecus* and *Sivapithecus. Nature,* **268,** 699–701.

Andrews, P. and Van Couvering, J. A. H. (1975). Palaeoenvironments in the East African Miocene. In *Approaches to primate paleobiology* (ed. F. S. Szalay). *Contributions to primatology,* Vol. 5, pp. 62–103. Karger, Basel.

Angelides, A. C. (1982). Ganglions of the hand and wrist. In *Operative Hand Surgery* (ed. D. P. Green) Vol. 2, pp. 1635–52. Churchill Livingstone, London.

Anson, B. J., Wright, R. R., Ashley, F. L., and Dykes, J. (1945). The fascia of the dorsum of the hand. *Surgery, Gynecology and Obstetrics,* **81,** 327–31.

Archer, M., Flannery, T. F., Ritchie, A., and Molnar, R. E. (1985). First Mesozoic mammal from Australia—an early Cretaceous monotreme. *Nature,* **318,** 363–6.

Archibald, J. D., Lovejoy, C. O., and Heiple, K. G. (1972). Implications of relative robusticity in the Olduvai metatarsus. *American Journal of Physical Anthropology,* **37,** 93–6.

Asiz, M. A. (1981). Muscular anomalies caused by delayed development in human aneuploidy. *Clinical Genetics,* **19,** 111–16.

Avis, V. (1962). Brachiation: the crucial issue for man's ancestry. *Southwestern Journal of Anthropology,* **18,** 119–48.

Barash, B. A., Freedman, L., and Opitz, J. M. (1970). Anatomic studies in the 18-trisomy syndrome. *Birth Defects: Original Article Series,* **4,** 3–15.

Barbour, R. A. (1963). The musculature and limb plexuses of *Trichosurus vulpecula. Australian Journal of Zoology,* **11,** 488–610.

Bardeen, C. R. (1905). Studies of the development of the human skeleton. *American Journal of Anatomy,* **4,** 265–302.

Bardeen, C. R. (1906). Development and variation of the nerves and the musculature of the inferior extremity and of the neighbouring regions of the trunk in man. *American Journal of Anatomy*, 6, 259–390.

Bardeleben, K. (1883*a*). Das Intermedium tarsi beim Menschen. *Sitzungsberichte der jena Gesellschaft für Medizin und Naturwissenschaften*, 17, 37–9.

Bardeleben, K. (1883*b*). Das Os intermedium tarsi der Säugethiere. *Sitzungsberichte der jena Gesellschaft für Medizin und Naturwissenschaften*, 17, 75–7.

Bardeleben, K. (1883*c*). Das Os intermedium tarsi der Säugethiere. *Zoologischer Anzieger*, 6, 278–80.

Bardeleben, K. (1885*a*). Zur Entwickelung der Fusswurzel. *Sitzungsberichte der jena Gesellschaft für Medizin und Naturwissenschaften*, 19, 27–32.

Bardeleben, K. (1885*b*). Zur Morphologie des Hand-und Fusskelets. *Sitzungsberichte der jena Gesellschaft für Medizin und Naturwissenschaften*, 19, 84–8.

Bardeleben, K. (1885*c*). Ueber neue Bestandteile der Hand-und Fusswurzel der Säugethiere, sowie die normale Anlage von Rudimenten 'Überzähliger' Finger und Zehen beim Menschen. *Sitzungsberichte der jena Gesellschaft für Medizin und Naturwissenschaften*, 19, 149–64.

Bardeleben, K. (1894). On the bones and muscles of the mammalian hand and foot. *Proceedings of the Zoological Society of London*, 1894, 354–76.

Barlow, T. E. (1949). An unusual anomaly of M. flexor digitorum longus. *Journal of Anatomy*, 83, 224–6.

Barlow, T. E. (1953). The deep flexors of the foot. *Journal of Anatomy*, 87, 308–10.

Barnett, C. H. (1954). The structure and function of fibrocartilages within vertebrate joints. *Journal of Anatomy*, 88, 363–8.

Barnett, C. H. (1954). Squatting facets on the European talus. *Journal of Anatomy*, 88, 509–13.

Barnett, C. H. (1955). Some factors influencing angulation of the neck of the mammalian talus. *Journal of Anatomy*, 89, 225–30.

Barnett, C. H. (1956). The phases of human gait. *Lancet*, 2, 617–21.

Barnett, C. H. (1966). The structure and function of joints. In *Clinical surgery* (ed. C. Robb and R. Smith), pp. 328–44. Butterworths, London.

Barnett, C. H. (1970). Talocalcaneal movements in mammals. *Journal of Zoology, London*, 160, 1–7.

Barnett, C. H. and Lewis, O. J. (1958). The evolution of some traction epiphyses in birds and mammals. *Journal of Anatomy*, 92, 593–601.

Barnett, C. H. and Napier, J. R. (1952). The axis of rotation at the ankle joint in man. Its influence upon the form of the talus and mobility of the fibula. *Journal of Anatomy*, 86, 1–9.

Barnett, Ç. H. and Napier, J. R. (1953). The form and mobility of the fibula in metatherian mammals. *Journal of Anatomy*, 87, 207–13.

Baur, G. (1884). Zur Morphologie des Tarsus der Säugethiere. *Morphologisches Jahrbuch*, 10, 458–61.

Baur, G. (1885*a*). On the morphology of the tarsus in mammals. *American Naturalist*, 19, 86–8.

Baur, G. (1885*b*). On the morphology of the carpus and tarsus of vertebrates. *American Naturalist*, 19, 718–20.

Baur, G. (1885*c*). Zur Morphologie des Carpus und Tarsus der Wirbelthiere. *Zoologischer Anzieger*, 8, 326–29.

Baur, G. (1885*d*). Zur Morphologie des Carpus und Tarsus der Reptilien. *Zoologischer Anzieger*, 8, 631–8.

Baur, G. (1886). Bemerkungen über den 'Astragalus' und das 'Intermedium tarsi' der Säugethiere. *Morphologisches Jahrbuch*, 11, 468–83.

Beard, K. C. and Godinot, M. (1988). Carpal anatomy of *Smilodectes gracilis* (Adapiformes, Notharctinae) and its significance for lemuriform phylogeny. *Journal of Human Evolution*, 17, 71–92.

Beard, K. C., Teaford, M. F., and Walker, A. (1986). New wrist bones of *Proconsul africanus* and *P. nyanzae* from Rusinga Island, Kenya. *Folia Primatologia*, 47, 97–118.

Bensley, B. A. (1903). On the evolution of the Australian Marsupialia; with remarks on the relationship of the marsupials in general. *Transactions of the Linnean Society of London*, 9, 83–217.

Benton, M. J. (1985). First marsupial fossil from Asia. *Nature*, 318, 313.

Biegert, J. (1971). Dermatoglyphics in the chimpanzee. In *The chimpanzee*, Vol. 4 (ed. G. H. Bourne), pp. 273–324. Basel: Karger.

Bingold, A. C. (1964). An extensor indicis brevis. *British Journal of Surgery*, 51, 236–237.

Bishop, A. (1964). Use of the hand in lower Primates. In *Evolutionary and genetic biology of primates*, Vol. 2 (ed. J. Buettner-Janusch), pp. 132–225. Academic Press, New York.

Bishop, W. W. (1978). Geochronological framework for African Plio-Pleistocene Hominidae: as Cerberus sees it. In *Early hominids of Africa* (ed. C. J. Jolly), pp. 255–65. Duckworth, London.

Bock, W. J. and von Wahlert, G. (1965). Adaptation and the form-function complex. *Evolution*, 19, 269–99.

Bojsen-Møller, F. (1976). Osteoligamentous guidance of the movements of the human thumb. *American Journal of Anatomy*, 147, 71–80.

Bojsen-Møller, F. (1978). Extensor carpi radialis longus muscle and the evolution of the first intermetacarpal ligament. *American Journal of Physical Anthropology*, 48, 177–84.

Bojsen-Møller, F. (1979). Calcaneocuboid joint and stability of the longitudinal arch of the foot at high and low gear push off. *Journal of Anatomy*, 129, 165–76.

Bowden, R. E. M. (1967). The functional anatomy of the foot. *Physiotherapy*, 53, 120–6.

Bown, T. M. and Gingerich, P. D. (1973). The Paleocene Primate *Plesiolestes* and the origin of the Microsyopidae. *Folia primatologica*, **19**, 1–8.

Bown, T. M. and Kraus, M. J. (1979). Origins of the tribosphenic molar and metatherian and eutherian dental formulae. In *Mesozoic mammals. The first two-thirds of mammalian history* (ed. J. A. Lillegraven, Z. Kielan-Jaworowska and W. A. Clemens), pp. 172–81. University of California Press, Berkeley.

Brace, C. L., Nelson, H., and Korn, N. (1971). *Atlas of fossil man*. Holt, Rinehart and Winston, New York.

Brailsford, J. F. (1953). *The radiology of bones and joints*. J. and A. Churchill, London.

Braithwaite, F., Channell, S. D., Moore, F. T., and Whillis, J. (1948). The applied anatomy of the lumbrical and interosseous muscles of the hand. *Guy's Hospital Reports*, **97**, 185–95.

Brand, P. W. (1985). *Clinical mechanics of the hand*. C. V. Mosby, St Louis.

Brandell, B. R. (1965). Innervation of the hand muscles of *Didelphis marsupialis virginiana*. *Journal of Morphology*, **116**, 133–48.

Brindley, H. H. (1969*a*). Homology of the extremities. Part I. *American Journal of Orthopaedic Surgery*, **1969**, 40–3.

Brindley, H. H. (1969*b*). Homology of the extremities. Part 2. *American Journal of Orthopaedic Surgery*, **1969**, 70–8.

Brizon, J. and Castaing, J. (1953). *Les feuillets d'anatomie*, Fascicule III. Librairie Maloine, Paris.

Brock, A. and Isaac, G. L. (1974). Paleomagnetic stratigraphy and chronology of hominid-bearing sediments east of Lake Rudolf, Kenya. *Nature*, **247**, 344–8.

Brooks, H. St. John. (1886). On the morphology of the intrinsic muscles of the little finger, with some observations on the ulnar head of the short flexor of the thumb. *Journal of Anatomy*, **20**, 645–61.

Brooks, H. St. John. (1887). On the short muscles of the pollex and hallux of the anthropoid apes, with special reference to the opponens hallucis. *Journal of Anatomy*, **22**, 78–95.

Brooks, H. St. John. (1889). On the morphology of the muscles of the extensor aspect of the middle and distal segments of the limbs: with an account of the various paths which are adopted by the nerve-trunks in these segments. *Studies from the Museum of Zoology in University College, Dundee*, **1**, 1–17.

Broom, R. (1901). On structure and affinities of *Udenodon*. *Proceedings of the Zoological Society of London*, **2**, 162–90.

Broom, R. (1904). The origin of the mammalian carpus and tarsus. *Transactions of the South African Philosophical Society*, **15**, 89–96.

Broom, R. (1921). On the structure of the reptilian tarsus. *Proceedings of the Zoological Society of London*, **1**, 143–55.

Broom, R. (1930). *The Origin of the human skeleton*. Witherby, London.

Broom, R. and Robinson, J. T. (1949). Thumb of the Swartkrans ape-man. *Nature*, **164**, 841–2.

Broom, R. and Schepers, G. W. H. (1946). *The South African fossil ape-men: the Australopithecinae. Transvaal Museum No. 2*. Transvaal Museum, Pretoria.

Bryce, M. B. (1897). On certain points in the anatomy and mechanisms of the wrist joint reviewed in the light of a series of Rontgen ray photographs of the living hand. *Journal of Anatomy*, **31**, 59–79.

Bunnell, S. (1942). Surgery of the intrinsic muscles of the hand other than those producing opposition of the thumb. *Journal of Bone and Joint Surgery*, **24**, 1–31.

Bunning, P. S. C. and Barnett, C. H. (1965). A comparison of adult and foetal talocalcaneal articulations. *Journal of Anatomy*, **99**, 71–6.

Bush, M. E., Lovejoy, C. O., Johanson, D. C., and Coppens, Y. (1982). Hominid carpal, metacarpal and phalangeal bones recovered from the Hadar Formation: 1974–1977 collections. *American Journal of Physical Anthropology*, **57**, 651–77.

Cahill, D. R. (1965). The anatomy and functions of the contents of the human tarsal sinus and canal. *Anatomical Record*, **153**, 1–18.

Campbell, B. (1939). The comparative anatomy of the dorsal interosseous muscles. *Anatomical Record*, **73**, 115–25.

Campbell, B. G. and Bernor, R. L. (1976). The origin of the Hominidae: Africa or Asia. *Journal of Human Evolution*, **5**, 441–54.

Cannon, H. G. (1958). *The evolution of living things*. Manchester University Press.

Carleton, A. (1941). A comparative study of the inferior tibiofibular joint. *Journal of Anatomy*, **76**, 45–55.

Carlsöö, S. (1972). *How man moves. Kinesiological studies and methods*. Heinemann, London.

Cartmill, M. (1974). Pads and claws in arboreal locomotion. In *Primate locomotion* (ed. F. A. Jenkins), pp. 45–83. Academic Press, New York.

Cartmill, M. (1980). Morphology, function and evolution of the anthropoid postorbital septum. In *Evolutionary biology of the new world monkeys and continental drift* (ed. R. L. Ciochon and A. B. Chiarelli), pp. 243–74. Plenum Press, New York.

Cartmill, M., MacPhee, R. D. E., and Simons, E. L. (1981). Anatomy of the temporal bone in early anthropoids, with remarks on the problem of anthropoid origins. *American Journal of Physical Anthropology*, **56**, 3–21.

Cartmill, M. and Milton, K. (1974). The lorisine wrist joint. *American Journal of Physical Anthropology*, **41**, 471.

Cartmill, M. and Milton, K. (1977). The lorisine wrist joint and the evolution of brachiating adaptations in the Hominoidea. *American Journal of Physical Anthropology*, **47**, 249–72.

Cassiliano, M. L. and Clemens, W. A. (1979). Symmetrodonta. In *Mesozoic mammals. The first two-thirds of mammalian history* (ed. J. A. Lillegraven, Z. Kielan-Jaworowska and W. A. Clemens), pp. 150–61. University of California Press, Berkeley.

Cave, A. J. E. (1968). Mammalian olecranon epiphyses. *Journal of Zoology*, **156**, 333–50.

Charles-Dominique, P. and Martin, R. D. (1970). Evolution of lorises and lemurs. *Nature*, **227**, 257–60.

Charlesworth, B. and Laude, R. (1982). Morphological stasis and developmental constraint: no problem for neo-Darwinism. *Nature*, **296**, 610.

Chauveau, A. (1891). *The comparative anatomy of the domesticated animals.* Churchill, London.

Cihak, R. (1972). Ontogenesis of the skeleton and intrinsic muscles of the human hand and foot. *Ergebnisse der Anatomie und Entwicklungsgeschichte*, **46**, 1–194.

Cihak, R. (1977). Differentiation and rejoining of muscular layers in the embryonic human hand. In *Morphogenesis and malformations of the limb* (ed. D. Bergsma and W. Lenz), pp. 97–108. Alan R. Liss, New York.

Ciochon, R. L. and Chiarelli, A. B. (1980). Paleobiogeographic perspectives on the origin of the Platyrrhini. In *Evolutionary biology of the new world monkeys and continental drift* (ed. R. L. Ciochon and A. B. Chiarelli), pp. 459–94. Plenum Press, New York.

Clark, W. E. Le Gros (1924). The myology of the treeshrew. *Proceedings of the Zoological Society of London*, **1924**, 461–97.

Clark, W. E. Le Gros (1926). On the anatomy of the pen-tailed tree-shrew (*Ptilocercus lowii*). *Proceedings of the Zoological Society of London*, **85**, 1179–1309.

Clark, W. E. Le Gros (1947). Observations on the anatomy of the fossil Australopithecinae. *Journal of Anatomy*, **81**, 300–33.

Clark, W. E. Le Gros (1952). Report on fossil hominoid material collected by the British–Kenya Miocene expedition, 1949–1951. *Proceedings of the Zoological Society*, **122**, 273–86.

Clark, W. E. Le Gros (1959). *The antecedents of man.* Edinburgh University Press.

Clark, Sir W. E. Le Gros (1967). *Man-apes or ape-men?* Holt, Rinehart and Winston, New York.

Clark, W. E. Le Gros and Leakey, L. S. B. (1951). The Miocene Hominoidea of Africa. *Fossil Mammals of Africa*, no. 1 pp. 1–117, British Museum (Natural History), London.

Clemens, W. A. (1971). Mammalian evolution in the Cretaceous. In *Early mammals* (ed. D. M. Kermack and

K. A. Kermack). *Supplement to the Zoological Journal of the Linnean Society*, **50**, pp. 165–80.

Clemens, W. A. (1977). Phylogeny of the marsupials. In *The biology of marsupials* (ed. B. Stonehouse and D. Gilmore), pp. 51–68. The Macmillan Press, London.

Clemens, W. A. (1979a). Notes on the Monotremata. In *Mesozoic mammals. The first two-thirds of mammalian history* (ed. J. A. Lillegraven, Z. Kielan-Jaworowska and W. A. Clemens), pp. 309–11. University of California Press, Berkeley.

Clemens, W. A. (1979b). Marsupialia. In *Mesozoic mammals. The first two-thirds of mammalian history* (ed. J. A. Lillegraven, Z. Kielan-Jaworowska and W. A. Clemens), pp. 192–220. University of California Press, Berkeley.

Clemens, W. A. and Kielan-Jaworowska, Z. (1979). Multituberculata. In *Mesozoic mammals. The first two-thirds of mammalian history* (ed. J. A. Lillegraven, Z. Kielan-Jaworowska and W. A. Clemens), pp. 99–149. University of California Press, Berkeley.

Clemens, W. A., Lillegraven, J. A., Lindsay, E. H., and Simpson, G. G. (1979). Where, when, and what—a survey of known Mesozoic mammal distribution. In *Mesozoic mammals. The first two-thirds of mammalian history* (ed. J. A. Lillegraven, Z. Kielan-Jaworowska, and W. A. Clemens), pp. 7–58. University of California Press, Berkeley.

Conroy, G. C. (1976a). Primate postcranial remains from the Oligocene of Egypt. *Contributions to Primatology* (ed. F. S. Szalay), vol. 8, pp. 1–134. Karger, Basel.

Conroy, G. C. (1976b). Hallucial tarsometatarsal joint in an Oligocene anthropoid *Aegyptopithecus zeuxis*. *Nature*, **262**, 584–688.

Conroy, G. C. and Fleagle, J. G. (1972). Locomotor behaviour in living and fossil pongids. *Nature*, **237**, 103–4.

Conroy, G. C. and Rose, M. D. (1983). The evolution of the primate foot from the earliest primates to the Miocene hominoids. *Foot and Ankle*, **3**, 342–64.

Cope, E. D. (1884). Fifth contribution to the knowledge of the fauna of the Permian formation of Texas and the Indian Territory. *Palaeontological Bulletin*, **39**, 28–47.

Cope, E. D. (1885). The relations between the theromorphous reptiles and the monotreme mammalia. *Proceedings of the American Association for the Advancement of Science*, **33**, 471–82.

Corner, E. M. (1898). The morphology of the triangular cartilage of the wrist. *Journal of Anatomy*, **32**, 272–7.

Corruccini, R. S. (1978). Comparative osteometrics of the hominoid wrist joint with special reference to knuckle-walking. *Journal of Human Evolution*, **7**, 307–21.

Corruccini, R. S., Ciochon, R. L., and McHenry, H. M. (1975). Osteometric shape relationships in the wrist joint of some anthropoids. *Folia primatologica*, **24**, 250–74.

Corruccini, R. S., Ciochon, R. L., and McHenry, H. M. (1976). The postcranium of Miocene hominoids: were Dryopithecines merely 'dental apes'? *Primates*, **17**, 205–23.

Coues, E. (1871). On the myology of Ornithorhynchus. *Proceedings of the Essex Institute*, **6**, 127–73.

Coues, E. (1872). On the osteology and myology of *Didelphis virginiana*. *Memoirs of the Boston Society of Natural History*, **2**, 11–154.

Cox, B. (1982). New branches for old roots. *Nature*, **298**, 321.

Crompton, A. W. and Jenkins, F. A. (1979). Origin of mammals. In *Mesozoic mammals. The first two-thirds of mammalian history* (ed. J. A. Lillegraven, Z. Kielan-Jaworowska and W. A. Clemens), pp. 59–73. University of California Press, Berkeley.

Cronin, J. E., Boaz, N. T., Stringer, C. B., and Rak, Y. (1981). Tempo and mode in hominid evolution. *Nature*, **292**, 113–22.

Cunningham, D. J. (1878). The intrinsic muscles of the hand of the thylacine (*Thylacinus cynocephalus*), cuscus (*Phalangista maculata*) and phascogale (*Phascogale calura*). *Journal of Anatomy*, **12**, 434–44.

Cunningham, D. J. (1882). Report on some points in the anatomy of the thylacine (*Thylacinus cynocephalus*), cuscus (*Phalangista maculata*) and phascogale (*Phascogale calura*) collected during the voyage of H.M.S. Challenger in the years 1873–1876; with an account of the comparative anatomy of the intrinsic muscles and nerves of the mammalian pes. *Challenger Reports*, **5**, 1–192.

Darlington, C. D. (1974). The origin of Darwinism. In *Biological anthropology* (ed. S. H. Katz) pp. 138–44. W. H. Freeman, San Francisco.

Daubenton, L. (1766). *Histoire naturelle générale et particulière avec la description du cabinet du roi*. Vol. 14. Impremierie Royale, Paris.

Davis, D. Dwight (1954). Primate evolution from the viewpoint of comparative anatomy. *Human Biology*, **26**, 211–19.

Davis, P. R. (1964). Hominid fossils from Bed I, Olduvai Gorge, Tanganyika. *Nature*, **210**, 967–8.

Day, M. H. (1986). *Guide to fossil man*. Cassell, London.

Day, M. H. and Napier, J. R. (1961). The two heads of flexor pollicis brevis. *Journal of Anatomy*, **95**, 123–30.

Day, M. H. and Napier, J. R. (1963). The functional significance of the deep head of flexor pollicis brevis in primates. *Folia primatologia*, **1**, 122–34.

Day, M. H. and Napier, J. R. (1964). Fossil foot bones. *Nature*, **201**, 969–70.

Day, M. H. and Scheuer, J. L. (1973). SKW14147: a new hominid metacarpal from Swartkrans. *Journal of Human Evolution*, **2**, 429–38.

Day, M. H. and Wood, B. A. (1968). Functional affinities of the Olduvai Hominid 8 talus. *Man*, **3**, 440–45.

Day, M. H. and Wood, B. A. (1969). Hominoid tali from East Africa. *Nature*, **222**, 591–2.

Dean, M. C. and Wood, B. A. (1982). Basicranial anatomy of Plio-Pleistocene Hominids from East and South Africa. *American Journal of Physical Anthropology*, **59**, 157–74.

De Beer, G. R. [1951]. *Embryos and ancestors*. Clarendon Press, Oxford.

Decker, R. L. and Szalay, F. S. (1974). Origin and function of the pes in the Eocene Adapidae (Lemuriformes, Primates). In *Primate locomotion* (ed. F. A. Jenkins, Jr.), pp. 261–91. Academic Press, New York.

Deloison, Y. (1985). Comparative study of calcanei of Primates on *Pan-Australopithecus-Homo* relationship. In *Hominoid evolution: past, present and future* (ed. P. V. Tobias). pp. 143–7. Alan R. Liss, New York.

Delson, E. (1975a). Evolutionary history of the Cercopithecidae. In *Approaches to primate palaeobiology* (ed. F. S. Szalay). *Contributions to Primatology*, Vol. 5, pp. 167–217. Karger, Basel.

Delson, E. (1975b). Paleoecology and Zoogeography of the Old World monkeys. In *Primate functional morphology and evolution* (ed. R. H. Tuttle), pp. 37–64. Mouton, The Hague.

Delson, E. (1977). Catarrhine phylogeny and classification: principles, methods and comments. *Journal of Human Evolution*, **6**, 433–59.

Delson, E. (1978). Models of early hominid phylogeny. In *Early hominids of Africa* (ed. C. J. Jolly), pp. 518–41. Duckworth, London.

Delson, E. (1979). *Prohylobates* (Primates) from the early Miocene of Libya: a new species and its implication for cercopithecid origins. *Geobios*, **12**, 725–33.

Delson, E. (1980). Fossil macaques, phyletic relationships and a scenario of deployment. In *The macaques: studies in ecology, behaviour and evolution* (ed. D. G. Lindburg), pp. 11–30. Van Nostrand, New York.

Delson, E. and Andrews, P. (1975). Evolution and interrelationships of the catarrhine Primates. In *Phylogeny of the primates* (ed. W. P. Luckett and F. S. Szalay), pp. 405–66. Plenum Press, New York.

Delson, E., Eldredge, N., and Tattersall, I. (1977). Reconstruction of hominid phylogeny: a testable framework based on cladistic analysis. *Journal of Human Evolution*, **6**, 263–78.

Delson, E. and Rosenberger, A. L. (1980). Phyletic perspectives on platyrrhine origins and anthropoid relationships. In *Evolutionary biology of the new world monkeys and continental drift* (ed. R. L. Ciochon and A. B. Chiarelli), pp. 445–58. Plenum Press, New York.

Deniker, J. (1886). *Recherches anatomiques et embryologiques sur les singes anthropoides*. Oudin, Poitiers.

Dobson, G. E. (1883a). On the homologies of the long

flexor muscles of the feet of mammalia, with remarks on the value of their leading modifications in classification. *Journal of Anatomy*, 17, 142–79.

Dobson, G. E. (1883b). *A monograph of the insectivora* Part II. John Van Voorst, London.

Douglas, J. (1750). *Myographiae comparatae*. Kincaid and Crawfurd, Edinburgh.

Dwight, T. (1910). A criticism of Pfitzner's theory of the carpus and tarsus. *Anatomischer Anzeiger*, 35, 366–70.

Dykes, J. and Anson, B. J. (1944). The accessory tendon of the flexor pollicis longus muscle. *Anatomical Record*, 90, 83–7.

Eckhardt, R. B. (1975). *Gigantopithecus* as a hominid. In *Paleoanthropology, morphology and paleoecology* (ed. R. H. Tuttle), pp. 105–29. Mouton, The Hague.

Effendy, W., Graf, J., and Khaledpour, C. (1985). Der seltene M. flexor carpi radialis brevis mit besonderem Verlauf. *Handchirurgie*, 17, 111–12.

Eldredge, N. and Gould, S. J. (1972). Punctuated equilibria: an alternative to phyletic gradualism. In *Models in paleobiology* (ed. T. J. M. Schopf) pp. 82–115. Freeman, Cooper and Co., San Francisco.

Eldredge, N. and Tattersall, I. (1975). Evolutionary models, phylogenetic reconstruction and another look at hominid phylogeny. In *Approaches to primate paleobiology* (ed. F. S. Szalay) pp. 218–42. Karger, Basel.

Elftman, H. (1960). The transverse tarsal joint and its control. *Clinical Orthopaedics*, 16, 41–6.

Elftman, H. and Manter, J. (1935). The evolution of the human foot, with special reference to the joints. *Journal of Anatomy*, 70, 56–67.

Elliott, D. H. (1965). Structure and function of mammalian tendon. *Biological Reviews*, 40, 392–421.

Emery, C. (1897). Beiträge zur Entwicklungsgeschichte und Morphologie des Hand-Fussketels der Marsupialier. *Semons Zoologische Forschungsreisen*, 2, 371–400.

Emery, C. (1901). Hand-und Fussskelet von *Echidna hystrix*. *Semon's Zoologische Forschungsreisen in Australien und dem Malayischen Archipel*, 3, 663–76.

Erikson, G. E. (1963). Brachiation in new world monkeys and in anthropoid apes. *Symposia of the Zoological Society of London No. 10* (ed. J. Napier and N. A. Barnicot), pp. 135–64.

Etter, H. F. (1973). Terrestrial adaptations in the hands of the Cercopithecinae. *Folia primatoligica*, 20, 331–50.

Etter, H. F. (1974). Morphologisch-und metrisch-Vergleichende Untersuchung am Handskelet rezeuter Primaten. Teil I, II, III, IV. *Morphologisches Jahrbuch*, 120, 1–21, 153–71, 299–322, 457–84.

Evans, E. M. N., Vancouvering, J. A. H., and Andrews, P. (1981). Palaeoecology of Miocene sites in Western Kenya. *Journal of Human Evolution*, 10, 99–116.

Eyler, D. L. and Markee, J. E. (1954). The anatomy and function of the intrinsic musculature of the fingers. *Journal of Bone and Joint Surgery*, 36A, 1–9.

Fick, L. (1857). Hand und Fuss. *Archiv für Anatomie und Physiologie*, 1857, 435–58.

Fick, R. (1904). *Handbuch der Anatomie und Mechanik der Gelenke*. Erster Teil: *Anatomie der Gelenke*. Gustav Fischer, Jena.

Fick, R. (1911). *Handbuch der Anatomie und Mechanik der Gelenke*, Dritter Teil: *Spezielle Gelenk-und Muskelmechanik*. Gustav Fischer, Jena.

Fleagle, J. G. (1983). Locomotor adaptations of Oligocene and Miocene hominoids and their phyletic implications. In *New interpretations of ape and human ancestry* (ed. R. L. Ciochon and R. S. Corruccini), pp. 301–24. Plenum Press, New York.

Fleagle, J. G. (1984). Are there any fossil gibbons? In *The lesser apes* (ed. M. Preuschoft, D. J. Chivers, W. Y. Brockelman and N. Creel), pp. 431–47. University Press, Edinburgh.

Fleagle, J. G. and Kay, R. F. (1983). New interpretations of the phyletic position of Oligocene hominoids. In *New interpretations of ape and human ancestry* (ed. R. L. Ciochon and R. S. Corruccini), pp. 181–210. Plenum Press, New York.

Fleagle, J. G. and Kay, R. F. (1985). The paleobiology of catarrhines. In *Ancestors: the hard evidence* (ed. E. Delson), pp. 23–36. Alan R. Liss, New York.

Fleagle, J. G. and Simons, F. L. (1978). *Micropithecus clarki*, a small ape from the Miocene of Uganda. *American Journal of Physical Anthropology*, 49, 427–40.

Fleagle, J. G. and Simons, F. L. (1982). The humerus of *Aegyptopithecus zeuxis*: a primitive anthropoid. *American Journal of Physical Anthropology*, 59, 175–93.

Fleagle, J. G. and Simons, E. L. (1983). The tibio-fibular articulation in *Apidium phiomense*, an Oligocene anthropoid. *Nature*, 301, 238–9.

Fleagle, J. G., Stern, J. T., Jungers, W. L., Susman, R. L., Vangor, A. K., and Wells, J. P. (1981). Climbing: a biomechanical link with brachiation and bipedalism. *Symposia of the Zoological Society of London*, 48, 359–75.

Flecker, H. (1933). Roentgenographic observations of the times of appearance of epiphyses and their fusion with the diaphyses. *Journal of Anatomy*, 67, 118–64.

Flower, W. H. (1876). *An introduction to the osteology of the mammalia*. Macmillan, London.

Forster, A. (1916). Die Mm. contrahentes and interossei manus in der Saugetierreihe und beim Menschen. *Archiv fur Anatomie und Physiologie*, 1916, 101–378.

Forster, A. (1933). Considerations sur l'os central du carpe dans l'éspèce humaine. *Archives d'Anatomie, d'Histologie et d'Embryologie*, 17, 85–98.

Forster, A. (1934). Mu cas de 'vrai' os centrale chez l'homme. *Archives d'Anatomie, d'Histologie et d'Embryologie*, 17, 385–95.

Frayer, D. W. (1973). *Gigantopithecus* and its relationship to *Australopithecus*. *American Journal of Physical Anthropology*, **39**, 413–26.

Frazer, J. E. (1908). The derivation of the human hypothenar muscles. *Journal of Anatomy*, **42**, 326–34.

Frazer, J. E. (1946). *The anatomy of the human skeleton*. Churchill, London.

Frets, G. P. (1908). Die Varietäten der Musculi peronaei beim Menschen und die Mm. peronaei bei den Säugetieren. *Morphologisches Jahrbuch*, **38**, 135–93.

Froimson, A. I. (1982). Tenosynovitis and tennis elbow. In *Operative hand surgery* (ed. D. P. Green) Vol. 2, pp. 1507–21. Churchill, London.

Furst, C. M. (1903). Der Musculus Popliteus und seine Sehne. *Lunds Universitets Arsskrift*, **39**, 1–134.

Gad, P. (1967). The anatomy of the volar part of the capsules of the finger joints. *Journal of Bone and Joint Surgery*, **49B**, 362–67.

Gardiner, B. G. (1982). Tetrapod classification. *Zoological Journal of the Linnean Society*, **74**, 207–32.

Gebo, D. L. (1985). The nature of the primate grasping foot. *American Journal of Physical Anthropology*, **67**, 269–77.

Gebo, D. L. (1986). Anthropoid origins—the foot evidence. *Journal of Human Evolution*, **15**, 421–30.

Gebo, D. L. and Simons, E. L. (1987). Morphology and locomotor adaptations of the foot in early Oligocene anthropoids. *American Journal of Physical Anthropology*, **74**, 83–101.

Geddes, A. C. (1912). The origin of the vertebrate limb. *Journal of Anatomy*, **46**, 350–83.

Gegenbaur, C. (1864). *Untersuchung zur vergleichenden Anatomie der Wirbeltiere. I. Carpus und Tarsus*. Englemann, Leipzig.

Gegenbaur, C. (1870). *Grundzuge der Vergleichenden Anatomie*. Engelman, Leipzig.

Gidley, J. W. (1919). Significance of divergence of the first digit in the primitive mammalian foot. *Journal of the Washington Academy of Sciences*, **9**, 273–80.

Giles, K. W. (1960). Anatomical variations affecting the surgery of de Quervain's disease. *Journal of Bone and Joint Surgery*, **42B**, 352–5.

Gingerich, P. D. (1975). Systematic position of *Plesiadapis*. *Nature*, **253**, 111–13.

Gingerich, P. D. (1980). Eocene Adapidae, Paleobiogeography, and the origin of South American Platyrrhini. In *Evolutionary biology of the new world monkeys and continental drift* (ed. R. L. Ciochon and A. B. Chiarelli), pp. 123–38. Plenum Press, New York.

Gingerich, P. D. (1981). Early Cenozoic Omomyidae and the evolutionary history of tarsiform primates. *Journal of Human Evolution*, **10**, 345–74.

Gingerich, P. D. (1984). Primate evolution: evidence from the fossil record, comparative morphology and molecular biology. *Yearbook of Physical Anthropology*, **27**, 57–72.

Gingerich, P. D. and Schoeninger, M. (1977). The fossil record and primate phylogeny. *Journal of Human Evolution*, **6**, 483–505.

Glaesmer, E. (1908). Untersuchung über die Flexorengruppe am Unterschenkel und Fuss der Säugetiere. *Morphologisches Jahrbuch*, **38**, 36–90.

Glaesmer, E. (1910). Die Beugemuskeln am Unterschenkel und Fuss bei den Marsupialia, Insectivora, Edentalta, Prosimiae und Simiae. *Morphologisches Jahrbuch*, **41**, 149–336.

Godinot, M. and Jouffroy, F. K. (1984). La main d'*Adapis* (Primate, Adapidae). In *Actes du symposium paléontologique G. Cuvier* (ed. E. Buffetant, J. M. Magin and E. Salmon), pp. 221–42. Montbeliard, France.

Gould, S. J. (1977). *Ontogeny and phylogeny*. Harvard University Press, Cambridge.

Gould, S. J. (1982). Change in developmental timing as a mechanism of macroevolution. In *Evolution and Development. Life Sciences Report No. 22* (ed. J. T. Bonner), pp. 333–46. Springer-Verlag, Berlin.

Gould, S. J. and Eldredge, N. (1977). Punctuated equilibria: the tempo and mode of evolution reconsidered. *Paleobiology*, **3**, 115–51.

Grand, T. I. (1972). A mechanical interpretation of terminal branch feeding. *Journal of Mammalogy*, **53**, 198–201.

Grasse, P-P. (1977). *Evolution of living organisms*. Academic Press, London.

Gregory, W. K. (1910). The orders of Mammals. *Bulletin of the American Museum of Natural History*, **27**, 1–524.

Gregory, W. K. (1916). Studies on the evolution of the Primates. *Bulletin of the American Museum of Natural History*, **35**, 239–355.

Gregory, W. K. (1947). The monotremes and the palimpsest theory. *Bulletin of the American Museum of Natural History*, **88**, 1–52.

Gregory, W. K. (1951). *Evolution emerging*. Macmillan, New York.

Haas, O. and Simpson, G. G. (1946). Analysis of some phylogenetic terms with attempts at redefinition. *Proceedings of the American Philosophical Society*, **90**, 319–49.

Haeckel, E. (1910). The evolution of man. Watts and Co, London.

Hafferl, A. (1929). Bau und Funktion des Affenfusses. *Zeitschrift für Anatomie und Entwicklunggeschaft*, **88**, 749–83.

Haines, R. W. (1932). The laws of muscle and tendon growth. *Journal of Anatomy*, **66**, 578–85.

Haines, R. W. (1935). A consideration of the constancy of muscular nerve supply. *Journal of Anatomy*, **70**, 33–55.

Haines, R. W. (1939). A revision of the extensor muscles of the forearm in tetrapods. *Journal of Anatomy*, **73**, 211–33.

Haines, R. W. (1940). Note on the independence of

sesamoid and epiphyseal centres of ossification. *Journal of Anatomy*, **75**, 101–5.

Haines, R. W. (1942). The tetrapod knee joint. *Journal of Anatomy*, **76**, 270–301.

Haines, R. W. (1944). The mechanism of rotation at the first carpo-metacarpal joint. *Journal of Anatomy*, **78**, 44–6.

Haines, R. W. (1950). The flexor muscles of the forearm and hand in lizards and mammals. *Journal of Anatomy*, **84**, 13–29.

Haines, R. W. (1951). The extensor apparatus of the finger. *Journal of Anatomy*, **85**, 251–9.

Haines, R. W. (1958). Arboreal or terrestrial ancestry of the placental mammals. *Quarterly Review of Biology*, **33**, 1–23.

Halstead, L. B. (1980). Museum of Errors. *Nature*, **288**, 208.

Harrison, T. (1981). New finds of small fossil apes from the Miocene locality of Koru in Kenya. *Journal of Human Evolution*, **10**, 129–37.

Hartman, C. G. and Straus, W. L. (1933). *The Anatomy of the rhesus monkey*. Hafner, New York.

Hast, M. H. and Perkins, R. E. (1986). Secondary tensor and supinator muscles of the human proximal radioulnar joint. *Journal of Anatomy*, **146**, 45–51.

Hay, R. L. (1980). The KBS tuff controversy may be ended. *Nature*, **284**, 401.

Henke, W. (1859). Die Bewegungen der Handwurzel. *Zeitschrift für rationell Medizin*, **7**, 27–41.

Henle, J. (1855). *Handbuch der Systematischen Anatomie des Menschen*, Erster Band. F. Vieweg und Sohn, Braunschweig.

Henle, J. (1856). *Handbuch der Systematischen Anatomie des Menschen*. Erster Band (Zweite Abtheilung). Vieweg und Sohn, Braunschweig.

Hennig, W. (1965). Extracts from: Phylogenetic Systematics. *Annual Review of Entomology*, **10**, 97–116.

Hennig, W. (1966). *Phylogenetic systematics*. University of Illinois Press, Urbana.

Hepburn, D. (1892). The comparative anatomy of the muscles and nerves of the superior and inferior extremities of the anthropoid apes. Part I. *Journal of anatomy*, **26**, 149–85.

Hesser, C. (1926). Beitrag zur Kenntnis der Gelenkentwicklung beim Menschen. *Morphologisches Jahrbuch*, **55**, 489–67.

Hicks, J. H. (1951). The function of the plantar aponeurosis. *Journal of Anatomy*, **85**, 414–15.

Hicks, J. H. (1953). The mechanics of the foot. I. The joints. *Journal of Anatomy*, **87**, 345–57.

Hicks, J. H. (1954). The mechanics of the foot. II. The plantar aponeurosis and the arch. *Journal of Anatomy*, **88**, 25–30.

Higgins, H. (1895a). The semilunar fibro-cartilages and transverse ligament of the knee joint. *Journal of Anatomy*, **29**, 390–8.

Higgins, H. (1895b). The popliteus muscle. *Journal of Anatomy*, **29**, 569–73.

Hildebrand, M. (1978). Insertions and functions of certain flexor muscles in the hind leg of rodents. *Journal of Morphology*, **155**, 111–22.

Hildebrand, M. (1982). *Analysis of vertebrate structure*. John Wiley, New York.

Hinchliffe, J. R. (1977). The chondrogenic pattern in chick limb morphogenesis: a problem of development and evolution. In *Vertebrate limb and somite morphogenesis* (ed. D. A. Ede, J. R. Hinchliffe and M. Balls), pp. 293–309. Cambridge University Press.

Hofstetter, R. (1974). Phylogeny and geographical deployment of the Primates. *Journal of Human Evolution*, **3**, 327–50.

Holmgren, N. (1952). An embryological analysis of the mammalian carpus and its bearing upon the question of the origin of the tetrapod limb. *Acta Zoologica*, **33**, 1–115.

Howell, A. Brazier (1936). Phylogeny of the distal musculature of the pectoral appendage. *Journal of Morphology*, **60**, 287–315.

Howell, A. B. and Straus, W. L. (1961). The muscular system. In *The anatomy of the rhesus monkey* (ed. C. G. Hartman and W. L. Straus). Hafner, New York.

Howells, W. W. (1980). *Homo erectus*—who, when and where: a survey. *Yearbook of Physical Anthropology* **23**, 1–23.

Humphry, G. M. (1858). *On the human skeleton*. Macmillan, Cambridge.

Humphry, G. M. (1872). *Observations in myology*. Macmillan, London.

Huntington, G. S. (1903). Present problems of myological research and the significance and classification of muscular variations. *American Journal of Anatomy*, **2**, 159–75.

Huxley, J. (1942). *Evolution. The modern synthesis*. Allen and Unwin, London.

Huxley, T. H. (1880). On the application of the laws of evolution to the arrangement of the Vertebrata, and more particularly of the Mammalia. *Proceedings of the Zoological Society of London*, **1880**, 649–69.

Inman, V. T. (1976). *The joints of the ankle*. Williams and Wilkins, Baltimore.

Isaac, G. L. (1978). Early man reviewed. *Nature*, **273**, 588–9.

Isaac, G. L. (1981). Emergence of human behaviour patterns. In *The emergence of man* (ed. J. Z. Young, E. M. Jope and K. P. Oakley), pp. 177–81. The Royal Society and the British Academy, London.

Jacob, T. (1973). Palaeoanthropological discoveries in Indonesia with special reference to the finds in the last two decades. *Journal of Human Evolution*, **2**, 473–85.

Jager, K. W. and Moll, J. (1951). The development of the human triceps surae. *Journal of Anatomy*, **85**, 338–49.

Jarvik, E. (1980). *Basic structure and evolution of vertebrates*, Vol. 2. Academic Press, London.

Jenkins, F. A. (1970*a*). Limb movements in a monotreme (*Tachyglossus aculeatus*): a cineradiographic analysis. *Science*, **168**, 1473–5.

Jenkins, F. A. (1970*b*). Cynodont postcranial anatomy and the 'prototherian' level of mammalian organisation. *Evolution*, **24**, 230–52.

Jenkins, F. A. (1971*a*). The postcranial skeleton of African cynodonts. *Bulletin of the Peabody Museum of Natural History*, **36**, 1–216.

Jenkins, F. A. (1971*b*). Limb posture and locomotion in the Virginia opossum (*Didelphys marsupialis*) and in other non-cursorial mammals. *Journal of Zoology, London*, **165**, 303–15.

Jenkins, F. A. (1974). Tree shrew locomotion and the origins of primate arborealism. In *Primate locomotion* (ed. F. A. Jenkins), pp. 85–115. Academic Press, New York.

Jenkins, F. A. (1981). Wrist rotation in primates: a critical adaptation for brachiators. *Symposia of the Zoological Society of London*, **48**, 429–51.

Jenkins, F. A. and Crompton, A. W. (1979). Tricondonta. In *Mesozoic mammals. The first two-thirds of mammalian history* ed. J. A. Lillegraven, Z. Kielan-Jaworowska and W. A. Clemens), pp. 74–90. University of California Press, Berkeley.

Jenkins, F. A. and Fleagle, J. G. (1975). Knuckle-walking and the functional anatomy of the wrists in living apes. In *Primate functional morphology and evolution* (ed. R. H. Tuttle), pp. 213–27. Mouton, The Hague.

Jenkins, F. A. and Krause, D. W. (1983). Adaptations for climbing in North American Multituberculates (Mammalia). *Science*, **220**, 712–15.

Jenkins, F. A. and McClearn, D. (1984). Mechanisms of hind foot reversal in climbing mammals. *Journal of Morphology*, **182**, 197–219.

Jenkins, F. A. and Parrington, F. R. (1976). The postcranial skeletons of *Eozostrodon*, *Megazostrodon* and *Erythrotherium*. *Royal Society of London Philosophical Transactions, B*, **273**, 387–431.

Johanson, D. C. (1978). Our roots go deeper. *Science Year. The world science annual 1979*. Field Enterprises Educational Corporation, Chicago.

Johanson, D. C., Lovejoy, C. O., Kimbel, W. H., White, T. D., Ward, S. C., Bush, M. E., Latimer, B. M., and Coppens, Y. (1982). Morphology of the Pliocene partial hominid skeleton (A.L. 288-1) from the Hadar formation, Ethiopia. *American Journal of Physical Anthropology*, **57**, 403–51.

Johanson, D. C. and Taieb, M. (1976). *Plio-Pleistocene hominid discoveries in Hadar, Ethiopia*. *Nature*, **260**, 293–7.

Johanson, D. C. and White, T. D. (1979). A systematic assessment of early African hominids. *Science*, **202**, 321–30.

Johanson, D. C., White, T. D., and Coppens, Y. (1978). A new species of the genus *Australopithecus* (Primates: Hominidae) from the Pliocene of eastern Africa. *Kirtlandia*, **28**, 1–14.

Johnston, H. M. (1907*a*). Varying positions of the carpal bones in the different movements of the wrist. Part I. *Journal of Anatomy*, **41**, 109–22.

Johnston, H. M. (1907*b*). Varying positions of the carpal bones in the different movements of the wrist. Part II. *Journal of Anatomy*, **41**, 280–92.

Jolly, C. J. (1970). The seed-eaters: a new model of hominid differentiation based on a baboon analogy. *Man*, **5**, 5–26.

Jones, F. Wood (1917). The genitalia of *Tupaia*. *Journal of Anatomy*, **51**, 118–26.

Jones, F. Wood (1929). The distinctions of the human hallux. *Journal of Anatomy*, **63**, 408–11.

Jones, F. Wood (1944). *Structure and function as seen in the foot*. Baillière, Tindall and Cox, London.

Jones, F. Wood (1948). *Hallmarks of mankind*. Baillière, Tindall and Cox, London.

Jones, F. Wood (1949*a*). *The principles of anatomy as seen in the hand*. Baillière, Tindall and Cox, London.

Jones, F. Wood (1949*b*). The study of a generalized marsupial (*Dasycercus cristicauda* Krefft). *Transactions of the Zoological Society of London*, **26**, 409–501.

Jones, F. Wood (1953). Some readaptations of the mammalian pes in response to arboreal habits. *Proceedings of the Zoological Society of London*, **123**, 33–41.

Jones, J. S. (1981). An uncensored page of fossil history. *Nature*, **293**, 427–8.

Jones, R. L. (1945). The functional significance of the declination of the axis of the subtalar joint. *Anatomical Record*, **93**, 151–9.

Jones, R. T. (1967). The anatomical aspects of the baboon's wrist joint. *South African Journal of Science*, **63**, 291–6.

Joseph, J. (1951). Further studies of the metacarpophalangeal and interphalangeal joints of the thumb. *Journal of Anatomy*, **85**, 221–9.

Jouffroy, F. K. and Lessertisseur, J. (1959). La main des Lemuriens Malgache comparee a celle des autres Primates. *Memoires de l'Institut Scientifique de Madagascar*, **13A**, 195–219.

Jouffroy, F-K. (1971). Musculature des membres. In *Traité de zoologie*, (ed. P-P. Grassé), Vol. 16. Masson, Paris.

Kadanoff, D. (1958). Uber die Erscheinungen des Umbildungsprozesses der Finger-und Zehenstrecker beim Menschen. *Morphologisches Jahrbuch*, **99**, 613–61.

Kaneff, A. and Cihak, R. (1970). Die Umbildung des M. extensor digitorum lateralis in der Phylogenese und in der menschlichen Ontogenese. *Acta anatomica*, **77**, 583–604.

Kapandji, I. A. (1966). *Physiologie articulaire*. Fascicule I. *Membre supérieur*. Librairie Maloine, Paris.

Kapandji, I. A. (1968). *Physiologie articulaire*. Fascicule II. *Membre inférieur*. Librairie Maloine, Paris.

Kaplan, E. B. (1955). The tibialis posterior muscle in relation to hallux valgus. *Bulletin of the Hospital for Joint Diseases*, **16**, 88–93.

Kaplan, E. B. (1958). Comparative anatomy of the extensor digitorum longus in relation to the knee joint. *Anatomical Record*, **131**, 129–50.

Kaplan, E. B. (1959). Morphology and function of the muscle quadratus plantae. *Bulletin of the Hospital for Joint Diseases*, **20**, 84–95.

Kaplan, E. B. (1961). The fabellofibular and short ligaments of the knee joint. *Journal of Bone and Joint Surgery*, **43A**, 169–79.

Kauer, J. M. G. (1974). The interdependence of carpal articulation chains. *Acta anatomica*, **88**, 481–501.

Kauer, J. M. G. (1986). The mechanism of the carpal joint. *Clinical Orthopaedics*, **202**, 16–26.

Kay, R. F. (1977a). Diets of early Miocene hominids. *Nature*, **268**, 628–30.

Kay, R. F. (1977b). Post-Oligocene evolution of catarrhine diets. *American Journal of Physical Anthropology*, **47**, 141–2.

Kay, R. F. (1981). The nut-crackers—a new theory of the adaptations of the *Ramapithecinae*. *American Journal of Physical Anthropology*, **55**, 141–51.

Kay, R. F. and Cartmill, M. (1977). Cranial morphology and adaptations of *Palaechthon nacimienti* and other Paromomyidae (Plesiadapoidea, Primates), with a description of a new genus and species. *Journal of Human Evolution*, **6**, 19–53.

Kay, R. F., Fleagle, J. G., and Simons, E. L. (1981). A revision of the Oligocene apes of the Fayum Provence, Egypt. *American Journal of Physical Anthropology*, **55**, 293–322.

Keast, A. (1977). Historical biogeography of the marsupials. In *The biology of marsupials* (ed. B. Stonehouse and D. Gilmore), pp. 69–95. Macmillan, London.

Keith, A. (1894a). The ligaments of the catarrhine monkeys with reference to corresponding structures in man. *Journal of Anatomy*, **28**, 149–68.

Keith, A. (1894b). Notes on a theory to account for the various arrangements of the flexor profundus digitorum in the hand and foot of primates. *Journal of Anatomy*, **28**, 335–9.

Keith, A. (1929). The history of the human foot and its bearing on orthopaedic practice. *Journal of Bone and Joint Surgery*, **11**, 10–32.

Keith, A. (1948). *Human embryology and morphology*. Edward Arnold, London.

Kemp, T. S. (1978). Stance and gait in the hindlimb of a therocephalian mammal-like reptile. *Journal of Zoology, London*, **186**, 143–61.

Kemp, T. S. (1982). *Mammal-like reptiles and the origin of mammals*. Academic Press, London.

Ker, R. F., Bennett, M. B., Bibby, S. R., Kester, R. C., and Alexander, R. McN. (1987). The spring in the arch of the human foot. *Nature*, **325**, 147–9.

Kesner, M. H. (1986). The myology of the manus of microtine rodents. *Journal of Zoology, London*, **210**, 1–22.

Kessler, I. and Silberman, Z. (1961). An experimental study of the radiocarpal joint by arthrography. *Surgery, Gynecology and Obstetrics*, **112**, 33–40.

Kidner, F. C. (1929). The prehallux (accessory scaphoid) in its relation to flatfoot. *Journal of Bone and Joint Surgery*, **11**, 831–7.

Kielan-Jaworowska, Z. (1975). Possible occurrence of marsupial bones in Cretaceous eutherian mammals. *Nature*, **255**, 698–9.

Kielan-Jaworowska, Z. (1979). Pelvic structure and nature of reproduction in Multituberculata. *Nature*, **277**, 402–3.

Kielan-Jaworowska, Z., Eaton, J. G., and Bown, T. M. (1979a). Theria of Metatherian-Eutherian Grade. In *Mesozoic mammals. The first two thirds of mammalian history* (ed. J. A. Lillegraven, Z Kielan-Jaworowska and W. A. Clemens), pp. 182–91. University of California Press, Berkeley.

Kielan-Jaworowska, Z., Bown, T. M., and Lillegraven, J. A. (1979b). Eutheria. In *Mesozoic mammals. The first two-thirds of mammalian history* (ed. J. A. Lillegraven, Z. Kielan-Jaworowska and W. A. Clemens), pp. 221–58. University of California Press, Berkeley.

Kielan-Jaworowska, Z., Crompton, A. W., and Jenkins, F. A. (1987). The origin of egg-laying mammals. *Nature*, **318**, 363–6.

Kingdon, J. (1971). *East African mammals*. Academic Press, London.

Kirk, T. S. (1924). Some points in the mechanism of the human hand. *Journal of Anatomy*, **58**, 228–30.

Koebke, J., Thomas, W., and Winter, H-J. (1982). Da Ligamentum metacarpeum dorsale I und die Arthrose des Daumensattelgelenkes. *Morphologische Medizin*, **2**, 1–8.

Koenigswald, G. H. R. (1969). Miocene Cercopithecoidea and Oreopithecoidea from the Miocene of East Africa. In *Fossil vertebrates of Africa*, Vol. 1, pp. 39–52.

Koenigswald, G. H. R. (1973). *Australopithecus, Meganthropus and Ramapithecus*. *Journal of Human Evolution*, **2**, 487–91.

Koenigswald, G. H. R. (1975). Early man in Java: catalogue and problems. In *Palaeoanthropology, morphology and palaeecology* (ed. T. S. Tuttle), pp. 304–9. Mouton, The Hague.

Kohlbrugge, J. H. F. (1890). Versuch einer Anatomie des Genus Hylobates. Erster Teil. In *Zoologische Ergebnisse einer Reise in Niederländisch Ost-Indien* (ed. M. Weber) Vol. 1, pp. 211–354.

Kootstra, G., Huffstadt, A. J. C., and Kauer, J. M. G. (1974). The styloid bone. A clinical and embryological study. *The Hand*, **6**, 185–9.

Kraus, M. J. (1979). Eupantotheria. In *Mesozoic mammals. The first two-thirds of mammalian history* (ed. J. A. Lillegraven, Z. Kielan-Jaworowska and W. A. Clemens), pp. 162–71. University of California Press, Berkeley.

Kron, D. G. (1979). Docodonta. In *Mesozoic mammals. The first two-thirds of mammalian history* (ed. J. A. Lillegraven, Z. Kielan-Jaworowska and W. A. Clemens), pp. 91–8. University of California Press, Berkeley.

Kuczynski, K. (1974). Carpometacarpal joint of the human thumb. *Journal of Anatomy*, **118**, 119–26.

Kuhne, W. G. (1973). The systematic position of monotremes reconsidered (Mammalia). *Zeitschrift für Morphologie der Tiere* **75**, 59–64.

Laidlaw, P. P. (1904). The varieties of the os calcis. Part I. *Journal of Anatomy*, **38**, 133–43.

Laidlaw, P. P. (1905). The os calcis. Parts II–IV. *Journal of Anatomy*, **39**, 161–77.

Lambert, E. H. (1969). The accessory deep peroneal nerve. *Neurology*, **19**, 1169–76.

Lamy, P. (1986). The settlement of the longitudinal plantar arch of some African Plio-Pleistocene hominids: a morphological study. *Journal of Human Evolution*, **15**, 31–46.

Landsmeer, J. M. F. (1949). The anatomy of the dorsal aponeurosis of the human finger and its functional significance. *Anatomical Record*, **104**, 31–44.

Landsmeer, J. M. F. (1955). Anatomical and functional investigations on the articulation of the human fingers. *Acta Anatomica, Supplement*, **24**.

Landsmeer, J. M. F. and Ansingh, H. R. (1957). X-ray observations on rotation of the fingers in the metacarpo-phalangeal joints. *Acta anatomica*, **30**, 404–10.

Langdon, J. H. (1986). Functional morphology of the Miocene hominoid foot. *Contributions to Primatology*, Vol. 22. Karger, Basel.

Langelaan, E. J. V. (1983). A kinematical analysis of the tarsal joints. Decor Davids, Alblasserdam.

Laporte, L. F. and Zihlman, A. L. (1983). Plates, climates and hominoid evolution. *South African Journal of Science*, **79**, 96–110.

Last, R. J. (1951). Specimens from the Hunterian collection. *Journal of Bone and Joint Surgery*, **33B**, 626–8.

Last, R. J. (1954). *Anatomy, regional and applied*. Churchill, London.

Latimer, B. M., Lovejoy, C. O., Johanson, D. C., and Coppens, Y. (1982). Hominid tarsal, metatarsal, and phalangeal bones recovered from the Hadar formation: 1974–1977 collections. *American Journal of Physical Anthropology*, **57**, 701–19.

Latimer, B., Ohman, J. C., and Lovejoy, C. O. (1987). Talocrural joint in African hominoids: implications for *Australopithecus afarensis*. *American Journal of Physical Anthropology*, **74**, 155–75.

Lavocat, R. (1980). The implications of rodent paleontology and biogeography to the geographical sources and origins of the platyrrhine Primates. In *Evolutionary biology of the new world monkeys and continental drift* (ed. R. L. Ciochon and A. B. Chiarelli), pp. 93–102. Plenum Press, New York.

Leakey, L. S. B. (1964). East African fossil Hominoidea and the classification within the superfamily. In *Classification and human evolution* (ed. S. L. Washburn), pp. 32–49. Methuen, London.

Leakey, M. D. (1970). Stone artefacts from Swartkrans. *Nature*, **225**, 1222–5.

Leakey, M. D. (1971). *Olduvai Gorge*. Vol. 3: Excavations in Beds I and II, 1960–1963. Cambridge University Press.

Leakey, M. D., Hay, R. L., Curtis, G. H., Drake, R. E., Jackes, M. K., and White, T. D. (1976). Fossil hominids from the Laetolil beds. *Nature*, **262**, 460–6.

Leakey, R. E. F. (1969). New Cercopithecidae from the Chemeron beds of Lake Baringo, Kenya. In *Fossil vertebrates of Africa*, Vol. I, pp. 53–69.

Leakey, R. E. F. (1972). Further evidence of lower Pleistocene hominids from East Rudolf, North Kenya, 1972. *Nature*, **242**, 170–3.

Leakey, R. E. F. (1973). Evidence for an advanced Plio-Pleistocene hominid from East Rudolf, Kenya. *Nature*, **242**, 447–50.

Leakey, R. E. F. (1974). Further evidence of lower Pleistocene hominids from East Rudolf, North Kenya, 1973. *Nature*, **248**, 653–6.

Leakey, R. E. F. and Lewin, R. (1977). *Origins*. Macdonald and Jones, London.

Leakey, R. E. F. and Walker, A. C. (1976). *Australopithecus*, *Homo erectus* and single species hypothesis. *Nature*, **261**, 572–4.

Lebboucq, H. (1884) Recherches sur la morphologie du carpe chez les mammifères. *Archives Biologie, Paris*, **5**, 35–102.

Lebboucq, H. (1886). Sur la morphologie du carpe et du tarse. *Anatomischer Anzeiger*, **1**, 17–21.

Leche, W. (1900). *Bronn's Klassen und Ordnungen des Their-Reichs*, Vol. 6, part 5A. Leipzig.

Le Double, A. F. (1897). *Traité des variations du système musculaire de l'homme*. Schleicher Frères, Paris.

Lee, M. M. C. and Garn, S. M. (1967). Pseudoepiphyses or notches in the non-epiphyseal end of metacarpal bones in healthy children. *Anatomical Record*, **159**, 263–72.

Legrand, J. J. (1983). The lunate bone: a weak link in the

articular column of the wrist. *Anatomia clinica*, **5**, 57–64.

Leonard, M. A. (1974). The inheritance of tarsal coalition and its relationship to spastic flat foot. *Journal of Bone and Joint Surgery*, **56B**, 520–6.

Lewis, O. J. (1959). The coraco-clavicular joint. *Journal of Anatomy*, **93**, 296–303.

Lewis, O. J. (1962a). The phylogeny of the crural and pedal flexor musculature. *Proceedings of the Zoological Society of London*, **138**, 77–109.

Lewis, O. J. (1962b). The comparative morphology of M. flexor accessorius and the associated long flexor tendons. *Journal of Anatomy*, **96**, 321–33.

Lewis, O. J. (1963). The monotreme cruro-pedal flexor musculature. *Journal of Anatomy*, **97**, 55–63.

Lewis, O. J. (1964a). The homologies of the mammalian tarsal bones. *Journal of Anatomy*, **98**, 195–208.

Lewis, O. J. (1964b). The tibialis posterior tendon in the primate foot. *Journal of Anatomy*, **98**, 209–18.

Lewis, O. J. (1964c). The evolution of the long flexor muscles of the leg and foot. In *International review of general and experimental zoology* (ed. W. J. L. Felts and R. J. Harrison), pp. 165–85. Academic Press, New York.

Lewis, O. J. (1965a). Evolutionary change in the primate wrist and inferior radio-ulnar joints. *Anatomical Record*, **151**, 275–86.

Lewis, O. J. (1965b). The evolution of the Mm. interossei in the primate hand. *Anatomical Record*, **153**, 275–88.

Lewis, O. J. (1966). The phylogeny of the cruropedal extensor musculature with special reference to the Primates. *Journal of Anatomy*, **100**, 865–80.

Lewis, O. J. (1969). The hominoid wrist joint. *American Journal of Physical Anthropology*, **30**, 251–68.

Lewis, O. J. (1970). The development of the human wrist joint during the fetal period. *Anatomical Record*, **166**, 499–516.

Lewis, O. J. (1971a). The contrasting morphology found in the wrist joints of semi-brachiating monkeys and brachiating apes. *Folia primatoligica*, **16**, 248–56.

Lewis, O. J. (1971b). Brachiation and the early evolution of the Hominoidea. *Nature*, **230**, 577–8.

Lewis, O. J. (1972a). Evolution of the hominoid wrist. In *Functional and evolutionary biology of the primates* (ed. R. H. Tuttle), pp. 207–22. Aldine-Atherton, Chicago.

Lewis, O. J. (1972b). Osteological features characterizing the wrists of monkeys and apes, with a reconsideration of this region in *Dryopithecus (Proconsul) africanus*. *American Journal of Physical Anthropology*, **36**, 45–58.

Lewis, O. J. (1972c). The evolution of the hallucial tarsometatarsal joint in the Anthropoidea. *American Journal of Physical Anthropology*, **37**, 13–34.

Lewis, O. J. (1973). The hominoid os capitatum, with special reference to the fossil bones from Sterkfontein and Olduvai Gorge. *Journal of Human Evolution*, **2**, 1–11.

Lewis, O. J. (1974). The wrist articulations of the Anthropoidea. In *Primate locomotion* (ed. F. A. Jenkins, Jr.), pp. 143–69. Academic Press, New York.

Lewis, O. J. (1977). Joint remodelling and the evolution of the human hand. *Journal of Anatomy*, **123**, 157–201.

Lewis, O. J. (1980a). The joints of the evolving foot. Part I. The ankle joint. *Journal of Anatomy*, **130**, 527–43.

Lewis, O. J. (1980b). The joints of the evolving foot. Part II. The intrinsic joints. *Journal of Anatomy*, **130**, 833–57.

Lewis, O. J. (1980c). The joints of the evolving foot. Part III. The fossil evidence. *Journal of Anatomy*, **131**, 275–98.

Lewis, O. J. (1981). Functional morphology of the joints of the evolving foot. *Symposia of the Zoological Society of London*, **46**, 169–88.

Lewis, O. J. (1983). The evolutionary emergence and refinement of the mammalian pattern of foot architecture. *Journal of Anatomy*, **137**, 21–45.

Lewis, O. J. (1985a). Derived morphology of the wrist articulations, and theories of hominoid evolution. Part I. The lorisine joints. *Journal of Anatomy*, **140**, 447–60.

Lewis, O. J. (1985b). Derived morphology of the wrist articulations, and theories of hominoid evolution. Part II. The midcarpal joints of higher Primates. *Journal of Anatomy*, **142**, 151–72.

Lewis, O. J., Hamshere, R. J., and Bucknill, T. M. (1970). The anatomy of the wrist joint. *Journal of Anatomy*, **106**, 539–52.

Lewis, W. H. (1910). The development of the muscular system. In *Manual of Human Embryology*, Vol. 1 (ed. F. Keibel and F. P. Mall). Lippincott, London.

Lillegraven, J. A. (1974). Biogeographical considerations of the marsupial–placental dichotomy. *Annual Review of Ecology and Systematics*, **5**, 263–83.

Lillegraven, J. A. (1975). Biological considerations of the marsupial-placental dichotomy. *Evolution*, **29**, 707–22.

Lillegraven, J. A. (1979a). Introduction. In *Mesozoic mammals, The first two-thirds of mammalian history* (ed. J. A. Lillegraven, Z. Kielan-Jaworowska and W. A. Clemens), pp. 1–6. University of California Press, Berkeley.

Lillegraven, J. A. (1979b). Reproduction in Mesozoic Mammals. In *Mesozoic mammals. The first two-thirds of mammalian history* (ed. J. A. Lillegraven, Z. Kielan-Jaworowska and W. A. CLemens), pp. 259–76. University of California Press, Berkeley.

Lillegraven, J. A., Kraus, M. J., and Bown, T. M. (1979). Paleogeography of the world of the Mesozoic. In *Mesozoic mammals. The first two-thirds of mammalian history* (ed. J. A. Lillegraven, Z. Kielan-Jaworowska and W. A. Clemens), pp. 277–308. University of California Press, Berkeley.

Linscheid, R. L. (1986). Kinematic considerations of the wrist. *Clinical Orthopaedics*, **202**, 27–39.

Lipson, S. and Pilbeam, D. (1982). *Ramapithecus* and hominoid evolution. *Journal of Human Evolution*, **11**, 545–8.

Lisowski, F. P. (1967). Angular growth changes and comparisons in the primate talus. *Folia primatologica*, **7**, 81–97.

Lisowski, F. P., Albrecht, G. H., and Oxnard, C. E. (1974). The form of the talus in some higher Primates: a multivariate study. *American Journal of Physical Anthropology*, **41**, 191–216.

Lisowski, F. P., Albrecht, G. H., and Oxnard, C. E. (1976). African fossil tali: further multivariate morphometric studies. *American Journal of Physical Anthropology*, **45**, 5–18.

Lorenz, R. (1974). On the thumb of the Hylobatidae. In *Gibbon and siamang*, Vol. 3 (ed. D. M. Rumbaugh), pp. 157–75. Karger, Basel.

Loth, E. (1908). Die Aponeurosis plantaris in der Primatenreihe. *Morphologisches Jahrbuch*, **38**, 194–322.

Lovejoy, C. O. (1978). A biomechanical review of the locomotor diversity of early hominids. In *Early hominids of Africa* (ed. C. J. Jolly), pp. 403–29. Duckworth, London.

Lovejoy, J. F. and Harden, T. P. (1971). Popliteus muscle in man. *Anatomical Record*, **169**, 727–30.

Lucien, M. (1907). Note sur le développement du ligament annulaire antérieur du carpe chez l'homme. *Compte Rendu de la Societe de Biologie, Paris* **62**, 169–71.

Lucien, M. (1909). Sur les connexions etre le pédieux et les muscles interosseux dorsaux chez l'homme. *Bibliotheca anatomica, Basel* **19**, 229–37.

Lucien, M. (1947). L'aponévrose dorsale moyenne de la main. *Acta anatomica*, **4**, 188–92.

Luckett, W. P. (1977). Ontogeny of amniote fetal membranes and their application to phylogeny. In *Major patterns in vertebrate evolution* (ed. M. K. Hecht, P. C. Goody and B. M. Hecht), pp. 439–516. Plenum, New York.

Luckett, W. P. (1980). *Comparative biology and evolutionary relationships of tree shrews* (ed. W. P. Luckett). Plenum, New York.

Luckett, W. P. (1980). The suggested evolutionary relationships and classification of tree shrews. In *Comparative biology and evolutionary relationships of tree shrews* (ed. W. P. Luckett), pp. 3–31. Plenum, New York.

Macalister, A. (1889). On the arrangement of the pronator muscles in the limbs of vertebrate animals. *Journal of Anatomy*, **3**, 335–40.

MacConaill, M. A. (1941). The mechanical anatomy of the carpus and its bearing on some surgical problems. *Journal of Anatomy*, **75**, 166–75.

MacConaill, M. A. (1945). The postural mechanism of the human foot. *Proceedings of the Royal Irish Academy*, **50**, 265–78.

MacConaill, M. A. (1953). The movements of bones and joints. The significance of shape. *Journal of Bone and Joint Surgery*, **35B**, 290–7.

MacConaill, M. A. (1973). A structuro-functional classification of synovial articular units. *Irish Journal of Medical Science*, **142**, 19–26.

MacConaill, M. A. and Basmajian, J. V. (1969). *Muscles and movements. A basis for kinesiology*. Williams and Wilkins, Baltimore.

McHenry, H. M. (1983). The capitate of *Australopithecus afarensis* and *A. africanus*. *American Journal of Physical Anthropology*, **62**, 187–98.

McHenry, H. M. (1986). The first bipeds: a comparison of the *A. afarensis* and *A. africanus* postcranium and implications for the evolution of bipedalism. *Journal of Human Evolution*, **15**, 177–91.

McHenry, H. M. and Corruccini, R. S. (1983). The wrist of *Proconsul africanus* and the origin of hominoid postcranial adaptations. In *New interpretations of ape and human ancestry* (ed. R. L. Ciochon and R. S. Corruccini), pp. 353–69. Plenum, New York.

McKee, P. R. and Bagnall, K. M. (1987). Skeletal relationships in the human embryonic foot based on three-dimensional reconstructions. *Acta anatomica*, **129**, 34–42.

McKenna, M. (1966). Paleontology and the origin of the Primates. *Folia primatologica*, **4**, 77–83.

McKenna, M. (1980). Early history and biogeography of South American's extinct land mammals. In *Evolutionary biology of the new world monkeys and continental drift* (ed. R. L. Ciochon and A. B. Chiarelli), pp. 43–78. Plenum, New York.

McMahon, T. A. (1987). The spring in the human foot. *Nature*, **325**, 108–9.

McMaster, M. (1972). A natural history of the rheumatoid metacarpophalangeal joint. *Journal of Bone and Joint Surgery*, **54**, 687–97.

McMurrich, J. P. (1903). The phylogeny of the forearm flexors. *American Journal of anatomy*, **2**, 177–209.

McMurrich, J. P. (1903). The phylogeny of the palmar musculature. *Americal Journal of Anatomy*, **2**, 463–500.

McMurrich, J. P. (1904). The phylogeny of the crural flexors. *American Journal of Anatomy*, **4**, 33–76.

McMurrich, J. P. (1907). The phylogeny of the plantar musculature. *American Journal of Anatomy*, **6**, 407–37.

McMurrich, J. P. (1927). The evolution of the human foot. *American Journal of Physical Anthropology*, **10**, 165–71.

Maderson, F. A. (1982). The role of development in macroevolutionary change. In *Evolution and develop-*

ment. *Life sciences report No. 22* (ed. J. T. Bonner), pp. 279–312. Springer-Verlag, Berlin.

Mann, R. and Inman, V. T. (1964). Phasic activity of intrinsic muscles of the foot. *Journal of Bone and Joint Surgery,* **46A**, 469–81.

Manners-Smith, T. (1894). On some points in the anatomy of *Ornithorhynchus paradoxus. Proceedings of the Zoological Society of London,* **1894**, 694–722.

Manners-Smith, T. (1907). A study of the navicular in the human and anthropoid foot. *Journal of Anatomy,* **41**, 255–79.

Manter, J. T. (1941). Movements of the subtalar and transverse tarsal joints. *Anatomical Record,* **80**, 397–410.

Manter, J. T. (1945). Variations of the interosseous muscles of the human foot. *Anatomical Record,* **93**, 117–24.

Martin, B. F. (1958*a*). The oblique cord of the forearm. *Journal of Anatomy,* **92**, 609–15.

Martin, B. F. (1958*b*). The tendons of flexor digitorum profundus. *Journal of Anatomy,* **92**, 602–9.

Martin, R. D. (1979). Phylogenetic aspects of prosimian behaviour. In *The study of prosimian behaviour* (ed. G. A. Doyle and R. D. Martin), Ch. 2, pp. 45–77. Academic Press, New York.

Marzke, M. W. (1983). Joint function and grips of the *Australopithecus afarensis* hand, with special reference to the region of the capitate. *Journal of Human Evolution,* **12**, 197–211.

Marzke, M. W. and Marzke, R. F. (1987). The third metacarpal styloid process in humans: origin and functions. *American Journal of Physical Anthropology,* **73**, 415–31.

Marzke, M. W. and Shackley, M. S. (1986). Hominid hand use in the Pliocene and Pleistocene: evidence from experimental archaeology and comparative morphology. *Journal of Human Evolution,* **15**, 439–60.

Masquelet, A. C., Salama, J., Outrequin, G., Serrault, M., and Chevrel, J. P. (1986). Morphology and functional anatomy of the first dorsal interosseous muscle of the hand. *Surgical and Radiologic Anatomy,* **8**, 19–28.

Matthew, W. D. (1904). The arboreal ancestry of the mammalia. *American Naturalist,* **38**, 811–18.

Mayfield, J. K., Johnson, R. P., and Kilcoyne, R. F. (1976). The ligaments of the human wrist and their functional significance. *Anatomical Record,* **186**, 417–28.

Mayr, E. (1942). *Systematics and the origin of species.* Columbia University Press, New York.

Mayr, E. (1960). The emergence of evolutionary novelties. In *Evolution after Darwin.* Vol. 1 (ed. Sol Tax), pp. 349–80. University of Chicago Press.

Mayr, E. (1974). Cladistic analysis or cladistic classification. *Zeitschrift fur Zoologische Systematik und Evolutionforschung,* **12**, 94–128.

Mayr, E. (1982). Questions concerning speciation. *Nature,* **296**, 609.

Mayr, E. and Provine, W. B. (1980). *The evolutionary synthesis.* Harvard University Press, Cambridge.

Meckel, J. F. (1826). *Ornithorhynchi paradoxi.* Descriptio anatomica. Folio, Leipzig.

Merrick, H. V., De Heinzelin, J., Haessaerts, P., and Howell, F. C. (1973). Archaeological occurrence of early Pleistocene age from the Shungura Formation, Lower Omo Valley, Ethiopia. *Nature,* **242**, 572–5.

Meschan, I. (1975). *An atlas of anatomy basic to radiology.* Saunders, Philadelphia.

Mestdagh, H., Bailleul, J. P., Vilette, B., Bocquet, F., and Depreux, R. (1985). Organisation of the extensor complex of the digits. *Anatomia Clinica,* **7**, 49–53.

Mivart, St. G. (1866). On some points in the anatomy of *Echidna hystrix. Transactions of the Linnean Society of London,* **25**, 379–403.

Morbeck, M. E. (1975). *Dryopithecus africanus* forelimb. *Journal of Human Evolution,* **4**, 39–46.

Morbeck, M. E. (1977). The use of casts and other problems in reconstructing the *Dryopithecus (Proconsul) africanus* wrist complex. *Journal of Human Evolution,* **6**, 65–78.

Mörike, K. D. (1964). Zur Herkunft und Funktion des ulnaren Diskus am Handgelenk. *Morphologisches Jahrbuch,* **105**, 365–74.

Morris, J. M. (1977). Biomechanics of the foot and ankle. *Clinical Orthopaedics and Related Research,* **122**, 10–17.

Morrison, A. (1953). The os paracuneiforme. *Journal of Bone and Joint Surgery,* **35B**, 254–5.

Morton, D. J. (1924*a*). Mechanism of the normal foot and of flat foot. Part I. *Journal of Bone and Joint Surgery,* **6**, 368–406.

Morton, D. J. (1924*b*). Evolution of the human foot. *American Journal of Physical Anthropology,* **7**, 1–52.

Morton, D. J. (1935). *The human foot: its evolution, physiology and functional disorders.* Columbia University Press, New York.

Murie, J. and Mivart, St. G. (1872). On the anatomy of the Lemuroidea. *Transactions of the Zoological Society of London,* **7**, 1–113.

Napier, J. R. (1952). The attachments and function of the abductor pollicis brevis. *Journal of Anatomy,* **86**, 335–41.

Napier, J. R. (1955). The form and function of the carpometacarpal joint of the thumb. *Journal of Anatomy,* **89**, 362–9.

Napier, J. R. (1956). The prehensile movements of the human hand. *Journal of Bone and Joint Surgery,* **38B**, 902–13.

Napier, J. R. (1959). Fossil metacarpals from Swartkrans. *Fossil Mammals of Africa, No. 17.* British Museum (Natural History), London.

Napier, J. R. (1961). Prehensility and opposability in the

hands of primates. *Symposia of the Zoological Society of London*, **5**, 115–32.

Napier, J. R. (1962). Fossil hand bones from Olduvai Gorge. *Nature*, **196**, 409–11.

Napier, J. R. (1964). Locomotor functions of hominids. In *Classification and human evolution* (ed. S. L. Washburn), pp. 178–89. Methuen, London.

Napier, J. R. (1970). Paleaecology and catarrhine evolution. In *Old world monkeys. Evolution, systematics and behaviour* (ed. J. R. Napier and P. H. Napier). Academic Press, London.

Napier, J. R. and Davis, P. R. (1959). The forelimb skeleton and associated remains of *Proconsul africanus*. *Fossil Mammals of Africa No. 16*. British Museum (Natural History), London.

Novacek, M. J. (1980). Cranioskeletal features in Tupaiids and selected Eutheria as phylogenetic evidence. In *Comparative Biology and evolutionary relationships of tree shrews* (ed. W. P. Luckett), pp. 35–93. Plenum Press, New York.

Oakley, K. P. (1972). Skill as a human possession. In *Perspectives on human evolution*, vol. 2 (ed. S. L. Washburn and P. Dolhinow), pp. 14–50. Holt, Rinehart and Winston, New York.

O'Connor, B. L. (1975). The functional morphology of the cercopithecoid wrist and inferior radioulnar joints and their bearing on some problems in the evolution of the Hominoidea. *American Journal of Physical Anthropology*, **43**, 113–22.

O'Connor, B. L. (1976). *Dryopithecus (Proconsul) africanus*: quadruped or non-quadruped? *Journal of Human Evolution*, **5**, 279–83.

Olivier, G. (1962). *Formation du squelette des membres chez l'homme*. Vigot Frères, Paris.

O'Rahilly, R., Gray, D. J., and Gardner, E. (1957). Chondrification in the hands and feet of staged human embryos. *Contributions to Embryology of the Carnegie Institution, No. 250*, **36**, 183–92.

Orlosky, F. J. (1980). Dental evolution trends of relevance to the origin and dispersion of the platyrrhine monkeys. In *Evolutionary biology of the new world monkeys and continental drift* (ed. R. I. Ciochon and A. B. Chiarelli), pp. 189–200. Plenum Press, New York.

Osman Hill, W. C. (1953). *Primates. Comparative anatomy and taxonomy. I. Strepsirhini*. University Press, Edinburgh.

Osman Hill, W. C. (1955). *Primates. Comparative anatomy and taxonomy. II. Haplorhini*. University Press, Edinburgh.

Owen, R. (1841). On the osteology of the marsupialia. *Transactions of the Zoological Society of London*, **2**, 379–408.

Owen, R. (1848). *On the archetype and homologies of the vertebrate skeleton*. R. and J. E. Taylor, London.

Owen, R. (1866). *Anatomy of vertebrates*. Vol. 2. Longmans, Green, London.

Owen, R. (1874). On the osteology of the marsupialia. *Transactions of the Zoological Society of London*, **8**, 483–500.

Oxnard, C. E. (1972). Some African fossil foot bones: a note on the interpolation of fossils into a matrix of extant species. *American Journal of Physical Anthropology*, **37**, 3–12.

Oxnard, C. E. (1975). The place of the australopithecines in human evolution: grounds for doubt? *Nature*, **258**, 389–95.

Oxnard, C. E. and Lisowski, F. P. (1980). Functional articulation of some hominoid foot bones: implications for the Olduvai (Hominid 8) foot. *American Journal of Physical Anthropology*, **52**, 107–17.

Palmer, A. K. (1984). The distal radioulnar joint. *Orthopaedic Clinics of North America*, **15**, 321–35.

Palmer, A. K. and Werner, F. W. (1981). The triangular fibrocartilage complex of the wrist—anatomy and function. *Journal of Hand Surgery*, **6**, 153–62.

Pankovich, A. M. and Shivaram, M. S. (1979). Anatomical basis of variability in injuries of the medial malleolus and the deltoid ligament. *Acta Orthopaedica Scandinavica*, **50**, 217–23.

Parsons, F. G. (1894a). On the morphology of the tendo-Achillis. *Journal of Anatomy*, **28**, 414–18.

Parsons, F. G. (1894b). On the myology of the sciuromorphine and hystricomorphine rodents. *Proceedings of the Zoological Society of London*, **1894**, 251–96.

Parsons, F. G. (1898a). The limb myology of *Gymnura rafflesii*. *Journal of Anatomy*, **32**, 312–24.

Parsons, F. G. (1898b). The muscles of mammals with special relation to human myology. *Journal of Anatomy*, **32**, 721–52.

Parsons, F. G. (1904). Observations on traction epiphyses. *Journal of Anatomy*, **38**, 248–58.

Parsons, F. G. (1905). On pressure epiphyses. *Journal of Anatomy*, **39**, 402–12.

Parsons, F. G. (1908). Further remarks on traction epiphyses. *Journal of Anatomy*, **42**, 388–96.

Partridge, T. C. (1973). Geomorphological dating of cave openings at Makapansgat, Sterkfontein, Swartkrans and Taung. *Nature*, **246**, 75–9.

Pearson, K. and Davin, A. G. (1921a). On the sesamoids of the knee joint. Part I. *Biometrika*, **13**, 133–75.

Pearson, K. and Davin, A. G. (1921b). On the sesamoids of the knee joint. Part II. Evolution of the sesamoids. *Biometrika*, **13**, 350–400.

Pfitzner, W. (1895). Beiträge zur Kenntniss des menschlichen Extremitätenskelets. *Morphologische Arbeiten*, **4**, 347–570.

Pfitzner, W. (1896). Beiträge zur Kenntnis des menschlichen Extremitätenskelets. *Morphologisches Arbeiten*, **6**, 245–527.

Pfitzner, W. (1900a). Beiträge zur Kenntniss des menschlichen Extremitätenskelets. VIII. Die morphologischen Elemente des menschlichen Handskelets. Abschmitt I: Allgemeiner Theil. *Zeitschrift für Morphologie und Anthropologie*, **2**, 77–157.

Pfitzner, W. (1900b). Beiträge zur Kentniss des menschlichen Extremitätenskelets. VIII. Die morphologischen Elements des menschlichen Handskelets. Abschmitt II. Specieller Theil. *Zeitschrift für Morphologie und Anthropologie*, **2**, 365–678.

Pilbeam, D. (1982). New hominoid skull material from the Miocene of Pakistan. *Nature*, **295**, 232–4.

Pilbeam, D. (1985). Patterns of hominoid evolution. In *Ancestors: the hard evidence* (ed. E. Delson), pp. 51–9. Alan R. Liss, New York.

Pocock, R. I. (1921). The external characters of the koala (*Phascolarctos*) and some related marsupials. *Proceedings of the Zoological Society of London*, pp. 591–607.

Pocock, R. I. (1925). Additional notes on the external characters of some platyrrhine monkeys. *Proceedings of the Zoological Society of London*, pp. 27–47.

Poirier, P. and Charpy, A. (1901). *Traité d'anatomie humaine*. Vol. 2. Masson et Cie, Paris.

Poirier, P. and Charpy, A. (1911). *Traité d'anatomie humaine*. Vol. 1. Masson et Cie, Paris.

Presley, R. (1981). Alisphenoid equivalents in placentals, marsupials, monotremes and fossils. *Nature*, **94**, 668–70.

Pridmore, P. A. (1985). Terrestrial locomotion in monotremes (Mammalia: Monotremata). *Journal of Zoology, London*, **205**, 53–73.

Rak, Y. (1983). *The australopithecine face*. Academic Press, New York.

Rang, M. (1969). *The growth plate and its disorders*. Livingstone, London.

Ratjova, V. (1967). The development of the skeleton in the guinea-pig. II. The morphogenesis of the carpus in the guinea-pig (*Cavia porcellus*) *Folia morphologica*, **15**, 132–9.

Raven, H. C. (1936). Comparative anatomy of the sole of the foot. *American Museum Novitates*, **871**, 1–9.

Raven, H. C. (1950). *The anatomy of the gorilla*. Columbia University Press, New York.

Reed, C. A. (1967). The generic allocation of the hominid species *habilis* as a problem in systematics. *South African Journal of Science*, **63**, 3–5.

Reimann, R. and Anderhuber, F. (1980). Kompensationsbewegungen der Fibula, die durch die keilform der Trochlea tali etzungen werden. *Acta anatomica*, **108**, 60–7.

Retterer, E. and Neuville, H. (1918). Des articulations metacarpophalangioennes de quelques singes. *Compte rendu de la Société de biologie*, **81**, 960–3.

Rewcastle, S. C. (1981). Stance and gait in tetrapods: an evolutionary scenario. *Symposia of the Zoological Society of London*, **48**, 239–67.

Rhodes, F. H. T. (1983). Gradualism, punctuated equilibrium and the origin of species. *Nature*, **305**, 269–72.

Ribbing, L. (1907). Die distale Armmuskulatur der Amphibien, Reptilien und Säugetiere. *Zoologisches Jahrbuch*, **23**, 587–682.

Ribbing, L. (1909). Die Unterschenkel- und Fussmuskulatur der Tetrapoden und ihr Verhalten zu der entsprechenden Arm- und Handmuskulatur. *Acta Universitatis Lund*, **5**, 1–158.

Robinson, J. T. (1972). *Early hominid posture and locomotion*. University of Chicago Press.

Romer, A. S. (1922). The locomotor apparatus of certain primitive and mammal-like reptiles. *Bulletin of the American Museum of Natural History*, **46**, 517–606.

Romer, A. S. (1955). *Vertebrate paleontology*. University of Chicago Press.

Romer, A. S. and Price, L. (1940). Review of the Pelycosauria. *Geological Society of America, Special Papers*, No. 28.

Rose, M. D. (1984). Hominoid postcranial specimens from the middle Miocene Chingi formation, Pakistan. *Journal of Human Evolution*, **13**, 503–16.

Rose, M. D. (1986). Further hominoid postcranial specimens from the late Miocene Nagri formation of Pakistan. *Journal of Human Evolution*, **15**, 333–67.

Rosenberg, E. (1876). Ueber die Entwicklung der Wirbelsaule und das Centrale carpi des Menschen. *Morphologisches Jahrbuch*, **1**, 83–197.

Rosenberger, A. L. (1977). *Xenothrix* and ceboid phylogeny. *Journal of Human Evolution*, **6**, 461–81.

Rosenberger, A. L. (1986). Platyrrhines, catarrhines and the anthropoid transition. In *Major topics in primate and human evolution* (ed. B. Wood, L. Martin and P. Andrews), pp. 66–88. Cambridge University Press.

Roth, V. L. (1984). On homology. *Biological Journal of the Linnean Society*, **22**, 13–29.

Ruge, G. (1878a). Untersuchung über die Extensorengruppe am Unterschenkel und Fusse der Säugethiere. *Morphologisches Jahrbuch*, **4**, 592–643.

Ruge, G. (1878b). Entwicklungsvorgänge an der Muskulatur des menschlichen Fusses. *Morphologisches Jahrbuch*, **4**, (Suppl), 117–52.

Russell, D. E. (1975). Paleoecology of the Paleocene-Eocene transition in Europe. In *Approaches to primate paleobiology* (ed. F. S. Szalay) *Contributions to Primatology*, Vol. 5, pp. 28–61. Karger, Basel.

Salsbury, C. R. (1937). The interosseous muscles of the hand. *Journal of Anatomy*, **71**, 395–403.

Santono, S. (1975). Implications arising from *Pithecanthropus VIII*. In *Paleoanthropology, morphology and paleoecology* (ed. R. S. Tuttle), pp. 327–60. Mouton, The Hague.

Sarich, V. M. (1968). The origin of the hominids: an immunological approach. In *Perspectives on human evolution* (ed. S. L. Washburn and P. C. Jay) Vol. 1, pp. 94–121. Rinehart and Winston, New York.

Sarrafian, S. K. (1983). *Anatomy of the foot and ankle.* Lippincott, Philadelphia.

Sarrafian, S. K. and Topouzian, L. K. (1969). Anatomy and physiology of the extensor apparatus of the toes. *Journal of Bone and Joint Surgery,* **51A,** 669–79.

Savage, D. E. (1975). Cenozoic—the Primate episode. In *Approaches to Primate paleobiology* (ed. F. S. Szalay). *Contributions to Primatology,* Vol. 5, pp. 2–27. Karger, Basel.

Sawalischin, M. (1911). Der musculus flexor communis brevis digitorum pedis in der Primatenreihe mit spezieller Berucksichtigung der menschlichen varietäten. *Morphologisches Jahrbuch,* **42,** 557–663.

Schaeffer, B. (1941*a*). The morphological and functional evolution of the tarsus in amphibians and reptiles. *Bulletin of the American Museum of Natural History,* **78,** 395–472.

Schaeffer, B. (1941*b*). The pes of *Bauria cynops* Broom. *American Museum Novitates,* **1103,** 1–7.

Schaeffer, B. (1947). Notes on the function of the artiodactyl tarsus. *American Museum Novitates,* **1356,** 1–24.

Schaeffer, B. (1948). The origin of a mammalian ordinal character. *Evolution,* **2,** 164–75.

Schäfer, E. A. and Thane, G. D. (1892). *Quain's Elements of Anatomy,* Vol. 2, Part 2. Longmans, Green, London.

Schäfer, E. A. and Thane, G. D. (1899). *Quain's Elements of Anatomy,* Vol. 2, Part 1. Longmans, Green, London.

Schenck, R. R. (1964). Variations of the extensor tendons of the fingers. *Journal of Bone and Joint Surgery,* **46A,** 103–10.

Schmidt, H-M. (1981). Die Artikulationsflächen der menschlichen Sprunggelenke. *Advances in Anatomy, Embryology and Cell Biology,* **66,** 1–81.

Schneider, L. H. and Hunter, J. M. (1982). Flexor tendons—late reconstruction. In *Operative hand surgery* (ed. D. P. Green) Vol. 2, pp. 1375–1440. Churchill Livingstone, London.

Schön, M. A. and Ziemer, L. K. (1973). Wrist mechanism and locomotor behaviour of *Dryopithecus (Proconsul) africanus. Folia primatoligica,* **20,** 1–11.

Schreiber, H. (1934). Zur morphologie der Primatenhand. I. Röntgenologische Ungersuchungen un der Handwurzel der Affen. *Anatomischer Anzeiger,* **78,** 369–429.

Schultz, A. H. (1930). The skeleton of the trunk and limbs in higher primates. *Human Biology,* **2,** 303–438.

Schultz, A. H. (1936). Characters common to higher primates and characters specific to man. *Quarterly Review of Biology,* **11,** 259–83.

Schultz, A. H. (1950). The physical distinctions of man. *Proceedings of the American Philosophical Society,* **94,** 428–49.

Schwartz, J. H. (1984). The evolutionary relationships of man and orang-utans. *Nature,* **308,** 501–5.

Schwartz, J. H. and Tattersal, I. (1979). The phylogenetic relationship of Adapidae (Primates, Lemuriformes). *Anthropological Papers of the American Museum of Natural History,* **55,** 271–83.

Sennwald, G. (1987). *The wrist.* Springer-Verlag, Berlin.

Sewell, R. B. S. (1904*a*). A study of the astragalus. Part I. *Journal of Anatomy,* **38,** 233–47.

Sewell, R. B. S. (1904*b*). A study of the astragalus. Part II. *Journal of Anatomy,* **38,** 423–34.

Sewell, R. B. S. (1904*c*). A study of the astragalus. Part III. *Journal of Anatomy,* **39,** 74–88.

Sewell, R. B. S. (1906). A study of the astragalus. Part IV. *Journal of Anatomy,* **40,** 152–61.

Sewertzoff, A. N. (1929). Directions of evolution. *Acta Zoologica,* **10,** 59–141.

Shellswell, G. B. and Wolpert, L. (1977). The pattern of muscle and tendon development in the chick wing. In *Vertebrate limb and somite morphogenesis* (ed. D. A. Ede, J. R. Hinchliffe, and M. Balls), pp. 71–86. Cambridge University Press.

Shephard, E. (1951). Tarsal movements. *Journal of Bone and Joint Surgery,* **33B,** 258–63.

Shipman, P., Walker, A., van Couvering, J. A., Hooker, P. J., and Miller, J. A. (1981). The Fort Ternan hominoid site, Kenya: geology, age, taphonomy and paleoecology. *Journal of Human Evolution,* **10,** 49–72.

Shrewsbury, M. M. and Sonek, A. (1986). Precision holding in humans, non-human primates and Plio-Pleistocene hominids. *Human Evolution,* **1,** 233–42.

Sidebotham, E. J. (1885). On the myology of the water-opossum. *Proceedings of the Zoological Society of London,* **1885,** 6–22.

Simons, E. L. (1972). *Primate evolution.* Macmillan, New York.

Simons, E. L. (1977). *Ramapithecus. Scientific American,* **236,** 28–35.

Simons, E. L. (1978). Diversity among the early hominids: a vertebrate paleontologist's viewpoint. In *Early hominids of Africa* (ed. C. J. Jolly), pp. 543–66. Duckworth, London.

Simons, E. L. (1981). Man's immediate forerunners. In *The emergence of man* (ed. J. Z. Young, E. M. Jope, and K. P. Oakley), pp. 21–41. The Royal Society and the British Academy, London.

Simons, E. L. (1985). Origins and characteristics of the first hominoids. In *Ancestors: the hard evidence* (ed. E. Delson), pp. 37–41. Alan R. Liss, New York.

Simons, E. L. and Delson, E. (1978). Cercopithecidae and Parapithecidae. In *Evolution of African mammals* (ed. V. J. Maglio and H. B. S. Cooke), pp. 100–19. Harvard University Press, Cambridge.

Simpson, G. G. (1935). The Tiffany fauna, upper Paleocene. II. Structure and relations of *Plesiadapis*. *American Museum Novitates*, **816**, 1–30.

Simpson, G. G. (1945). *The principles of classification and a classification of mammals*. American Museum of Natural History, New York.

Singh, I. (1959). Variations in the metacarpal bones. *Journal of Anatomy*, **93**, 262–7.

Smith, E. Barclay (1896). The astragalo-calcaneonavicular joint. *Journal of Anatomy*, **30**, 390–412.

Sonntag, C. F. (1923). On the anatomy, physiology and pathology of the chimpanzee. *Proceedings of the Zoological Society, London*, **1**, 323–429.

Sonntag, C. F. (1924). *The morphology and evolution of the apes and man*. John Bale, Sons and Danielsson, London.

Stack, H. G. (1962). Muscle function in the fingers. *Journal of Bone and Joint Surgery*, **44B**, 899–909.

Stack, H. G. (1963). A study of muscle function in the fingers. *Annals of the Royal College of Surgeons, England*, **33**, 307–22.

Stains, H. (1959). Use of the calcaneum in studies of taxonomy and food habits. *Journal of Mammalogy*, **40**, 392–401.

Stamm, T. T. (1931). The constitution of the ligamentum cruciatum cruris. *Journal of Anatomy*, **66**, 80–3.

Stanley, S. M. (1979). *Macroevolution. Pattern and process*. W. H. Freeman, San Francisco.

Stein, A. H. (1951). Variations of the tendons of insertion of the abductor pollicis longus and the extensor pollicis brevis. *Anatomical Record*, **110**, 49–55.

Stein, B. R. (1981). Comparative limb myology of two opossums, *Didelphis* and *Chironectes*. *Journal of Morphology*, **169**, 113–40.

Stern, J. T. (1971). Functional myology of the hip and thigh of cebid monkeys and its implications for the evolution of erect posture. *Bibliotheca primatologica*, **14**, 1–318.

Stern, J. T. (1975). Before bipedality. *Yearbook of Physical Anthropology*, **19**, 59–68.

Stern, J. T. and Susman, R. L. (1983). The locomotor anatomy of *Australopithecus afarensis*. *American Journal of Physical Anthropology*, **60**, 279–317.

Stopford, J. S. B. (1914). The supracondyloid tubercles of the femur and the attachment of the gastrocnemius muscle to the femoral diaphysis. *Journal of Anatomy*, **44**, 80–84.

Strauch, B. and de Moura, W. (1985). Digital flexor tendon sheath: an anatomic study. *Journal of Hand Surgery*, **10A**, 785–9.

Straus, W. L. (1930). The foot musculature of the Highland Gorilla (*Gorilla beringei*). *Quarterly Review of Biology*, **5**, 261–317.

Straus, W. L. (1941a). The phylogeny of the human forearm extensors. *Human Biology*, **13**, 23–50.

Straus, W. L. (1941b). The phylogeny of the human forearm extensors. *Human Biology*, **13**, 203–38.

Straus, W. L. (1942). The homologies of the forearm flexors: urodeles, lizards, mammals. *American Journal of Anatomy*, **70**, 281–317.

Susman, R. L. (1983). Evolution of the human foot: evidence from Plio-Pleistocene hominids. *Foot and Ankle*, **3**, 365–76.

Susman, R. L., Badrian, N. L., and Badrian, A. J. (1980). Locomotor behaviour of *Pan paniscus* in Zaire. *American Journal of Physical Anthropology*, **53**, 69–80.

Susman, R. L. and Creel, N. (1979). Functional and morphological affinities of the subadult hand (OH7) from Olduvai Gorge. *American Journal of Physical Anthropology*, **51**, 311–32.

Susman, R. L., Stern, J. T., and Jungers, W. L. (1985). Locomotor adaptations in the Hadar hominids. In *Ancestors: the hard evidence* (ed. E. Delson), pp. 184–92. Alan R. Liss, New York.

Susman, R. T. and Stern, J. T. (1982). Functional morphology of *Homo habilis*. *Science*, **217**, 931–4.

Swindler, D. R. and Wood, C. D. (1973). *An atlas of primate gross anatomy. Baboon, chimpanzee, and man*. University of Washington Press, Seattle.

Szalay, F. S. (1973). New Paleocene Primates and a diagnosis of the new suborder Paromomyiformes. *Folia primatologica*, **19**, 73–87.

Szalay, F. S. (1975). Haplorhine phylogeny and the status of the Anthropoidea. In *Primate functional morphology and evolution* (ed. R. H. Tuttle), pp. 3–22. Mouton, The Hague.

Szalay, F. S. (1977). Constructing Primate phylogenies: a search for testable hypotheses with maximum empirical content. *Journal of Human Evolution*, **6**, 3–18.

Szalay, F. S. (1981). Functional analysis and the practice of the phylogenetic method as reflected by some mammalian studies. *American Zoologist*, **21**, 37–45.

Szalay, F. S. (1984). Arboreality: is it homologous in metatherian and eutherian mammals? In *Evolutionary biology*, Vol. 18 (ed. M. K. Hecht, B. Wallace, and G. T. Prance), pp. 215–58. Plenum Press, New York.

Szalay, F. S. and Decker, R. L. (1974). Origins, evolution and function of the tarsus in late Cretaceous Eutheria and Paleocene primates. In *Primate locomotion* (ed. F. A. Jenkins), pp. 223–59. Academic Press, New York.

Szalay, F. S. and Delson, E. (1979). *Evolutionary history of the primates*. Academic Press, London.

Szalay, F. S. and Drawhorn, G. (1980). Evolution and diversification of the Archonta in an arboreal milieu. In *Comparative biology and evolutionary relationships of tree shrews* (ed. W. P. Luckett), pp. 133–69. Plenum Press, New York.

Szalay, F. S. and Katz, C. C. (1973). Phylogeny of lemurs, galagos, and lorises. *Folia Primatologica*, **19**, 88–103.

Szalay, F. S. and Li, C. (1986). Middle Paleocene

Euprimate from Southern China and the distribution of Primates in the Paleogene. *Journal of Human Evolution,* **15**, 387–97.

Szalay, F. S., Tattersall, I., and Decker, R. L. (1975). Phylogenetic relationships of *Plesiadapis*—postcranial evidence. In *Approaches to primate paleobiology. Contributions to Primatology,* Vol. 5 (ed. F. S. Szalay), pp. 136–66. Karger, Basel.

Szarski, H. (1949). The concept of homology in the light of the comparative anatomy of vertebrates. *Quarterly Review of Biology,* **24**, 124–31.

Taleisnik, J. (1976). The ligaments of the wrist. *Journal of Hand Surgery,* **1**, 110–18.

Tarling, D. H. (1980). The geologic evolution of South America with special reference to the last 200 million years. In *Evolutionary biology of the new world monkeys and continental drift* (ed. R. L. Ciochon and A. B. Chiarelli), pp. 1–42. Plenum Press, New York.

Tattersall, I. and Eldredge, N. (1977). Fact, theory and fantasy in human paleontology. *American Scientist,* **65**, 204–11.

Taylor, G. T. and Bonney, V. (1905). On the homology and morphology of the popliteus muscle: a contribution to comparative myology. *Journal of Anatomy,* **40**, 34–50.

Testut, L. (1884). *Les anomalies musculaires.* Masson et Cie, Paris.

Testut, L. (1904). *Traité d'anatomie humaine.* Vol. 1. Octave Doin, Paris.

Thane, G. D. (1892). *Quain's Elements of Anatomy.* Vol. 2, Pt. 2 (ed. E. A. Schäfer and G. D. Thane). Longmans, Green, London.

Thilenius, G. (1896). Untersuchungen über die morphologische Bedeutung accessorischer Elemente am menschlichen Carpus (und Tarsus). *Morphologische Arbeiten* **5**, 462–554.

Thomas, H. (1985). The early and middle Miocene land connection of the Afro-Arabian plate and Asia: a major event for hominoid dispersal? In *Ancestors: the hard evidence* (ed. E. Delson), pp. 42–50. Alan R. Liss, New York.

Thompson, D'Arcy W. (1952). *On growth and form.* Cambridge University Press.

Tobias, P. V. (1968). Cranial capacity in anthropoid apes, *Australopithecus* and *Homo habilis,* with comments on skewed samples. *South African Journal of Science,* **64**, 81–91.

Tobias, P. V. (1971). *The brain in hominid evolution.* Columbia University Press, New York.

Tobias, P. V. (1973). Implications of the new age estimates of the early South African hominids. *Nature,* **246**, 79–83.

Tobias, P. V. (1978). The South African australopithecines in time and hominid phylogeny, with special reference to the dating of the Taung skull. In *Early hominids of Africa* (ed. C. J. Jolly), pp. 45–84. Duckworth, London.

Tobias, P. V. (1981). The emergence of man in Africa and beyond. In *The emergence of man* (ed. J. Z. Young, E. M. Jope, and K. P. Oakley), pp. 45–56. The Royal Society and The British Academy, London.

Turner, W. (1864). On variability in human structure, with illustrations from the flexor muscles of the fingers and toes. *Transactions of the Royal Society of Edinburgh,* **24**, 175–89.

Tuttle, R. H. (1967). Knuckle-walking and the evolution of hominoid hands. *American Journal of Physical Anthropology,* **26**, 271–306.

Tuttle, R. H. (1972). Functional and evolutionary biology of hylobatids hands and feet. In *Gibbon and siamang,* Vol. 1 (ed. D. M. Rumbaugh), pp. 136–206. Karger, Basel.

Vallois, H. V. (1914). *L'Articulation du genou chez les primates.* Imprimerie Cooperative, Montpellier.

Van Dam, G. (1934). On active rotation of the metacarpalia, e.g. in spreading of the fingers. *Acta radiologica,* **15**, 304–9.

Van Valen, L. (1965). Treeshrews, Primates and fossils. *Evolution,* **19**, 137–51.

Van Valen, L. (1967). New Paleocene insectivores and insectivore classification. *Bulletin of the American Museum of Natural History,* **135**, 217–84.

Van Valen, L. and Sloan, R. E. (1965). The earliest primates. *Science,* **150**, 743–745.

Virchow, H. (1929). Das Os centrale des Menschen. *Morphologisches Jahrbuch,* **63**, 480–530.

Volkmann, R. (1973). Zur Anatomie und Mechanick des Lig. calcaneonaviculare plantare sensu strictiori. *Anatomischer Anzeiger,* **134**, 460–70.

Vrba, E. S. (1975). Some evidence of chronology and palaeoecology of Sterkfontein, Swartkrans, and Kromdraai from the fossil Bovidae. *Nature,* **254**, 301–4.

Waddington, C. H. (1957). *The strategy of the genes.* Allen and Unwin, London.

Walker, A. (1972). The dissemination and segregation of early Primates in relation to continental configuration. In *Calibration of hominoid evolution* (ed. W. W. Bishop and J. A. Miller), pp. 195–218. Scottish Academic Press, Edinburgh.

Walker, A. and Leakey, R. E. F. (1978). The hominids of East Turkana. *Scientific American,* **239**, 44–56.

Walker, A. C. and Pickford, M. (1983). New postcranial fossils of *P. africanus* and *P. nyanzae.* In *New interpretations of ape and human ancestry* (ed. R. L. Ciochon and R. S. Corruccini), pp. 325–51. Plenum Press, New York.

Weidenreich, F. (1922). Der Menschenfuss. *Zeitschrift für Morphologie und Anthropologie,* **22**, 51–282.

Weidenreich, F. (1940). The external tubercle of the

human tuber calcanei. *American Journal of Physical Anthropology*, **23**, 473–87.

Weinberg, E. (1929). Recherches d'anatomie comparee sur l'appareil fibreux de la face posterieure du genou. *Archives d'Anatomie, d'Histologie et d'Embryologie*, **9**, 253–307.

Weinert, C. R., McMaster, J. H., and Ferguson, R. J. (1973). Dynamic function of the human fibula. *American Journal of Anatomy*, **138**, 145–50.

Weitbrecht, J. (1742). *Syndesmologia*. Academy of Sciences, St. Petersburg.

Westling, C. (1889). Anatomische Untersuchungen über Echidna. *Bihang Svenska vetensakademiens handlingar*, **15**, 1–71.

Wiedersheim, R. (1895). *The structure of man*. Macmillan, London.

Wikander, R., Covert, H. H., and Deblieux, D. D. (1986). Ontogenetic intraspecific, and interspecific variation of the prehallux in Primates: implications for its utility in the assessment of phylogeny. *American Journal of Physical Anthropology*, **70**, 513–23.

Williamson, P. G. (1981*a*). Palaeontological documentation of speciation in Cenozoic molluscs from Turkana Basin. *Nature*, **293**, 437–43.

Williamson, P. G. (1981*b*). Morphological stasis and developmental constraint: real problems for neo-Darwinism. *Nature*, **294**, 214–15.

Wilson, A. C. and Sarich, V. M. (1969). A molecular time scale for human evolution. *Proceedings of the National Academy of Science*, **63**, 1088–93.

Winckler, G. and Giacomo, G. (1955). La veritable terminaison de la chair carrée de Sylvius (musc. quadratus plantae). *Archives d'Anatomie, Strasbourg* **38**, 47–66.

Windle, B. C. A. (1883). On the embryology of the mammalian muscular system. No. I—the short muscles of the human hand. *Transactions of the Royal Irish Academy*, **28**, 211–40.

Windle, C. A. (1889). The flexors of the digits of the hand. *Journal of Anatomy*, **24**, 72–84.

Windle, B. C. A. and Parsons, F. G. (1897). On the myology of the terrestrial Carnivora—Part I. *Proceedings of the Zoological Society of London*, **1897**, 370–409.

Windle, B. C. A. and Parsons, F. G. (1898). On the myology of the terrestrial Carnivora. Part II. *Proceedings of the Zoological Society of London*, **1898**, 152–86.

Windle, B. C. A. and Parsons, F. G. (1899). On the myology of the Edentata. *Proceedings of the Zoological Society of London*, **1899**, 990–1017.

Windle, B. C. A. and Parsons, F. G. (1903). On the muscles of the Ungulata. Part II. *Proceedings of the Zoological Society of London*, **1903**, 261–98.

Wirtschafter, Z. T. and Tsujimura, J. K. (1961). The sesamoid bones in the C3H mouse. *Anatomical Record*, **139**, 399–408.

Wise, K. S. (1975). The anatomy of the metacarpo-phalangeal joints, with observations on the aetiology of ulnar drift. *Journal of Bone and Joint Surgery*, **57B**, 485–90.

Wolpoff, M. H. (1978). Analogies of interpretation in paleoanthropology. In *Early hominids of Africa* (ed. C. J. Jolly), pp. 461–503. Duckworth, London.

Wood, A. E. (1980). The origins of the caviomorph rodents from a source in Middle America: a clue to the area of origin of the platyrrhine Primates. In *Evolutionary biology of the new world monkeys and continental drift* (ed. R. L. Ciochon and A. B. Chiarelli), pp. 79–92. Plenum Press, New York.

Wood, B. A. (1973). Locomotor affinities of hominoid tali from Kenya. *Nature*, **246**, 45–6.

Wood, B. A. (1974). Olduvai Bed I postcranial fossils: a reassessment. *Journal of Human Evolution*, **3**, 373–8.

Wood, J. (1866). Variations in human myology observed during the winter session of 1865–66 at King's College, London. *Proceedings of the Royal Society*, **15**, 229–44.

Wood, J. (1867). Variations in human myology observed during the winter session of 1866–67 at King's College, London. *Proceedings of the Royal Society*, **15**, 518–46.

Wood, J. (1868). Variations in human myology observed during the winter session of 1867–68 at King's College, London. *Proceedings of the Royal Society*, **16**, 483–525.

Wright, D. G., Desai, S. M., and Henderson, W. J. (1964). Action of the subtalar and ankle joint complex during the stance phase of walking. *Journal of Bone and Joint Surgery*, **46A**, 361–82.

Wright, D. G. and Rennels, D. C. (1964). A study of the elastic properties of plantar fascia. *Journal of Bone and Joint Surgery*, **46**, 482–92.

Wright, J. D. (1935). An analysis of movements of various bones of the carpus during movements of the thumb and wrist joints. *Journal of Anatomy*, **70**, 137–42.

Zapfe, H. (1960). Die Primatenfunde aus der Miozänenspaltenfüllung von Neudorf an der March. *Schweizerische Palaeontologische Abhundlungen*, **78**, 1–293.

Zihlman, A. and Brunker, L. (1979). Hominid bipedalism: then and now. *Yearbook of Physical Anthropology*, **22**, 132–62.

Zwell, M. and Conroy, G. C. (1973). Multivariate analysis of the *Dryopithecus africanus* forelimb. *Nature* **244**, 373–5.

Index

Items in this index are listed under the nouns rather than under the descriptive adjectives. For example, the various ligaments will be found under the general heading ligament(s). Page references to descriptions alone are in roman type. Page references in italic type indicate figures, often in addition to descriptions on the same page.